Lecture Notes in Earth Sciences

Vol. 1: Sedimentary and Evolutionary Cycles. Edited by U. Bayer and A. Seilacher. VI, 465 pages. 1985

Vol. 2: U. Bayer, Pattern Recognition Problems in Geology and Paleontology. VII, 229 pages. 1985.

Vol. 3: Th. Aigner, Storm Depositional Systems. VIII, 174 pages. 1985.

Vol. 4: Aspects of Fluvial Sedimentation in the Lower Triassic Buntsandstein of Europe. Edited by D. Mader. VIII, 626 pages. 1985

Vol. 5: Paleogeothermics. Edited by G. Buntebarth and L. Stegena. II, 234 pages. 1986.

Vol. 6: W. Ricken, Diagenetic Bedding. X, 210 pages. 1986

Vol. 7: Mathematical and Numerical Techniques in Physical Geodesy. Edited by H. Sünkel. IX, 548 pages. 1986.

Lecture Notes in Earth Sciences

Edited by Somdev Bhattacharji, Gerald M. Friedman, Horst J. Neugebauer and Adolf Seilacher

7

Mathematical and Numerical Techniques in Physical Geodesy

Lectures delivered at the
Fourth International Summer School in the Mountains
on Mathematical and Numerical Techniques in Physical Geodesy
Admont, Austria, August 25 to September 5, 1986

Edited by Hans Sünkel

Springer-Verlag
Berlin Heidelberg GmbH

Editor

Prof. Dr. Dipl.-Ing. Hans Sünkel
Institute of Mathematical Geodesy, Technical University Graz
Technikerstr. 4, A-8010 Graz, Austria

ISBN 978-3-540-16809-6 ISBN 978-3-540-47059-5 (eBook)
DOI 10.1007/978-3-540-47059-5

Originally published by Springer-Verlag Berlin Heidelberg New York in 1986

2132/3140-543210

THE AUTHORS

COLOMBO, O.L., EG&G Washington Analytical Services Center, Inc., 5000 Philadelphia Way, Suite J, Lanham, Maryland 20706, U.S.A.

HEIN, G.W., Institute of Astronomical and Physical Geodesy, University FAF Munich, Werner-Heisenberg-Weg 39, D-8014 Neubiberg, F.R. Germany

HOFMANN-WELLENHOF, B., Institute of Theoretical Geodesy, Division of Physical Geodesy, Technical University Graz, Rechbauerstraße 12, A-8010 Graz, Austria

MORITZ, H., Institute of Theoretical Geodesy, Division of Physical Geodesy, Technical University Graz, Rechbauerstraße 12, A-8010 Graz, Austria

RAPP, R.H., Department of Geodetic Science and Surveying, The Ohio State University, 1958 Neil Avenue, Columbus, Ohio 43210, U.S.A.

RUMMEL, R., Department of Geodesy, Delft University of Technology, Thijsseweg 11, NL-2600 GA Delft, The Netherlands

SANSO, F., Institute of Topography, Photogrammetry and Geophysics, Politecnico Milano, Piazza Leonardo da Vinci 32, I-20133 Milano, Italy

SCHWARZ, K.-P., Division of Surveying Engineering, The University of Calgary, 2500 University Drive N.W., Calgary, Alberta T2N 1N4, Canada

SÜNKEL, H., Institute of Theoretical Geodesy, Division of Mathematical Geodesy, Technical University Graz, Rechbauerstraße 12, A-8010 Graz, Austria

TSCHERNING, C.C., Geodetic Institute, Gamlehave Allé 22, DK-2920 Charlottenlund, Denmark

PREFACE

This volume comprises the main lectures delivered at the Fourth International Summer School in the Mountains on "**Mathematical and Numerical Techniques in Physical Geodesy**", held from August 25 to September 5, 1986 in Admont, Austria.

The School was organized by the Institute of Theoretical Geodesy of the Technical University Graz, Austria under the auspices of the International Association of Geodesy. All five continents were represented by 70 participants from over 20 countries.

The purpose of the Summer School was to provide an introduction to advanced techniques which represent the mathematical vehicle for the treatment of modern geodetic problems, to familiarize participants with the present state of the art of global and local gravity field determination methods, ranging from orbit theory, the key satellite techniques, to inertial and standard terrestrial methods, and to discuss future scientific developments.

The arrangement of this volume matches the sequence of lectures given at the School. The theoretical PART A represents the mathematical framework of modern physical geodesy, the application PART B deals with the key satellite and surface techniques, providing the detailed structure of the earth's gravity field.

PART A:

One of the main goals in physical geodesy, global and local gravity field determination, is pursued by extensively applying functional analytic methods. Recently special attention is being given to the base function and norm choice problem, and to the establishment of a sound link between density distributions inside the earth as the source and observed or estimated gravity field quantities as the effect. The lectures by C.C. Tscherning focus on this topic.

Space and time dependent problems of discrete and continuous type are encountered in modern geodesy nowadays and dealt with in the lectures by F. Sansò. Estimation theory either in its stochastic or statistic formulation plays a key role in the processing of processes like the earth's gravity field.

The consistent processing of large structured data sets calls for equally structured numerical algorithms. Spectral analysis with its powerful fast Fourier transform has become a common tool for the treatment of such problems. An introduction to spectral methods, supplemented by numerous examples, is provided by B. Hofmann-Wellenhof and H. Moritz.

PART B:

The theory of orbit dynamics, tailored to the near circular orbits of most geodetic satellites, is fundamental to modern geodetic satellite techniques and discussed in the lectures by O.L. Colombo. Particular emphasis is put on the interplay between orbit perturbations and the earth's disturbing gravity field and its mapping by satellite techniques like satellite altimetry, satellite-to-satellite tracking and satellite gradiometry.

Satellite gradiometry, which is discussed in the lectures by R. Rummel in detail, with regard to the geometric structure of the gravitational field, the observability of the gradients, and the mathematical model underlying the gravity field recovery problem, promises to provide particularly detailed information about the gravity field of our planet.

The global structure of the earth's gravity field is described in terms of earth gravity field models which are derived from both satellite and surface data. The many delicate, mathematically as well as numerically challenging problems, related to the consistent processing of very large space distributed data sets, and proposed solutions are presented in the lecture by R.H. Rapp.

For many years various attempts have been made to explain the shorter wavelength part of the earth's anomalous gravity field by isostatic phenomena. Recently several high resolution topographic-isostatic earth models have been computed based on global digital terrain data using different techniques for the estimation of the parameters of the chosen isostatic model. A declared goal is the maximum smoothing of the observed gravity field by removing the contribution of the topography and its isostatic compensation. This topic is discussed in the lectures by H. Sünkel.

Inertial methods are steadily gaining importance, power and application. This is not only due to hardware improvements in terms of precision and reliability, but also due to recent advances in the mathematical and numerical modelling of the system's performance. An investigation of the error characteristics of inertial survey systems and their interaction with the anomalous gravity field, studied in the framework of dynamic system analysis, is the topic of the lectures by K.-P. Schwarz and the key issue for further improvements and possible integrations with other positioning systems.

Geodetic data have both geometric and physical ingredients of various nature. Standard geodetic processing procedures aim at a separation of geometry from physics. Integrated geodesy, in contrast, has been designed as a very sophisticated melting pot which handles practically all available geodetic data in a consistent and optimal way.It handles surface and satellite data with either geometrically or gravity field dominated content, and geophysical data in terms of density and seismic information just as well and represents as such the great synthesis of mathematical modelling in connexion with geodetic data processing techniques; these advanced ideas are presented in the lectures by G. Hein.

This volume presents highlights of modern geodetic activity and takes the reader to the frontiers of current research. It is not a textbook on a closed and limited subject, but rather a reference book for graduates and scientists working in the vast and beautiful, demanding but rewarding field of earth science in general and physical geodesy in particular.

The editor expresses his appreciation to all authors of this volume for their advice and help in formulating and designing the scientific program of the Summer School, for providing typewritten lecture notes, and for their excellent cooperation.

Graz, Austria
September 1986

Hans Sünkel
Editor

CONTENTS

PART A

FUNCTIONAL METHODS
FOR GRAVITY FIELD APPROXIMATION

by

C.C. Tscherning

Geodetic Institute
Gamlehave Allé 22
DK-2920 Charlottenlund, Denmark

Lecture Notes in Earth Sciences, Vol. 7
Mathematical and Numerical Techniques in Physical Geodesy
Edited by H. Sünkel

1. Introduction

Many methods have been proposed (and a few used in practice) for gravity field approximation:

-series expansion in orthogonal functions,

-linear combinations of potentials of point masses, multipoles, or mass lines,

-linear combinations of harmonic splines, kernel functions or finite elements,

-least squares or minimum norm collocation.

The methods have in common that the approximation to the gravity potential is a linear combination of a number of base functions. They differ by their choice of base functions and by the method used to determine the coefficients of the linear combination.

Some of the methods are based on clear mathematical principles, even admitting proofs of convergence for the situation where the number of base functions increase. But in practice they do not always give good results. Conversely, methods have been proposed based on mainly heuristic arguments, which in practice works very well, and for wich proofs of convergence are not known.

Functional analysis is the branch of mathematics, which provides us with the tools to analyse, and hopefully better understand, the functioning of the different methods. The main propose of these lecture notes is therefore to give an introduction to the basic concepts in functional analysis. However, large parts of basic functional analysis is now well known to geodesists and geophysists, at least as it is presented in basic mathematical texts. Also potential theory, and physical geodesy is well known, but I feel that the close connection between functional analysis and physical geodesy maybe is not valued well enough.

As in my lecture notes prepared for earlier summerschools (Tscherning, 1975, 1978a,1985) (from which much material presented here has been taken) I will skip nearly all proofs and concentrate on examples. Some of these are the solutions to excercises contained in earlier lecture notes. This might interest the reader, who has not been able to verify his solutions to these excercises.

In the following we will mainly deal with objects which are functions defined in a subset of a three-dimensional enclidian space $\mathbf{R}^3$, which we will denote Ω. The boundary will be denoted σ or ω. Functions will denoted f, g, h or l and a, b, c are reserved for real or complex numbers. Points in $\mathbf{R}^n$ will be denoted x and y with coordinates $\{x_i\}$ and $\{y_i\}$. A fat "period" ● will denote end of example.

2. Linear vector spaces, dual spaces, norms and inner products

2.1 Linear vector spaces.

A linear vector space over the real (or complex) numbers is a set of elements (or vectors) f, g, h, ..., for which it is possible to add two elements or to multiply an element with a scalar. There must also exist a unique zero element, 0, and every element must have an inverse, -f with $f + (-f) = 0$. The following rules must hold: $f+g = g+f$, $f+(g+h) = (f+g)+h$, $0+f = f$, $a(f+g) = af+ag$, $a(bf) = (ab)f$, $(a+b)f = af+bf$, and $1 \cdot (f) = f$, where 1 as written is the real number one.

The vectors which form a linear combination,

$$f = \sum_{i=1}^{n} a_i f_i, \qquad a_i \in \mathbf{R}$$

are called linear dependent if there exist constants $\{a_i\} \in \mathbf{R}^n$, so that $f = 0$. Otherwise they are called independent. A vector space is said to have dimension n, if there exist n linearly independent vectors, while any set of n + 1 vectors are dependent. If no n exist, the space is infinite dimensional.

A set of elements f_1, f_2,...., is a basis for the vector space, if they are independent and every element can be expressed uniquely as a linear combination of the elements.

Well known examples of finite dimensional spaces are the real enclidian space $\mathbf{R}^n$ and polynomials p(x) of maximal degree n. In general functions $f : \Omega \subseteq \mathbf{R}^n \rightarrow \mathbf{R}$ form a linear vector space since we are able to add functions and multiply functions by a constant. The space of functions with derivatives continuous up to order m is denoted $C^m(\Omega)$.

Example 2.1 The harmonic functions outside a sphere i $\mathbf{R}^3$.
Let Ω be the open set in $\mathbf{R}^3$ outside a sphere with center of the origin and radius R, and consider the vector space of functions harmonic in Ω and regular at infinity,

$$\Delta f = \sum_{i=1}^{3} \frac{\partial^2 f}{\partial x_i^2} = 0 , \qquad \lim_{|x|\to\infty} |f(x)|x|| < \infty .$$

A finite dimensional sub space of this vector space is formed by talking all linear combinations of the solid spherical harmonics,

$$V_{ij}(P) = \left(\frac{R}{r}\right)^{i+1} P_{ij}(\sin\varphi) \begin{Bmatrix} \cos j\lambda \\ \sin|j|\lambda \end{Bmatrix} \begin{matrix} 0 \le j < i \\ -i \le j < 0, \end{matrix} \tag{2.1}$$

where $0 \le i \le n$. The point P has spherical coordinates φ (latitude), λ (longitude) and r(distance from the origin), and $P_{ij}(\sin\varphi)$ are the associated Legendre functions of degree i and order $|j|$. The dimension of the subspace is $(n+1)^2$. ●

In general, linear vector spaces may be formed by taking all (finite) linear combinations of given elements. This is called the space spanned by the elements,

$$\text{span}\{f_i , i \in I\},$$

where I is an index-set. Note , that span $\{V_{ij}, i = 0, \ldots\ldots, \infty , |j| \le i\}$, does not contain all functions harmonic in Ω.

2.2 Dual spaces.

In practice we encounter, not functions, but measurements. Forgetting measurement errors, we generally regard the measurements as values of mappings from a function into the real numbers, for example the value of one of its derivatives in a point. An important class of functionals are the linear functionals, which we will denote by L, sometimes with a subscript affixed. The functional just mentioned is an example of a linear functional. In general they must fulfil

$$L(af + bg) = aL(f) + b L(g).$$

For linear vector spaces of functions the most important linear functional is the evaluation functional. $L_P(f) = f(P)$. It should be obvious that the set of linear functionals associated with a vector space, H, will form a new linear vector space H^* , the dual space. (We will later restrict H^* to the space only containing the bounded functionals).

Example 2.2 Linear functionals in $\mathbf{R}^n$.

In $\mathbf{R}^n$ the linear functionals are the linear mappings

$$L(x) = \sum_{i=1}^{n} a_i x_i$$

Note, that also $\{a_i\}$ is an element of $\mathbf{R}^n$, and that all linear functionals have a unique representation as the scalar product of a fixed vector $\{a_i\}$ with the variable vector x. We shall see later that this nice property is valid in a large class of function spaces. ●

Example 2.3 The coordinate functionals for functions harmonic outside or inside a sphere. It is well known that functions harmonic outside a sphere and regular at ∞ permit a representation through a convergent series

$$f(P) = \sum_{i=0}^{\infty} \sum_{j=-i}^{i} a_{ij} V_{ij}(P). \tag{2.2}$$

The functions harmonic inside a sphere and regular at 0 have a similar expansion with $V^o_{ij}(P)$ substituted for $V_{ij}(P)$, where $V^o_{ij}(P)$ are the inner solid spherical harmonics,

$$V^o_{ij}(P) = \frac{r^{2i+1}}{R^{2i+1}} V_{ij}(P) \tag{2.3}$$

The coordinate functionals $L_{ij}(f) = a_{ij}$ are evaluated by

$$L_{ij}(f) = \frac{1}{4\pi R} \int_\sigma f(P) V_{ij}(P) d\sigma = a_{ij}, \tag{2.4}$$

where σ is the surface of the sphere. (The integral should be taken as the limit over concentric spheres with radius converging towards R from above or from below.)●

Example 2.4 The basic measurement functionals.
The basic measurements functionals of physical geodesy are (with W denoting the gravitational potential and T the anomalous potential):

$$\text{Gravity:} \quad g(P) = \left(\sum_{i=1}^{3} \left(\frac{\partial W}{\partial x_i}\right)^2 \right)^{\frac{1}{2}} \tag{2.5}$$

$$\text{Gravity anomaly and disturbance:} \quad \Delta g(P) = -\frac{\partial T}{\partial r}\Big|_P - \frac{2}{r} T(P) \tag{2.6}$$

$$\delta g(P) = -\frac{\partial T}{\partial r}\Big|_P \tag{2.7}$$

Deflections of the vertical (meridian and prime vertical components):

$$\xi(P) = -\frac{1}{r_\gamma} \frac{\partial T}{\partial \varphi}\Big|_P, \quad \eta(P) = -\frac{1}{r_\gamma} \frac{\partial T}{\partial \lambda} \frac{1}{\cos\gamma} \tag{2.8}$$

Gradiometry: linear combinations of $T_{ij}(P) = \frac{\partial^2 T}{\partial x_i \partial x_j}\big|_P$ (2.9)

Height anomaly: $\zeta(P) = L_p(T)/\gamma = T(P)/\gamma,$ (2.10)

where γ is normal gravity. (The indication of the point of evaluation may in the following be dropped.) The functionals (2.6)-(2.10) are linear(ized), while (2.5) is non-linear. ●

Example 2.5 The linear space spanned by a random function.
In this example one should keep in mind a set of (harmonic) functions all with a certain probality being equal to the anomalous gravity potential of the Earth. However, the basic probability space to be studied consist of the (linear) functionals giving for a fixed point P the values of the anomalous potential or the gravity anomaly in this point.

In general the starting point is a probability space (H, A, P), where H is an arbitrary set (generally denoted Ω in statistical literature), A is a Boolean σ-algebra of subsets of H and P is a probability measure. The stochastic or random variables, X, are mappings from H to **R**. If these mappings are parameterized through an index set, T, a stochastic process $\{X_t, t \in T\}$ is defined, provided the joint probabilities are defined, cf. Parzen (1959). The index set is frequently the real line (t is time), but here we should think of T as being the set $\mathbf{R}^3$.

The stochastic process is called a random function provided the second order moments are finite, i.e. $\int_H X_t^2 \, dP < \infty$ for all t. In this case we will define a linear vector space as the one spanned by the random function,

$$L\{X_t, t \in T\} = \text{span } \{X_t, t \in T\}$$ (2.11)

If H is the linear vector space of example 2.1, a stochastic process is formed by the evaluation functionals $L_p(f) = f(P)$, with Ω as the index set. The probability measure may be defined by requiring all L_p, $P \in \Omega$ to be normally distributed with zero mean and covariance

$$\int_H L_p \cdot L_Q \, dP = \text{cov}(P, Q),$$ (2.12)

where $\text{cov}(P,Q)$ is a positive definite function, $\Omega \times \Omega \to \mathbf{R}$.

Note, that the linear vector space (2.11) will not automatically be equal to H*. We would also like that this space should contain the functionals, which give the value of the Laplace operator in a point, and that these stochastic variables should be equal to zero with probility one. This will be achieved, if $\text{cov}(P,Q)$ is a harmonic function in each of its variables, and if we add to the space (2.11) all limits of sequences of stochastic variables which have finite variance. ●

2.3 Inner product and normed space.

Example 2.5 was in fact an example of what we now will define, namely a linear vector space with an inner product. In general it is a mapping $(\cdot,\cdot)$: $H \times H \to R$, which is symmetric, linear, homogeneous and positive, i.e. $(f,g) = (g,f)$, $(f_1 + f_2, g) = (f_1, g) + (f_2, g)$, $(af,g) = a(f,g)$ and $(f,f) \geq 0$, with zero only occuring for $f \equiv 0$. The most simple example is the inner product in $\mathbf{R}^n$, $(x,y) = \sum_{i=1}^{n} x_i y_i$.

From the inner product a norm may be defined, $\|f\| = (f, f)^{\frac{1}{2}}$, which also will be positive, definite and homogeneous. The norm will furthermore fulfil the triangle inequality,

$$\|f + g\| \leq \|f\| + \|g\| \tag{2.13}$$

A norm need not to be an offspring of an inner product. The most simple example of a norm, which is not related to an inner product is the maximum norm,

$$\|f\| = \max |f(x)| , x \in \Omega, f \in C^0(\Omega).$$

Note that a normed vector space is also a metric space, since we to all elements f, g may associate their distance, $d(f,g) = \|f - g\|$. ●

Example 2.6 The Sobolev norms. Regard the linear vector space $C^m(\Omega), \Omega \subseteq \mathbf{R}^n$. Define

$$D^\alpha f = \frac{\partial^{\alpha_1}}{\partial x_1^{\alpha_1}} \frac{\partial^{\alpha_2}}{\partial x_2^{\alpha_2}} \cdots\cdots \frac{\partial^{\alpha_n}}{\partial x_n^{\alpha_n}} f, \quad \sum_{i=1}^{n} \alpha_i = |\alpha| \leq s. \tag{2.14}$$

Then inner products may be defined for positive integer values of s on subsets of $C^m(\Omega)$, $m \geq s$ by

$$(f,g)_s = \sum_{|\alpha| \leq s} \int D^\alpha f \cdot D^\alpha g \, d\Omega \tag{2.15}$$

Norms may be defined as the p'th root of the integral of $(D^\alpha f)^p$. Then the subsets we will consider are these for which the corresponding norm is finite. The corresponding complete spaces (see next capter) are denoted Sobolev or $H^{s,p}$-spaces. We will here only consider spaces with p = 2, so we will drop the second superscript.

Norms may also be defined for non-integer (and even negative) values of s. Non-integer values are defined using Fourier-transforms of the functions, while norms corresponding to negative values of s are defined using

$$\|f\|_{-s} = \max_{g \in H^s} \frac{\int f \cdot g \, d\Omega}{\|g\|_s} \tag{2.16}$$

The Sobolev spaces are used in modern theory of boundary value problems, because the spaces are used to characterize for which spaces boundary value problems have a unique solution. They are also much used in practice, because solutions to for example the Laplace-equation, which are elements of $C^\infty(\Omega)$, may be approximated arbitrarity well using elements of the H^s-spaces, see e.g. Aubin (1972). ●

In a linear vector space with an inner product the Schwarz inequality is valid,

$$|(f,g)|^2 \leq \|f\|^2 \cdot \|g\|^2 \tag{2.17}$$

The equality sign is only valid when f and g are linearily dependent. This enables us to define an angle θ between two non-zero vectors f,g, because

$$-1 \leq \cos\theta = \frac{(f,g)}{\|f\| \cdot \|g\|} \leq 1 \tag{2.18}$$

With $\theta = 0$, f, g are parallel and when $\theta = 90^0$, the two elements are orthogonal, i.e. $(f,g) = 0$.

Example 2.7 Correlation as angle between vectors.
In the space (2.11) the covariance expresses an inner product,

$$E(x,y) = (x,y) = \int_H x \cdot y \; dP$$

$$\|x\| = E(x^2)^{\frac{1}{2}}$$

and the correlation is cosine of the angle

$$\cos(\theta) = E(x,y)/(E(x^2) \cdot E(y^2))^{\frac{1}{2}}$$ ●

From a set of linearly independent elements, a new set of orthogonal elements may be constructed using the so-called Gram-Schmidt orthonomalization procedure.

From the set $\{f_i, \; i = 1,\ldots\ldots,n\}$ we generate a new set $\{g_i, \; i = 1,\ldots, n\}$ by

$$g_1 = f_1/\|f_1\|,$$

and for $k = 2,\ldots\ldots, n$

$$g_k = (f_k - \sum_{i=1}^{k-1} (f_k, g_i)g_i)/\|f_k - \sum_{i=1}^{k-1} (f_k, g_i)g_i\| .$$

In fact in each step we first subtract from f_k its projection on the space spanned by the k-1 first vectors and then divide by the norm of this vector. (Hereby we do not only arrive to an orthogonal set, but all vectors have norm equal to 1).

Example 2.8 The first Legendre polynomials.
Regard the 3-dimensional vector-space spanned by the polynomials $f_o=1$, $f_1=x$, $f_2=x^2$

on the interval from -1 to 1, with inner product

$$(f,g) = \tfrac{1}{2} \int_{-1}^{1} f(x)g(x)\,dx$$

Using the Gram-Schmidt procedure we get

$$\| f_o \|^2 = \tfrac{1}{2}\int_{-1}^{1} 1\cdot dx = 1, \text{ i.e. } g_o = 1.$$

Since $(g_o, f_1) = 0$ and $\|f_1\|^2 = \tfrac{1}{2}\int_{-1}^{1} x^2 dx = 1/3$, then

$$g_1(x) = \sqrt{3}\cdot x$$

From

$$(g_o, f_2) = \tfrac{1}{2}\int_{-1}^{1} x^2\,dx = 1/3, (g_1, f_2) = \tfrac{1}{2}\int_{-1}^{1} x^2 x\sqrt{3}\,dx = 0$$

and

$$\| f_2 - 1/3\cdot 1 \|^2 = \tfrac{1}{2}\int_{-1}^{1} (x^2 - 1/3)^2 dx = (2/(3\sqrt{5}))^2$$

we get

$$g_2(x) = (x^2 - 1/3)/(2/(3\sqrt{5})) = \sqrt{5}(3/2\,x^2 - 1/2).$$

A continuation of the process with x^3, x^4, etc. will produce all the normalized Legendre polynomials, $\bar{P}_n(x) = \sqrt{2n+1}\ P_n(x)$. ●

Let then g_o, g_1....., be a finite or infinite sequence of orthonormal elements. The series

$$\tilde{f} = \sum_{n=1}^{\infty} (f, g_n)g_n = \sum_{n=1}^{\infty} a_n g_n \tag{2.19}$$

is called the orthogonal expansion or Fourier expansion of f with the coefficients (or coordinates) $a_n = (f, g_n)$.

Remember that $\tilde{f}(P)$ not necessarily is equal to $f(P)$. It is well known that the harmonic functions of example 2.1 with boundary values in $H^o(\sigma)$ may be expressed as a series in (normalized) spherical harmonics,

$$\left.\begin{aligned} \bar{V}_{ij}(P) &= \left(2(2i+1)\frac{(i-j)!}{(i+j)!}\right)^{\frac{1}{2}} V_{ij}(P) \qquad j \neq 0 \\ \text{and} \\ \bar{V}_{io}(P) &= (2i+1)^{\frac{1}{2}} V_{io}(P). \end{aligned}\right\} \tag{2.20}$$

Example 2.9 The function $g(x) = -2x+1$ is an element of the space of example 2.8. The Fourier-expansion of g is $g(x) = 1\cdot g_0(x) - \frac{2}{\sqrt{3}} g_1(x)$. The norm of $g(x)$ is

$$\|g(x)\|^2 = \frac{1}{2}\int_{-1}^{1}(-2x+1)^2 dx = \frac{7}{3} = 1 + (2/\sqrt{3})^2. \bullet$$

Example 2.10 Use of the orthogonality concept to characterize the set of density distributions with zero exterior potential.

We regard as possible density distributions the set of all square-integrable functions in Ω_0. Let $H(\Omega_0)$ be the set of all harmonic functions regular in Ω_0, zero outside Ω_0. Regard then the functions which are equal to the Laplace operator applied on a two times differentiable function, which is zero in a set enclosed in Ω_0. Then $f = \Delta(g)$. This function will generate g as its potential, i.e. have potential zero outside Ω_0, and it will be orthogonal on all functions harmonic in Ω_0. Then suppose $\Delta(h) = 0$ in Ω_0, and using Greens first identity:

$$\int h\cdot f\, d\Omega_0 = \int h\Delta(g) d\Omega_0 = \int \Delta(h) g\, d\Omega_0 + \int_\omega h \frac{\partial g}{\partial n} d\omega - \int_\omega g\frac{\partial h}{\partial n} d\omega = 0.$$

Hence, f and h are orthogonal. ●

3. Hilbert spaces and reproducing kernels

3.1 Banach and Hilbert spaces

A sequence $\{f_i, i=1,\dots,\infty\}$ in a normed (linear) space is called a Cauchy sequence if it has the property that for all $\varepsilon > 0$ there is an integer n so that $d(f_i,f_j) \le \varepsilon$ for $i,j \ge n$. The space is called complete if every Cauchy sequence has a limit which belongs to the space. A complete, normed linear space is called a Banach space, and if the norm is derived from an inner product, then it is a Hilbert space.

Example 3.1. $C^0(\Omega)$, with Ω being the closed interval from a to b, may be equipped with the norm $\| f \| = \max_{a\le x\le b} |f(x)|$, and it is complete in this norm. But using $\| f \|^2 = \int_a^b f(x)^2 dx$ makes the space incomplete because the limits of continuous functions need not to be continuous using this norm . ●

Since the anomalous gravity potential, T, always may be considered an element of a vector space with a countable basis, we need only consider so-called separable Hilbert spaces. In such a space the expansion of an element f with respect to an orthonormal basis $\{f_1, f_2,\dots\}$ converge in the norm towards f, $\lim_{n\to\infty} \| f - \Sigma_{i=1}^{n} a_i f_i \| = 0$, (but we still do not necessarily have $f(P) = \sum_{i=1}^{\infty} a_i f(P)$.) The spaces have a lot of nice properties, which makes them nearly look like a finite dimensional Euclidean space:

The inner product of two elements may be calculated as scalar products of their coordinates,

$$(f,g) = \sum_{i=1}^{\infty} (f,f_i)\cdot(g,f_i).$$

Also an element is uniquely determined by its coordinates

$$(g,f_i) = (f,f_i),\ i=1,\dots,\infty \quad\Rightarrow\quad g = f$$

$$(f_i,f) = 0,\quad i=1,\dots,\infty \quad\Rightarrow\quad f = 0\ .$$

Example 3.2. The space of harmonic functions introduced in example 2.1 may as mentioned in Chapter 2 be equipped with the inner product

$$(f,g)_0 = \frac{1}{4\pi R}\int_\sigma f\cdot g\, d\sigma \tag{3.1}$$

The subset for which $\| f \| < \infty$ will form a separable Hilbert space which we will denote $H^o(\sigma)$. (The integral in eq. (3.1) is to be understood as the limit of the integrals over concentric spheres with radii $R + \varepsilon$, $\varepsilon > 0$ for $\varepsilon \to 0$. This is because we only require the boundary values to be square integrable). ●

Exercise 3.1. The potential of a point mass located at $P_1 = (\varphi_1, \lambda_1, R_1)$ with mass M_1 is at $P = (\varphi, \lambda, r)$ equal to $G \cdot M_1/L_1$, $L_1 = \sqrt{r^2 - 2rR_1\cos\psi_1 + R_1^2}$, with ψ_1 being the spherical distance between P and P_1. Use the inner product (3.1) to show that the inner product of two potentials in P_1 and $P_2 = (\varphi_2, \lambda_2, R_2)$, mass M_2, is equal to

$$\left(\frac{GM_1}{L_1}, \frac{GM_2}{L_2}\right)_o = \sum_{n=o}^{\infty} \frac{GM_1 \cdot GM_2}{(2n+1)R^2} \left(\frac{R_1R_2}{R^2}\right)^n P_n(\cos\psi_{12}). \quad ●$$

3.2 Representers of functionals

One of the most useful properties of Hilbert spaces (and certain Banach-spaces) is that the continuous linear functionals may be represented "analytically" as the inner product of the function on which the functional is to be applied with a fixed function. This is as pointed out in example 2.2 a well known property in $\mathbb{R}^n$.

A functional, L, is bounded if there exist a constant, M, so that $|L(f)| \leq M \cdot \|f\|$ for all $f \in H$. If $\|f_i - f\| \to 0$ then

$$|L(f_i) - L(f)| = |L(f_i - f)| \leq M \cdot \|f_i - f\|, \tag{3.2}$$

so $|L(f_i) - L(f)| \to 0$, i.e. a bounded functional is continuous. Also a continuous functional is bounded. For these functionals a norm may be introduced being equal to the maximal value M for which eq. (3.2) holds for all $f \in H$, i.e.

$$\|L\|_* = \sup_{f \in H} \frac{|L(f)|}{\|f\|} \qquad (f \neq 0)$$

and then consequently

$$|L(f)| \leq \|f\| \cdot \|L\|_* \tag{3.2a}$$

Example 3.3. The norm of the coordinate functionals. Regard L_i, $L_i(f) = (f, f_i) = a_i$. Then it is obvious that

$$\|L_i\|_* = \sup \frac{|a_i|}{(\Sigma a_i^2)^{\frac{1}{2}}} = 1. \quad ●$$

This introduction of a norm of the bounded linear functionals makes this linear subspace of the space of all linear functionals, a Hilbert space, H^*.

To each $L \in H^*$ there exist a unique element $\ell \in H$ so that

$$L(f) = (\ell, f) .$$

The element ℓ is denoted the (Riesz) representer of L. Using eq. (3.2)

$$|L(f)| \leq |(f,\ell)| \leq \|f\| \cdot \|\ell\|$$

i.e. $\|L\|_* \leq \|\ell\|$. Also

$$\|L\|_* = \sup_{f \in H} \frac{|L(f)|}{\|f\|} \geq \frac{|L(\ell)|}{\|\ell\|} = \frac{|(\ell,\ell|}{\|\ell\|} = \|\ell\|,$$

so $\|L\|_* \geq \|\ell\|$ and consequently $\|L\|_* = \|\ell\|$.

The inner product in H^* of two functionals L_1 and L_2 may be defined directly as the inner product of the representers, ℓ_1 and ℓ_2,

$$(L_1, L_2)_* = (\ell_1, \ell_2).$$

3.3 Reproducing kernels

Let us now regard a separable Hilbert space of functions, $f: \Omega \rightarrow \mathbf{R}$. If the evaluation functionals $L_P(f) = f(P)$ are bounded, then there exist a <u>reproducing kernel</u>. Let us denote the representer of L_P by $K_P(Q)$. Then

$$(K_P(Q), f(Q)) = f(P) \qquad P,Q \in \Omega .$$

If P varies, we have a mapping

$$K(P,Q): \Omega \times \Omega \rightarrow \mathbf{R},$$

with the reproducing property

$$(K(P,Q), f(Q)) = f(P). \tag{3.3}$$

It is called a kernel, since it, when used in spaces with inner products like in example 3.2, looks like the kernel of an integral equation.

The reproducing kernel may be expressed explicitly using a known orthonormal base, f_i, $i=1,\ldots,\infty$,

$$K(P,Q) = \sum_{i=1}^{\infty} f_i(P)f_i(Q) \ . \tag{3.4}$$

This is easily seen by inserting this expression in eq. (3.3) and using the expansion of $f=\sum_{i=1}^{\infty} a_i f_i$,

$$(f(P), K(P,Q)) = (\sum_{i=1}^{\infty} a_i f_i(P), \sum_{j=1}^{\infty} f_j(P)f_j(Q)) = \sum_{i=1}^{\infty} a_i f_i(Q) = f(Q),$$

where we have used the linearity of the inner product and the orthonormality of the base functions.

Example 3.4. Regard the 3-dimensional space of example 2.8. The base functions are 1, $\sqrt{3}x$ and $\sqrt{5}(\frac{3}{2}x^2-\frac{1}{2})$. Hence,

$$K(x,y) = 1\cdot 1 + 3xy + 5(\tfrac{3}{2}x^2 - \tfrac{1}{2})(\tfrac{3}{2}y^2 - \tfrac{1}{2}) = \tfrac{9}{4} + 3xy - \tfrac{15}{4}(x^2+y^2) + \tfrac{45}{4}x^2y^2 . \bullet \tag{3.5}$$

Exercise 3.2. Check that the function $g(x) = -2x+1$ used in example 2.9 is reproduced by the kernel (3.5). ●

Exercise 3.3. Show using eq. (2.20), (3.1) and (3.4) that the Hilbert space of harmonic functions $H_0(\sigma)$ has the reproducing kernel

$$K_0(P,Q) = \sum_{i=0}^{\infty} (2i+1) \frac{R^{2i+2}}{(rr')^{i+1}} P_i(\cos\psi), \tag{3.6}$$

and that the reproducing property eq. (3.3) then becomes identical to Poissons integral

$$f(\varphi,\lambda,r) = \frac{1}{4\pi} \int_{-\pi/2}^{\pi/2} \int_{-\pi}^{\pi} \frac{(r^2-R^2)}{(r^2+R^2-2Rr\cos\psi)^{3/2}} f(\varphi',\lambda',R)\cos\varphi' d\lambda' d\varphi' \tag{3.7}$$

with $Q = (\varphi',\lambda',r')$. Show also, that the evaluation functionals at the boundary does not belong to $H_0(\sigma)^*$.●

Example 3.5. The Krarup kernel. Regard the space of harmonic functions in example 2.1. A Hilbert space is formed by the functions for which the norm

$$\| f \|^2 = \frac{1}{4\pi} \int_\Omega \frac{1}{r}(\nabla f)^2 d\Omega \tag{3.8}$$

is finite .

The reproducing kernel of this space is of special interest because it is very simple. In the space the solid spherical harmonics will still be orthogonal, but they must be normalized with respect to the new norm. Since

$$\frac{1}{r}\nabla\bar{V}_{ij} = \begin{Bmatrix} -\frac{(i+1)}{r^2}\bar{V}_{ij} \\ \frac{1}{r^2}\frac{\partial}{\partial\varphi}\bar{V}_{ij} \\ \frac{1}{r^2}\frac{1}{\cos\varphi}\frac{\partial}{\partial\lambda}\bar{V}_{ij} \end{Bmatrix} = \begin{Bmatrix} -\frac{(i+2)}{r}(\frac{1}{r}\bar{V}_{ij}) \\ \frac{1}{r}\frac{\partial}{\partial\varphi}(\frac{1}{r}\bar{V}_{ij}) \\ \frac{1}{r\cos\varphi}\frac{\partial}{\partial\lambda}(\bar{V}_{ij}) \end{Bmatrix} + \begin{Bmatrix} \frac{1}{r^2}\bar{V}_{ij} \\ 0 \\ 0 \end{Bmatrix}$$

we see that

$$\|\bar{V}_{ij}\|^2 = \frac{1}{4\pi}\int \nabla(\frac{1}{r}\bar{V}_{ij})\nabla\bar{V}_{ij}d\Omega + \frac{1}{4\pi}\int\frac{1}{r^2}\bar{V}_{ij}(-i-1)\frac{1}{r}\bar{V}_{ij}d\Omega$$

$$= \frac{1}{4\pi}\int_\sigma \frac{1}{r}\bar{V}_{ij}\frac{\partial}{\partial r}|_R V_{ij}R^2\cos\varphi d\lambda d\varphi + \int_\Omega (\frac{R}{r})^{2i+2}\frac{(-i+1)}{r^3}$$

$$\bullet(\bar{P}_{ij}(\sin\varphi)\begin{Bmatrix}\cos j\lambda \\ \sin j\lambda\end{Bmatrix})^2 r^2 d\lambda\cos\varphi\, d\varphi dr$$

$$= i+1 + [\frac{-1}{2i+2}(\frac{R}{r})^{2i+2}(-i-1)]_R^\infty = \frac{2i+1}{2R}\ .$$

Hence

$$K(P,Q) = \sum_{i=0}^{\infty}\sum_{j=-i}^{i}\frac{2R}{2i+1}\bar{V}_{ij}(P)\bar{V}_{ij}(Q) = 2R\sum_{i=0}^{\infty}(\frac{R^2}{rr'})^{i+1}P_i(\cos\psi)\ .$$

Using $L = (R^2-2rr'\cos\psi + (rr'/R)^2)^{\frac{1}{2}}$

$$K(P,Q) = \frac{2R}{L}\ . \tag{3.9}$$

Note that this kernel may be interpreted as the potential of a point mass located at $Q' = (\varphi',\lambda',\frac{R^2}{r'})$ with mass $2R^2/(r'\cdot G)$,

$$K(P,Q) = (2R^2/r')/|P-Q'|\ .\ \bullet \tag{3.10}$$

<u>Example 3.6.</u> The Hilbert space of exploration geophysics.

Suppose a geological structure has been divided in n non-overlapping blocks with volume v_k, k=1,...,n. Then regard the linear vector space spanned by the indicator functions I_k, equal to 1 inside the k'th block and zero outside. The functions span an n-dimensional space. Using the inner product equal to the integral of the product of two functions, we see that the functions are orthogonal, with norm equal to the square-root of the volume. The reproducing kernel is then

$$K(P,Q) = \sum_{k=1}^{n} I_k(P)\, I_k(Q)/v_k \; . \quad \bullet \tag{3.11}$$

<u>Example 3.7.</u> The covariance function of a stohastic process as a reproducing kernel.

A family of vectors $\{L_t,\ t \in T\}$ in a Hilbert space H^* is said to be a representation of the stochastic process $\{X(t),\ t \in T\}$ with covariance kernel cov(s,t) if

$$(L_t, L_s)_{H^*} = \mathrm{cov}(s,t) = E(X_s X_t).$$

Let now T be a "nice" subset of $\mathbf{R}^n$. Then by the so-called Mercer's theorem, there exist a sequence of normalized eigenfunctions g_i,

$$\int_T \mathrm{cov}(s,t) g_i(s) ds = \lambda_i g_i(t) ,$$

$$\int_T g_n(t) g_m(t)\, dt = \delta_{nm} ,$$

so that

$$\mathrm{cov}(s,t) = \sum_{n=1}^{\infty} \lambda_n g_n(s) g_n(t), \quad \lambda_n \geq 0. \tag{3.12}$$

Then let H be the space of all functions spanned by the set $g_n/\sqrt{\lambda_n}$, which we define as being orthonormal. Hereby the inner product of H is fixed, and the space consist of all the functions for which the square-sum of the coordinates (i.e. the norm in H) is finite. The dual space H^* will then be a representation of the process, and cov(s,t) the reproducing kernel. In this manner may all stochastic processes (random functions), be regarded as spanning a Hilbert space with reproducing kernel. The converse is also true . ●

<u>Exercise 3.4.</u> The reproducing kernel eq. (3.11) may be interpreted as a covariance function for the related density structure. What is the covariance of two gravity anomalies in two points P and Q outside the structure ? ●

<u>Exercise 3.5.</u> Show that the reproducing kernel for the subset of the space in example 2.1 equipped with the $H^0(\Omega)$ norm (multiplied by $\frac{1}{4\pi R^3}$), but with the subspace of degree zero excluded is

$$K_0(P,Q) = \sum_{i=1}^{\infty} (2i-1)(2i+1)\left(\frac{R^2}{rr'}\right)^{i+1} P_i(\cos\psi) . \bullet \qquad (3.13)$$

Like in the Sobolev space $H^1(\Omega)$ and in the space of example 3.6 we may introduce a norm involving the integral of the product of the first order derivatives. This is called the Dirichlet inner product,

$$(f,g)_2 = \int \nabla f \cdot \nabla g \, d\Omega. \qquad (3.14)$$

In this case the reproducing kernel is the sum of the Green's and Neumann's functions for Ω , see Garabedian (1964, section 7.3),

$$K_1(P,Q) = G(P,Q) + N(P,Q).$$

Exercise 3.6. Use Green's identity, and the fact that the normal derivative, $\frac{\partial N}{\partial n}$, and G are zero at the boundary to show that the reproducing property is equivalent to the properties that the two kernels solve the boundary value problems where either $\frac{\partial f}{\partial n}$ or f are known. ●

Exercise 3.7. Show that the reproducing kernel for the functions harmonic outside a sphere with radius R, equipped with the Dirichlet inner product (multiplied by $\frac{1}{4\pi R^3}$), is

$$K_2(P,Q) = \frac{1}{R} \sum_{i=0}^{\infty} \frac{2i+1}{i+1} \left(\frac{R^2}{rr'}\right)^{i+1} P_i(\cos\psi) . \qquad (3.15)$$

Use the result in example 3.6 and

$$\int^{\infty} \frac{ds}{L} = -s\,\ell n\left(\frac{1-\cos\psi}{L+s+\cos\psi}\right), \qquad s = \frac{R^2}{rr'}$$

to construct a closed expression for the kernel . ●

A Hilbert space may also be constructed based on a norm on the space of bounded (density) functions, with the same support Ω_0 as the Earth's density distribution. Since the purpose of using such a space must be that we subsequently should be able to evaluate (compute) density estimates, then the evaluation functionals should be bounded, i.e. the space should have a reproducing kernel. This may be achieved like in example 3.6 by working in a finite dimensional space, but also spaces of functions which fulfil an elliptic (and certain parabolic) partial differential equations can be used. Here the most simple example is the space of harmonic density functions, cf. example 2.10. Let us use the inner product $H^0(\Omega_0)$,

$$(\rho_1,\rho_2)_D = \int_{\Omega_o} \rho_1(P)\rho_2(P)\, d\,\Omega_o . \tag{3.16}$$

The value of the potential T_1 generated by ρ_1 in a point Q is the value of the (Newton) functional, N_Q, applied on ρ_1, which may be expressed very simply by the inner product

$$N_Q(\rho_1) = T_1(Q) = G\int \frac{\rho_1(P)}{|P-Q|}\, d\,\Omega_o . \tag{3.17}$$

Note, that the harmonic function $k(P,Q) = G/|P-Q|$, (with singularity in Q outside Ω_o) is the Riesz representer of N_Q in $H^o(\Omega_o)$.

If Q varies, we have an operator N from $H^o(\Omega_o)$ to the set of functions harmonic in Ω. This will be a linear vector space, and we may introduce an inner product by $(T_1,T_2)_H = (\rho_1,\rho_2)_D$. Since

$$T(P) = (\rho(Q),\, k(P,Q))_D = (N(\rho),\, N(k(P,Q)))_H = (T(Q),\, N_P(k(S,Q)))_H ,$$

we see that (cf. Krarup (1978))

$$N_P(k(S,Q)) = G^2 \int_{\Omega_o} \frac{1}{|P-S|\,|Q-S|}\, d\,\Omega_o = K_D(P,Q) \tag{3.18}$$

is the reproducing kernel in H.

<u>Example 3.8.</u> If Ω_o is a sphere with radius R, K_D may be computed explicitly.

$$\int \frac{1}{|P-S|}\,\frac{1}{|S-Q|}\, d\,\Omega_o =$$

$$\int_\varphi\int_\lambda\int_o^R \sum_{n=o}^{\infty} \frac{r_S^{\,n}}{r_P^{\,n+1}} P_n(\cos\psi_{PS}) \sum_{i=o}^{\infty} \frac{r_S^{\,i}}{r_Q^{\,i+1}} P_i(\cos\psi_{QS})\, r_S^{\,2}\cos\varphi_S\, d\varphi_S\, d\lambda_S\, dr_S$$

$$= \sum_{j=o}^{\infty} \frac{R^{2j+3}}{2j+3}\, \frac{4\pi}{(r_P r_Q)^{j+1}}\, \frac{1}{2j+1}\, P_j(\cos\psi_{PQ})$$

$$= \sum_{j=o}^{\infty} \frac{4\pi R}{(2j+3)(2j+1)} \left(\frac{R^2}{r_P r_Q}\right)^{j+1} P_j(\cos\psi_{PQ}) = K_{\Omega_o}(P,Q) \tag{3.19}$$

Note that the evaluation functionals at the boundary have finite norm . ●

We have now seen two examples of Hilbert spaces, where the set of harmonicity is the actual set of harmonicity for T, and where the inner product is computed by integration over the actual set over harmonicity. Hence, T may be supposed to be an element of such a space. This is important if one tries to assure the convergence of an approximation algorithm, where more and more base functions (or data) are used. On the other hand, T may be approximated arbitrarily well using functions which are not even harmonic, or which belongs to spaces having functions with sets of harmonicity larger than the set outside the solid earth, cf. the Runge-Krarup theorem, (Krarup, 1969).

We will now consider an important space, where the inner product in some way has been selected so that it suits T optimally, but as we will see does not contain T itself. The idea is to use the metric used in practice to evaluate the quality of an approximation, namely the integral of the square of the difference between the true values and the estimated values. The basic operator, is the mean value operator

$$M(f(P)) = \frac{1}{A} \int f(P)\, d\sigma \tag{3.20}$$

where A is the area or volume over which the integral is calculated. In statistics, the mean value is calculated as the integral with respect to a probability measure, but the actual estimate is the usual mean value of the observations (assuming they all have the same weight). When computing the so-called empirical covariances, we will use the same mean value operator, but introduce (in some cases) a further mean over all directions, so that the mean is formed over all point pairs with the same spherical distance and having the same distance from the origin. (See Goad et al. (1985) and Tscherning (1985, section 3.2) for a discussion of the estimation and modelling of empirical covariance functions.) Let us evaluate this covariance function on the sphere with radius R. Then

$$C(P,Q) = M(T(P)\cdot T(Q)) =$$

$$\frac{1}{8\pi^2} \int_0^{2\pi} \int_{-\pi/2}^{\pi/2} \int_{\alpha=0}^{2\pi} T(\varphi,\lambda,R)T(\varphi',\lambda',R)\cos\varphi \, d\varphi d\lambda d\alpha \,. \tag{3.21}$$

If we work in sperical approximation, we may suppose that T can be developed in normalized solid spherical harmonics,

$$T(\varphi,\lambda,r) = \frac{GM}{R} \sum_{i=2}^{\infty} \sum_{j=-i}^{i} \bar{a}_{ij} \bar{V}_{ij} (\varphi,\lambda,r) \,. \tag{3.22}$$

Then as shown in Heiskanen and Moritz (1967, section 7.3) we have

$$C(P,Q) = C(\psi_{PQ}) = \sum_{i=2}^{\infty} \sigma_i P_i(\cos\psi_{PQ}), \sigma_i = (\frac{GM}{R})^2 \sum_{j=-1}^{i} \bar{a}_{ij}{}^2 \,, \tag{3.23}$$

where σ_i are the so-called (potential) degree-variances.

The general expression for r or r' ≠ R contains a factor $(\frac{R^2}{rr'})^{i+1}$. It is well known that the degree-variances tend to zero for $i \to \infty$ somewhat faster than i^{-3}, see (Tscherning, and Rapp 1974), Forsberg (1984).

The covariance function may as explained in example 3.8 (in a slightly different context) be interpreted as a reproducing kernel, where the norm has been fixed, by defining which vectors form an orthogonal system of eigenfunctions. For the function (3.23) the eigenfunctions are the solid spherical harmonics, and the eigenvalues are $(2i+1)\sigma_i$ (where we suppose $\sigma_i > 0$, i=1). The new base vectors are then

$$V^*_{ij} = (\sigma_i/(2i+1))^{\frac{1}{2}}\bar{V}_{ij}$$

and

$$T(P) = \sum_{i=2}^{\infty}\sum_{j=-i}^{i} \frac{\bar{a}_{ij}(2i+1)^{\frac{1}{2}}}{\sigma_i^{\frac{1}{2}}} V^*_{ij}(P).$$

Hence

$$\| T \|_C^2 = \sum_{i=2}^{\infty} \frac{2i+1}{\sigma_i} \sum_{j=-i}^{i} \bar{a}_{ij}^2 = \sum_{i=2}^{\infty} 2i+1 = \infty ,$$

which shows that T does not belong to the space. (This paradox has been solved in F. Sansò: Statistical Methods in Physical Geodesy, published also in these Lecture Notes).

3.4 Representers of functionals in a reproducing kernel Hilbert space

Using the reproducing kernel, the representer of a linear functional is easily obtained,

$$L(f) = L(f(P), K(P,Q)) = (f(P), L(K(P,\cdot)),$$

i.e. the representer is L(K(P,·)), which we in the following will denote K(P,L). If two functionals are applied on K(P,Q) we will use

$$L_1L_2(K(P,Q)) = K(L_1,L_2) .$$

This is equal to the inner product of the two functionals, because it is defined as the inner product of their representers:

$$(L_1,L_2)_* = (K(P,L_1), K(P,L_2)) = L_1(K(P,Q), K(P,L_2)) = K(L_1,L_2) \qquad (3.24)$$

And then

$$\| L \|_*^2 = K(L,L).$$

We have earlier shown how a covariance function may be interpreted as a reproducing kernel in a Hilbert space, and also conversely explained how a reproducing kernel may be regarded as a covariance function of the evaluation functionals, again being equal to the inner product of the two functionals. From eq.(3.24) we may conclude, that also the inner product of two arbitrary functionals (in H^*) may be interpreted as covariances, obtained by applying the linear functionals on the basic covariance function (reproducing kernel). In this manner we can find the covariance function of the gravity anomaly functionals, by applying eq. (2.6) and (3.23):

$$C(\Delta g_P, \Delta g_Q) = \sum_{i=2}^{\infty} \sigma_i \frac{(i-1)^2}{R^2} \left(\frac{R^2}{rr'}\right)^{i+2} P_i(\cos\psi), \tag{3.25}$$

However, let us look at some simple examples.

Example 3.9. Let us find the representers of some simple functionals in the space with the reproducing kernel given in example 3.4

(a) $L(f) = f(\tfrac{1}{2}) \Rightarrow \ell(x) = K(x,\tfrac{1}{2}) = -\frac{15}{16}x^2 + \frac{3}{2}x + \frac{21}{16}$

(b) $L(f) = f(0) \Rightarrow \ell(x) = K(x,0) = -\frac{15}{4}x^2 + \frac{9}{4}$

(c) $L(f) = \frac{\partial f}{\partial x}\Big|_0 \Rightarrow \ell(x) = \frac{\partial}{\partial x}\Big|_0 K(x,y) = 3x$ ●

Example 3.10. A subspace of the space described in example 2.1 has reproducing kernel

$$K(P,Q) = \sum_{i=2}^{\infty} \frac{A}{(i-1)^2} \left(\frac{R^2}{rr'}\right)^{i+1} P_i(\cos\psi), \tag{3.26}$$

where A is in units of m^4/s^4. Having selected the kernel on this specific form makes the inner product of the gravity anomaly functionals very simple:

$$K(\Delta g_P, \Delta g_Q) = \left(\left(-\frac{\partial}{\partial r}\Big|_P - \frac{2}{r}L_P\right), \left(-\frac{\partial}{\partial r}\Big|_Q - \frac{2}{r'}L_Q\right)\right)_*$$

$$= \sum_{i=2}^{\infty} \frac{A}{rr'}\left(\frac{R^2}{rr'}\right)^{i+1} P_i(\cos\psi) = \frac{A\,R^2/(rr')^2}{\left(1-2\frac{R^2}{rr'}\cos\psi + \left(\frac{R^2}{rr'}\right)^2\right)^{\frac{1}{2}}} - \frac{AR^2}{(rr')^2}$$

$$- \frac{AR^4}{(rr')^3}\cos\psi\;. \tag{3.27}$$

Let us evaluate the covariance for $r = r' = \sqrt{2}R$. Then

$$K(\Delta g_P,\Delta g_Q) = \frac{A/(4R^2)}{(5/4-\cos\psi)} - \frac{A}{4R^2} - \frac{A}{8R^2}\cos\psi \quad .$$

For $P = Q$ we get

$$\| \frac{\partial}{\partial r} - \frac{2}{r}L_P \|_*^2 = \frac{A}{4R^2}(2-1-\tfrac{1}{2}) = \frac{A}{8R^2} \quad .$$

Note, that if the summation started from $i=0$, then $K(\Delta g_P,\Delta g_Q)$ would have been positive everywhere. By subtracting two terms, the function will get two zero-points, i.e. for two values of the spherical distance will the gravity anomaly functionals be independent.

The "trick" of selecting a model for the degree-variances (3.23) which makes the derived inner products for the most commonly used functional very simple has been used in Tscherning and Rapp (1974), where

$$\sigma_i \simeq \frac{A}{(i-1)(i-2)(i+B)} \quad , \quad B \in \mathbf{Z}_+ \ . \tag{3.28}$$

The reason for not using the denominator $(i-1)^2$ like in eq. (3.27), is that a closed expression can be found for a reproducing kernel using (3.28), while no closed expression is known for eq. (3.26).●

We conclude this chapter by pointing out how the norms and degree-variances of the associated reproducing kernels are related. In general the degree variances behave like i^{-2k+2} or i^{-2k+1} if we use a $H^k(\Omega)$, $H^k(\sigma)$-norm, respectively. This is important, if we want to select a space where the evaluation functionals, or gravity anomaly functionals have finite norm at the boundary. However by selecting a Hilbert space of functions harmonic in a set just slightly larger than the set outside the Earth, makes all functionals of geodetic interest nice and continuous. We only miss the possibility to estimate the variance of density which correspond to a certain estimated variance of T(P) in the set between the Earths surface and the new set of harmonicity. And this may unfortunately be exactly where we would like to know something about the density variation.

4. Approximation in a Hilbert space

4.1. Approximation in a vector space

The most simple form for an approximation is constructed as the linear combination ($\tilde{f}$) of a given set of m functions, f_i, which gives the best least squares fit to the data, $y_i = L_i(f)$, $i=1,\ldots,n$,

$$\sum_{i=1}^{n} (L_i(\sum_{j=1}^{m} f_j a_j) - y_i)^2 \cdot w_i = \min_{\{a_j\}} \tag{4.1}$$

where w_i are (positive) weights, depending e.g. on the quality of the data. There are then three different situations, depending on whether $n > m$, $n = m$ or $n < m$.

If we have an overdetermined problem, the solution is well known

$$\{a_j\} = (A^T W A)^{-1} A^T W y \;,\; A = \{L_i f_j\} \tag{4.2}$$

if $A^T W A$ is of full rank. (W is a diagonal matrix with w_i in the diagonal).

Note, that if the observations are values of f, $L_i(f) = f(P_i)$, regularily distributed with unit weights, then we here have a discretization of the minimum norm condition

$$\int_\Omega (\sum_{j=1}^{m} a_j f_j(P) - f(P))^2 d\Omega = \min_{a_j} \tag{4.3}$$

In case n=m, the minimum is simply found using

$$\{a_j\} = A^{-1} \{y_j\}. \tag{4.4}$$

These solutions are also in certain cases equivalent to mimimum norm solutions, where $\|f-\tilde{f}\|$ or $\|\tilde{f}\|$ is minimalized, as we shall see in section 4.3. In these cases A will be symmetric and positive definite.

The underdetermined problem is generally reduced to the problem n=m by finding the eigenvalues of the $m \times m$ matrix AA^T which has non-zero eigenvalues (singular value decomposition). These (n) eigenvectors are then used as new base vectors, see (Sansó et al., 1986, Appendix 1). Other alternatives are found, if we work inside a Hilbert space, see the following sections.

4.2 Best linear approximation in an inner product space, H

Given a set of linear independent elements $g_i \in H$, $i=1,\ldots,n$ it is possible to find a unique "best" linear approximation $\tilde{f}$ to a function $f \in H$,

$$\tilde{f} = \sum_{i=1}^{n} a_i g_i$$

in the sense that $\tilde{f}$-f has the smallest possible norm. This means that for any set b_i, i=1,...,n

$$\| f-\tilde{f} \| = \| f - \sum_{i=1}^{n} a_i g_i \| \leq \| f - \sum_{i=1}^{n} b_i g_i \| .$$

Using Gram-Schmidt orthonormalization, we can findan orthonormal set g_i^*, and it is easily seen that a best linear approximation is given by

$$\tilde{f} = \sum_{i=1}^{n} (f_i, g_i^*) g_i^*,$$

the projection of f on the subspace spanned by $\{g_i\}$. The difference between $\tilde{f}$ and f is orthogonal on all g_i^* because

$$(f - \sum_{j=1}^{n} (f,g_j^*) g_j^*, g_i^*) = (f,g_i^*) - \sum_{k=1}^{n} (f,g_k^*)(g_k^*, g_i^*) = (f,g_i^*) - (f,g_i^*) = 0$$

and must therefore always be orthogonal in all g_k, see Fig. 1.

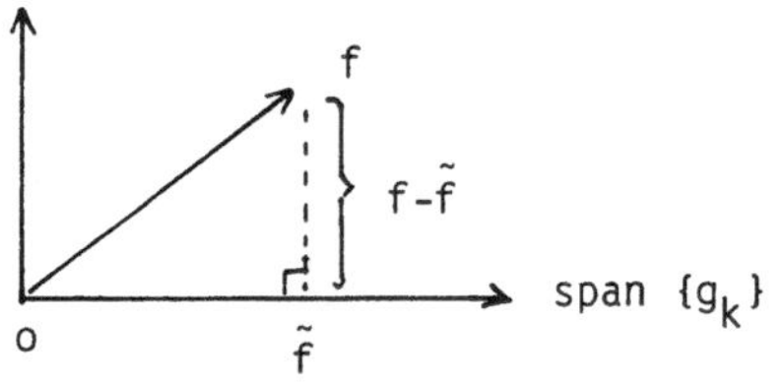

Fig. 1. Construction of $\tilde{f}$ as a projection.

We can use this to write down a system of <u>normal equation</u> which directly will determine the constants $\{a_i\}$. We must have

$$(f-\tilde{f}, g_i) = (f - a_1 g_1 \ldots - a_n g_n, g_i) = 0$$

or

$$a_1(g_1,g_i) + a_2(g_2,g_i) + \ldots a_n(g_n,g_i) = (f,g_i) .$$

On matrix form

$$\{ (g_i,g_j) \} \{a_j\} = \{(f,g_i)\} \tag{4.5}$$

Note, that in order to find $\tilde{f}$ we must require that f is an element of the <u>same</u> Hilbert space as the functions g_i so that (f,g_i) can be calculated. Also data must be available permitting the calculation. Or we must select the inner product, so that it uses the available data. When approximating T using gravity data, for example, an inner product

$$(f,g)_{\Delta g} = \int_\omega \left(\frac{\partial f}{\partial r}\Big|_P + \frac{2}{r}f(P)\right)\left(\frac{\partial g}{\partial r}\Big|_P + \frac{2}{r}g(P)\right)d\omega_P \quad (4.6)$$

could "easily" be used.

<u>Exercise 4.1:</u> Calculate the reproducing kernel corresponding to this norm used on the space of example 2.1 with the subspaces of degree 0 and 1 removed. Show that the gravity anomaly functionals evaluated at the boundary will have infinite norm, and hence do not belong to the dual space.●

4.3 Approximation in a reproducing kernel Hilbert space (RKHS)

A method to circumvent the actual evaluation of the inner product in eq. (4.5) is to use for g_i the representers of the functionals corresponding to the observations, $g_i = K(P,L_i)$. Eq. (4.5) becomes

$$\{(g_i,g_j)\}\ \{a_j\} = \{K(L_i,L_j)\}\ \{a_j\} = \{(f,g_i)\} = \{(f,K(P,L_i))\} = \{L_i f\}\ ,$$

where $L_i f$ are the observations. In this case we also have

$$L_i\tilde{f} = \sum_{k=1}^{n} a_k L_i(g_k) = \sum_{k=1}^{n} \{L_k f\}^T \{K(L_k,L_j)\}^{-1} \{K(L_i,L_j)\} = L_i f \quad (4.7)$$

i.e. there is an exact agreement between observations and values computed using the best approximation $\tilde{f}$.

The fulfilment of this condition is the basis for minimum norm collocation. Here a function $\tilde{f}$ is obtained so that eq. (4.7) is fulfilled and so that $\tilde{f}$ has the minimum norm in between the elements of a RKHS which fulfil this equation. Since we do not require that $\|f-\tilde{f}\|$ is minimal (or that it can be computed) we do not even need to require that f is an element of the RKHS, H. Only the linear functionals associated with the observations must be elements of H^*. (On the other hand if $f \in H$, then $\|f-\tilde{f}\|$ will be minimal).

The situation is shown in Fig. 2, where $\tilde{f}$ must be an element of an affine subspace, $\tilde{f} \in A = \{g \mid L_i g = L_i f,\ i=1,\dots,n\}$, where again $L_i f$ are the observed values.

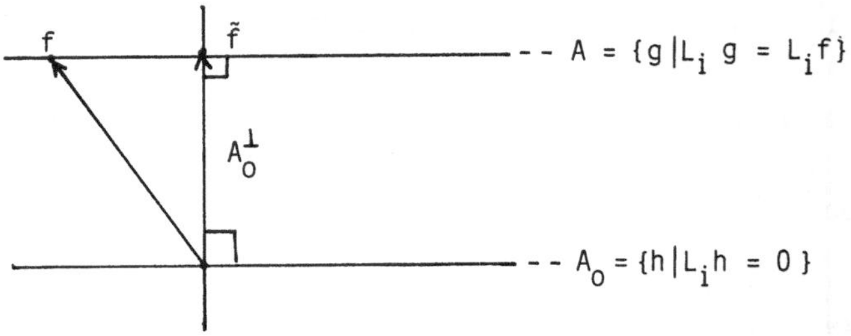

Fig. 2. The construction of $\tilde{f}$ as the intersection of two subspaces. Note that f is supposed to be in H in the figure.

$\tilde{f}$ must be the element in the affine subspace, which has the shortest distance from zero, i.e. it must be located on the n-dimensional subspace orthogonal to the affine subspace. This subspace must also be orthogonal on the subspace $A_0 = \{ h \mid L_i h = 0, i=1,\ldots, n \}$, parallel to the affine subspace A.

If we regard the representers $K(L_i,P)$, then if $h \in A_0$

$$(h(P),K(L_i,P)) = L_i(h(P),K(P,Q)) = L_i h = 0,$$

i.e. the functions $K(L_i,P)$ span the subspace orthogonal to A_0,

$$A_0^{\perp} = \{g = \sum_{i=1}^{n} a_i K(L_i P),\ \{a_i\} \in \mathbf{R}^n \} .$$

The function $\tilde{f}$ must therefore be equal to the intersection between $A_0^{\perp}$ and A, so

$$\tilde{f}(P) = \sum_{i=1}^{n} a_i\ K(L_i,P) \tag{4.8}$$

and

$$\{ L_i\tilde{f} \} = \{K(L_i,L_j)\} \ \{a_j\} = \{ L_i f \},$$

$$\{ a_j \} = \{K(L_i,L_j)\}^{-1} \ \{L_i f\} \tag{4.9}$$

A further condition for finding $\tilde{f}$ is then naturally that $\{K(L_i,L_j)\}$ is positive definite, or that the functionals as elements of H^* are linearily independent.

It can easily be proved rigorously (cf. Tscherning, 1975, p. 89) that $\tilde{f}$ has the minimum norm. In fact the norm is

$$\| \tilde{f} \|^2 = (\sum_{i=1}^{n} a_i \; K(L_i,P), \; \sum_{j=1}^{n} a_j \, K(L_j,P))$$

$$= \sum_{i=1}^{n} \sum_{j=1}^{n} a_i a_j (K(L_i,P), \; K(L_j,P))$$

$$= (\{ L_i f \}^T \{K(L_i,L_k)\}^{-1})^T \{K(L_k,L_p)\} \{K(L_p,L_j)\}^{-1} \{L_j f\}$$

$$= \{ L_i f \}^T \{K(L_i,L_j)\}^{-1} \{ L_j f \} \tag{4.10}$$

Using $\tilde{f}$, the approximate value corresponding to any functional $L \varepsilon H^*$ can be calculated,

$$L\tilde{f} = \sum_{i=1}^{n} a_i \; K(L,L_i) \; . \tag{4.11}$$

If $f \varepsilon H$ we may also calculate an upper limit for the error of prediction using eq. (3.2)

$$| Lf - L\tilde{f} \,| = |L(f-\tilde{f})| = |L(f) - \{K(L,L_i)\}^T \{ K(L_i,L_j) \}^{-1} \{ L_j f \} |$$

$$= | (L - \{K(L,L_i)\}^T \; \{K(L_i,L_j)\}^{-1} \{ L_j \}) (f) \, |$$

$$\leq \| f \| \cdot \| L - \{K(L,L_i)\}^T \{K(L_i,L_j)\}^{-1} \{ L_j \} \|_*$$

or

$$| L(f) - L(\tilde{f}) | \leq \|F\| \, (\|L\|_*^2 \; - \{ K(L,L_i) \}^T \{K(L_i,L_j)\}^{-1} \{ K(L,L_j) \})^{\frac{1}{2}} \; . \tag{4.12}$$

An alternative expression for the upper limit of the error is found in (Krarup, 1978, eq. (12)). However, such equations have a limited use since in all cases $\|T\|$ must be known.

<u>Exercise 4.2</u>. Consider the case where we have only one observation $L_P f = f(P)$. Write down the minimum norm collocation solution and the expression for the error bound. ●

<u>Exercise 4.3</u>. Consider the case where the n observations are the n first coefficients $\{b_i\}$ with respect to an orthogonal base f_i. Show that the minimum norm collocation solution is the Fourier expansion

$$\tilde{f}_n(P) = \sum_{i=1}^{n} b_i f_i(P)$$

as we should expect. Then suppose we have a second set of observations as well, $y_i = L_i(f)$, $i=n+1,\ldots,N$. Show that the collocation solution in this case is identical to the determination of an approximation $\tilde{f}_N$, using a reproducing kernel with the first n terms removed, and new data values $x_i = y_i - L_i(\tilde{f}_n)$, $\tilde{f} = \tilde{f}_n + \tilde{f}_N$. ●

Example 4.1. Cholesky decomposition of the normal equations. Let $C = \{K(L_i, L_j)\}$, $C_o = K(L,L) = \|L\|_*^2$ and $C_P = \{K(L, L_i)\}$. Then the solution to the normal equations $\{a_i\} = C^{-1}\{y_i\}$ may be calculated using Cholesky decomposition of C,

$$C = U^T U$$

where U is an upper triangular matrix, called the Cholesky factor. {a} is found by solving the simple system of equations

$$U\,a = (U^T)^{-1} y \ .$$

The Cholesky factor of the matrix

$$\begin{Bmatrix} C & C_P \\ C_P^T & C_o \end{Bmatrix}$$

is

$$\begin{Bmatrix} U\,(U^T)^{-1} C_P \\ 0 \ (C_o - C_P^T C^{-1} C_P)^{\frac{1}{2}} \end{Bmatrix} ,$$

which shows that Cholesky reduction produces the quantity needed in the estimate of the error bound eq. (4.12) as its diagonal element no. n+1.

As a part of the solution a new vector of observations $(U^T)^{-1}y$ is computed. It is easily seen that the representers of the linear functionals corresponding to these new observations are orthonormal. Hence, the addition of new observations does not require a new calculation of the Cholesky factor of the first n columns of the new normal equation matrix. (This is the basis for the re-start possibility implemented in the program GEOCOL, (Tscherning, 1985a), and corresponds to the so-called permanency property, cf. (Freeden, 1982).) ●

Example 4.2. Approximation of the function $g(x) = -2x+1$, using minimum norm collocation, and computation of error bound. We use the Hilbert space of example 3.4, and suppose we have two observations $g(0) = 1$ and $g(1) = -1$. We use the results of the examples 2.9, 3.9 and 4.1.

The representer of L_o and L_1 are f_o and f_1, where

$$f_o = -\frac{15}{4}x^2 + \frac{9}{4} \text{ and } f_1 = -\frac{15}{2}x^2 + 3x - \frac{3}{2}.$$

The approximation is then

$$\tilde{g}(x) = b_o f_o + b_1 f_1 \quad \text{with} \quad \{b\} = \{K(L_i, L_i)\}^{-1} \left\{ \begin{matrix} 1 \\ -1 \end{matrix} \right\}.$$

Then

$$\begin{Bmatrix} \frac{9}{4} & -\frac{3}{2} \\ -\frac{3}{2} & 9 \end{Bmatrix} \begin{Bmatrix} b_o \\ b_1 \end{Bmatrix} = \begin{Bmatrix} 1 \\ -1 \end{Bmatrix}$$

and from this $b_o = \frac{5}{12}$ and $b_1 = -\frac{1}{24}$, hence

$$\tilde{g}(x) = -\frac{15}{8}x^2 - \frac{1}{8}x + 1.$$

The difference $|\tilde{g}(x) - g(x)| = \frac{15}{8}|(x-1)x|$, is equal to zero for $x = 0,1$ as expected. The error expression (4.12) gives for the evaluation functional $L_x(f) = f(x)$

$$|g(x) - \tilde{g}(x)| \leq \|g\| \left(K(x,x) - \begin{Bmatrix} K(0,x) \\ K(1,x) \end{Bmatrix}^T \begin{Bmatrix} \frac{9}{4} & -\frac{3}{2} \\ -\frac{3}{2} & 9 \end{Bmatrix}^{-1} \begin{Bmatrix} K(0,x) \\ K(1,x) \end{Bmatrix} \right)^{\frac{1}{2}}$$

The error expression is according to example 4.1 the Cholesky factor of the 3×3 matrix

$$\begin{Bmatrix} \frac{9}{4} & -\frac{3}{2} & K(0,x) \\ -\frac{3}{2} & 9 & K(1,x) \\ K(0,x) & K(1,x) & K(x,x) \end{Bmatrix}$$

which is easily seen to be

$$\begin{Bmatrix} \frac{3}{2} & -1 & \left(-\frac{5}{2}x^2 + \frac{3}{2}\right) \\ 0 & 2\sqrt{2} & \frac{\sqrt{2}}{4}(5x^2 + 3x) \\ 0 & 0 & \left(\frac{15}{8}x^4 - \frac{30}{8}x^3 + \frac{15}{8}x^2\right)^{\frac{1}{2}} \end{Bmatrix}$$

Hence, since $\|g\| = (7/3)^{\frac{1}{2}}$,

$$|g(x) - \tilde{g}(x)| \le (\frac{7}{3})^{\frac{1}{2}} (\frac{15}{8})^{\frac{1}{2}} |x(x-1)| = (\frac{35}{8})^{\frac{1}{2}} |x(x-1)|$$

which is slightly more pessimistic than the true error. ●

Exercise 4.4. Repeat the calculations in example 4.2 with the empirical covariance function of g(x) as reproducing kernel,

$$K(x,y) = 1^2 \cdot 1 \cdot 1 + (\frac{2}{\sqrt{3}})^2 3xy = 1+4xy . \quad ●$$

4.4. Approximation using the empirical covariance function

Besides the reqirement of minimum norm of $\tilde{f}$, another (similar) requirement may lead to a collocation solution. The requirement is, that we want to find a function, so that the overall mean square prediction error becomes minimal. Let the error of prediction be e_P , and suppose our observations are $T(Q_i)$ $i=1, \ldots, n$, then

$$e_P = \tilde{T}(P) - T(P) = \sum_{i=1}^{n} b_{Pi} T(Q_i) - T(P)$$

then

$$e_P^2 = T(P)^2 - 2 \sum_i b_{Pi} T(P) T(Q_i) + \sum_i \sum_k T(Q_i) T(Q_k) b_{Pi} b_{Pk} .$$

We want $M(e_P^2)$ to be minimal. But with

$$M(T(Q_i)T(Q_k)) = C_{ik}$$
$$M(T(P)\, T(Q_i)) = C_{Pi} , \qquad M(T(P)^2) = C_0$$

we see

$$m_P^2 = M(e_P^2) = C_0 - 2 \sum b_{Pi} C_{pi} + \sum \sum b_{Pi} b_{Pk} C_{ik} .$$

If we want to minimize this expression, then

$$\frac{\partial m_P^2}{\partial b_{Pi}} = - 2 C_{Pi} + 2 \sum_{k=1}^{n} b_{Pk} C_{ik} = 0, \quad i=1,\ldots,n ,$$

or

$$\sum_{k=1}^{n} C_{ik} b_{Pk} = C_{Pi} , \tag{4.13}$$

so that

$$\tilde{T}(P) = \sum_{i=1}^{n} a_i \, C_{Pi}$$

with

$$\{a_i\} = \{C_{ik}\}^{-1} \{T(Q_k)\} \tag{4.14}$$

We see this is the collocation solution with K(P,Q) = C(P,Q). Unfortunately $\| T \|$ is infinite in this norm, but we may compute a mean square error

$$m_L^2 = C(L,L) - \{C(L,L_i)\}^T \{C(L_i,L_j)\}^{-1} \{C(L,L_j)\}. \tag{4.15}$$

4.5 Treatment of noise and parameters in minimum norm collocation

The data we have will generally not be directly related to T through a linear functional. It contain measurement errors (e_i) and may be affected by parameters such as an incompletely known relationship between a local geodetic datum and a geocentric, correctly oriented datum. Let us denote the observations y_i, the m parameters $\{X_j\} = X$ and suppose the observations are related to the parameters through a vector A_L. Then

$$y_i = L_i(T) + e_i + A_L^T X . \tag{4.16}$$

Let us suppose, that the noise vector e has the variance-covariance matrix D. A minimum norm collocation solution may then be obtained so that the sum of three quantities are minimized, namely $\| \tilde{T} \|^2$, $e^T D^{-1} e$ and $X^T P X$, where P is a positive definite matrix expressing some a-priori weights of the parameters,

$$\|\tilde{T}\|^2 + e^T D^{-1} e + X^T P X = \min . \tag{4.17}$$

The solution may be found using Lagrange multipliers as described in (Moritz, 1980, Chapter 29 and 30).

$$\begin{aligned} \tilde{X} &= P^{-1} A^T \bar{C}^{-1} (y + A\tilde{X}) \\ &= (A^T \bar{C}^{-1} A + P)^{-1} A^T \bar{C}^{-1} y \end{aligned} \tag{4.18}$$

$$\tilde{T} = \{K(L_i,P)\}^T \bar{C}^{-1} (y - A\tilde{X}) , \tag{4.19}$$

with $\bar{C} = \{K(L_i,L_j) + D_{ij}\}$.

The fulfilment of the last two equations are only a necessary condition for a minimum. However, it is possible to show that we <u>have</u> obtained a minimum.

It is still possibly to write down expressions for maximal error bounds similar to eq. (2.34), now using both the norm in the reproducing kernel Hilbert space, and the norms implicitly introduced in the n-dimensional space of measurements by D and the m-dimensional space of parameters by P, see Moritz (1980, p. 128). For the mean square error of prediction we find with

$$H = \{ K(L,L_i) \}^T \bar{C}^{-1}$$

that

$$m_x^2 = (A^T\bar{C}^{-1} A + P)^{-1} \tag{4.20}$$

and

$$m_L^2 = K(L,L) - H\{K(L,L_i)\} + HAm_x^2 (HA)^T \tag{4.21}$$

We will end this section with two examples showing that the addition of parameters makes no problem from the computational standpoint, but that the estimated results maybe should be evaluated very carefully using the error estimates.

<u>Example 4.3.</u> Modified Cholesky factorization.

Like in example 4.1 we will now modify the Cholesky decomposition of C, so that we directly may obtain the estimates of the parameters and the error estimates. The solution is obtained from the extended matrix

$$\begin{Bmatrix} \bar{C} & A & y \\ A^T & P & 0 \\ y & 0 & 0 \end{Bmatrix}$$

by defining the Cholesky factor as

$$\begin{Bmatrix} U & (U^T)^{-1}A & (U^T)^{-1}y \\ 0 & V & (V^T)^{+1} A^T\bar{C}^{-1}y \\ 0 & 0 & (y^TC^{-1}y - y^T\bar{C}^{-1}A(VV^T)^{-1}A^T\bar{C}^{-1}y)^{\frac{1}{2}} \end{Bmatrix}$$

where V is the Cholesky reduced of $(P + A^T\bar{C}^{-1}A)$. The modification with respect to an ordinary Cholesky reduction is that this would produce $P - A^T\bar{C}^{-1}A$, which might be negative definite, since P might be the 0-matrix.

It is then easily seen, that the X vector is obtained by solving

$$VX = (V^T)^{-1} A^T C^{-1} y$$

and that the error estimate is obtained by computing the Cholesky factor of

$$\begin{Bmatrix} \bar{C} & A & C_P \\ A^T & P & 0 \\ C_P^T & 0 & C_o \end{Bmatrix} \bullet$$

Example 4.4. Suppose we have observed two sea-surface heights (ζ_1 and ζ_2) by satellite radar altimeter measurements. However we will treat the measurements as if they were geoid heights biased by a constant X due to orbit error and sea surface topography,

$$L_i(T) + X + e_i = \zeta_i \quad , \qquad i=1,2,$$

where e_i is the error and L_i is given by eq. (2.10). Then suppose $K(L_1,L_2) = 0.60\ m^2$, $K(L_i,L_i) = 0.99\ m^2$, and the variance of the (supposedly uncorrelated) noice is $0.01\ m^2$. Suppose P = 0 in eq. (4.17). Then we will show how an estimate of T and of the bias may be obtained, and we will predict the geoid height in a point Q where $K(L_1, \zeta_Q) = K(L_2, \zeta_Q) = 0.8\ m^2$, and compute the error estimate.

The estimate of T is

$$\tilde{T}(P) = b_1 K(P,L_1) + b_2 K(P,L_2).$$

As shown in example 4.3 we will get the solution (b_1,b_2,X) by modified Cholesky reduction of the matrix

$$\begin{Bmatrix} \bar{C} & A^T & y \\ A & 0 & 0 \\ y & 0 & 0 \end{Bmatrix} = \begin{Bmatrix} 1 & 0.6 & 1 & \zeta_1 \\ 0.6 & 1 & 1 & \zeta_2 \\ 1 & 1 & 0 & 0 \\ \zeta_1 & \zeta_2 & 0 & 0 \end{Bmatrix}$$

and then execute the back-substitution using the last column as the reduced right-hand side. The Cholesky factor U is

$$U = \begin{Bmatrix} 1 & 0.6 & 1 & \zeta_1 \\ 0 & 0.8 & 0.5 & (\zeta_2 - 0.6\zeta_1)\frac{5}{4} \\ 0 & 0 & \sqrt{5}/2 & (\zeta_1+\zeta_2)\sqrt{5}/4 \\ 0 & 0 & 0 & ((\zeta_1-\zeta_2)\frac{5}{4})^{\frac{1}{2}} \end{Bmatrix}$$

The back-substitution is then a solution of the equations

$$\begin{Bmatrix} 1 & 0.6 & 1 \\ 0 & 0.8 & 0.5 \\ 0 & 0 & \sqrt{5}/2 \end{Bmatrix} \begin{Bmatrix} b_1 \\ b_2 \\ X \end{Bmatrix} = \begin{Bmatrix} \zeta_1 \\ (\zeta_2 - 0.6\,\zeta_1)\frac{5}{4} \\ (\zeta_1+\zeta_2)\sqrt{5}/4 \end{Bmatrix}$$

or

$$X = (\zeta_1 + \zeta_2)/2 \qquad \text{with} \quad (A^T C^{-1} A)^{\frac{1}{2}} = \frac{\sqrt{5}}{2} \simeq 1.12 \text{ m}$$

$$b_2 = \frac{5}{4}(\zeta_2 - \zeta_1), \quad b_1 = \frac{5}{4}(\zeta_1 - \zeta_2) .$$

Note that the prediction back in P_1 gives

$$\tilde{\zeta}_1 = K(L_1,L_1)b_1 + K(L_1,L_2)\,b_2 = 0.99\frac{5}{4}(\zeta_1 - \zeta_2) + \frac{3}{5}\frac{5}{4}(\zeta_2 - \zeta_1) \simeq \frac{1}{2}(\zeta_1 - \zeta_2)$$

i.e. the bias has been removed. The prediction in Q becomes

$$\tilde{\zeta}_Q = \frac{4}{5}\frac{5}{4}(\zeta_2 - \zeta_1 + \zeta_1 - \zeta_2) = 0 \quad .$$

The error of prediction may be computed from the Cholesky reduced of the right-hand side with elements $K(L_i, \zeta_Q)$, 0 and $K(\zeta_o, \zeta_o)$:

$$\left\{\begin{array}{ccc|c} \multicolumn{3}{c|}{\text{reduced part}} & \\ 1 & 0.6 & 1 & 0.8 \\ 0 & 0.8 & 0.5 & 0.8 \\ 0 & 0 & \sqrt{5}/2 & 0 \\ \hline 0.8 & 0.8 & 0 & -0.99 \end{array}\right\} \xrightarrow{\text{reduction of last column}} \left\{\begin{array}{c} 0.8 \\ 0.4 \\ 2.4/\sqrt{5} \\ \sqrt{0.99} \end{array}\right\}$$

i.e. a standard deviation of nearly 1 m. Note that the bias is estimated to be the mean value of the observations as could be expected. ●

<u>Exercise 4.5</u>. Repeat the example (1) without assuming a bias and (2) with only one observation, but with a bias. ●

We have here seen that noise and parameters may be easily treated in the minimum norm collocation set-up. The same is true if $\tilde{T}$ is estimated using least squares with fewer base functions than observations. For the best approximation methods (section 4.2) it seems more difficult.

4.6 Density estimation

We know (cf. example 2.10) that it in principle is impossible to determine the Earth's density distribution from information on the outer potential only. But we will here show that an estimate may be made anyway using principles which might be given a physical interpretation in the form of implied elastic properties.

If potentials of point masses or rectangular boxes (examples 3.5,3.6) are used, then the interpretation is immediate using the coefficients of the linear combination obtained. Volume distributions filling the whole space inside the Earth may be estimated using the harmonic density functions discussed in section 3.3.

The estimation of the harmonic density may be made if the inverse of the Newton-functional eq. (3.17) can be constructed. This is easily done if the boundary surface is a sphere, because the exterior spherical harmonic functions are mapped into the interior spherical harmonic functions, multiplied by a constant:

$$N_P(V_{ij}^{o}) = \frac{4\pi G R^2}{(2i+3)(2i+1)} V_{ij}(P) \; ,$$

cf. (Tscherning, 1977). Using the inverse of this equation, we can find the representer of the functional N^{-1} for a reproducing kernel belonging to a space of functions harmonic outside a sphere with radius R by

$$K(N_Q^{-1},P) = \sum_{i=2}^{\infty} \sigma_i \frac{(2i+3)(2i+1)}{4\pi G R} \frac{(r')^i}{r^{i+1}} P_i(\cos\psi), \quad r'<R \; .$$

For a non-spherical boundary a Hilbert space may be constructed using the product of two Hilbert spaces. One must consist of functions harmonic outside a sphere enclosed in the Earth and the other must be the finite dimensional space of potentials of indicator functions filling the space between the sphere and the actual boundary used, for details see Sansó and Tscherning (1980).

Hence, not only point mass models, but (nearly) all methods can be given a geophysical interpretation in the form of an implied <u>volume</u> density distribution for the Earth. Further possibilities for determining a unique density distribution (besides the use of harmonic densities) are discussed in Tscherning (1977) and Tscherning and Sünkel (1981).

5. Choice of base functions and inner product

5.1 Introductory remarks.

Suppose we have a set of observations $y_i = L_i(T)$, $i = 1,....n$. Then we have identified 3 situations, namely corresponding to whether we use a number of base functions m with $n > m$, $n = m$ and $n < m$. The last, underdetermined case, is generally reduced to one of the two other cases, using a singular value decomposition of the normal equations and then selecting as base functions the eigenfunctions with eigenvalues numerically larger than a fixed constant.

The selection of base functions may then be done using one of the following considerations:

(1) An approximation is wanted, which describe the gravity field down to a certain resolution, e.g. 500 km, over a given area, despite that maybe the data contains information on even shorter wavelengths.

(2) An approximation is wanted, which represents the data in the best possible manner or which extracts the maximal possible information from the data.

In the first case can we normally have $m < n$ and in the last case $n \geq m$, (which we here will consider equivalent to $n = m$).

5.2 Choice of base functions.

If a global approximation is needed, then case (1) may lead us to use (linear combinations of) spherical harmonics to wavelengths corresponding to the resolution. Or an intermediate solution, containing shorter wavelengths, is constructed and the longer wavelength information is then extracted from this.

In both cases large systems of equations may have to be solved. But if data are globally distributed in a regular manner (equal spacing in longitude for each fixed degree of latitude), the equations may have a Toplitz structure, see Colombo (1979), or be strongly diagonal dominat see Wenzel (1985). In both cases very large systems of equations may be solved with a relatively minor effort.

Also it is possible to use finite elements, taking advantage of the sparceness of the normal equations as discussed in Meissl (1981). However finite elements are not harmonic, so additional equations have to be solved in order to model this property of T.

If a local area is considered, methods similar to spline function or finite elements have been proposed. A set of bell-shaped (generally harmonic) functions, regularity distributed with a spacing equal to the resolution may be used. If harmonic functions are used, we loose some of the advantages which made the original spline functions and finite elements a good choice, namely the property that they were zero outside a finite interval. The systems of equations to be solved will be full, but a quasi-ortogonallity may be achieved by a careful selection of the functions, which permit large systems of equations to be solved iteratively in very few steps.

Example 5.1 Potentials of point masses may be used as base functions. However they are positive everywhere. If a number of the first terms in the Legendre expansions are eliminated, then the new function will have several zero-points, cf. example 3.11. ●

As actual base functions potentials of point masses or harmonic kernel functions (see example 3.1, 3.3, 3.5, 3.10) have been proposed and used, see Lelgemann (1981). Depending on the data available, it was proposed to use representers of the associated functionals, distributed in a regular grid covering the actual area. If for example the component of the gravity vector is used, then the three functions are orthogonal in each point, if a rotational invariant kernel is used, see Tscherning (1970).

In a first choice of base functions generally only a class of functions are selected. For point mass potentials or kernel functions the depth to the point mass, or the radius of a sphere bounding the set of harmonicity, respectively, may then be determined subsequently. Here a generally non-linear optimization problems must be solved (see (Barthelmes,1986)). For regularly spaced data and certain classes of kernels rules to determine a best "depth" has been found, see Hardy and Göpfert (1975) and Lelgemann (1981). These rules solve the problem that most of the kernels used have strong singularities, i.e. they take on infinite values. Typically, if a point mass is located too close to the Earth's surface, its generated potential will go too fast to ∞ and the determined approximation will look as illustrated in Fig. 3.

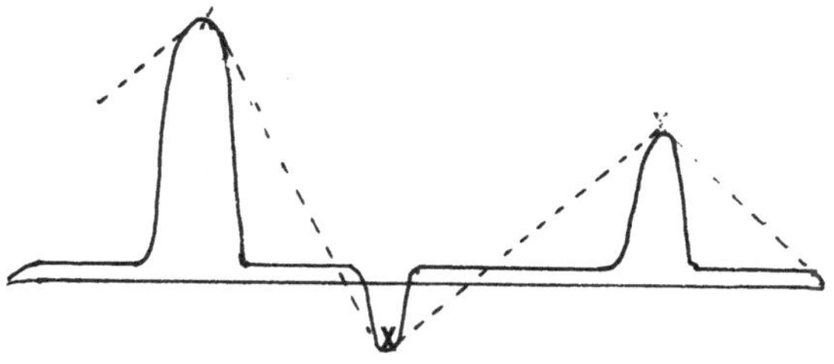

Fig. 3. Approximation obtained using too singular base functions. - and with base functions assuring an exact mid-point interpolation... .

Naturally also the opposite phenomenon may occur, well known from polynomial approximation, that the function has oscillations with large amplitudes between the data-points. On the other hand we also know how to counteract this phenomenon, namely by using a minimum norm condition, and base functions of a different kind.

In the second case, $n = m$, the choice of base functions is fixed, if we use minimum norm collocation. Here the representers $K(L_i, P)$ of the linear functionals, L_i, must be used. In some other cases the base functions are selected, because they occur when discretizising an integral formula like Poissons or Stokes equation relating data at the Earth's surface with data in space, cf. some of the methods proposed by Bjerhammar. However, the kernels in most of these integral formulae are reproducing kernels (see section 3.3), so we have in practice a very little difference to the use of minimum norm collocation in the space where these kernels are reproducing kernels or covariance functions for that sake.

Since the integral formulae normally are solutions to boundary value problems with data on a sphere, then the base functions used will be associated with points on the sphere . This sphere will, when being used for selecting base functions, generally have a radius smaller than the Earth's mean radius, and the coefficients then look like "artificial" data which must be estimated explicitly on the sphere, (see e.g. Bjerhammar, 1976). In this way we avoids the problem of the possible singularity of the kernels. If data are not on a sphere, then a non-symmetric system of equations must be solved in order to find the coefficients of the linear combination. But if they are on a sphere, then we have symmetry and an exact equivalence to a minimum norm collocation solution.

<u>Example 5.2</u> Point mass modelling with carrier points right under the data points on a sphere with radius R.

Then $\tilde{T} = \Sigma a_i / |P - P_i'|$, where P_i' is right under the data point P_i. Suppose the observations are the values $T(P_i)$. Then the equations to be solved have the coefficients

$$1/|P_i' - P_j| = \frac{1}{R} \sum_{k=0}^{\infty} \left(\frac{R}{r_j}\right)^{k+1} P_k(\cos\psi_{ij}) \tag{5.1}$$

If $r_i = r_j = r$ (all data points on a sphere), and $R_B^2 = R \cdot r$, then

$$1/|P_i' - P_j| = \frac{1}{R} \sum_{k=0}^{\infty} \left(\frac{R_B^2}{r^2}\right)^{k+1} P_k(\cos\psi_{ij}). \tag{5.2}$$

In this case the solution is equivalent to the minimum norm collocation solution in the space with the Krarup-kernel, example (3.5). This gives a nice interpolation if the point mass modelling technique as a general approximation method also for non-gravity field related data. The solution will have a maximal smoothness in terms of minimal first order derivatives. ●

Since we in many cases will have a similar close equivalence between the two kinds of collocation solutions, we will in the following only consider the minimum-norm type of solutions.

5.3 Choice of inner product.

For the overdetermined problem n > m, a solution is found as the "best" approximation of T on the subspace spanned by the base vectors. This involves, cf. eq. (4.5) an evaluation of several inner products, requiring that T or some of its derivatives are known. Hence, the inner product must depend on the data available. If only gravity data are available, then

$$(f, g)_{\frac{1}{2}} = \frac{1}{4\pi} \int_{\omega} \Delta g_f \cdot \Delta g_g \, d\omega \tag{5.3}$$

can be used. The corresponding base vectors are the normalized solid spherical harmonics multiplied by $\frac{R}{i-1}$ and the reproducing kernel

$$K(P,Q)_{\frac{1}{2}} = R^2 \sum_{i=2}^{\infty} \frac{2i+1}{(i-1)^2} \left(\frac{R^2}{rr'}\right)^{i+1} P_i(\cos\psi) \tag{5.4}$$

In practice the integrals of type (5.3) are not evaluated over the whole Earths surface, but only over a limited area. Also sometimes in practice, not the discretized integral is used, but one simply calculates the products of the available data with weights according to their error variances. It is easy to imagine how this can go wrong, if the data are not in some manner regularity distributed. The difficulty is obvious in the case where low order potential coefficients have to be estimated from satellite orbit pertubations. These pertubations are of a similar character as the potential itself (integrals along the orbit of accelerations = velocities) and integrals of the potential (a further integration of the velocities). The minimalization of the differences between observed and computed pertubations must then correspond to one of the negative Sobolev norms. Seen from a functional-analytic standpoint is such a norm much too weak, permitting functions with wildly behaving boundary values. In practice (Lerch et al., 1977, section 3.2) we then also see that the H_0-norm is minimalized simultaneously (the square-sum of the estimated potential coefficients). In the minimum norm collocation situation, n = m, the choice of inner product has a consequence, that the base functions are fixed, and the calculation of the inner products does not involve actual observed data. Then, we might use the advice given by for example approximation by spline functions, cf. (Moritz, 1978, p. 44), and select a norm minimalizing the second order derivatives. Now in spline function theory, only values of the approximating function are considered, while we in geodesy frequently work with first or several order derivatives. We should consequently work with a norm involving the fourth-order derivatives at or close to the Earth's surface. Now, our primary goal in geodesy is to have a good representation of the anomalous potential itself and the gravity anomaly vector. Hence, maybe minimatization of third order derivatives would be reasonable. If the goal is a further geophysical interpretation in terms of density estimates, then the fourth order derivatives might be the right choice. On the other hand, if we as basic function space use a space of functions harmonic in a larger set than the set outside the Earth, then all derivatives are in principle minimalized.

The norms implied using the experiences from spline functions are strong norms. The linear functionals will have increasing correlations and numerical instabilities will be encountered when solving the normal equations. However, this may be counteracted by using a small depth to the Bjerhammar sphere and by removing or decreasing the weight of low degree subspaces. If a subspace corresponding to spherical harmonic of maximal degree i are removed, then the corresponding i first terms in the reproducing kernel are also removed. This has a tremendous effect on the kernel and thereby on the correlation of the various linear functionals. It is illustrated in (Tscherning ,1985, Fig. 5).

If we want a best least-squares approximation, then we also only have one choice: the empirical covariance function. The computation of this function is discussed at many places: Tscherning and Rapp (1974), Lachapelle and Schwarz (1980), Goad at al. (1984), Forsberg (1984) and Tscherning (1985). It is, fortunately, possible to model the covariance function using reproducing kernels, cf. Tscherning (1972), thus giving a compromise between least-squares and minimum norm collocation.

The reproducing kernels, which may be used to model the empirical covariance function(s), correspond to norms minimatizing the second order derivatives on the boundary of a (Bjerhammar) sphere a few kilometers inside the Earth, or third order derivatives over the space outside such a sphere, see Tscherning (1985, Table 1).

It would be interesting to investigate, whether the application of even stronger norms would give better results. Such norms would make it possible to skip the Bjerhammar sphere and we would work directly with the true boundary. In this case T would be an element of the same space as $\tilde{T}$, and the convergence of the collocation solutions towards T is assured for increasing amounts of data, see Tscherning (1978a).

Finally two arguments should be mentioned, which have been put forward, (but which I think are wrong). One argument is that the base functions should be as easy to compute as possible. However, all functions may be tabulized once and for all, and then later evaluated with the same speed as any other functions, cf. Sünkel (1979). The other argument is (see e.g. Barthelmes (1986)) that some basefunctions have an immediate geophysical interpretation as giving estimates of the density. But as we have seen in section 4.6, then all harmonic functions may be given such an interpretation.

6. Conclusion - Choice of method

We have in chapter 4 described a number of methods and in chapter 5 discussed various choices of base functions and inner products. A comparison of the methods show that several satisfy the following criteria:

all kinds of data can be used,

all kinds of quantities can be predicted,

the methods are fairly easy to implement on a computer,

numerical stability problems can be avoided by proper selections of base functions (Bjerhammar sphere, removal of subspaces).

data noise can be taken into account,

non-gravity field related parameters may be estimated simultaneous with the approximation of T and density estimates may be derived from $\tilde{T}$.

However, some of the methods require a considerable numerical effort, especially when solving the normal equations. But also this effort may be reduced, taking in into account repretitive pattens of the data, or by limiting the extend of a solution to a local area, and then using an integration method outside the area, see e.g.(Lachapelle and Tscherning, 1978). Also data selection can be used as described in (Goad et al., 1983) and (Barthelmes, 1986).

Furthermore, some methods delives error-estimates. Here it has been argued that such estimates may be found for all methods, simply by comparing a solution to data not used to construct the solution. But this does not solve the problem we want to solve, namely of knowing error estimates of other data types than these used or available.

Methods are primarily compared in terms of mean square differences between observed and computed quantities. If this is to be taken as the standard for comparison, then least-squares collocation must be the best method, because it has be proved that it (section 4.4) minimalizes exactly this error. But the proper use of this methods requires that a good estimate of the empirical covariance function is available. And this is a difficult quantity to estimate. Lauritzen (1973) has found that it under certain conditions is more difficult to estimate the covariance function than to estimate $\tilde{T}$.
However also this paradox has been solved in the Lecture Notes by F. Sansò!

Hence, if it in practical comparisons is found that least squares collocation does not give the best results (see e.g. Bjerhammar, 1985), then the explanation must be that a wrong covariance function was used.

From theoretical considerations and practical experiences I am convinced that least squares collocation is the best method for local gravity field approximation. The difficulties accounted when using the method with a large number of data will also be solved, cf. the work by Colombo (1979), Jekeli (1985). For global approximation, I would judge, that least square collocation also would work very well, but comparisons between methods are lacking.

This does not mean that all problems have been solved in global and local gravity field approximation. In Tscherning (1986) I have listed a large number of current problems: how to do we fit local solutions together, why is there a height-dependent error in geoid heights computed from spherical harmonic expansions, is it possible to give the error estimates a statistical interpretation?

There are still many problems to be solved. Their solution requires a good understanding not only of the theory of functional analysis, but also numerical experiments and a careful comparison of the results of several methods as done e.g. in Kearsly et al. (1985).

Appendix: Hints to the solution of the exercises.

Exercise	Hint
3.1	See Barthelmes (1986).
3.3	See Tscherning (1985, example 2.2)
3.4	Solution: The product sum of the gravity values produced by each block evaluated in P and Q divided by the volume of the block.
3.5	See Tscherning (1985, example 2.6)
3.6	See Tscherning (1985, example 2.7)
3.7	See Tscherning (1985, example 2.7)
4.1	Use eq. (3.24).
4.2	See Tscherning (1985, example 2.3)
4.3	See Tscherning (1985, example 2.4 and 3.1).

References:

Aubin, J.-P.: Approximation of elliptic boundary-value problems. Wiley-Interscience, 1972.

Barthelmes,F.: Untersuchungen zur Approximation des aeusseren Gravitationsfeldes der Erde durch Punktmassen mit optimierten Positionen. Akademie der Wissenschaften der DDR, ZIPE, Potsdam 1986.

Bjerhammar, A.: A Dirac Approach to Physcial Geodesy. Z. f. Vermessungswesen, 101 Jg., no. 2, pp. 41-44, 1976.

Bjerhammar,A.: A robust approach to global problems in physical geodesy. Bulletin Geodesique, Vol. 59, pp. 303-315, 1985.

Colombo, O.: Optimal estimation from data regularly sampled on a sphere with applications in geodesy. Reports of the Dep. of Geodetic Science, No. 291, The Ohio State University, Columbus, 1979.

Forsberg, R.: Local Covariance Functions and Density Distributions. Reports of the Department of Geodetic Science and Surveying No. 356, The Ohio State University, Columbus, 1984.

Freeden, W.: On the Permanence Property in Spherical Spline Interpolation. Reports of the Department of Geodetic Science and Surveying, No. 341, The Ohio State University, Columbus, Ohio, 1982.

Garabedian, P.R.: Partial Differential Equations. John Wiley & Son, New York, 1964.

Goad, C.C., C.C.Tscherning and M.M.Chin: Gravity Empirical Covariance values for the Continental United States. J.Geophys.Res., Vol. 89, No. B9, pp. 7962-7968, 1984.

Hardy,R.L. and W.M.Goepfert: Least squares prediction of gravity anomalies, geoid undulations and deflections of the vertical with multiquadratic harmonic functions. Presented XVI General Assembly IUGG, Grenoble 1975.

Heiskanen, W.A. and H. Moritz: Physical Geodesy. W.H. Freeman & Co, San Francisco, 1967.

Jekeli, C.: On Optimal Estimation of Gravity from Gravity Gradients at Aircraft Altitude. Rev. of Geophysics, Vol. 23, pp. 301-311, 1985.

Kearsley, A.H.W., M.G.Sideris, J.Krynski, R.Forsberg and K.P.Schwarz: White Sands Revisited - A Comparison of Techniques to Predict Deflections of the Vertical. Report 30007, Division of Syrveying Engineering, University of Calgary, 1985.

Krarup, T.: A Contribution to the Mathematical Foundation of Physical Geodesy. Meddelelse no. 44, Geodætisk Institut, København 1969.

Krarup, T.: Some remarks about collocation. In: H.Moritz and H.Suenkel (Ed.): Approximation methods in Geodesy, pp. 193-209, H.Wichmann Verlag, Karlsruhe, 1978.

Lachapelle, G. and K.-P.Schwarz: Empirical Determination of the gravity anomaly covariance function in mountainous areas. The Canadian Surveyor, Vol. 34, no. 3, pp. 251-264, 1980.

Lachapelle, G. and C.C. Tscherning: Use of Collocation for Predicting Geoid Undulations and related Quantities over Large Areas. Proceedings Int. Symposium on the Geoid in Europe and Mediterranean Area, Ancona, Italy, September 25-29, 1978, pp. 1-19, 1978.

Lauritzen, S.L.: The Probabilistic Background of Some Statistical Methods in Physical Geodesy. Meddelelse no. 48, Geodætisk Institut, 1973.

Lelgemann, D.: On the numerical Properties of Interpolation with Harmonic Kernel Functions. Manuscripta Geodaetica, Vol. 6, no. 2, pp. 157-191, 1981.

Lerch, F.J., S.M.Klosko, R.E.Laubscher and C.A.Wagner: Gravity Model Improvement using GEOS-3 (GEM 9 & 10), Goddard Spece Flight Center, X-921-77-246, 1977.

Meissl, P.: The use of finite elements in physical geodesy. Reports of the Dep. of Geodetic Science and Surveying, No. 313, The Ohio State University, Columbus, 1981.

Moritz,H.: Introduction to Interpolation and Approximation. In: Moritz and Suenkel (Ed.): Approximation methods in Geodesy, pp. 1-45, H.Wichmann Verlag, Karlsruhe, 1978.

Moritz, H.: Advanced Physical Geodesy. H.Wichmann Verlag, Karlsruhe, 1980.

Parzen, E.: Statistical Inference on Time Series by Hilbert Space Methods, I. 1959. (Reprinted in "Time Series Analysis Papers", Holden-Day, San Francisco, 1967, pp. 251-282).

Sanso',F., R.Barzaghi and C.C.Tscherning: Choice of Norm for the Density Distribution of the Earth. In print Geoph.Jour.Royal Astr. Soc., 1986.

Sanso, F. and C.C.Tscherning: Mixed Collocation: A proposal. Quaterniones Geodasiae, Vol. 3, no. 1, pp. 1-15, 1982.

Suenkel, H.: A Covariance Approximation Procedure. Reports of the Department of Geodetic Science, No. 286, The Ohio State University, Columbus, Ohio, 1979.

Tscherning,C.C.: Bestemmelse af lodafvigelser ved collokation. (In Danish, "Determination of Deflections of the Vertical using Collocation", Masters thesis, University of Copenhagen, 1970.

Tscherning, C.C.: Representation of Covariance Functions Related to the Anomalous Potential of the Earth using Reproducing Kernels. The Danish Geodetic Institute Internal Report No. 3, 1972.

Tscherning, C.C.: Application of Collocation: Determination of a Local Approximation to the Anomalous Potential of the Earth using "Exact" Astro-Gravimetric Collocation. In: Brosowski, B. and E. Martensen (Ed's): Methoden und Verfahren der Mathematischen Physik, Vol. 14, pp. 83-110, 1975.

Tscherning, C.C.: Models for the Auto- and Cross Covariances between Mass Density Anomalies and First and Second Order Derivatives of the Anomalous Potential of the Earth. Proceedings 3rd. Int. Symposium "Geodesy and Physics of the Earth", Weimar, October, 1976, pp. 261-268, Potsdam, 1977.

Tscherning, C.C.: On the Convergence of Least Squares Collocation. Bolletino di Geodesia e Scienze Affini, Vol. XXXIII, No. 2-3, pp. 507-516, 1978.

Tscherning, C.C.: Introduction to Functional Analysis with a View to its Application in Approximation Theory. In: Moritz, H. and H.Suenkel (Ed's): Approximation Methods in Geodesy, H.Wichmann Verlag, Karlsruhe, pp. 157-192, 1978a.

Tscherning, C.C.: Local Approximation of the Gravity Potential by Least Squares Collocation. In: K.P.Schwarz (Ed.): Proceedings of the International Summer School on Local Gravity Field Approximation, Beijing, China, Aug. 21 - Sept. 4, 1984. Publ. 60003, Univ. of Calgary, Calgary, Canada, pp. 277-362, 1985.

Tscherning, C.C.: GEOCOL - A FORTRAN-program for Gravity Field Approximation by Collocation. Technical Note, Geodætisk Institut, 3.ed., 25 MAR 1985a.

Tscherning, C.C.: Current Problems in Gravity Field Approximation. Proceedings 1. Hotime-Marussi Symposium, Rome, June 3-6, 1985, pp. 363-384. Politecnico di Milano, 1986.

Tscherning, C.C. and R.H.Rapp: Closed Covariance Expressions for Gravity Anomalies, Geoid Undulations, and Deflections of the Vertical Implied by Anomaly Degree-Variance Models. Reports of the Department of Geodetic Science No. 208, The Ohio State University, Columbus, Ohio, 1974.

Tscherning, C.C. and H. Suenkel: A Method for the Construction of Spheroidal Mass Distributions consistent with the harmonic Part of the Earth's Gravity Potential. Manuscripta Geodaetica, Vol. 6, pp. 131-156, 1981.

Wenzel,H.-G.: Hochaufloesende Kugelfunktionsmodelle fuer das Gravitationspotential der Erde. Wiss. Arb. Fachrichtung Vermessungswesen der Universitaet Hannover, (in print), 1985.

STATISTICAL METHODS IN PHYSICAL GEODESY

by

F. Sansò

Institute of Topography, Photogrammetry
and Geophysics
Politecnico Milano
Piazza Leonardo da Vinci 32
I-20133 Milano, Italy

Lecture Notes in Earth Sciences, Vol. 7
Mathematical and Numerical Techniques in Physical Geodesy
Edited by H. Sünkel

1. Introduction

The method of least squares and the related theory has dominated the scene of the application of statistical methods to geodesy, for two centuries. However, when more complex problems like the approximation of functions representing physical fields (typically the anomalous gravity field but also fields of temperature, of the refraction index, strain fields, etc.), have been clearly focussed, it was realized that a simple least squares approach as data reduction tool was insufficient and unsatisfactory specially when limited to a purely finite dimensional environment. At the same time, a new powerful tool was created by mathematicians by introducing the theory of weakly stationary stochastic processes of the second order and the related estimation theory. It turned out that this theory can be viewed as a kind of least squares approach in an infinite dimensional space, where however we take advantage, built in the metric,of some statistical information on our unknown function through its so called covariance function.

These methods have become more and more important in many applications in geodetic sciences, quite similarly to a lot of other disciplines, primarily to electronics where the theory was born.

We can classify them according to whether we refer to time dependent phenomena or space dependent phenomena as well as whether we use continuous methods or discrete ones, depending on whether we consider the data as a continuous or a discrete set. The situation is tentatively illustrated in Tab. 1.1. The first classification is quite essential, since for time dependent phenomena an extra effort is done in the estimation theory in order to apply the principle of causality (i.e. the future can depend on the past and not viceversa), while in space dependent phenomena only certain invariance properties, like homogeneity or isotropy, are usually accounted for.

TAB. 1.1	Time dependent	Space dependent
Discrete	Integrated, moving average Autoregressive processes Discrete Kalman filters	Least squares collocation Collocation in harmonic Spaces
Continuous	Stochastic differential equations Continuous Kalman filters	Continuous collocation Overdetermined boundary value problems

Just to fix the ideas, to the first class belongs the treatment of time series of control data as well as the real time elaboration of inertial or spatial data; to the second class belongs the estimation of global or local gravity models, the solution of overdetermined b.v.p.'s, etc.
The second classification in terms of discrete or continuous methods might be perceived as unjustified since up-to-day it is only in few cases, like in inertial geodesy, that within numerical applications a continuous noise process is really modelled, although even in this case the measurement process is described as discrete.
Nevertheless, I believe that the increase in the density of available data, both time-wise and space-wise, produced by modern technology makes it desirable that geodesists be acquainted with both approaches, preserving and improving their historical skilfulness in switching, forth and back, from a discrete to a continuous description of large sets of data.
No doubt, this could be a possible breakthrough for the problem of a full exploitation of the available data.
In these lecture notes however we shall concentrate only on discrete methods for space dependent fields, namely we shall analyse that complex of stochastic and functional methods which is known in geodesy as collocation theory.
As the reader might know, there has been a long discussion among geodesists onto whether a stochastic interpretation of the collocation method was acceptable or not: one argument against this, at the time supported by the author, was that we are in reality work-

ing with one field only (specifically the gravity field of the earth) and we cannot see any physical mechanism which could generate other sample functions (there is one earth only!)(*).

Under the pressure of this discussion it was clarified that a stochastic interpretation becomes particularly acceptable when the family of field functions on which we perform the averaging operation, is generated from the actual field by applying to it the elements of a transformation group: by assuming simple distributions on the transformation group one gets estimators which are invariant under the action of the group itself and this furnishes the bridge for a mathematical-physical interpretation of the results.

Driven by the need of clarifying in a precise way what is the role of the stochastic interpretation in the whole approximation process, we shall try in these lectures to reconstruct such a process step by step from the more general principles to the specific estimation formulas.

At the same time we shall solve some contradictions which were up to now left unexplained in the theory, like the infinite norm paradox (cfr. C.C.Tscherning /20/).

2. Recalls on Hilbert spaces with reproducing kernel (RKHS)

For the reader's sake we try to collect in this paragraph the main facts about RKHS which will be used throughout the text. In order to be more comprehensive a short review of the properties of Hilbert spaces is added in Appendix 1, where also many examples relevant to these lectures are to be found.

Def. 2.1: given a closed set A and a real Hilbert space H of functions defined on A, we say that $k(x,y)$ $(x,y \in A)$ is a reproducing kernel in H if

a) $k(x,y) \in H$ as function of y, $\forall x \in A$

b) $\langle k(x,y), u(y)\rangle_H = u(x) \qquad \forall u \in H \qquad (2.1)$

(*) The idea that one could consider other planets as different realizations of the same process, though fascinating, is not correct in this context since in this case we should mix also the gravity data of all the planets to derive "collective" estimates, while here we are interested only in the gravity field of our planet as such.

Remark 2.1: if $k(x,y)$ is the RK of H and if this function is continuous in a neighborhood of the diagonal $(|x-y| < \varepsilon \ ; \ x,y \in A)$ then the functions of H are continuous. In fact by Schwarz inequality

$$|u(x) - u(x')| = |\langle k(x,\cdot) - k(x',\cdot), u\rangle| \leq \|k(x,\cdot) - k(x',\cdot)\| \, \|u\|$$

and by the RK property, when $x' \to 0$

$$\|k(x,\cdot) - k(x',\cdot)\|^2 = k(x,x) + k(x',x') - 2k(x,x') \to 0 \quad .$$

From now on we shall assume that the above hypothesis is always satisfied. This means that as a set (of functions) $H \subset \mathcal{C}(A)$: this embedding however is also topological in the sense that

$$\sup_{x \in A} |u(x)| \leq \sup_{x} \|k(x,\cdot)\| \cdot \|u\|_H = \sup_{x} \sqrt{k(x,x)} \, \|u\|_H$$

Remark 2.2: we list some fundamental properties of reproducing kernels.

1) RK is a symmetric function.
(According to the definition,

$$k(x,y) = \langle k(y,\cdot),k(x,\cdot)\rangle = \langle k(x,\cdot),k(y,\cdot)\rangle = k(y,x) \quad) \tag{2.2}$$

2) The RK of a Hilbert space is unique.
(If $k,\bar{k}$ are both RK's in H, then

$$\bar{k}(x',x) = \langle k(x,\cdot),\bar{k}(x',\cdot)\rangle = \langle \bar{k}(x',\cdot),k(x,\cdot)\rangle = k(x',x) \quad)$$

3) A RK is a positive definite function.
(Let $\{\lambda_i\}$ be any finite sequence in R and $\{x_i\}$ any sequence in A; then we can write

$$\sum_{1}^{N}{}_{i,j} \lambda_i\lambda_j \, k(x_i,x_j) = \sum_{1}^{N}{}_{i,j} \lambda_i\lambda_j \, \langle k(x_i,\cdot),k(x_j,\cdot)\rangle = \Big\|\sum_{1}^{N}{}_{i} \lambda_i k(x_i,\cdot)\Big\| \geq 0 \ )$$

4) Let us take any sequence of points $\{x_i\}$, dense in A, then the linear manifold

$$V = \text{Span}\,\{k(x_i,\cdot)\} \tag{2.3}$$

is densely embedded in H.
(Suppose $u \in H$, $u \perp V$; then $u(x_i) = \langle k(x_i,\cdot), u\rangle = 0$ and since u is continuous by Remark 2.1, then $u = 0$ on A)

From the properties 2) and 4) in particular it follows that there is a one to one correspondence between a RKHS and its reproducing kernel: in fact if H admits a (continuous) RK then this is unique and viceversa if K is the (continuous) RK of H, as it is stated in the following Theorem 2.1, H is uniquely defined. Whence from now on we shall use the symbol H_k to mean the Hilbert space with reproducing kernel k.

Theorem 2.1: if both H and $\bar{H}$ admit the same reproducing kernel $k(x,y)$, then they are identical.
Let $u \in H$; take $u_N = \sum_1^N{}_i \lambda_i\, k(x_i,\cdot) \in V$ and such that $u_N \to u$ in H, so that $\{u_N\}$ is also a Cauchy sequence in H. Since for any $v \in V$, $v = \Sigma_1^n{}_i\, \mu_i\, k(x_i,\cdot)$

$$\|v\|_H^2 = \sum_1^n{}_{i,j}\, \mu_i\mu_j\, k(x_i,x_j) = \|v\|_{\bar{H}}^2 ,$$

the sequence $\{u_N\}$ mentioned above has to be a Cauchy sequence in $\bar{H}$ too: but then there is a limit of u_N in $\bar{H}$ ($u_N \to \bar{u}$ in $\bar{H}$).
Set

$$\bar{u} = Ju$$

we have

$$\|\bar{u}\|_{\bar{H}} = \lim_{N\to\infty} \|u_N\|_{\bar{H}} = \lim_{N\to\infty} \|u_N\|_H = \|u\|_H$$

and J is an isometry of H in $\bar{H}$. Since H and $\bar{H}$ enter symmetrically in the proof, there is also an isometry of $\bar{H}$ in H and therefore J is an isomorphism.

On the other hand

$$u(x) = \langle k(x,\cdot),u\rangle_H = \langle k(x,\cdot),\bar{u}\rangle_{\bar{H}} = \bar{u}(x)$$

and $H,\bar{H}$ are seen to be done by the same functions with the same norms, i.e. they are identical.

Remark 2.3: as we have observed in Remark 2.2 if H has a (continuous) RK then $H \subset \mathcal{C}(A)$; the converse is also true. In fact if $H \subset \mathcal{C}(A)$ we must have $\forall u \in H$, $\|u\|_{\mathcal{C}(A)} = \sup_{x\in A} |u(x)| \leq c \|u\|_H$; but then the evaluation functional

$$ev_x(u) = u(x) \quad ,$$

which is trivially linear, is also bounded and therefore continuous so that, applying the Riesz theorem (see Appendix 1), for every x there is in H a representer of ev_x , i.e.

$$u(x) = \langle k(x,\cdot),u\rangle$$

q.e.d.

Remark 2.4: one of the main advantages of working in RKHS is that when $k(x,y)$ is explicitely known it is relatively simple to find the representers in H of the linear functionals, the existence of which is guaranteed by the Riesz theorem. In fact let F be any bounded functional; one has (F_x meaning that the functional F acts on functions of x)

$$F_x(u(x)) = F_x(\langle k(x,\cdot),u(\cdot)\rangle) \overset{(*)}{=} \langle F_x k(x,\cdot),u(\cdot)\rangle \tag{2.4}$$

Whence the Riesz representer of F_x is the function (of y) $F_x k(x,y)$, which is in many cases easy to compute, specially when F_x is a local functional (e.g. $ev_x u = u(x)$, $ev_x(\nabla u) = \nabla u(x)$, etc.).
It is to be noticed that according to (2.4) we have a criterion to judge

(*) This step can be directly justified by a computation when $u \in V$ (cfr. (2.3)) and then extended to all of H_k .

whether a given linear functional F is bounded in H_k or not, namely this happens if

$$F_x\, k(x,y) \in H_k \tag{2.5}$$

Furthermore, if we want to perform the scalar product between two functionals, or equivalently between their representers, we see from (2.4) that

$$\langle F,G\rangle_{H_k^*} = \langle F_x k(x,\cdot), G_y k(y,\cdot)\rangle_{H_k} = F_x\, G_y\, k(x,y) \tag{2.6}$$

Since (2.5) means that the norm of $F_x k(x,y)$ in H_k, which is the same as the norm of F in H_k^*, must be finite, we see that (2.5) is equivalent to

$$|F_x\, F_y\, k(x,y)| < +\infty \tag{2.7}$$

This is again a criterion of boundedness of the functional F.

Now we turn ourselves to explore the relation between RKHS and CONS.(*) In this respect the following theorem holds.

<u>Theorem 2.2</u>: a Hilbert space H is endowed with a reproducing kernel k, if for some CONS, $\{e_n(x)\}$, we have

$$\sum_{1}^{+\infty}{}_n\, e_n^2(x) < +\infty \qquad \forall x \in A \tag{2.8}$$

If H has the RK $k(x,y)$, we must have for any CONS and $\forall x$

$$k(x,y) = \Sigma \langle k(x,\cdot), e_n(\cdot)\rangle_{H_k}\, e_n(y) = \Sigma_n\, e_n(x)\, e_n(y)$$

then (2.8) has to hold since, from Parseval's identity,

$$\Sigma\, e_n^2(x) = \|k(x,\cdot)\|^2 = k(x,x) < +\infty \quad .$$

(*) Complete Orthonormal System.

Viceversa if (2,8) holds for a CONS, we can put

$$k(x,y) = \sum_{1}^{+\infty}{}_n \, e_n(x) \, e_n(y) \quad , \tag{2.9}$$

the series being H convergent for every x . Whence, $\forall u \in H$

$$\langle k(x,\cdot),u\rangle = \Sigma \, e_n(x) \, \langle e_n(\cdot),u\rangle = u(x)$$

and k is a RK.

Remark 2.5: the representation (2.9) is quite characteristic for reproducing kernels: so much so that if we know that (2.9) holds, the series being convergent in H_k , and if $\{e_n\}$ is known to be a σ-independent sequence, i.e.(*)

$$\sum_{1}^{+\infty}{}_i \, \lambda_i \, e_i(x) = 0 \;\rightarrow\; \lambda_i = 0 \tag{2.10}$$

then $\{e_n\}$ is a CONS in H_k .
In fact $\{e_n\}$ is total in H_k , as

$$u \perp \text{Span } \{e_n\} \;\rightarrow\; k = \langle k(x,\cdot),u\rangle = \Sigma \, e_n(x) \, \langle e_n,u\rangle_{H_k} = 0 \quad .$$

Furthermore from

$$e_m(x) = \langle k(x,y),e_m\rangle_{H_k} = \Sigma \, \langle e_n,e_m\rangle_{H_k} \, e_n(x)$$

and recalling (2.10), we see that

$$\langle e_n,e_m\rangle_{H_k} = \delta_{nm} \quad ,$$

q.e.d.

(*) the notion of σ- independence is quite different from that of linear independence of finite combinations, as one can easily verify with the sequence:

$\{\sin t, 1, \sin 2nt, \cos 2nt, \ldots\}$ in $L^2(0\pi)$.

We note explicitely that (2.10) is certainly satisfied if $\{e_i\}$ is orthonormal in some space $H_o \supset H_k$.

This remark opens the way to an explicit construction of the scalar product $\langle,\rangle_{H_k}$ when we know only the function $k(x,y)$.
In fact assume that A is finite, so that

$$H_k \subset \mathcal{C}(A) \subset L^2(A) \quad :$$

defines the integral operator k in $L^2(A)$ with kernel $k(x,y)$. If $\{\bar{e}_n\}$ is the sequence of eigenfunctions of k orthonormalized in L^2 and $\{k_n\}$ the corresponding sequence of (necessarily positive) eigenvalues, from

$$\langle k(x,\cdot),\bar{e}_n\rangle_{L^2} = k_n \bar{e}_n(x)$$

we see that

$$k(x,y) = \Sigma k_n \bar{e}_n(x) \bar{e}_n(y) \quad : \tag{2.11}$$

whence

$$\{e_n(x)\} = \{k_n^{\frac{1}{2}} \bar{e}_n(x)\} \tag{2.12}$$

is a CONS in H_k .
Moreover the scalar product in H_k can be defined as

$$\langle u,v\rangle_{H_k} = \Sigma \langle u,\bar{e}_n\rangle_{L^2} \langle v,\bar{e}_n\rangle_{L^2} k_n^{-1} \quad , \tag{2.13}$$

since in this way the reproducing property is verified.
One has to notice that the above approach is correct only if $k_n \neq 0$: in this case we see also that H_k is densely embedded in L^2 . If on the contrary $k_n = 0$ is an eigenvalue of k , then H_k is dense in a proper subspace of L^2 and in (2.13) the corresponding value of n should be skipped.

We make now some examples of RKHS: they follow the definitions and the symbolism presented in Appendix 1.

Ex. 2.1 - The space $H_0^{1,2}(0,1)$, endowed with scalar product

$$\langle u,v\rangle_{H_0^{1,2}} = \int_0^1 u'(t)\, v'(t)\, dt$$

is a RKHS. In fact:

$$|u(t)|^2 = \left|\int_0^t u'(\tau)\, d\tau\right|^2 \le t \int_0^t |u'(t)|^2\, dt \le \|u\|^2_{H_0^{1,2}} \quad ,$$

so that

$$H_0^{1,2} \subset \mathcal{C}(0,1)$$

and, by dint of Remark 2.3, this is a RKHS.

Exercise 2.1: prove that the kernel

$$k(t,\tau) = \begin{cases} \tau - t\tau & \tau \le t \\ t - t\tau & \tau \ge t \end{cases} \qquad (\ 0 \le t,\tau \le 1\ )$$

is the RK of $H_0^{1,2}$.

(Hint: prove a) that $k(t,\cdot)$ belongs to $H_0^{1,2}(0,1)$; b) that it satisfies the reproducing property, exploiting the fact that $u \in H_0^{1,2} \to \quad u(1) = u(0) = 0$).

Exercise 2.2: find the sum of the series

$$2 \sum_1^{+} {}_n \frac{\sin n\pi t \sin n\pi\tau}{n^2 \pi^2} \quad .$$

(Hint: $\{\frac{\sqrt{2}\sin n\pi t}{n\pi}\}$ is according to Appendix 1 a CONS in $H_0^{1,2}$. Use Theorem 2.2).

Ex. 2.2 - $L^2(A)$ is not a RKHS: in fact on the contrary hypothesis we would have $L^2(A) \subset \mathcal{C}(A)$ which is obviously false. As a matter of fact the evaluation functional, which is continuous on $\mathcal{C}(A)$, cannot be extended to L^2 since for any point x we can always construct a sequence $f_n \in \mathcal{C}(A), f_n(x) = 1$, $f_n \to 0$ in $L^2(A)$; i.e. ev_x is not continuous on $\mathcal{C}(A)$ as a subset of $L^2(A)$.

It follows from these remarks that if $\{e_n(x)\}$ is a CONS in L^2, then $\Sigma e_n^2(x)$ is almost everywhere divergent, otherways after changing $e_n(x)$ on a set of zero measure, (2.8) would be satisfied.

Ex. 2.3 - Let $H^{s,2}(R^m)$ be the space of functions square integrable over R^m, with all the derivatives up to order s : $H^{s,2}(R^m)$ is a RKHS for $s > m/2$ (Sobolev lemma).

In fact we note that taking advantage of the Fourier identities

$$\hat{u}(p) = \frac{1}{2\pi^{m/2}} \int e^{j\underline{p}\cdot\underline{x}}\, u(x)\, d_m x$$

$$u(x) = \frac{1}{2\pi^{m/2}} \int e^{-j\underline{x}\cdot\underline{p}}\, \hat{u}(p)\, d_m p$$

We can create an isometry between $H^{S,2}$ and $\hat{H}^{S,2}$, i.e.

$$||u||^2_{H^{S,2}} = \int \{u^2 + |\nabla^s u|^2\}\, d_m x = \int (1+p^{2S})\, |\hat{u}(p)|^2\, d_m p = ||\hat{u}||_{\hat{H}^{S,2}} \; .$$

It follows that the evaluation functional in $H^{S,2}$ can be represented as

$$u(x) = ev_x\, u = \int e^{-j\underline{p}\cdot\underline{x}}\, \hat{u}(p)\, d_m p = \langle \hat{u}, e^{j\underline{p}\cdot\underline{x}}\, (1+p^{2S})^{-1} \rangle_{\hat{H}^{S,2}} \; .$$

But then ev_x is a bounded functional on $H^{S,2}$ if $e^{+j\underline{p}\cdot\underline{x}}\,(1+p^{2S})^{-1} \in \hat{H}^{S,2}$ i.e. if

$$\int \frac{(1+p^{2S})\, |e^{+j\underline{p}\cdot\underline{x}}|^2}{(1+p^{2S})^2}\, d_m p = A_m \int d\sigma \int_0^{+\infty} dp\, \frac{p^{m-1}}{1+p^{2S}} < +\infty \qquad (2.14)$$

where $d\sigma$ is the element of surface of the unit sphere in R^m and $d_m p = A_m p^{m-1}\, d\sigma\, dp$.

Now (2.14) is verified if $2s-(m-1) > 1$, i.e. $s > m/2$.

Exercise 2.3: consider a Hilbert space of functions defined on the circle and assume that it is endowed with a reproducing kernel of the form $k(\theta'-\theta)$: show that this is possible if k can be expanded in the Fourier series

$$k(\theta) = \frac{1}{\pi}\sum_{0}^{+\infty}{}_n\, k_n \cos n\theta \qquad k_n \geq 0 \tag{2.15}$$

If $k_o = 0$, what is the property of the functions of H_k ?

(Hint: develop k in Fourier series and use (2.2) and Remark (2.5). If $k_o = 0$, $\int_o^{2\pi} d\theta = 0$).

Exercise 2.4: assume H_k as in Exercise 2.3 with $k_n = A/n^6$ $(n = 1,2,\dots)$. How many derivatives are continuous for all the functions of H_k ? Compute the norms of the functionals $ev_{\bar\theta}$, $(ev_{\bar\theta}\, \partial_\theta)$.

(Hint: use (2.7). Note that after application to each of the two variables θ,θ' , the functionals imply evaluation in $\bar\theta$).

Ex. 2.4 - Let us consider the Sobolev spaces $H^{s,2}(\sigma)$ of functions defined on the unit sphere, with norms

$$\|u\|^2_{H^{s,2}} = \int \{u^2 + |\nabla_\sigma^s u|^2\}\, d\sigma \quad : \tag{2.16}$$

by ∇_σ^s we mean the tensor product of ∇_σ s times and by $|\nabla_\sigma^s u|^2$ we mean the contraction of the tensor $\nabla_\sigma^s u$ with itself. Whence, if u is sufficiently regular, by applying the Green's identity s times we see that

$$\int |\nabla_\sigma^s u|^2\, d\sigma = \int u(-\Delta_\sigma)^s u\, d\sigma \quad . \tag{2.17}$$

We claim that $H^{s,2}$ is a RKHS for $s \geq 2$. In fact $\{Y_{lm}\}$ is a COS in $H^{s,2}$, with norm

$$\|Y_{lm}\|^2_{H^{s,2}} = 1 + l^s(l+1)^s \quad :$$

whence $\left\{\dfrac{Y_{lm}}{\sqrt{1 + l^s(l+1)^s}}\right\}$ is a CONS.

From (2.8) we see that $H^{s,2}$ is a RKHS if

$$\sum_{0}^{+\infty}{}_{l}\sum_{-l}^{l}{}_{m}\frac{Y_{lm}^{2}(\sigma)}{1+l^{s}(l+1)^{s}}<+\infty\quad : \tag{2.18}$$

but

$$\sum_{-l}^{l}{}_{m}Y_{lm}^{2}(\sigma)=(2l+1)\,P_{l}(1)=2l+1\quad ,$$

so that (2.18) is verified only from $s=2$ on.

Exercise 2.5: let H be a space of functions harmonic outside the unit sphere: assume that the trace of $u\in H$ on the unit sphere is an $H^{2,2}$ function and therefore define the norm of u as in (2.16) with $s=2$:

a) prove that this is a Hilbert space;

b) write its reproducing kernel in terms of a spherical harmonic series;

c) prove that inside the domain of harmonicity ($r_P>1$) all the functionals $ev_P(\partial/\partial r)^s$, $ev_P(-\Delta_\sigma)^s$ are bounded.

(Hint: a) it is enough to reason on $\|u\|=0\to u=0$ and on the completeness; b) write the RK $k_\sigma(P,Q)$ for $H^{2,2}(\sigma)$ of Ex. 2.4 and compute its harmonic continuation. To verify the reproducing property note that in the definition of $<,>$ we must first take $r=1$ and then apply $H^{2,2}(\sigma)$ product).

Exercise 2.6: assume that in a RKHS of functions defined on the sphere σ, the kernel k has the form

$$k(P,Q)=k(\theta_{PQ})\qquad(\cos\theta_{PQ}=\underline{r}_P\cdot\underline{r}_Q)\quad ;$$

prove that then necessarily

$$k(P,Q)=\sum_{0}^{+\infty}{}_{l}\sum_{-l}^{l}{}_{m}k_{l}\,Y_{lm}(P)\,Y_{lm}(Q)\qquad k_l\geq 0\quad . \tag{2.19}$$

What are the relations satisfied by $u\in H_k$ if $k_o=0$, $k_1=0$?

(Hint: make the development $k(\theta)=\Sigma\,\bar{k}_l\,P_l(\cos\theta)$, $\bar{k}_l=(2l+1)\,k_l$, apply the theorem of summation of spherical harmonics and Remark 2.5. $k_0 = k_1 = 0 \rightarrow \int u \, d\sigma =$ $= \int u \, Y_{1m} \, d\sigma = 0$).

Remark 2.6: consider a Lie group G of transformations acting on the points of the set A on which are defined the functions of a RKHS , H_k . We shall call g_ω the elements of the group depending on the multidimensional parameter ω ; g_ω sends the point P of A to another point $Q = g_\omega P$ always in A. For instance we can consider translations on R^1 , rotations around a circle, rototranslations in R^2 , rotations around the origin of a sphere in R^3 or of the exterior domain of a sphere etc.

We define an operator G_ω acting on functions of H_k or on $k(P,Q)$, according to

$$G_\omega u(P) = u(g_\omega P)$$
$$G_\omega k(P,Q) = k(g_\omega P, g_\omega Q) \tag{2.20}$$

We shall say that the scalar product in H_k is invariant under G if G_ω is unitary in H_k , i.e.

$$< G_\omega u, G_\omega v >_{H_k} = < u,v >_{H_k} \quad ; \tag{2.21}$$

we shall say that k is invariant under G if

$$G_\omega k(P,Q) = k(P,Q)$$

the following remarkable theorem holds.

Theorem 2.3: the scalar product $<,>_{H_k}$ is invariant iff k is invariant. (Sketch of the proof: if $<,>$ is invariant then $u(g_\omega P) = <k(g_\omega P,Q),u(Q)>_{H_k} =$ $= <k(g_\omega P,g_\omega Q),u(g_\omega Q)>_{H_k}$ and $k(g_\omega P,g_\omega Q) = k(P,Q)$ by the uniqueness of the RK: if k is invariant you can verify for finite combinations $u = \Sigma \lambda_i k(p_i,\cdot)$ that $<,>$ is invariant.)

Ex. 2.5 - A RKHS is invariant under:

1) translations in R^1; 2) rotations of the circle; 3) rototranslations

of R^2; 4) rotations of the sphere; 5) rotations of the exterior of a
sphere, if the corresponding RK's have the form

1) $k(t,t') = k(|t-t'|)$
2) $k(\theta,\theta') = k(|\theta-\theta'|)$
3) $k(\underline{x},\underline{y}) = k(|\underline{x}-\underline{y}|)$ (2.22)
4) $k(P,Q) = k(\theta_{PQ})$
5) $k(P,Q) = k(r_P,r_Q,\theta_{PQ})$

Ex. 2.6 -   Let's consider a stochastic process $x(t,\omega)$ with finite covariance function

$$C(t,t') = E\{x(t)\ x(t')\} \tag{2.23}$$

The Hilbert space of r.v.'s, H_x spanned by the process is defined in Appendix 1.
At the same time, since $C(t,t')$ is symmetric, positive definite, and,
we assume, continuous, we can define the corresponding RKHS, H_c. If
$x(t,\omega)$ is not deterministic, i.e. if C is a strictly positive function,
there is a congruence between H_x and H_c, namely the one implied by
the position

$$\Psi(x(t,\cdot)) = C(t,\cdot)$$
$$\tag{2.24}$$
$$\Psi^{-1}(C(t,\cdot)) = x(t,\cdot)$$

Let us note first that since the manifolds

$$V = \text{Span}\ \{C(t_i,\cdot)\quad ;\quad (t_i)\ \text{dense in A}\}$$
$$U = \text{Span}\ \{x(t_i,\cdot)\quad ;\quad (t_i)\ \text{dense in A}\}$$

are dense respectively in H_c and H_x, once (2.24) are proved to be isometries between U and V, we can extend by continuity Ψ, Ψ^{-1} to all
of H_c, H_x.
On the other hand, we see that:

$$\langle x(t,\cdot),x(t',\cdot)\rangle_{H_k} = C(t,t') = \langle C(t,\cdot),C(t',\cdot)\rangle_{H_c} \qquad (2.25)$$

what proves the isometry.

Moreover Ψ is invertible, since

$$\Psi\left[\Sigma \lambda_i\, x(t_i)\right] = 0 \;\rightarrow\; \Sigma \lambda_i\, C(t_i,t) = 0 \;\rightarrow\; \Sigma \lambda_i \lambda_j\, C(t_i,t_j) = 0 \;\rightarrow\; \lambda_i = 0$$

by the strict positiveness of C . This completes the proof.

This congruence is useful in several respects: for instance if we want to know whether a functional F is bounded in H_x it is enough to verify that $(F\ \Psi^{-1})$ is bounded in H_c , by applying (2.7).

For example, given a certain process $x(t,\omega)$, we can ask: does it exist the variable $\partial_t x(t,\omega)$ in H_x ?

This is the same as: is the functional $E\ \{Y\ \partial_t x(t,\omega)\}$ bounded for any $Y \in H_x$?

By applying Ψ the question becomes: is the functional $\langle \Psi Y,\ \partial_t C(t,\cdot)\rangle_{H_c}$ bounded for any Y ?

Since when Y spans H_x , ΨY spans H_c , the answer is in the affirmative if $\partial_t C(t,\cdot) \in H_c$, i.e. if

$$\partial_t\, \partial_{t'}\, C(t,t')\Big|_{t'=t} < +\infty \qquad (2.26)$$

Exercise 2.7 (Solution of the "best linear estimation" problem): recall that for any separable Hilbert space the "best" approximation of an element y by an element of V_n = Span $\{x_i,\ i=1,2,\dots n\}$ is given by the projection of y onto V_n , characterized by the equations

$$P_n y = \Sigma \lambda_i\, x_i \;\;;\;\; \langle P_n y, x_j\rangle = \langle y, x_j\rangle \quad .$$

Prove that the best approximation of $x(t,\cdot)$ in U_n = Span $\{x(t_i,\cdot),\ i=1,2,\dots n\}$ is given by the same linear combination as the best approximation of $C(t,t')$ in V_n = Span $\{C(t_i,t'),\ i=1,2,\dots n\}$.

Generalize the result to the case in which instead of $x(t_i,\cdot)$, linear functionals of the process $L_i x(t,\cdot)$ are given.

Ex. 2.7 - Beyond the isometry Ψ described in Ex. 2.6, there is another isometry of fundamental importance for stochastic processes.
We treat first the case in which A is a bounded set so that

$$H_C \subset \mathcal{C}(A) \subset L^2(A) \quad .$$

In this case in fact, following the Remark 2.5, we can represent C as

$$C(t,t') = \sum_{1}^{+\infty}{}_n \, c_n \, e_n(t) \, e_n(t') \tag{2.27}$$

where $\{e_n(t)\}$ is a CONS in $L^2(A)$ and $c_n > 0, \forall n$ since we suppose C to be strictly positive.
Now, let us define the weighted l^2 space which is done by sequences $\{a_n\}$ with finite norm

$$\| \{a_n\} \|^2_{l^2(c_n)} = \Sigma \, a_n^2 \, c_n < +\infty \tag{2.28}$$

Obviously we have $\{a_n(t)\} \in l^2(c_n), \forall t \in A$.
We then define the congruence $\Phi : H_X \to l^2(c_n)$

$$\Phi(x(t,\cdot)) = \{e_n(t)\} \quad . \tag{2.29}$$

Let us verify that Φ is an isometry: in fact we have, by exploiting (2.27),

$$\langle x(t,\cdot), x(t',\cdot) \rangle_{H_X} = C(t,t') = \langle \{e_n(t)\}, \{e_n(t')\} \rangle_{l^2(c_n)} \quad . \tag{2.30}$$

Since $\{x(t_i, \ ), t_i$ dense in $A\}$ is total in H_X and $\{\{e_n(t_i)\}, t_i$ dense in $A\}$ is total in $l^2(c_n)$(*), (2.30) can be extended to the whole spaces H_X , $l^2(c_n)$.

(*) In fact $\{a_n\} \in l^2(c_n)$, $\Sigma \, a_n c_n e_n(t_i) = 0 \to \Sigma \, a_n c_n e_n(t) = 0$ since the series is in reality convergent in H_C and hence also uniformly; but then $a_n c_n = 0 \to a_n = 0$.

Then the situation is that we have three spaces congruent each other according to the scheme

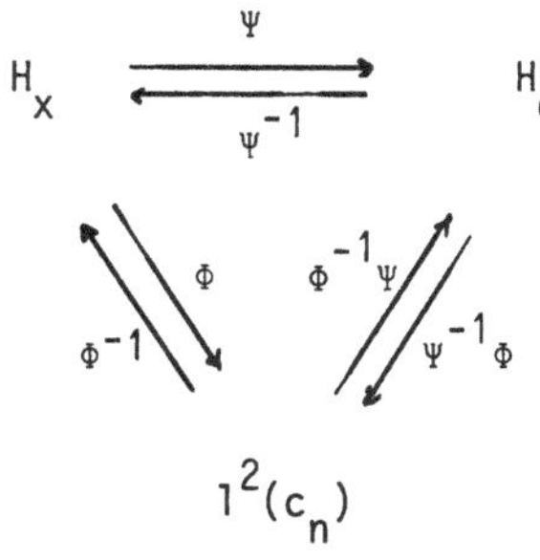

Now consider in particular the system of elements of $1^2(c_n) = \{\{\delta_{nm}\}, n=1,2,\dots\}$: this obviously is a COS of $1^2(c_n)$. It follows that there is a corresponding COS in H_x , namely

$$\{x_n\} = \{\Phi^{-1}(\{\delta_{nm}\})\} \quad ; \tag{2.31}$$

Moreover

$$\|x_n\|^2_{H_x} = \|\{\delta_{nm}\}\|^2_{1^2(c_n)} = c_n \quad .$$

We observe now that we can write

$$x(t,\cdot) = \Phi^{-1}(\{e_m(t)\}) = \Phi^{-1}(\Sigma_n\{\delta_{nm}\}\, e_n(t)) = \Sigma_n\, x_n\, e_n(t) \quad :$$

in this way we have proved the expansion (39) of Appendix 1.
Incidentally we must stress the conclusion that, also taking Ex. 2.5 and Exercises 2.3 and 2.6 into account, any rotationally invariant process on the circle, respectively on the sphere, admits the representation

$$x(\theta,\omega) = \sum_{n=-\infty}^{+\infty} x_n\, e^{jn\theta} \quad ,$$

respectively

$$x(\sigma,\omega) = \sum_{0}^{+\infty}{}_{1} \sum_{-1}^{1}{}_{m} x_{1m} Y_{1m}(\sigma)$$

where x_n (resp. x_{1m}) are incorrelated r.v.s with variance

$$\|x_n\|^2 = c_n \quad , \quad (c(\theta) = \Sigma_n c_n \cos n\theta)$$

resp.

$$\|x_{1m}\|^2 = c_1 \quad , \quad (c(\theta_{PQ}) = \Sigma_{1,m} c_1 Y_{1m}(P) Y_{1m}(Q))$$

If A has not finite measure, for instance A equals all of R^m, the situation is more complicated: however a similar result can be easily derived for translationally invariant processes, i.e. those for which the covariance function has the form $C(x,y) = C(x-y)$.

By taking advantage of the Bochner theorem, stating that

$$C(x-y) = \int_{R^m} e^{j\underline{p}\cdot(\underline{x}-\underline{y})} F(d_m p) \quad ,$$

where $F(d_m p)$ is a measure called the spectral distribution, it is not difficult to see that the correspondence $\Phi(X(x,\cdot)) = e^{j\underline{p}\cdot\underline{x}}$ extends to all H_X into $L^2(F)$, i.e. the space of functions f with bounded norm

$$\int |f(p)\|^2 F(d_m p) < +\infty \quad .$$

In analogy with the bounded case, one proves afterwords that there is a random measure $Z(A;\omega)$ with orthogonal increments on R^m, i.e. $E\{Z(A)Z(B)\} = F(A \cap B)$, such that the spectral representation holds

$$X(\underline{x},\cdot) = \int e^{j\underline{p}\cdot\underline{x}} Z(d_m p) \quad .$$

3. The observation equations of physical geodesy and their linearization

An observation equation is a functional equation trying to model the relation between the quantity Q, on which the measure has been performed, and the unknown quantities on which we want to draw some information. Since we are never able to model completely this relation, as there are always small factors which influence the measurement process but are out of control of the experimenter, the observation equation must contain a discrepancy ν which we treat as a stochastic variable since we have only a collective information on it, namely distributions, averages, variances, correlations estimated from repetitions: ν is usually called the measurement error or noise.

The unknowns of the problem can be distinguished in two cathegories: the finite dimensional parameters and the fields. The first unknowns will be collected in a (real) vector x of dimension n ; $x \in R^n$.

As for the fields, these are characterized by some functions which will range in suitable functional spaces; as such they are infinite dimensional unknowns.

For the sake of simplicity, we shall suppose, without spoiling the generality of our procedure, that we have only one unknown field v . The typical observation equation then reads

$$Q_{oi} = F_i(x,v) + \nu_i \tag{3.1}$$

where Q_{oi} is the observed value of Q_i , F_i is the (generally) non linear observational functional, ν_i the noise.

Since we want to treat here the discrete case we shall assume that we have a discrete and finite set of observations indexed by $i = 1, 2, \ldots m$.

If we put all together the Q_{oi} in a vector $Q_o \in R^m$, the functionals F_i in a vector functional F , i.e. a non linear operator sending (x,v) in R^m , and the ν_i in a vector noise ν , we can write

$$Q_o = F(x,v) + \nu \quad . \tag{3.2}$$

We shall assume that the vector ν has zero average, i.e. that the average of Q_o is $F(x,v)$, and we say that the observation equations are correct:

(3.3)

$$E(\nu) = 0 \tag{3.3}$$

It is better to stress that in (3.3) E means average with respect to the population of ν's: would it not be clear from the context what is the variable (i.e. the probability space) on which we take the expectation, we could specify it by adding an index to the symbol E . Moreover we assume that the covariance structure of ν is given, i.e. to know the matrix

$$C_{\nu\nu} = E\{\nu\nu^+\} \quad ; \tag{3.4}$$

in many examples for the sake of simplicity we shall assume

$$C_{\nu\nu} = \sigma_\nu^2 I \tag{3.5}$$

i.e. incorrelated errors with the same variance, although this will never be essential in our conclusions.

Now we try to put some elementary requirement on the space H in which v is supposed to range. Obviously F is determined from the physics of the measurement and we don't control it directly: in most cases F_i are smooth functionals, however they include the evaluation of the field v or of some of its derivatives at a point P .

Take one for all, the observation equation of an absolute gravity measurement: after a number of "corrections" and simplifications are performed, the observation equation can be reduced to its main part

$$g(P) = |\nabla v(P)| \quad . \tag{3.6}$$

Since it would be a nonsense to try to derive information on v by observing functionals of v which depend on it in an unstable way, we put on v the constraint to belong to a suitable space where functionals like (3.6) are continuous.

If the measurement is physically meaningful this hypothesis should be satisfied by the field v ; for instance in the case (3.6) we know that the earth's potential has continuous derivatives since it is generated by a bounded density.

Since for many reasons it is convenient to work in Hilbert spaces, it becomes straightforward to choose that v belongs to a RKHS, H_k : the advantage is that we work with reasonably smooth functions, that linear functionals are more easily represented in such spaces and that by means of scalar products we can construct in a plane way projections on subspaces, what is a key issue for a "best approximation" theory (cfr. Exercise 2.7).

We leave unspecified for the moment what H_k , i.e. what $k(P,Q)$, we choose, and we send this decision to the moment in which a stochastic meaning will be given to our estimates.

At present we only formulate our problem as, to define a suitable "optimality" principle which permits us to estimate $\hat{x},\hat{v}$ from the vector Q_o, from the observation equations (3.2), the covariance matrix $C_{\nu\nu}$, the hypothesis (3.3) and the hypothesis that $v \in H_k$.

Concerning this last point, we observe only, for the moment, that the tighter is H_k, i.e. the more regular is k compatibility with the hypothesis $v \in H_k$, the better it is from the estimation point of view since we have to chose among a smaller "number" of possible estimates $\hat{v}$: from this remark we can already understand that the problem will be to model k on the function v. Since a direct analysis of the non-linear problem is too difficult, we have to linearize (3.2) around approximate values $\tilde{x}$, $\tilde{v}$ which we assume to know. In order to do that we must use the concept of Frechet differential (cfr. Appendix 1, Def. 7, 8, 9). Assuming that F is suitably regular, we can write

$$x = \tilde{x} + \xi$$
$$v = \tilde{v} + u$$

$$F(x,v) = F(\tilde{x},\tilde{v}) + \partial_x F(\tilde{x},\tilde{v})\xi + \langle F_v(\tilde{x},\tilde{v}),u\rangle_{H_k} + r \tag{3.7}$$

where the residual $r = r(\tilde{x},\tilde{v},\xi,u)$ satisfies

$$|r| = o(|\xi| + \|u\|_{H_k}) \quad , \tag{3.8}$$

in a neighborhood of $\tilde{x}$, $\tilde{v}$.

We note that $\partial_x F$ is a matrix, the design matrix

$$\partial_x F = A = \left| \frac{\partial F_i(x,v)}{\partial x_k} \right| \quad , \tag{3.9}$$

and F_v is a vector of elements of H_k

$$F_v(\tilde{x},\tilde{v}) = \underline{f} = \begin{vmatrix} f_1 \\ \vdots \\ f_m \end{vmatrix}$$

so that

$$\langle F_v,u\rangle_{H_k} = \begin{vmatrix} \langle f_1,u\rangle_{H_k} \\ \vdots \\ \langle f_m,u\rangle_{H_k} \end{vmatrix} = \langle \underline{f},u\rangle_{H_k} \tag{3.10}$$

Substituting in (3.2) we can write the linearized model as

$$q_o = A\xi + \langle \underline{f},u\rangle_{H_k} + \nu \tag{3.11}$$

where

$$q_o = Q_o - \tilde{Q} = Q_o - F(\tilde{x},\tilde{v}) \tag{3.12}$$

is, according to the geodetic tradition, the (observed) anomaly of the quantity Q . In (3.11) we are implicitely admitting that the non-linearity error, r , is so small as to be negligible in comparison with $\sigma(\nu_i)$: would this hypothesis be unsatisfactory, we should think of iterating our procedure, as customary for instance in least squares problems.

Remark 3.1: in general it is not necessary that in each equation enter together parameters and field: for instance if u represents the anomalous gravity field and ξ the coordinates of station points we have that the equation of a distance does depend on ξ , but not on u , while the equation of a gravity measurement (3.6), if we know the coordinates of P , does depend on u but not on ξ .
In any way we shall always assume that the design matrix is of full rank, i.e.

$$A\xi = 0 \rightarrow \xi = 0 \quad . \tag{3.13}$$

As for the vector $\underline{f}$, apart from those components which might be identically zero, we shall assume that $\{f_i\}$ are linearly independent functions: this is equivalent to maintain that the Hessian matrix

$$H = \langle \underline{f},\underline{f}^+\rangle_{H_k} = \{\langle f_i,f_j\rangle_{H_k}\} \tag{3.14}$$

is strictly positive definite. In fact for any vector $\lambda \in R^m$, $\lambda \neq 0$

$$\lambda^+ H\lambda = \langle \lambda^+ \underline{f}, \underline{f}^+ \lambda \rangle_{H_k} = \|\lambda^+ \underline{f}\|^2_{H_k} > 0 \quad . \tag{3.15}$$

We shall call V_m , the manifold generated by $\underline{f}$, i.e.

$$V_m = \{u \in H_k \; ; \; u = \lambda^+ \underline{f}\} \quad . \tag{3.16}$$

In case some of the equations do not contain the field u , after a suitable reordering we shall have a Hessian H of the form

$$H = \begin{vmatrix} 0 & 0 \\ 0 & H_o \end{vmatrix} \quad , \tag{3.17}$$

and we can assume again that H_o is positive definite; in this case $\underline{f}$ generates a manifold $V_{m'}$, $m' < m$.

Remark 3.2: although our unknowns (ξ,u) are infinite in number, once the set of observation equations (3.11) is fixed, the estimation problem becomes necessarily finite dimensional. In fact q_o does depend on u only through the vector $\langle \underline{f},u \rangle_{H_k}$ and in turn this depends only on the projection of u on V_m ,

$$u_m = P_m u$$

$$P_m u = u, \; \forall u \in V_m, \; P_m u = 0, \; \forall u \in V_m^{\perp} \quad :$$

as a matter of fact we have obviously $\underline{f} = P_m \underline{f}$ and

$$\langle \underline{f},u \rangle_{H_k} = \langle P_m \underline{f},u \rangle_{H_k} = \langle \underline{f},P_m u \rangle_{H_k} = \langle \underline{f},u_m \rangle_{H_k} \quad . \tag{3.18}$$

Hence, unless we have different information on u other than that implied by (3.11), there is no sense in considering as unknown u , since the only knowledge we can achieve from (3.11) is u_m .
Since $u_m \in V_m$, we can write it in the form

$$u_m = \lambda^+ \underline{f} = \underline{f}^+ \lambda \quad :$$

introducing in (3.11) and recalling (3.14)(3.18), we see that the observation equations take the form

$$q_o = A\xi + H\lambda + \nu \quad , \tag{3.19}$$

where our unknowns are the parameter vector $\xi \in R^m$, the parameter vector $\lambda \in R^m$ and the stochastic vector $\nu \in R^m$.
Even more customary is the form of (3.19) if we set $s = H\lambda$, i.e.

$$q_o = A\xi + s + \nu \tag{3.20}$$

where by s we mean "signal" and for the moment we understand a pure deterministic vector.
We note, for future use, that

$$s^+ H^{-1} s = \lambda^+ H \lambda = ||u_m||^2_{H_k} \quad . \tag{3.21}$$

In case H has the form (3.17) we should specify that

$$s = \begin{vmatrix} 0 \\ s_o \end{vmatrix} = H \begin{vmatrix} 0 \\ \lambda_o \end{vmatrix}$$

and (3.21) would become simply

$$s_o^+ H_o^{-1} s_o = \lambda_o^+ H_o \lambda_o = ||u_{m'}||^2_{H_k} \quad .$$

As it is obvious, finding ξ, s, ν from (3.20) is always an undetermined problem and we have to state an optimality criterion in order to derive estimates: this will be the object of the next paragraph.

Remark 3.3: it is a typical situation in geodesy that the coordinates of the point where the measurement is performed are among the unknowns of the problem; in our terminology this means that they are included in the vector x . Particular care should be taken of this fact when the linearization (3.7) is practically computed.

Ex. 3.1 - we want to linearize the observation equation (3.6), when both the point P and the field v are unknowns. Since

$$F(\bar{P},v) = ev_{\bar{P}} \, |\nabla v(P)|$$

we have,

$$F(\tilde{P} + \tau\delta P, \tilde{v} + tu) = ev_{\tilde{P}+\tau\delta P} \, |\nabla\tilde{v}(P) + tu(P)| =$$

$$= |\nabla\tilde{v}(\tilde{P} + \tau\delta P) + t\nabla u(\tilde{P} + \tau\delta P)|$$

so that, setting $\underline{\tilde{\gamma}} = \nabla\tilde{v}(\tilde{P})$, $\tilde{\gamma} = |\underline{\tilde{\gamma}}|$, $\tilde{V} = |\partial_{ik}\tilde{v}(\tilde{P})|$ we get

$$\partial_\tau F\Big|_{\tau=t=0} = \frac{\underline{\tilde{\gamma}}.\tilde{V}\delta P}{\tilde{\gamma}} \tag{3.22}$$

$$\partial_t F\Big|_{\tau=t=0} = \frac{\underline{\tilde{\gamma}}.\nabla u(\tilde{P})}{\tilde{\gamma}} \tag{3.23}$$

The vector $V(\underline{\tilde{\gamma}}/\tilde{\gamma})$, according to (3.22), corresponds to one raw of the design matrix, while (3.23) has to be set in the form $\langle f,u\rangle_{H_k}$. Since (3.23) can be written as

$$ev_{\tilde{P}}\left(\frac{\underline{\tilde{\gamma}}}{\tilde{\gamma}} . \nabla u\right) = \langle k(\tilde{P},\cdot), \frac{\underline{\tilde{\gamma}}}{\tilde{\gamma}} . \nabla \langle k(\cdot,Q),u(Q)\rangle \rangle =$$

$$= \langle \frac{\underline{\tilde{\gamma}}}{\tilde{\gamma}} . \nabla_{\tilde{P}} k(\tilde{P},Q),u(Q)\rangle \quad ,$$

we see that the sought functional f is

$$f = \frac{\underline{\tilde{\gamma}}}{\tilde{\gamma}} . \nabla_{\tilde{P}} k(\tilde{P},Q) \quad . \tag{3.24}$$

In spherical approximation $\tilde{v} = \mu/r$, $\underline{\tilde{\gamma}}/\tilde{\gamma} = -\underline{r}/r = -\underline{e}_r$, $\tilde{V} = -\frac{\mu}{r_p^3}[I-3P_r]$, where $P_r = \underline{e}_r\,\underline{e}_r^+$ is the radial projection, so that

$$\tilde{V}\frac{\underline{\tilde{\gamma}}}{\tilde{\gamma}} = -\frac{2\mu}{r_P^3}\underline{e}_r \qquad , \qquad f = -\frac{\partial}{\partial r_P} k(P,Q) \quad :$$

accordingly,

$$q_o = g(P) - \tilde{\gamma} = -\frac{2\mu}{r^3}(\underline{e}_r \cdot \delta P) - \frac{\partial}{\partial r_P} u(\tilde{P}) \quad ,$$

which, in a different notation, coincides with the classical definition of gravity anomaly if $\tilde{P}$ is chosen so that $\tilde{v}(\tilde{P}) = v(P)$, $\Phi(P) = \phi(\tilde{P})$, $\Lambda(P) = \lambda(\tilde{P})$.

It is worth observing that in view of the above procedure the linearized observational functional $\underline{f}$ is always computed in the approximate points $\tilde{P}$, when these are not exactly known.

Remark 3.4: in order to be rigorous one should verify that the Gateaux differentials (3.22), (3.23) are continuous in their variables: in this way in fact one can control the linearization error, r . As for the differential

$$\partial_v F(x,v) = \frac{\nabla v}{|\nabla v|} \cdot \nabla k(P,\cdot) \tag{3.25}$$

we see that if k has continuous mixed derivatives, $\nabla_P \nabla_Q k(P,Q)$, then any function in H_k has continuous first order derivatives, so that

$$\frac{\nabla v}{|\nabla v|} = \frac{\nabla(\tilde{v}+u)}{|\nabla(\tilde{v}+u)|} \tag{3.26}$$

depends continuously on u in H_k if $|\nabla\tilde{v}| = \tilde{\gamma}(\tilde{P}) \geq \text{const} > 0$ when $\tilde{P}$ varies on a bounded domain, like the surface of the earth or a neighborhood of it.

More complicated seems the situation with the other differential, $\partial_x F(x,v)$, since according to (3.22) we have

$$\partial_x F(x,v) = \nabla(\nabla v(P)) \cdot \frac{\nabla v(P)}{|\nabla v(P)|} \quad : \tag{3.27}$$

this is continuous in v and P only if v has continuous second derivatives, what implies a loss of regularity.

This case would not be crucial if we knew the boundary S , since we can suppose that the actual potential admits continuous second derivatives on S , however already if S is assumed to be unknown, things become more complicated, as second derivatives must be discontinuous across S .

Fortunately one can see that writing (3.7) in the form

$$\begin{aligned} r &= F(x,v) - F(\tilde{x},\tilde{v}) - F_x(\tilde{x},\tilde{v})\xi - \langle F_v(\tilde{x},\tilde{v}),u\rangle = \\ &= \int_0^1 \langle F_v(x,\tilde{v} + tu) - F_v(x,\tilde{v}),u\rangle \, dt + \\ &+ \int_0^1 F_x(\tilde{x} + \tau\xi,\tilde{v}) - F_x(\tilde{x},\tilde{v}) \, . \, \xi \, d\tau \end{aligned} \qquad (3.28)$$

it is sufficient to prove that F_v is uniformly continuous in the two variables x,v in a neighborhood of $\tilde{x},\tilde{v}$, while F_x <u>needs only to be continuous in x</u> in a neighborhood of $\tilde{x}$ for $\tilde{v}$ fixed, in order that

$$r = o(|\xi| + \|u\|) \quad .$$

Whence it is only a matter of choosing a normal potential $\tilde{v}$ which is regular with all the derivatives in a relevant neighborhood of the earth's surface, in order to obtain a sound linearization and this hypothesis is certainly satisfied.

<u>Exercise 3.1</u>: verify and linearize with respect to δP and u the astrogeodetic observation equations

$$\Phi(P) = \text{arctg} \frac{g_z(P)}{\sqrt{g_x^2(P) + g_y^2(P)}} \qquad (3.29)$$

$$\Lambda(P) = \text{arctg} \frac{g_y(P)}{g_x(P)} \qquad (3.30)$$

where the directions of the geocentric axes x,y,z are supposed to be known.

<u>Exercise 3.2</u>: verify and linearize with respect to $\delta\underline{r}_P$, $\delta\underline{r}_Q$, u the theodolite observation equations

$$\text{tg}\,\alpha_{PQ} = \frac{\underline{r}_{PQ} \cdot (\underline{e}_o \wedge \underline{\nu}(P))}{\underline{r}_{PQ} \cdot \underline{e}_o} \qquad (3.31)$$

$$\cos z_{PQ} = \frac{\underline{r}_{PQ}}{r_{PQ}} \cdot \underline{\nu}(P) \tag{3.32}$$

where α_{PQ} = reading at the horizontal circle

z_{PQ} = reading at the vertical circle

$\underline{r}_{PQ} = \underline{r}_Q - \underline{r}_P$

$\underline{\nu}(P) = \frac{\underline{g}(P)}{g(P)}$ = actual vertical direction

$\underline{e}_o$ = direction of the origin of the horizontal circle

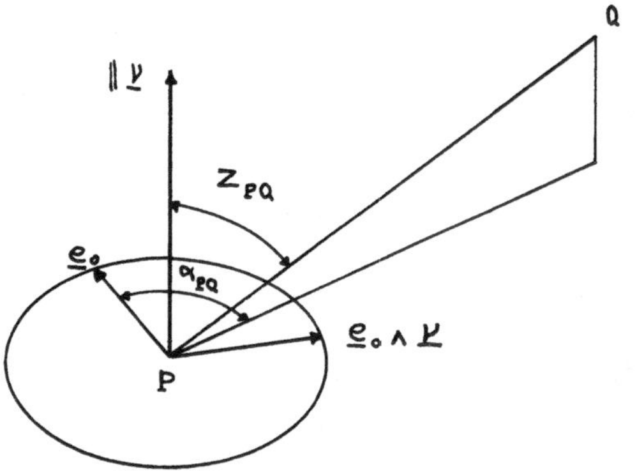

FIG. 3.1

(Hints: the approximate quantities $\tilde{\underline{r}}_P, \tilde{\underline{r}}_Q, \tilde{\underline{e}}_o$, $\tilde{\underline{n}} = -\frac{\underline{\gamma}(\tilde{P})}{\gamma(\tilde{P})}$, $\tilde{V}$ are supposed to be given. Note that

$$\delta\underline{\nu} = \underline{\nu}(P) - \tilde{\underline{n}} = -(I - P_{\tilde{n}})(\underline{g} - \tilde{\underline{\gamma}}) = -(I - P_{\tilde{n}})(\tilde{V}\ \delta\underline{r}_P + \nabla u)$$

where $P_{\tilde{n}}$ is the projection along the vertical $\tilde{n}$.

Note also that ($\delta\underline{e}_o \cdot \tilde{\underline{e}}_o = 0$ since $|\tilde{\underline{e}}_o + \delta\underline{e}_o| = 1$)

$$\delta\underline{e}_o = \underline{e}_o - \tilde{\underline{e}}_o = a\ \tilde{\underline{e}}_o \wedge \tilde{\underline{n}} + b\tilde{\underline{n}} \quad ;$$

a is the unknown orientation correction, while b is related to $\tilde{\underline{n}}$ thorugh the relation $\delta\underline{e}_o \cdot \tilde{\underline{n}} + \tilde{\underline{e}}_o \cdot \delta\underline{\nu} = 0$, which is derived from $(\tilde{\underline{e}}_o + \delta\underline{e}_o) \cdot (\tilde{\underline{n}} + \delta\underline{\nu}) = 0$)

Exercise 3.3: the observation equation of a levelling line between the end points P and Q , i.e. the sum of the levelling increments along the line, is (cfr. Heiskanen-Moritz /4//2/)

$$\Delta n_{PQ} = \int_L dn = \frac{1}{\gamma_0} |v(P) - v(Q)| - \int_L \frac{g(P')-\gamma_0}{\gamma_0} dn_{P'} \qquad (3.33)$$

Linearize (3.33) with respect to δP, δQ, δL, u asuming to know approximate values $\tilde{P}$, $\tilde{Q}$, $\tilde{L}$, $\tilde{v}$.

(Hint: write

$$\int_L \frac{g(P')-\gamma_0}{\gamma_0} dn_{P'} = \int_L \frac{g(P')-\gamma(\tilde{P}')}{\gamma_0} dn_{\tilde{P}'} + \int_L \frac{\gamma(P')-\gamma_0}{\gamma_0} dn_{\tilde{P}'}$$

and notethat only the first term in the second member has to be linearized).

4. Optimum criteria: the hybrid norm principle

We must now decide onto what principle we want to adopt in order to derive the estimates $\hat{\xi}$, $\hat{s}$, $\hat{\nu}$ from the equation (3.20)

$$q_0 = A\xi + s + \nu \qquad . \qquad (4.1)$$

Once these are calculated we have also automatically an estimate of the field u_m by

$$\hat{\lambda} = H^{-1}\hat{s} \ , \quad \hat{u}_m = \hat{\lambda}^+\underline{f} \quad : \qquad (4.2)$$

as it has been discussed in §3 u_m is the maximum we can hope to know about u on the basis of (4.1).

In order to decide about our optimum principle we first consider a simplified example of the form

$$q_o = s + \nu \qquad C_{\nu\nu} = \sigma_o^2 I \quad , \tag{4.3}$$

and we discuss it to the light of two main criteria which we already use separately in different contexts, namely:

I) the least squares principle, of stochastic character, which suggests to minimize the form

$$\nu^+ C_{\nu\nu}^{-1} \nu = \min \quad , \tag{4.4}$$

i.e. in our case $\nu^+\nu$; this is the usual approach when there is no signal in the model;

II) the minimum norm principle, of deterministic nature, which consists in minimizing the norm of the unknown function $\|u\|^2_{H_k}$, with the given information. When the model is not noisy (i.e. $\sigma_o=0$) this is just the "least squares" principle in H_k which leads from u to u_m : in fact it is a simple exercise to prove the equivalence of the following two formulations:

$$\min_{v \in V_m} \|u-v\|_{H_k} \rightarrow \quad v = u_m = P_m u \leftarrow \min_{(P_m v = P_m u)} \|v\|_{H_k} \tag{4.5}$$

This principle corresponds to find the smoothest function in H_k which agrees with the data; if we apply it to the function u_m , recalling (3.21), it becomes

$$s^+ H^{-1} s = \min \tag{4.6}$$

One point is immediately clear for the case (4.3), namely that we should not use (4.4) and (4.6) one after the other.

In fact, let us interpret (4.3) as an interpolation problem and see what happens if we apply first (4.4) and then (4.6) or viceversa.

I) If we impose $\nu^+\nu = \min$ we get the estimates

$$\hat{\nu}_I = 0 \qquad \hat{s}_I = q_o \tag{4.7}$$

which corresponds to interpret our data as all signal (cfr. Fig. 4.1): the second principle (4.6) becomes ineffective.

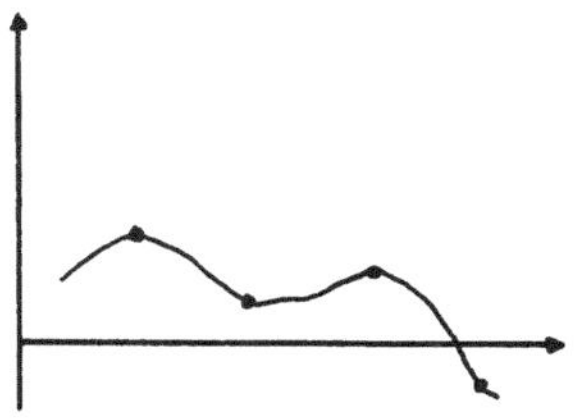

FIG. 4.1

This estimate $\hat{s}_I$ is unbiased since

$$E\{\hat{s}_I\} = E\{q_o\} = s \quad ,$$

the expectation being taken with respect to ν .
On the other hand the norm of $\hat{u}_m$ is unduly inflated, since

$$\|\hat{u}_m\|^2_{H_k} = \hat{s}^+H^{-1}\hat{s} = q_o^+H^{-1}q_o = s^+H^{-1}s + 2\, s^+H^{-1}\nu + \nu^+H^{-1}\nu$$

so that

$$E\{\|\hat{u}\|^2\} = s^+H^{-1}s + \sigma_o^2\mathrm{Tr}\, H^{-1} = \|u_m\|^2 + \sigma_o^2\mathrm{Tr}\, H^{-1} \tag{4.8}$$

For instance if the components of $\underline{f}$ form an ONS,

$$H = \{\langle f_i, f_j\rangle_{H_k}\} = \{\delta_{ij}\} = I$$

and (4.8) becomes

$$E\{\|\hat{u}\|^2\} = \|u_m\|^2 + \sigma_o^2\, m \tag{4.9}$$

We remark also that $\sigma_o^2\mathrm{Tr}\, H^{-1}$ is the m.s.e.e. (mean square estimation error) of $\hat{u}_m$, since

$$E\{\|\hat{u}_m - u_m\|^2\} = E\{(\hat{s}-s)^+H^{-1}(\hat{s}-s)\} = \sigma_o^2 \mathrm{Tr}\, H^{-1} \quad . \tag{4.10}$$

II) If we impose $s^+H^{-1}s = \min$, we find

$$\begin{aligned} \hat{\nu}_{II} &= q_o \\ \hat{s}_{II} &= 0 \end{aligned} \tag{4.11}$$

i.e. we explain all the data as noise (cfr. Fig. 4.2); in this case (4.4) remains ineffective.

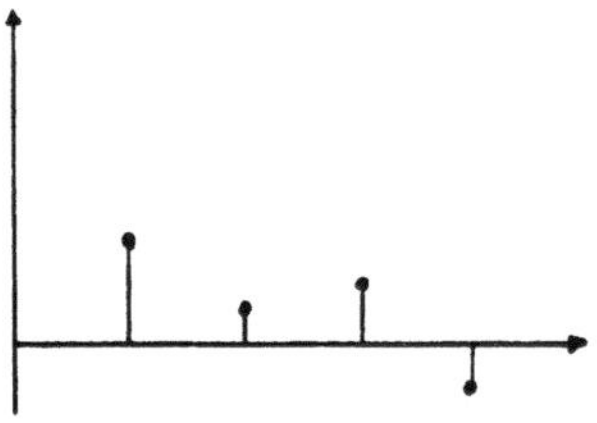

FIG. 4.2

Obviously this estimate is biased and in particular the estimate of σ_o derived from (4.11) is systematically larger than the true σ_o :

$$\hat{\sigma}_o^2 = \frac{1}{m}\hat{\nu}^+\hat{\nu} = \frac{1}{m} q_o^+ q_o = \frac{1}{m}(s^+s + 2\, s^+\nu + \nu^+\nu)$$

$$E\{\hat{\sigma}_o^2\} = \frac{1}{m}\{s^+s + m\,\sigma_o^2\} = \frac{s^+s}{m} + \sigma_o^2$$

Moreover it is clear that in this case the m.s.e.e. is given by

$$E\{(\hat{s}-s)^+H^{-1}(\hat{s}-s)\} = s^+H^{-1}s \quad . \tag{4.12}$$

Another possibility one could think of, is to use the two principles I) and II) , so to say, one into the other. Namely it is known that the least squares principle is also characterized as it gives the minimum variance linear estimator of the parameters: one could be inclined then, also for the case (4.3), to minimize the estimation error of u_m , i.e.

$$E\{\|\hat{u}_m - u_m\|^2_{H_k}\} = \min \quad , \tag{4.13}$$

among linear estimators. On the other hand

$$\|\hat{u}_m - u_m\|^2_{H_k} = (\hat{s} - s)\ H^{-1}(\hat{s} - s)$$

and if we take $\hat{s}$ to be a linear function of q_o

$$\hat{s} = Lq_o = Ls + L\nu$$

we have

$$\hat{s} - s = (L - I)\ s + L\ \nu \quad ,$$

so that going back to (4.13), we find the principle

$$\underset{L}{\mathrm{Min}}\ E\ \{(\hat{s} - s)^+H^{-1}(\hat{s} - s)\} = \underset{L}{\mathrm{Min}}\ s^+(L - I)^+H^{-1}(L - I)s + \sigma_o^2 \mathrm{Tr}\ L^+H^{-1}L \tag{4.14}$$

The minimum condition is easily handled and leads to the estimator

$$L = ss^+(\sigma_o^2 I + ss^+)^{-1} = \frac{1}{\sigma_o^2 + |s|^2}\ ss^+ \tag{4.15}$$

this estimator, as we see, depends quite strongly on s, i.e. on the quantity we want to estimate and the only way to apply it would be to write non-linear equations which, outside the present simple example, might be very hard to be solved. Moreover, yet in the very simple case (4.15), there exist more than one solution.

We are pushed therefore to try a different combination of (4.4) and (4.6); since it is intuitive to search for a principle in between I) and II), i.e. one which distributes in the same way the observed values between noise and signal, we look for the hybrid norm principle

$$\|\hat{u}_m\|^2_{H_k} + \hat{\nu}^+C^{-1}_{\nu\nu}\ \hat{\nu} = \min \quad , \tag{4.16}$$

or

$$\hat{s}^+H^{-1}\hat{s} + \hat{\nu}^+C^{-1}_{\nu\nu}\ \hat{\nu} = \min \quad . \tag{4.17}$$

Before solving (4.17) we notice that the objection that we could take a different positive linear combination of the two quadratic forms is meaningless here, since we have not yet fixed $k(P,Q)$, i.e. the matrix H , so that any positive proportionality constant could go into it.

Now, to solve (4.17) we observe that $\hat{s},\hat{\nu}$ must satisfy the condition (4.3) so that we can write

$$\hat{\nu} = q_o - \hat{s}$$

in (4.17) and minimize with respect to $\hat{s}$: we obtain, recalling that $C_{\nu\nu} = \sigma_o^2 I$,

$$\hat{s} = H\,(\sigma_o^2 I + H)^{-1}\, q_o \qquad ; \tag{4.18}$$

the corresponding estimate of $\hat{\nu}$ is then

$$\hat{\nu} = \sigma_o^2\,(\sigma_o^2 I + H)^{-1}\, q_o \qquad . \tag{4.19}$$

In this case to compute the m.s.e.e., we set $(\sigma_o^2 I + H) = C$ to shorten the formulas and we observe that

$$\hat{s} = HC^{-1}q_o = HC^{-1}s + HC^{-1}\nu$$

$$\hat{s} - s = -\,\sigma_o^2 C^{-1}s + HC^{-1}\nu \qquad ;$$

hence we find

$$E\,\{(\hat{s}-s)^+H^{-1}(\hat{s}-s)\} = \sigma_o^4\; s^+C^{-1}H^{-1}C^{-1}s + \sigma_o^2 \mathrm{Tr}\; C^{-1}H^{-1}C^{-1} \tag{4.20}$$

<u>Remark 4.1</u>: the estimate (4.18) is biased; in fact

$$E\,\{\hat{s}\} = H\,(\sigma_o^2 I + H)^{-1}s \qquad ,$$

so that we have the bias

$$b = s - E\,\{\hat{s}\} = \sigma_o^2\,(\sigma_o^2 I + H)^{-1}s \qquad . \tag{4.21}$$

However it is well known in statistics that the use of a biased estimator may reduce sometimes the estimation error.

Let us compare for example $\hat{s}$ with $\hat{s}_I$; recalling (4.10) and (4.20) we see that $\hat{s}$ has a smaller m.s.e.e. if

$$\sigma_o^4\, s^+C^{-1}H^{-1}C^{-1}s + \sigma_o^2 \mathrm{Tr}\, C^{-1}H^{-1}C^{-1} \leqq \sigma_o^2 \mathrm{Tr}\, H^{-1} \tag{4.22}$$

This relation is not trivial, however just to see that it might be verified, let us take the case that $H = I$: (4.22) then reduces to $(C = (1 + \sigma_o^2)I)$

$$\sigma_o^4 \frac{s^+s}{(H\,\sigma_o^2)^2} + \frac{m\,\sigma_o^2}{(1+\sigma_o^2)^2} = \frac{m\,\sigma_o^2}{(1+\sigma_o^2)^2}\{\sigma_o^2 \frac{s^+s}{m} + 1\} \leqq m\,\sigma_o^2 \tag{4.23}$$

and since

$$s^+s = \|u_m\|^2 \leqq \|u\|^2$$

we see that for m large enough (4.23) is certainly satisfied.
It is interesting however to observe that (4.23) and a-fortiori (4.22) is not automatically satisfied; this condition in fact depends on s as well as on m.
In an analogous way we can compare $\hat{s}$ with $\hat{s}_{II} = 0$; recalling (4.12) we see that $\hat{s}$ is better than $\hat{s}_{II}$ if

$$\sigma_o^4\, s^+C^{-1}H^{-1}C^{-1}s + \sigma_o^2 \mathrm{Tr}\, C^{-1}H^{-1}C^{-1} \leqq s^+H^{-1}s \quad . \tag{4.24}$$

As one could expect, (4.24) is certainly verified if the noise variance σ_o is small enough.
In the case $H = I$ we have

$$\sigma_o^4 \frac{s^+s}{(1+\sigma_o^2)^2} + \frac{m\,\sigma_o^2}{(1+\sigma_o^2)^2} < s^+s \tag{4.25}$$

Again we find that $\hat{s}$ is better than $\hat{s}_{II}$ under suitable conditions.
In particular we see that when $H = I$, $\hat{s}$ is better than both $\hat{s}_I$, $\hat{s}_{II}$ only if

$$\frac{s^+s}{2+\sigma_o^2} \leqq m \leqq \frac{1+2\sigma_o^2}{\sigma_o^2}\, s^+s \quad , \tag{4.26}$$

i.e., since $s^+s \to \|u\|^2$, in a quite definite interval of m .
We can conclude this Remark by saying that the hybride norm principle furnishes estimates where the data q_o are distributed between signal and noise according to the weighting formulas (4.18),(4.19); these estimates are generally biased. They can be better than other estimates but this happens only under suitable cirumstances. This situation will be improved only when the stochastic interpretation will be assumed, according to which we extend the averages also on a family of signals and the hybride norm principle becomes a complete minimum m.s.e.e. principle.

If we accept the hybrid norm principle, we can go back to our general formulation (4.1) and state that

$$\begin{cases} \hat{s}^+H^{-1}\hat{s} + \hat{\nu}^+C_{\nu\nu}^{-1}\hat{\nu} = \min \\ q_o = A\hat{\xi} + \hat{s} + \hat{\nu} \end{cases} \tag{4.27}$$

Taking $\hat{\nu}$ as function of $\hat{s}$, $\hat{\xi}$ from the second of these equations, we find the minimum conditions

$$(H^{-1} + C_{\nu\nu}^{-1})\ \hat{s} - C_{\nu\nu}^{-1}(q_o - A\hat{\xi}) = 0$$

$$A^+(q_o - A\hat{\xi} - \hat{s}) = 0$$

From the first, continuing to call $C = H + C_{\nu\nu}$, we get

$$\hat{s} = H\ C^{-1}(q_o - A\hat{\xi}) \quad ; \tag{4.28}$$

from the second, we get

$$\hat{\xi} = (A^+C^{-1}A)^{-1}A^+C^{-1}q_o \quad . \tag{4.29}$$

The two estimation formulas should be completed by the other

$$\hat{u}_m = \underline{f}^+\lambda = \underline{f}^+H^{-1}s = \underline{f}^+C^{-1}(q_o - A\hat{\xi}) \quad . \tag{4.30}$$

It is to be stressed once more that the estimates $\hat{s}$,$\hat{\xi}$ derived from (4.28),(4.29) are biased.

For future applications it is of crucial importance to make the following remark.

Remark 4.1: the role of $\hat{s}^{+}H^{-1}\hat{s}$ in (4.27) is essentially to smooth the estimate $\hat{u}_m$, (4.30), by controlling its norm in H_k : this norm has not yet been chosen, but we have already stated the principle that it should include some information on the degree of smoothness of the unknown function u (cfr. also the next Ex. 4.1). However it is also a common practice in approximation theory to use smoother functions to approximate u ; this is what we do for instance when using polynomials. In this case we have the following situation: we know that $u \in H_{k_o}$ (with reproducing kernel k_o) and we assume that the linearized observational functionals are

$$L_n u = \langle f_{on}, u\rangle_{H_{k_o}} \quad .$$

We now define a smoother RKHS, H_k (with reproducing kernel k) and assume that H_k is densely embedded in H_{k_o}, so that there are no functions in H_{k_o} which cannot be approximated by functions in H_k.

Take any function $\hat{u}$ in H_k, since this is also in H_{k_o}, we can write for it the observational functionals in both forms

$$L_n \hat{u} = \langle f_{on}, \hat{u}\rangle_{H_{k_o}} = \langle f_n, \hat{u}\rangle_{H_k} \quad ; \tag{4.31}$$

naturally $f_n \neq f_{on}$.

Now we are free to repeat for $\hat{u}$ the above reasonings and in particular to state the principle (4.27) and derive the solutions (4.28), (4.29), (4.30). Naturally in this way we open a new question of convergence of the solution to the true u : in fact at least if $\nu = 0$, $A = 0$, when $u \in H_k$ we have $\hat{u}_m = u_m \to u$ in H_k, if the sequence of the observational functionals f_n spans all of H_k.

The same cannot be maintained if $u \in H_{k_o}$, $u \notin H_k$ (cfr. /19/,/20/).

In spite of this theoretical complication the use of a smoother H_k to solve the hybride norm principle (4.27), is crucial in two respects: first of all the computation of the matrix H with the norm of H_k can be enormously simpler than with the norm of H_{k_o} (this is the case for instance for the anomalous potential of the earth); second, we usually don't know precisely the degree of smoothness of u, of which we know only what the measurements tell us, and the use of a smoother H_k opens the way to the stochastic interpretation which is capable of adapting k to u, but in such a way we are not sure that $u \in H_k$ (see next paragraph).

Ex. 4.1 - Assume $H_k = H_o^{1,2}(0,1)$ and that we observe, with noise, a number of harmonic coefficients of $u \in H_o^{1,2}$: then we have the linear observation equations

$$q_{on} = s_n + \nu_n = \int_o^1 \sqrt{2} \sin n\pi t \, u(t) \, dt + \nu_n \quad , \quad n=1,2,\ldots m \tag{4.32}$$

$$\{\nu_n\} \text{ independent } \quad \sigma^2(\nu_n) = \sigma_o^2$$

First we remind that for $u \in H_o^{1,2}$,

$$\int_o^1 \sqrt{2} \sin n\pi t \, u(t) \, dt = - \frac{1}{n^2\pi^2} \int_o^1 (\sqrt{2} \sin n\pi t)'' \, u(t) \, dt =$$

$$= \frac{1}{n^2\pi^2} \int_o^1 (\sqrt{2} \sin n\pi t)' \, u'(t) \, dt = \tag{4.33}$$

$$= \frac{1}{n^2\pi^2} < \sqrt{2} \sin n\pi t, u >_{H_o^{1,2}}$$

so that the components $f_n(t)$ of the vector $\underline{f}(t)$ are

$$f_n(t) = \frac{\sqrt{2} \sin n\pi t}{n^2\pi^2} \quad :$$

therefore the matrix H is, recalling that $\frac{\sqrt{2} \sin n\pi t}{n\pi}$ is ON in $H_o^{1,2}$,

$$H = \{ \frac{\delta_{nk}}{n^2\pi^2} \} \quad . \tag{4.34}$$

Subsequently, we compute

$$H \, C^{-1} = H \, (H + \sigma_o^2 I)^{-1} = \{ \frac{\delta_{nk}}{1 + \sigma_o^2 n^2 \pi^2} \}$$

and we find the estimate

$$\hat{s} = \{ \frac{q_{on}}{1 + \sigma_o^2 n^2 \pi^2} \}$$

which, according to (4.30), yields

$$\hat{u}_m(t) = \underline{f}^+ H^{-1} \hat{s} = \sum_1^m {}_n \frac{q_{on}}{1 + \sigma_o^2 n^2 \pi^2} \sqrt{2} \sin n\pi t \quad . \tag{4.35}$$

We observe first of all that if $\sigma_o^2 = 0$ we obtain $\hat{u}_m = u_m$, as it should be. Moreover, the estimation operator has the form of a low pass filter cutting high frequencies, but becoming more and more flat as $\sigma_o^2 \to 0$. The qualitative behaviour of the transfer function is illustrated in Fig. 4.3.

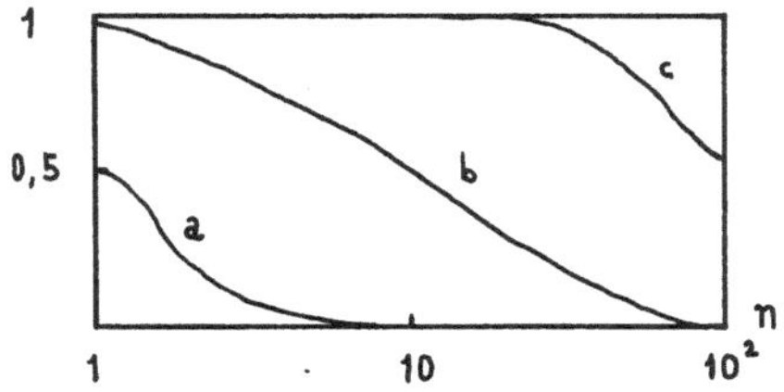

FIG. 4.3 - Transfer functions $(1 + \sigma_o^2 n^2 \pi^2)^{-1}$ for a) $\sigma_o = \pi^{-1}$; b) $\sigma_o = 10^{-1}\pi^{-1}$; c) $\sigma_o = 10^{-2}\pi^{-1}$.

This corresponds essentially to the fact that since $u \in H_o^{1,2}$ (recall $s_n = \int_o^1 \sqrt{2} \sin n\pi t \, dt$), we must have $\Sigma s_n^2 n^2 \pi^2 < +\infty$, so that the true s_n have to go to zero at least with a certain speed: it is this "gross" information which is transferred to the estimate (4.35) by the hybrid norm principle, since the higher is the frequency n, the more probable becomes that the observed value q_{on} is essentially noise.

Exercise 4.1: prove the following properties of the estimate (4.35):

1) $\hat{u}_m$ admits almost surely a limit in $H_o^{1,2}$.
(Hint: it is enough to prove that

$$\lim_{m\to\infty} E\,\{\|\hat{u}_{m+p} - \hat{u}_m\|^2_{H_0^{1,2}}\} = 0$$

since this imply convergence in probability, i.e.

$$\lim_{m\to\infty} P\,\{\|\hat{u}_{m+p} - \hat{u}_m\|_{H_0^{1,2}} > \varepsilon\} = 0 \quad) ;$$

this limit is obviously

$$\lim_{m\to\infty} u_m = \bar{u} = \sum_{1}^{+\infty}{}_n \frac{q_{on}}{1 + \sigma_0^2 n^2 \pi^2} \sqrt{2} \sin n\pi t \qquad ;$$

2) prove that $\bar{u}$ is a biased estimate of u,

$$b = E(\bar{u}) - u \neq 0$$

3) prove that

$$\lim_{\sigma_0 \to 0} b = 0 \quad \text{in} \quad H_0^{1,2}$$

(Hint: recall that $\Sigma\, s_n^2 n^2 \pi^2 < +\infty$)

4) prove that

$$E\,\{\|\bar{u} - E(\bar{u})\|^2_{H_0^{1,2}}\} = \sum_{1}^{+\infty}{}_n \frac{\sigma_0^2\, n^2\, \pi^2}{(1 + \sigma_0^2 n^2 \pi^2)^2}$$

and that this quantity tends to ∞ when $\sigma_0 \to 0$.
(Hint: use the minorization

$$\sum_{1}^{+\infty}{}_n \frac{\sigma_0^2\, n^2\, \pi^2}{1 + \sigma_0^2 n^2 \pi^2} > \sum_{1}^{[1/\sigma_0\pi]}{}_n \frac{\sigma_0^2\, n^2\, \pi^2}{4}$$

and the fact that $\sum_{1}^{N}{}_n\, n^2 > O(N^3)$.)

5) prove that

$$E\,\{\|\bar{u} - E(\bar{u})\|_{L^2}\} \xrightarrow[\sigma_0^2 \to 0]{} 0 \quad ,$$

so that at least in L^2 we have almost sure convergence of $\hat{u}_m$ to u.

Ex. 4.2 - Assume that $u \in H_k$, $k(P,Q)$ continuous, and to observe m values of u with noise: the linear observation equations are then

$$q_{ok} = u(P_k) + \nu_k = s_k + \nu_k \qquad (C_{\nu\nu} = \sigma_o^2 I) \quad . \tag{4.36}$$

Since

$$s_k = u(P_k) = \langle k(P_k,\cdot), u \rangle_{H_k}$$

we find

$$\underline{f} = \begin{vmatrix} \vdots \\ k(P_k,\cdot) \\ \vdots \end{vmatrix} \qquad k=1,\ldots m$$

and

$$H = \{\langle f_i, f_k \rangle_{H_k}\} = \{k(P_i,P_k)\}$$

Whence

$$\hat{s} = H\,(H + \sigma_o^2 I)^{-1}\, q_o$$

and

$$\hat{u}_m(\cdot) = \sum_{1}^{m}{}_{n,k}\; k(P_n,\cdot)\, \{k(P_n,P_k) + \sigma_o^2\, \delta_{nk}\}^{(-1)}\, q_{ok} \quad , \tag{4.37}$$

which is quite a classical estimation formula in collocation theory.
What we want to underline here is a difficulty which is also very typical of collocation when we really want to compute (4.37).
Namely while $\sigma_o^2 > 0$, the matrix $C = H + \sigma_o^2 I$ is always definite positive and its inverse is stable whatever is the configuration of measure points: on the contrary if $\sigma_o = 0$ and we let two points, say P_1 and P_2, to approach each other we see that

$$\det C = \det H = \det \{k(P_n,P_k)\} \to 0$$

since the first two columns of C tend to become identical. From the numerical point of view this means that if σ_0 is very small (with respect to the diagonal $k(P_n,P_n)$) and two points are too close (in the sense that the two functions $k(P_1,\cdot)$, $k(P_2,\cdot)$ are very close in H_k, or that $\langle k(P_1,\cdot),k(P_2,\cdot)\rangle = k(P_1,P_2)$ is very close to $k(P_1,P_1)$) we might have a numerical instability in computing (4.37).
This is certainly so in the computation of the inverse matrix C^{-1} ; but is the same true for the whole expression ?
The following exercise sheds some light on this question.

<u>Exercise 4.2</u>: assume that $u \in H_0^{1,2}(0,1)$ and that two values $u(t_1)$, $u(t_2)$ $(t_1 < t_2)$ are observed without noise.
Construct explicitely the solution $\hat{u}_2(\tau)$ and show that if we take $t_2 \to t_1$ $\hat{u}_2(\tau)$ tends in $H_0^{1,2}$ to the solution corresponding to one observed value only, namely

$$\hat{u}_1(\tau) = \frac{k(t_1,\tau)\; u(t_1)}{k(t_1,t_1)} \quad .$$

(Hint: use k given by Exercise 2.1. Set $u_1 = u(t_1)$, $u_2 = u(t_2)$, $k_i(\tau) = k(t_i,\tau)$, $k_{ik} = k(t_i,t_k)$ and prove that

$$\hat{u}_2(\tau) = \frac{1}{D}\,\{(u_2-u_1)\;[k_{22}k_2(\tau) - k_{12}k_1(\tau)] +$$

$$+\; u_1k_2(\tau)\;(k_{22}-k_{12}) - u_1k_1(\tau)\;(k_{12}-k_{11})\} \quad ,$$

where $D = \det H = C_{11}C_{22} - C_{12}^2$;
prove that $D = (t_2-t_1).(1-t_1)t_1 + o(t_2-t_1)$;
prove that the first term in $\hat{u}_2$ is $o(t_2-t_1)$,
the second is $u_1k_1(\tau)\;(\partial_{t_1}k_{12})\big|_{t_2=t_1} \cdot (t_2-t_1) + o(t_2-t_1)$,
the third is

$$u_1k_1(\tau)\;(\partial_{t_2}k_{12})\big|_{t_2=t_1} \cdot (t_2-t_1) + o(t_2-t_1) \; .$$

5. Collocation: the stochastic interpretation. Optimality of the estimates

Let us go back to the estimates established in the previous paragraph, i.e. $\hat{s},\hat{\xi}$ given by (4.28), (4.29), and to the principle (4.27) from which they arise.
The first obvious remark one can do is that in the estimation process the two vectors $\hat{s}$ and $\hat{\nu}$, apart from their a-priori interpretation, play a quite symmetrical role. It follows that we could derive exactly the same estimates if s were considered a stochastic vector independent from ν and obeying to the same rules, namely

$$E\{s\} = 0$$
$$E\{ss^+\} = H \quad ; \tag{5.1}$$

in this case in fact we could interpret (4.27) as a least squares principle applied to the vector $z = \begin{vmatrix} s \\ \nu \end{vmatrix}$ with mean and covariance

$$E\{z\} = 0$$
$$E\{zz^+\} = \begin{vmatrix} H & 0 \\ 0 & C_{\nu\nu} \end{vmatrix} \tag{5.2}$$

on which we put the observational constraint

$$q_o = A\xi + Bz \quad , \quad (B = |\; I \quad I \;|) \quad . \tag{5.3}$$

We shall pursue this line of thought with two main targets in mind: to find a criterion to choose advantageously the reproducing kernel k, modelling it on the problem we face; to prove that the estimates (4.28), (4.29), derived from the hybrid norm principle and having not many optimal property in that context, do become optimal when considered with an enlarged interpretation, i.e. in weaker sense; more precisely they become minimum m.s.e.e. estimates if we don't compute this quantity on a single s but we rather average it on a suitably defined family of s.
We shall do that in a rigorous way, although at the beginning we proceed formally to see what constraints we have to put on the expectation operator $E\{\;\}$ when applied to s; otherwise stated, we see what properties s must possess in order that some required conclusions can be drawn for the estimates (4.28), (4.29).

First of all, since we have elaborated a general scheme allowing to treat any set of functionals of u ,

$$s = \langle \underline{f}, u_m \rangle_{H_k} = \langle \underline{f}, u \rangle_{H_k} \quad ,$$

we see that (5.1) is equivalent to

$$\begin{cases} E\ \{\langle f,u\rangle_{H_k}\} = 0 \\ E\ \{\langle f_1,u\rangle_{H_k}\ \langle f_2,u\rangle_{H_k}\} = \langle f_1,f_2\rangle_{H_k} \end{cases} \tag{5.4}$$

for every choice of $f, f_1, f_2 \in H_k$.
If we take $f = k(P,\cdot)$, $f_i = k(P_i,\cdot)$, we see that (5.4) entails

$$\begin{cases} E\ \{u(P)\} = 0 \\ E\ \{u(P_1)\ u(P_2)\} = k(P_1,P_2) \end{cases} \quad , \tag{5.5}$$

implying that $u(P)$ has to be considered as a stochastic second order process with zero mean and covariance function $k(P,Q)$. Whence, when we shall specify the rules to compute E , we shall also arrive immediately at the identification of k as the covariance function of u .
The first difficulty we find in continuing along with this interpretation is the so-called infinite norm paradox which was pointed out by C.C. Tscherning (cfr. /20/).
The point is that if $u(P,\omega)$(*) is a process with covariance function $k(P,Q)$, then it might very well be that

$$P\ \{u(P,\omega) \notin H_k\} = 1 \tag{5.6}$$

i.e. that $\|u(P,\omega)\|_{H_k}$ is unbounded a.s. on Ω ; but then we are not sure to be able to interpret in any case expressions like $\langle f,u\rangle$, whatever is $f \in H_k$.
As a matter of fact (5.6) is always verified when we can estimate the covariance func-

---

(*) We make here explicit the dependence from the variable ω ranging in some probability space $(\Omega, \ ,P)$, although we still have to construct it.

tion a.s., as we shall see in the next paragraph; here, let us consider the example of a normal process. If $u(P,\omega)$ is normal, take any CONS in H_k and perform the series development

$$u(P,\omega) = \sum_{1}^{+\infty}{}_n u_n(\omega)\, e_n(P) \quad ; \tag{5.7}$$

as pointed out in Ex. 2.7, the series represents the variable $u(P,\omega)$, being convergent for every P in H_u (the Hilbert space spanned by u).
The variables u_n have zero mean, variance 1 and are incorrelated, whence, being normal, they are also independent.
On the other hand we see that the norms in H_k of the partial sums of (5.7) are r.v.'s given by

$$\left\|\sum_{1}^{N}{}_n u_n(\omega)\, e_n(P)\right\|_{H_k}^2 = \sum_{1}^{N}{}_n u_n^2(\omega) = \chi_N^2$$

and it is easy to ascertain that

$$P\,(\chi_N^2 < a) \xrightarrow[N\to\infty]{} 0 \qquad \forall a > 0 \quad .$$

Before proceeding we want to get rid of this contradiction whatever is the construction of $u(P,\omega)$.
If we go back to our discussion of (5.4) and (5.5) we see that what we need here is, as a matter of fact, a linear operator L , which we shall write

$$Lf = \langle f,u\rangle_{H_k} \tag{5.8}$$

acting from H_k into H_u

$$L:\ f \in H_k \to (Lf) \in H_u$$

and such that

$$L\ k(P,\cdot) = u(P) \quad . \tag{5.9}$$

In fact if $u(P,\omega) \in H_k$, $\langle f,u\rangle_{H_k}$, has to satisfy (5.9) and since Span $\{k(P,\cdot)\}$ is

dense in H_k (5.9) defines completely L ; it follows that if we can define L satisfying (5.9), whether $u(P,\omega) \in H_k$ or not, we find an extension of the usual definition of (5.8). But the construction of such an L has already been accomplished in Ex. 2.6, and is as a matter of fact the congruence Ψ^{-1}.
Therefore we can state by definition that

$$\langle f,u\rangle_{H_k} \stackrel{\mathrm{Def}}{=} \Psi^{-1}(f) \tag{5.10}$$

and we shall not need any more to care of the norm of u in H_k(*).
We are now in position to state a principle of stochastic equivalence.

Principle of stochastic equivalence: define a second order stochastic process $u(\cdot,\omega)$ whatever, with zero mean and covariance $k(P,Q)$, and such that the actual unknown function $u(\cdot) = u(\cdot,\bar{\omega})$ is one realization of the process; then we can construct the RKHS, H_k associated with the process $u(\cdot,\omega)$ and we assume that $u(\cdot) = u(\cdot,\bar{\omega})$ belongs to some other RKHS, H_{k_o} possibly larger than H_k but such that $H_k \subset H_{k_o}$ densely. The problem of estimating the vector of parameters ξ and the stochastic vector $s(\omega) = \langle \underline{f},u(\cdot,\omega)\rangle_{H_k}$ (defined as in (5.10)) in the observational model

$$q_o = A\xi + s(\omega) + \nu \qquad (C_{\nu\nu} \text{ given, } \nu \text{ indep. of } s) \tag{5.11}$$

by a least squares approach (i.e. the search of the best linear estimate) and the problem of estimating the parameter vector ξ and the non stochastic vector $s = \langle f_o,u\rangle_{H_{k_o}}$ by means of a vector $\hat{s}$ of the form (cfr. (4.31))

$$\hat{s} = \langle \underline{f}_o,\hat{u}_m\rangle_{H_{k_o}} = \langle \underline{f},\hat{u}_m\rangle_{H_k} \quad ,$$

applying the hybrid norm principle (4.27) in the space H_k give rise to the same solution.

Remark 5.1: also in the stochastic approach, as in the discussion of §4, we can substitute the problem of estimating $u(P,\omega)$ with that of estimating

(*) This procedure is well known in the theory of stochastic processes (cfr. Ito / 7 /) and is exactly the same that is used to define a Wiener integral of the type $\int_T f(t)\, dW(t,\omega)$, which is a variable in H_W , $\forall f \in L^2(T)$, although it cannot be

$$u_m(P,\omega) = \underline{f}^+(P)\ H^{-1}\ \langle\underline{f}(\cdot),\ u(\cdot,\omega)\rangle_{H_k} \quad . \tag{5.12}$$

In fact it is clear that

$$\langle\underline{f},u\rangle_{H_k} = \langle\underline{f},u_m\rangle_{H_k} \quad ,$$

so that in the observation equations (5.11) only $\langle\underline{f},u_m\rangle_{H_k}$ appear and we cannot draw more information on u than that which comes from u_m. On the other hand the vector $u_m(\cdot,\omega)$ is already the projection in H_u of u on the manifold spanned by $\underline{s}$, i.e.

$$E_u\ \{\underline{s}\ [u(P,\omega) - u_m(P,\omega)]\} \equiv 0 \tag{5.13}$$

so that $\underline{s}$ cannot give more information on u than u_m itself. To prove (5.13) one can for instance use the inverse of (5.10), namely

$$\Psi(\langle f(\cdot),\ u(\cdot,\omega)\rangle_{H_k}) = f(Q) \tag{5.14}$$

Whence

$$\Psi(\underline{s}) = \Psi\langle\underline{f},u\rangle_{H_k} = \underline{f}(Q)$$

$$\Psi(u(P,\omega)) = k(P,Q)$$

$$\Psi(\hat{u}_m(P,\omega)) = \underline{f}^+(P)\ H^{-1}\underline{f}(Q) = \underline{f}^+(Q)\ H^{-1}\underline{f}(P) \quad .$$

Through the isometry Ψ , (5.13) is equivalent to the other relation

$$\langle\underline{f}(Q),\ k(P,Q)\rangle_{H_k} = \langle f(Q),\ \underline{f}^+(Q)\ H^{-1}\underline{f}(P)\rangle_{H_k} = \underline{f}(P)$$

which is trivially true.

We deduce from this remark that u_m is already the best approximation of u in the

---

computed for each realization since $\frac{d}{dt}\, w(t,\omega)$ is known to be a.s. out of $L^2(T)$.

manifold spanned by $\underline{s}(\omega)$, so that a reasonable measure of the goodness of an estimate $\hat{u}_m$ is obtained by suitably defining an estimation error norm for $\hat{u}_m - u_m$, in analogy with what we did in the discussion of the deterministic case.

Remark 5.2: while $u(\cdot,\omega)$ may not belong to H_k ; we see that $u_m(\cdot,\omega)$, like any element of the form $\underline{f}^+\lambda$, does belong to H_k so that we can compute its norm in H_k .

We are able now to prove a first theorem concerning the optimality of the estimate (4.30)

$$\hat{u}_m(P,\omega) = \underline{f}^+(P)\, C^{-1}(q_o - A\hat{\xi})^{(*)} \quad . \tag{5.15}$$

Theorem 5.1: let $\tilde{s}$ be any unbiased estimate of s linear in q_o , i.e.

$$\tilde{s} = L\, q_o + l \tag{5.16}$$

and let

$$\tilde{u}_m = \underline{f}^+ H^{-1}\tilde{s} \quad , \tag{5.17}$$

then, if $\hat{u}_m$ is the collocation estimate (5.15),

$$E\,\{\|u_m - \hat{u}_m\|^2_{H_k}\} \leq E\,\{\|u_m - u_m\|^2_{H_k}\} \quad , \tag{5.18}$$

the expectation being taken on both populations of ν and s (i.e. on ω).

Before starting with the proof, we note that for any unbiased estimate $\tilde{u} \in H_k$ of u_m, if we call P_m the projector on Span $\{\underline{f}\}$, we have necessarily

$$\|\tilde{u} - u_m\|_{H_k} \geq \|P_m\tilde{u} - u_m\|_{H_k} \quad ,$$

(*) The estimate $\hat{u}_m$, though not explicitely written, depends on both the random variables ω and ν.

so that the choice of $\tilde{u}_m$ in the form (5.17) is by no means restrictive, since a minimum if it exists, has to be of this form.

Now, if $\tilde{s}$ is an unbiased estimate of s we must have, for any $\xi \in R^n$

$$E(\tilde{s}) = L\,A\xi + l = 0 \quad , \tag{5.19}$$

the expectation being taken on both ν and s.

The relation (5.19) implies

$$l = 0 \quad , \tag{5.20}$$

$$LA = 0 \quad . \tag{5.21}$$

Therefore we have

$$\tilde{s} = L\,q_o = L\,A\xi + L\,s + L\,\nu = L\,s + L\,\nu$$

$$\tilde{s} - s = (L - I)s + L\,\nu \quad .$$

Since on the other hand

$$||\tilde{u}_m - u_m||^2_{H_k} = ||\underline{f}^+ H^{-1}(\tilde{s} - s)||^2_{H_k} = (\tilde{s} - s)^+ H^{-1}(\tilde{s} - s) \quad ,$$

we find

$$\begin{aligned} \underset{s}{E}\,\{\underset{\nu}{E}\,||\tilde{u}_m - u_m||^2_{H_k}\} &= \underset{s}{E}\,\{s^+(L-I)^+H^{-1}(L-I)s + \mathrm{Tr}\,L^+H^{-1}\,L\,C_{\nu\nu}\} \\ &= \mathrm{Tr}(L - I)^+\,H^{-1}(L - I)H + \mathrm{Tr}\,L^+\,H^{-1}\,L\,C_{\nu\nu} \quad . \end{aligned} \tag{5.22}$$

The target function (5.22) should be minimized with the side condition (5.21), which can be accounted for by a Lagrange multiplier term of the type $2\mathrm{Tr}\,A^+L^+\Lambda$, to be added to (5.22).

Since the function to be minimized is quadratic, positive definite, we know that a minimum exists and that it satisfies the variation equation

$$H^{-1}(L - I)H + H^{-1}\,L\,C_{\nu\nu} + \Lambda A^+ = 0 \ .$$

This leads to (recall that $C = H + C_{\nu\nu}$)

$$L = H C^{-1} - H \Lambda A^{+} C^{-1} \qquad ;$$

after imposing (5.19) we receive

$$L = H C^{-1} - H C^{-1} A (A^{+}C^{-1}A)^{-1} A^{+} C^{-1} \qquad ,$$

so that the sought minimum $\hat{s}$ does coincide with (4.29)

$$\hat{s} = L q_o = H C^{-1} [q_o - A(A^{+}C^{-1}A)^{-1} A^{+} C^{-1} q_o] = H C^{-1} [q_o - A\hat{\xi}] \qquad ,$$

as it was to be proved.

Remark 5.3: the key step, to make a comparison between this theorem and the situation discussed in §4, is performed in (5.22). In fact by (5.22) we see that essentially our minimum principle (5.18) corresponds strictly to the minimum principle (4.14), with the difference that now the m.s.e.e. is computed by averaging also on s, in addition to taking the expectation with respect to ν.

This is why our optimality theorem is weaker than a deterministic optimality theorem; in fact we cannot prove a property of minimum of the m.s.e.e for the function u (i.e. for the vector s) considered as a single realization of the process $u(P,\omega)$, but we rather have a minimum of the m.s.e.e. averaged over all the family $u(P,\omega)$ $(\omega \in \Omega)$.

The estimate $\hat{u}_m$ is optimal still in another sense than that implied by Theorem 5.1; in fact in general we don't want to estimate u_m but rather some functionals of u and what we want to know is whether we can put

$$\widehat{\langle F,u\rangle}_{H_k} = \langle F,\hat{u}_m\rangle_{H_k} \qquad \forall F \in H_k$$

This is a classical issue in collocation theory and is in fact the way in which the collocation estimates were derived.

<u>Theorem 5.2</u>: let $F \in H_k$ and set

$$s_F = \langle F,u\rangle_{H_k} \quad ; \tag{5.23}$$

take now any unbiased estimator of s_F, linear in q_o, i.e.

$$\tilde{s}_F = \lambda^+ q_o \quad , \tag{5.24}$$

with λ fulfilling the condition

$$\lambda^+ A = 0 \qquad (A^+\lambda = 0) \quad ; \tag{5.25}$$

then

$$\hat{s}_F = \langle F,\hat{u}_m\rangle_{H_k} \tag{5.26}$$

is one estimator of this family corresponding to (call $N = A^+C^{-1}A$, the normal matrix)

$$\hat{\lambda} = C^{-1}\,[I - AN^{-1}A^+C^{-1}]\,\langle F,\underline{f}\rangle_{H_k} \tag{5.27}$$

and more precisely it is an estimator of minimum variance

$$\underset{\lambda}{\text{Min}}\; E\,\{(\tilde{s}_F - s_F)^2\} = E\,\{(\hat{s}_F - s_F)^2\} \quad . \tag{5.28}$$

Note first of all that the general form of a linear unbiased estimator (5.24),(5.25) is derived from a more general linear form with the same reasoning as in (5.19),(5.20), (5.21).

The equation (5.27) can be straightforwardly verified from the definition of $\hat{u}_m$ and from $\hat{\xi} = N^{-1}A^+C^{-1}q_o$. The minimum condition (5.28) can be proved for instance by showing that

$$E\,\{(\tilde{s}_F - \hat{s}_F)(\hat{s}_F - s_F)\} = 0 \qquad \forall\lambda\;(A^+\lambda = 0) \tag{5.29}$$

since in this case

$$E\{(\tilde{s}_F - s_F)^2\} = E\{(\tilde{s}_F - \hat{s}_F)^2\} + E\{(\hat{s}_F - s_F)^2\} \quad .$$

Now we have, by virtue of condition (5.25),

$$\tilde{s}_F = \lambda^+ q_o = \lambda^+(A\xi + s + \nu) = \lambda^+(s + \nu)$$

so that

$$\hat{s}_F = \hat{\lambda}^+(s + \nu)$$
$$\tilde{s}_F - \hat{s}_F = (\lambda - \hat{\lambda})^+ (s + \nu) \tag{5.30}$$

Whence, recalling that ν and s are independent,

$$E\{(\tilde{s}_F - \hat{s}_F)(\hat{s}_F - s_F)\} = (\lambda - \hat{\lambda})^+ |E\{(s + \nu)(s + \nu)^+\} \hat{\lambda} - E\{(s + \nu)s_F\} =$$
$$= (\lambda - \hat{\lambda})^+ |C \hat{\lambda} - E\{ss_F\}| \quad . \tag{5.31}$$

On the other hand

$$E\{ss_F\} = E\{<\underline{f},u>_{H_k} <F,u>_{H_k}\} = <\underline{f},F>_{H_k} \quad ,$$

so that recalling (5.27), we can rewrite (5.31) as

$$-(\lambda - \hat{\lambda})^+ A\, N^{-1}\, A^+\, C^{-1}\, <\underline{f},F>_{H_k} \quad ;$$

but this last expression is zero since

$$(\lambda - \hat{\lambda})^+ A = 0$$

by condition (5.25) and the theorem is proved.

Remark 5.4: if we apply the Theorem 5.2 to the evaluation functional we see that we can claim the property of minimum variance

$$E\ \{|u(P,\omega) - \hat{u}_m(P,\omega)|^2\} = \min \qquad ;$$

it is as a matter of fact by minimizing this quadratic functional that $\hat{u}_m$ was at first calculated (cfr. H. Moritz /13/).

Exercise 5.1: show directly that in the average $\hat{s}$ is a better estimator than $\hat{s}_I$, $\hat{s}_{II}$ of the previous paragraph.
(Hint: take the expectation with respect to s of (4.22), (4.24)).

6. Invariance principles and random fields with almost surely estimable covariance function

In order to make effective the stochastic interpretation illustrated in the previous paragraph from the theoretical point of view, we must decide upon a mechanism generating the stochastic process $u(P,\omega)$ on which for the moment we put as only constraints

$$u(P;\bar{\omega}) = u(P) \tag{6.1}$$

$$E\ \{u(\cdot,\omega)\} = 0 \tag{6.2}$$

$$E\ \{u(P,\omega)\ u(Q,\omega)\} = k(P,Q) \tag{6.3}$$

Since we want to use these relations to estimate from the data the covariance function k and on the other hand we are free to choose our process $u(P,\omega)$, we stipulate to build it in such a way that the covariance function is surely estimable from the single realization $u(P)$.
In other words let's assume that u belongs to some Hilbert space H_o. We want to set up a process such that, for some bounded family of operators M_{PQ} in H_o, the relation holds

$$\langle u(\cdot,\omega),\ M_{PQ}u(\cdot,\omega)\rangle_{H_o} = k(P,Q), \quad \forall\omega \in \Omega \quad : \tag{6.4}$$

in this way at least if we know exactly our function $u(P)$, we can compute the covariance function of the corresponding process, by estimating it through (*)

$$k(P,Q) = \langle u(\cdot), M_{PQ} u(\cdot)\rangle_{H_0} \quad . \tag{6.5}$$

Indeed in general we don't know u but in the most lucky case some finite values $u(P_i,\omega)$, also observed with noise: in this case we shall set up an approximation of (6.5) tending to the true $k(P,Q)$ when $m\to\infty$ and the noise tends to zero.

Remark 6.1: the general form of the processes generated with the above characteristics is

$$u(P,\omega) = R(P,\omega)\ u \tag{6.6}$$

with $R(P,\omega)$ a family of operators acting in H_0 , depending on some parameter ω ranging on a probability space $(\Omega, ,P)$ and continuous in P ; in this case in fact from the boundedness of the quadratic form

$$E\ \{R(P,\omega)\ u\ R(Q,\omega)u\} = \int R(P,\omega)\ u\ R(Q,\omega)\ u\ dP(\omega)$$

we deduce the existence of a bounded selfadjoint M_{PQ} for every P,Q via general theorems of functional analysis.

Ex. 6.1 - We show that there is always at least one process with the required characteristics. Namely, let $u(P)$ be the given function and let

$$u(P) = \Sigma\ u_n\ e_n(P) \tag{6.7}$$

be any Fourier series representation of u in H_0 .
Define a sequence $\{b_n\}$ of binomial variables assuming the values $-1,1$ with probability 1/2 and independent from each other. The product of the probability spaces of b_n is the probability space of $\omega = \{b_n\}$: each

(*) To be more precise we could slightly relax this condition by assuming that (6.4) holds almost surely and hope to be not so unlucky that (6.5) does not hold. This is equivalent to assume (6.4) is a r.v. with variance zero.

realization of ω is a sequence $\{b_1, b_2, \ldots\}$ $b_n = \pm 1$.
If we set

$$u(P,\omega) = \Sigma\, b_n\, u_n\, e_n(P)$$

we see that

$$E\,\{u(P,\omega)\} = \Sigma\, E(b_n)\, u_n\, e_n = 0$$

$$k(P,Q) = \Sigma\, u_n^2\, e_n(P)\, e_n(Q) \quad .$$

In this case, if we put π_n = projection in H_o on e_n , the operators $R(P,\omega)$, M_{PQ} mentioned above are

$$R(P,\omega) = \Sigma\, b_n\, \pi_n$$

$$M_{PQ} = \Sigma\, e_n(P)\, e_n(Q)\, \pi_n \quad .$$

This example introduces us to the spectral representation of the condition that $k(P,Q)$ be estimable with $P = 1$ from u only. Namely let us fix the CONS $\{e_n\}$ in H_o and assume a-priori that $k(P,Q)$ is diagonal on $\{e_n\}$, i.e.

$$k(P,Q) = \Sigma\, k_n\, e_n(P)\, e_n(Q) \quad . \tag{6.8}$$

Consider first the case that k is non degenerate, i.e. k_n are all different; since from

$$u(P,\omega) = \Sigma\, u_n(\omega)\, e_n(P)$$

we know that

$$\sigma^2(u_n) = E\,\{u_n^2(\omega)\} = k_n \tag{6.9}$$

we can see that k_n (and hence $k(P,Q)$) can be estimated from u_n almost surely if

$$u_n^2(\omega) = k_n \quad ; \quad P = 1 \tag{6.10}$$

if (6.10) is verified $\forall\omega \in \Omega$, k is surely estimable.
In case we admit that k can be degenerate so that k_n is repeated N_n times we can write

$$k(P,Q) = \sum_{1}^{+\infty}{}_n\, k_n \sum_{1}^{N_n}{}_m\, e_{nm}(P)\, e_{nm}(Q) \tag{6.11}$$

and

$$u(P,\omega) = \Sigma_{n,m}\, u_{nm}(\omega)\, e_{nm}(P) \quad ; \tag{6.12}$$

in this case the condition (6.10) can be relaxed to

$$\frac{1}{N_n} \sum_{1}^{N_n}{}_n\, u_{nm}^2(\omega) = k_n \qquad P = 1 \quad . \tag{6.13}$$

It is important to stress that conditions (6.10) or (6.13) are constraints to the construction of the r.v. ω . Namely if we choose a CONS $\{e_{nm}\}$, define ω and the r.v.'s $u_{nm}(\omega)$ in such a way that: a) u_{nm} have zero mean and are incorrelated; b) u_{nm} satisfy (6.13), then it is clear that (6.12) is a process of zero mean, and covariance of the form (6.11) almost surely estimable from one realization.
Why to choose a-priori one system $\{e_{nm}\}$ or another, is a problem that we shall look at, starting from Ex. 6.2.
Another important consequence of (6.10) or (6.13) is that, if the covariance function is a.s. estimable, then a.s. $u(P,\omega) \notin H_k$.
In fact in this case

$$\|u(P,\omega)\|_{H_k}^2 = \Sigma_{n,m}\, u_{nm}^2(\omega)\, k_n^{-1} = \Sigma_n\, N_n = +\infty \quad .$$

This difficulty however has already been treated and solved in §5.

<u>Ex. 6.2</u> - let $u(\theta)$ be given in $L^2(0,2\pi)$: assume to state that the reproducing kernel $k(\theta,\theta')$ be invariant under the rotation group: according to Exercise 2.3 and Remark (2.6) we must have

$$k(\theta,\theta') = \sum_{0}^{+\infty}{}_n\, k_n \cos n(\theta-\theta') =$$

$$= k_o + \sum_{1}^{+\infty}{}_n\, k_n \left[\cos n\,\theta \cos n\,\theta' + \sin n\,\theta \sin n\,\theta'\right]$$

As we see the requirement of invariance of k forces k to be of a particular form, fixing on the same time $\{e_n(\theta)\}$ and the fact that the eigenvalues k_n are twofold (corresponding to both eigenfunctions $\cos n\,\theta$, $\sin n\,\theta$), but for $n=0$.
In this case if we set

$$u(\theta) = \sum_{0}^{+\infty}{}_n \{c_n \cos n\,\theta + s_n \sin n\,\theta\}$$

and analogously for $u(\theta,\omega)$ the condition (6.13) reads

$$\frac{1}{2}\left[c_n^2(\omega) + s_n^2(\omega)\right] = k_n = c_n^2 + s_n^2 \quad .$$

By analyzing Ex. 6.2 an important feature of our construction is highlighted, namely the fact that the choice of an a-priori invariance group for the kernel k, determines a particular CONS in H_o, with respect to which k is diagonal, and the type of degeneracy of k in this base.
This is by no means a case, but rather a general fact as we shall try to illustrate shortly.
Let G be a Lie group of transformations acting within the domain A on which $u(P)$ is defined: let g_ω be an element of G, parametrized by the r dimensional parameter ω ; in an infinitesimal neighborhood of the identity, we can write $(g_{\omega=0}=1)$

$$g_\omega P = (1 + \omega G)P \qquad\qquad (\omega G = \sum_{1}{}_i\, \omega_i\, G_i) \tag{6.14}$$

where G_i are the infinitesimal generators of the group G.
Corresponding to every g_ω we can define an operator G_ω as

$$G_\omega\, u(P) = u(g_\omega P)$$

and from (6.14), when u is sufficiently regular, we get

$$G_\omega u(P) = u(P) + (\omega GP) \cdot \nabla u(P) \quad : \tag{6.15}$$

the operators

$$D_i = (G_i P) \cdot \nabla \tag{6.16}$$

are the generators of a Hilbert representation of the group G. Assume now to require that $k(P,Q)$ be invariant under the group G, i.e. according to Remark 2.6 that the scalar product of H_k be invariant under this group.

Some properties follow:

- since k represents the identity in H_k, we must have the operator identity

$$Dk = kD \qquad (D = \text{r-vector with components } D_i) \quad ; \tag{6.17}$$

- since for any operator B in H_k

$$\langle k(P,Q), B\cdot\rangle_{H_k} = \langle B_Q^+ k(P,Q), \cdot\rangle_{H_k}$$

  we have also

$$D_P\, k(P,Q) - D_Q^+\, k(P,Q) = 0 \quad ; \tag{6.18}$$

- from the invariance of $k(P,Q)$ we easily see that

$$k(g_\omega P, g_\omega Q) - k(P,Q) = D_P k(P,Q) + D_Q k(P,Q) = 0 \tag{6.19}$$

- comparing (6.18) with (6.19) we find

$$D = -D^+ \quad ,$$

  i.e. the generators are antisymmetric operators (they enjoy the same properties as the selfadjoint operators, however with purely imaginary spectrum instead of a real one);
- since generally the components of D do not commute (they rather follow the commutation rules implied by the structure constants of the group G) we cannot state from

the beginning that there is a system of eigenfunctions common to D_i , and to the operator k too.

Usually it is more useful the following relation

$$(D.D)_P\ k = \Sigma\ D^2_{iP}\ k(P,Q) = D_P k(P,Q) \cdot D_Q = -\ D_Q \cdot D_P\ k(P,Q) \quad (*)$$

which, being symmetrical in P,Q, shows that

$$D^2_P\ k = D^2_Q\ k \ ; \tag{6.20}$$

since D^2 is obviously selfadjoint, (6.20) is nothing but

$$\left[D^2,k\right] = 0 \ . \tag{6.21}$$

This equation entails the existence of a common set of eigenfunctions of D^2 and k; if they are still not completely specified (case of degenerate eigenvalues), and the components of D commute with D^2 , one can further use one such component to reduce the degeneracy.

Remark 6.2: it seems not unuseful to stress that when a property maybe trivial in terms of operators in H_k , like (6.21), is translated into a property of k(P,Q) as function of P and Q , like (6.20), then it can be extended from H_k into the larger Hilbert space H_O , for instance into L^2 , on condition that the scalar product in this space is invariant under G too. Whence when we say that k has a common set with D^2 we can look into the eigenfunctions of D^2 in H_0 and then deduce the particular form of k . So, for instance, since all finite transformations G can be approximated by products of infinitesimal transformations, we understand that (6.17) entails

$$G_\omega k = k\ G_\omega \qquad \forall \omega \in \Omega \ .$$

(*) Obviously D_P, D_Q commute since they are acting on different variables.

Exercise 6.1: consider the one-dimensional rotation operator

$$G_\omega u(\theta) = u(\theta+\omega)$$

and prove that

$$D = \frac{\partial}{\partial\theta} \quad , \quad D^2 = \frac{\partial^2}{\partial\theta^2} \quad ;$$

find the eigenfunctions common to D^2 and k for an invariant k .

Exercise 6.2: compute the generators of the 3-D rotation group

$$\underline{D} = \underline{r}_P \wedge \nabla_P \quad .$$

(Hint: every infinitesimal rotation can be represented by a vector product $G_\omega u(P) = u(P + \underline{\omega} \wedge \underline{r}_P)$ ...)

Derive

$$D^2 = \Delta_\sigma \qquad \text{(Laplace Beltrami operator)}$$

(Hint: $(\underline{r} \wedge \nabla).(\underline{r} \wedge \nabla)u = \underline{r} . \nabla \wedge [- \nabla \wedge (\underline{r}u)] = -\underline{r} . \{\nabla[\nabla . (\underline{r}u)] - \nabla \underline{r}u\}$...)

Show from (6.19) that, if P,Q are on the unit sphere,

$$k(P,Q) = k(\underline{r}_P \cdot \underline{r}_Q) = k(\theta_{PQ})$$

(Hint; from $\underline{r}_P \wedge \nabla_P\, k(P,Q) = -\underline{r}_Q \wedge \nabla_Q\, k(P,Q)$ derive

$\underline{r}_P \wedge \nabla_P k(P,Q) = \lambda\, \underline{r}_P \wedge \underline{r}_Q = -\, \underline{r}_Q \wedge \nabla_Q k(P,Q)$, and then

$\nabla_P k(P,Q) = \lambda \underline{r}_Q$, $\nabla_Q k(P,Q) = \lambda \underline{r}_P$, so that

$dk = d\underline{r}_P \cdot \nabla_P k(P,Q) + d\underline{r}_Q \cdot \nabla_Q k(P,Q) = \lambda d(\underline{r}_P \cdot \underline{r}_Q)$ ...)

Exercise 6.3: let $\underline{x}$ be 2-D vector and G be the rototranslation group with infinitesimal transformations

$g_{(\underline{\eta},\underline{\varepsilon})}\ \underline{x} = \underline{x} + \underline{\eta} + \underline{\varepsilon} \wedge \underline{x}$ ($\underline{\varepsilon} \perp$ to the plane of $\underline{x}$)

Find the generators of the translations ($\underline{\varepsilon} = 0$)

$$D_{\underline{\eta}}\ u = \underline{\eta} \cdot \nabla u$$

and the generators of the rotations ($\underline{\eta} = 0$)

$$D_{\underline{\varepsilon}}\ u = \underline{\varepsilon} \cdot \underline{x} \wedge \nabla u$$

Show from (6.19), namely

$$(\nabla_x + \nabla_y)\ k\ (\underline{x},\underline{y}) = 0$$

$$(\underline{x} \wedge \nabla_x + \underline{y} \wedge \nabla_y)\ k\ (\underline{x},\underline{y}) = 0$$

that

$$k(\underline{x},\underline{y}) = k(|\underline{x}-\underline{y}|) \quad .$$

Ex. 6.3 - We want to prove once more that any rotationally invariant kernel on the unit sphere admits the representation

$$k(P,Q) = \Sigma_{n,m}\ k_n\ Y_{nm}(P)\ Y_{nm}(Q) \quad ; \tag{6.22}$$

(cfr. also Exercise 2.6).
We observe that according to Exercise 6.2, k commutes with Δ_σ ; furthermore we see that k commutes with

$$D_z = \underline{e}_z \cdot \underline{r} \wedge \nabla = \frac{1}{r \sin\theta}\ \frac{\partial}{\partial\lambda}$$

by means of (6.17) and therefore it must have a common set of eigenfunctions with Δ_σ, jD_z, or in real form with Δ_σ, D_z^2 .
On the other hand these are the CONS $\{Y_{nm}(\sigma)\}$, so that we must have

$$k(P,Q) = \Sigma\ k_{nm}\ Y_{nm}(P)\ Y_{nm}(Q) \quad ;$$

finally since k is rotationally invariant, we can choose P to be the pole and Q such that $\lambda_Q = 0$, $\theta_Q = \theta_{PQ}$.
Since

$$Y_{n,m}(0,\lambda) = \sqrt{2n+1}\,\delta_{mo}$$

$$Y_{n,o}(\theta,\lambda) = \sqrt{2n+1}\,P_n(\cos\theta)$$

we find that (6.22) holds with $k_n = (2n+1)\,k_{no}$.

A partial conclusion for the moment is that given an invariance group and the corresponding base $\{e_{nm}\}$, if we are able to define a process satisfying (6.10), or (6.13) with incorrelated r.v.'s $u_{nm}(\omega)$, we can construct our process $u(P,\omega)$ with an a.s. estimable covariance function $k(P,Q)$, satisfying the same invariance principle too.

We restrict now our attention to compact groups since these enjoy particular properties of great importance in the study of the rotation group which is prominent in the applications to physical geodesy.
Let us consider k as an operator in H_o with kernel (6.11). Since k_n have to satisfy (6.13) we see that $k_n \to 0$ and k is therefore compact in H_o.
The operators

$$\pi_n = \sum_{1}^{N_n}{}_m\, e_{nm}(P)\,\langle e_{nm}(\cdot),\cdot\rangle_{H_o}$$

are nothing but the projectors on V_n, the invariant subspaces of k, which must be finite dimensional, $N_n < +\infty$, otherwise k could not be compact.
Since (6.19) holds, (6.17) holds too in the sense of the operators in H_o; but then

$$[\pi_n, D] = 0 \rightarrow [\pi_n, G_\omega] = 0 \quad ,$$

i.e. V_n are also invariant under the transformations G_ω. In particular we must have

$$e_{nm}(g_\omega P) = \sum_{1}^{N_n}{}_k\, c^n_{mk}(\omega)\, e_{nk}(P) \quad : \qquad (6.23)$$

developing twice $e_{nm}(g_{\omega 1} g_{\omega 2} P)$ by (6.22) it is easy to see that

$$c^n_{ml}(g_{\omega 1} g_{\omega 2}) = \Sigma_k \, c^n_{mk}(g_{\omega 1}) \, c^n_{mk}(g_{\omega 2}) \quad , \tag{6.24}$$

so that $[c^n_{ml}(\omega)]$ is a representation of the group G, in R^{N_n}. Since the scalar product H_o is invariant under G, we see also that

$$\langle e_{nm}(g_\omega P), e_{nl}(g_\omega P)\rangle_{H_o} = \delta_{ml} = \Sigma_k \, c^n_{mk}(\omega) \, c^n_{lk}(\omega) \quad , \tag{6.25}$$

i.e. $[c^n_{ml}(\omega)]$ is a unitary representation of G.

The space V_n can be further decomposed into irreducible components (i.e. subspaces which are completely spanned starting from one vector and applying to it all G_ω), or equivalently we let in the representation (6.11) that the same k_n is maybe repeated, but in such a way that the subspaces V_n are basis of irreducible representations of G. We obtain in this way a particular enunciation of a quite famous theorem of Peter-Weyl on Hilbert-unitary representations of continuous compact groups, claiming that the infinite dimensional unitary representation of G, done with the operators G_ω can be decomposed into the sum of finite dimensional irreducible representations, the underlying Hilbert space H_o being like-wise decomposed into the direct sum of orthogonal components, $H_o = \oplus V_n$, each of which is the base of one of the irreducible representations

The next step relies still on another general theorem on topological compact groups.

Theorem 6.1: let G be a topological compact group; we say that a measure $\mu(dg)$ is left (right-)-invariant if $\forall \bar{g} \in G$, $\mu(\bar{g}^{-1} dg) = \mu(dg)$ $(\mu(dg\bar{g}^{-1}) = \mu(dg))$: on G there is a measure which is both left and right invariant, unique up to a multiplicative constant, and such that

$$\mu(G) < +\infty \quad :$$

therefore on G there is a unique uniform probability distribution

$$P(dg) = \frac{\mu(dg)}{\mu(G)} \quad . \tag{6.26}$$

The following fact is also taken from the general group theory (cfr. W. Miller /12/, §6.2).

Theorem 6.2: let $H_o = \oplus V_n$ be a decomposition of H_o into invariant subspaces, bearing irreducible representations of G ; let $\{c^n_{jk}(g_\omega)\}$ be defined as in (6.23) and let $P(dg) = P(d\omega)$ be the uniform probability distribution on the group manifold G (or equivalently on the parameter space Ω); then the orthogonality relations hold

$$\int_G c^n_{jk}(g)\, c^m_{il}(g)\, P(dg) = c\, \delta_{nm}\, \delta_{ji}\delta_{kl} \quad ; \tag{6.27}$$

setting $n=m$, $j=i$, $k=l$ and adding on the last index we see that $c = \frac{1}{N_n}$.

Remark 6.3: let us consider a constant function e_{oo} on A ; since clearly

$$G_\omega e_{oo} = e_{oo} \quad ,$$

we see that the space spanned by e_{oo} is invariant under G and as a matter of fact it is the invariant subspace bearing the trivial representation

$$g_\omega \to 1 \quad .$$

But then one of c^n_{nm} , let us call it c^o_{oo} , must be

$$c^o_{oo} = 1 \quad .$$

From the orthogonality relations (6.27) we then derive

$$\int_G c^n_{jk}(\omega)\, P(dg) = 0 \tag{6.28}$$

unless $n=0$, $j=0$, and $k=0$.

We stipulate that in $k(P,Q)$ the e_{oo} component is null, i.e. $k_o = 0$; this is the same to say that

$$\langle u,e_{oo}\rangle_{H_o} = 0 \to u_{oo} = 0 \to k_o = 0 \quad . \tag{6.29}$$

For instance if $H_o = L^2(A)$, this condition means simply that

$$\int_A u(P)\, dA = 0 \quad , \tag{6.30}$$

i.e. it has a null average on A .

We are able now to show how to construct the stochastic process $u(P,\omega)$.

<u>Theorem 6.3:</u> let Ω be the parameter space of an invariance group G and let us define on it a uniform probability distribution, i.e. one which generates a left and right invariant probability measure on G ; let H_o be a given Hilbert space, with scalar product invariant under G , and let $u \in G$ be given, satisfying (6.29); then the process

$$u(P,\omega) = u(g_\omega P) \tag{6.31}$$

satisfies the relations

$$\begin{cases} E\,\{u(P,\omega)\} = f(P) = 0 \\ \\ E\,\{u(P,\omega)\, u(Q,\omega)\} = k(P,Q) = \Sigma_{nm}\, k_n\, e_{nm}(P)\, e_{nm}(Q) \\ k_n = \frac{1}{N_n} \sum_{1}^{N_n}{}_m\, u^2_{nm}(\omega) = \frac{1}{N_n} \sum_{1}^{N_n}{}_m\, u^2_{nm} \end{cases} \tag{6.32}$$

In fact, following the definition (6.31),

$$\begin{aligned} u(P,\omega) &= \sum_{1}^{\infty}{}_n \sum_{1}^{N_n}{}_m\, u_{nm}\, e_{nm}\,(g_\omega P) = \\ &= \sum_{nm} u_{nm} \sum_k c^n_{mk}(\omega)\, e_{nk}(P) = \\ &= \sum_{nk} (\Sigma_m\, c^n_{mk}(\omega)\, u_{nm})\, e_{nk}(P) \quad , \end{aligned}$$

so that

$$u_{nk}(\omega) = \sum_{1}^{N_n}{}_m\, c^n_{mk}(\omega)\, u_{nm} \quad .$$

On the other hand from (6.28) we see that

$$E\,\{u_{nm}(\omega)\} = 0 \;\rightarrow\; E\,\{u(P,\omega)\} = 0 \qquad ;$$

from (6.27) we find

$$E\,\{u_{nm}(\omega)\; u_{jl}(\omega)\} = \sum_{ki} E\,\{c^{n}_{km}(\omega)\; c^{j}_{il}(\omega)\}\; u_{nk}\; u_{ji} = \delta_{nj}\; \delta_{ml}\; \frac{1}{N_n} \sum\nolimits_k u^2_{nk}$$

so that the variables $\{u_{nm}(\omega)\}$ are mutually incorrelated and their variance is

$$\sigma^2(u_{nm}(\omega)) = \frac{1}{N_n} \sum\nolimits_k u^2_{nk} \qquad ;$$

then we have also

$$k(P,Q) = \sum\nolimits_{nmjl} e_{nm}(P)\; e_{jl}(Q) \quad E\,\{u_{nm}(\omega)\; u_{jl}(\omega)\} = \sum\nolimits_{nm} k_n\; e_{nm}(P)\; e_{nm}(Q)$$

with

$$k_n = \frac{1}{N_n}\, \Sigma\, u^2_{nm} \qquad ;$$

finally from the unitary relations (6.25) we see that

$$\frac{1}{N_n}\, \Sigma_m\, u^2_{nm}(\omega) = \frac{1}{N_n}\, \Sigma_m\, u^2_{nm} = k_n \qquad ,$$

so that $k(P,Q)$ is a.s., estimable.

<u>Remark 6.4</u>: due to the importance of this point let us see another proof of (6.31) which does not make use of the orthogonality relations of the coefficients $c^{n}_{mk}(\omega)$.

Let $\bar{g} \in G$; from the invariance of $P(dg_\omega)$, we see that

$$f(\bar{g}P) = E\,\{u(\bar{g}P,\omega)\} = \int_G u(g_\omega \bar{g}P)\; P(dg) = \int_G u(g_n P)\; P(dg g_n^{-1}) =$$

$$= E\,\{u(P,\omega)\} = f(P)$$

$$k(\bar{g}P,\bar{g}Q) = E\{u(\bar{g}P,\omega)u(\bar{g}Q,\omega)\} = \int_G u(g_\omega \bar{g}P)u(g_\omega \bar{g}Q)P(dg) =$$

$$= \int_G u(g_\eta P)u(g_\eta Q)P(dgg_\eta^{-1}) = k(P,Q) \quad .$$

The second relation already proves that k is invariant under G so that we can repeat the discussion of the first part of the paragraph and find a complete set $\{e_{nm}\}$ of eigenfunctions of k, reducing the Hilbert representation of G into irreducible components.

From the first relation it follows $f(P) = \text{const} = \mu$ and since we must have

$$\langle e_{oo},f\rangle = \langle e_{oo},e_{oo}\rangle = \int \langle e_{oo},u(g_\omega P)\rangle_{H_o} P(dg) =$$

$$= \int \langle e_{oo},u(P)\rangle_{H_o} P(dg) = 0 \quad ,$$

we see that $\mu = 0$ follows.

Ex. 6.4 - let G be the group of proper rotations in 3-D. As everybody knows G can be parametrized by 3 parameters, for instance 3 Euler angles, which we shall choose as $(\lambda_P,\theta_P,\alpha)$ in Fig. 6.1.

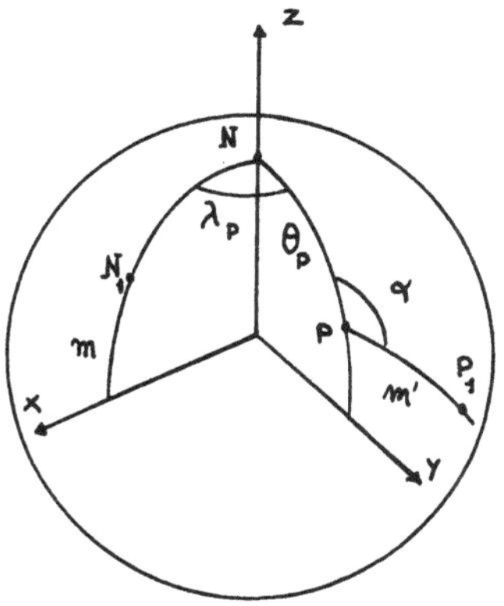

FIG. 6.1

Noting that a reference system on the sphere can be characterized by a pole and a meridian, e.g. N (the north pole) and m (the origin meridian), the rotation parametrized by $(\lambda_P,\theta_P,\alpha)$ can be viewed as the one which changes the pole from N to P and the origin meridian from m to m' : alternatively we could say that if two couple of points (N,N_1), (P,P_1) are given so that

$$\theta_{NN_1} = \theta_{PP_1} \quad ,$$

then the rotation with parameters $(\lambda_P,\theta_P,\alpha)$ is the one which sends (N,N_1) into (P,P_1) the relation of (λ,θ,α) to a set of Euler angles is understood by noting that the rotation from (N,m) to (P,m') can be obtained as the product of the three

$$g = g_3\, g_2\, g_1 \quad ;$$

g_1 = rotation around z of λ_P
g_2 = rotation around y of θ_P
g_3 = rotation around z of $\pi-\alpha$.

Whence the proper rotation group, also called $O(3)$ can be made an analytic manifold by transporting on it the product topologies of the unit sphere $\sigma \to P \to (\lambda_P,\theta_P)$ and of the unit circle $c \to \alpha$:

$$O(3) = \sigma \otimes c \quad .$$

From this we see that $O(3)$ is a compact group and the above theory applies to it. In particular there must be a measure on $O(3)$, invariant under left and right multiplication.
In fact an infinitesimal set of rotations dg around a fixed rotation $\bar{g}$ $(\bar{g} : (N,m) \to (P,m'))$ can be described as the product of an infinitesimal set in σ and an infinitesimal one on c : choosing neighborhoods as in Fig. 6.2, one can see that if

$$\mu(d\sigma_N) = \mu(d\sigma_P)$$

$$|d\alpha| = |d\alpha'| \quad ,$$

there is a one to one correspondence between these two sets, realized by the equation

$$dg_{Pm'}, \bar{g} = \bar{g}\ dg_{Nm} \qquad (\bar{g}:(N,m) \rightarrow (P,m'))$$

so that one is forced to set

$$\mu(dg_{Pm'}) = \mu(dg_{Nm}) = \mu(d\sigma_N)\ |d\alpha| \quad . \tag{6.33}$$

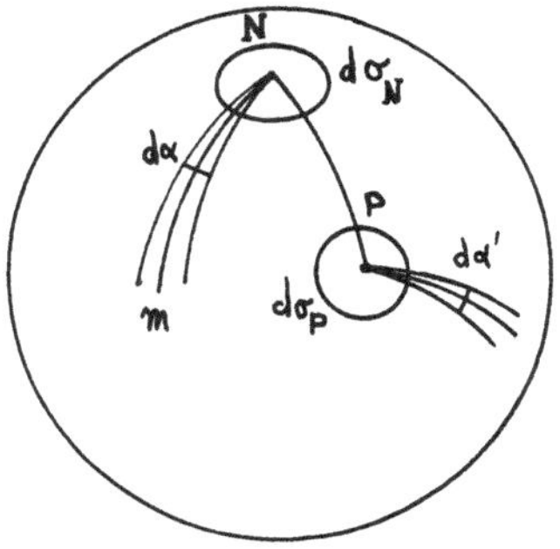

FIG. 6.2

Since both $\mu(d\sigma_N) \cong 2\pi\ d\theta$, $|d\alpha|$ are invariant under rotations, it follows that $\mu(dg)$ defined as in (6.33) is invariant too.
Since

$$\mu(G) = \mu(\sigma)\ \mu(c) = 8\ \pi^2 \quad ,$$

we have that the required uniform probability distribution can be written

$$P\ (\bar{\lambda},\bar{\theta},\bar{\alpha};d\lambda,d\theta,d\alpha) = \frac{\sin\bar{\theta}\ d\theta\ d\lambda\ d\alpha}{8\ \pi^2}$$

Two main consequences follow by specifying (6.32) in this case: let $\bar{g}$ be the rotation which brings P',Q' to $N,\bar{Q}$ $(\theta_{N\bar{Q}} = \theta_{P'Q'}\ ,\ \lambda_{\bar{Q}} = 0)$ and let g be specified by $\lambda_P,\theta_Q,\alpha_{PQ}$, i.e.

$$g : (N,\bar{Q}) \rightarrow (P,Q)\ ,\ \theta_{PQ} = \theta_{N\bar{Q}} = \theta_{P'Q'}$$

$$\int u(gP)\,P\,(dg) = \int u(gN)\,P\,(dg) = \frac{1}{8\pi^2}\int_0^{2\pi} d\alpha \int_0 u(P)\,d\sigma_P =$$

$$= \frac{1}{4\pi}\int_0 u(P)\,d\sigma_P = 0$$

$$k(P',Q') = \int u(gP')\,u\,(gQ')\,P\,(dg) = \int u(gN)\,u\,(g\bar{Q})\,P\,(dg) =$$

$$= \frac{1}{8\pi^2}\int d\sigma_P \int_0^{2\pi} d\alpha_{PQ}\,u(P)\,u(Q) \qquad (6.34)$$

$$(\theta_{PQ} = \theta_{P'Q'})$$

this last equation displaies explicitely the dependence of $k(P'Q')$ only on $\theta_{P'Q'}$. In particular from

$$k\,(\theta_{PQ}) = \Sigma\, k_n(2n+1)\,P_n(\cos\theta_{PQ}) =$$

$$= \Sigma\, k_n\, Y_{nm}(P)\, Y_{nm}(Q) \qquad (6.35)$$

we see that $\{Y_{nm}(P)\}$ is the set of eigenfunctions of k in $H_o = L^2$, decomposing $O(3)$ into irreducible components, and from

$$k_n = E\,\{\langle Y_{nm}, u(\cdot,\omega)\rangle^2_{H_o}\}$$

and the third of (6.32) we get

$$k_n = \frac{1}{2n+1}\sum_{-n}^{n}{}_m\, u^2_{nm} \qquad (6.36)$$

Remark 6.5: the construction of the process $\{u(P,\omega)\}$ we have proposed in this paragraph, is by no means the only possible one.
For instance we are not at all obliged to accept that $u(P,\omega)$ be constructed so that its covariance function is almost surely estimable.
For example, starting from empirical hystograms of the gravity anomaly, there has been the suggestion that, in the case of the anomalous gravity potential considered as a function on a sphere, $u(P,\omega)$ should be a normal process (cfr. H. Moritz /13/). But in this case u_{nm} should be inde-

pendent normal r.v.'s and $u(P,\omega)$ cannot be really fixed from one realization only.

Taking up this point of view S.L. Lauritzen (cfr. /11/) has shown that the covariance function $k(P,Q)$ in this case cannot be consistently estimated, throwing some serious shadows on the whole estimation process. Naturally, our answer to that approach is that we are free to choose $u(P,\omega)$; only, depending on our choice, we shall derive different average optimal properties of the estimator. Whence there is no sense in trying to make things more complicated.

On the other hand a normal-shaped empirical hystogram of $u(P)$ (or of some of its functionals), puts indeed a constraint on $u(P,\omega)$ but does not oblige this process to be normal.

Still another stochastic model has been proposed by H. Moritz (cfr./13/): namely, let $O(2n+1)$ represent the proper rotation group in R^{2n+1} and let g_n be a r.v. uniformly distributed in $O(2n+1)$. Then g_n is represented by a matrix taking any $(2n+1)$ vector x into another one, $g_n x$, with the same euclidean norm. Define ω as the direct product of g_n considered as independent r.v.'s. Finally define

$$u(P,\omega) = \sum_{nm} (g_n u_{nm}) Y_{nm}(P) \quad .$$

It is a simple exercise to verify that this process satisfies the conditions (6.1), (6.2), (6.3), (6.13) and furtheron that k is rotationally invariant. Nevertheless we continue to prefer our approach defining $u(P,\omega)$ through a transformation group acting only on $P \in A$, since this is simpler (we define a smaller number of r.v.'s), and has a simpler physical interpretation. Moreover this approach is equivalent to a more general invariant estimation principle which can be applied also to non compact groups, where a stochastic interpretation is not as simple.

Exercise 6.4: let $G = O(3)$ and $u(P)$ an L^2 function defined on the unit sphere σ and with zero average on it.

Let $\omega = \{g_n\}$; g_n independent r.v.'s uniformly distributed on G .

Set

$$u(P,\omega) = \sum_{nm} u_{nm} Y_{nm}(g_n P)$$

and show that $u(P,\omega)$ has an a.s. estimable function, invariant under G.

Remark 6.6: the stochastic interpretation cannot be so simply extended to non compact group of transformations: as an example one can think of translations of the real line. In this case in fact there is no invariant measure, such that $\mu(G) < +\infty$, whence we cannot define a uniform probability distribution.

The above theory however can survive if we accept to reduce our requirements. For instance, to treat the R^1 case, we can define a family of r.v.'s dependent on a parameter T, namely τ_T, uniformly distributed on $[-T,T]$.

Then we can define a family of stochastic processes

$$G_{\tau_T} u = u(t,\tau_T) = u(t + \tau_T) \qquad ,$$

requiring that only asymptotically for $T \to \infty$, they admit a shift invariant covariance function.

In this case if $\qquad \forall t,t'$

$$\frac{1}{2T} \int_{-T}^{T} u(t + \tau)\, u(t' + \tau)\, d\tau \to k(t - t') \qquad , \tag{6.37}$$

we are able to derive estimates which are also asymptotically optimal, or more precisely which satisfy an optimal asymptotic criterion.

In fact, take for instance a model without parameters and noise: if (6.37) is satisfied we see that on the class of shift invariant linear estimates, i.e. those fulfilling the commutation rule

$$G_\tau \hat{u} = G_\tau \Sigma \lambda_j(t;t_1 \ldots t_m) L_j u(t_j) =$$

$$= G_\tau \Lambda u = \Sigma \lambda_j(t + \tau; t_1 + \tau, \ldots) L_j u(t_j + \tau) =$$

$$= \Sigma \lambda_j(t;t_1 \ldots) L_j u(t_j + \tau) = \Lambda G_\tau u ,$$

it is possible to define the limit

$$\lim_{T\to\infty} \frac{1}{2T} \int_{-T}^{T} [G_\tau(\Lambda u - u)]^2 \, d\tau = \Sigma \lambda_i \lambda_j L_i L_j k(t_i - t_j) + - 2 \Sigma \lambda_i L_i k(t_i - t) + k(0) \quad : \tag{6.38}$$

by minimizing (6.38) with respect to λ_i we find again the collocation solution.

We can conclude this paragraph by summarizing the process which has led us to the procedure of constructing our estimates with the stochastic interpretation: assume you have a set observations $\{q_i\}$ which includes a dense set of values of $u(P_i)$ (*) in A ; define an invariance principle, depending on the geometry of A and on your physical intention; estimate the covariance function of u by approximating expressions like (6.5) or (6.13) with your discrete data; apply the "optimal" estimates (4.29) (4.30), coinciding with the minimum hybrid norm estimates.
The whole process is usually referred to as least squares collocation, though the author prefers to consider it as a collocation solution with stochastic (or group-invariant) interpretation. It seems interesting to stress that going from the deterministic to the stochastic interpretation, one fundamental step has been done, transforming a purely linear method into a non-linear one. Fortunately the non-linear part of the procedure can be performed separately, since the subsequent estimation formulas are not too strongly sensitive to the shape of $k(P,Q)$.

7. Applications to harmonic fields: the case of physical geodesy

We want to apply in this paragraph the discussed, general approach to the problem of estimating from geodetic measurements the anomalous gravity field of the earth, that we shall call u as in the preceding paragraphs.

(*) Instead of $u(P_i)$ we could have a set of $L_i u(P_i)$, derive first the covariance function of $L_P u$ and then convert it into the covariance function of u .

The first important specific characteristic of this case is due to the fact that u is now a function harmonic in the outer space, which we shall call A as opposed to the set of the earth's interior which we shall call B ; A and B are separated by their boundary, the earth's surface S .

Let us assume then that u belongs to soma RKHS, $H_{k_o}(A)$ and that in this space the Dirichlet problem has a unique solution. There is a natural correspondence (isomorphism) between $u(P)$, $P \in A$ and $u(P)$, $P \in S$, in the sense that to every $u \in H_{k_o}$ we can associate the restriction to S $Ru = u|_S$ and viceversa to any function of the form $u|_S$ we can associate only the original function $u = R^{-1}(u|_S)$ since the corresponding Dirichlet problem has a unique solution: as we see $\mathcal{D} = R^{-1}$ represents the resolvent of the Dirichlet problem.

By transporting isometrically the structure of $H_{k_o}(A)$ to the family of traces $u|_S$ $(u \in H_{k_o})$, namely defining

$$\langle Ru,Rv\rangle = \langle u|_S,v|_S\rangle = \langle u,v\rangle_{H_{k_o}} \tag{7.1}$$

we see that we can make of this family a RKHS too, which we can call $H_{k_o}(S)$. As a matter of fact the double restriction of $k_o(P,Q)$ to S , i.e. with a little abuse of notation

$$k_o(P,Q) = R(P,P')\, R(Q,Q')\, k_o(P',Q') \quad , \quad (P,Q \in S;\ P',Q' \in A) \tag{7.2}$$

satisfies the relation

$$\begin{aligned} &\langle R(P,P')\, R(Q,Q')\, k_o(P',Q'),\ R(Q,Q')\, u(Q')\rangle_{H_{k_o}(S)} = \\ &\quad = \langle R(P,P')\, k_o(P',Q'),\ u(Q')\rangle_{H_{k_o}(A)} = R(P,P')\, u(P') \quad , \end{aligned} \tag{7.3}$$

showing that it is the sought reproducing kernel of $H_{k_o}(S)$.

By inverting (7.2) we see that the reproducing kernel in space can be found from its trace by solving a double Dirichlet problem, i.e.

$$k_o(P',Q') = R^{-1}(P',P)\, R^{-1}(Q',Q)\, k_o(P,Q) \quad : \tag{7.4}$$

incidentally this shows that k_o has to be harmonic in the outer space.

It follows from these preliminary remarks that if we like to use a stochastic approach in estimating u , we can apply it to the trace only of u on S and then extend harmonically the reproducing kernel, found on S , to all of A .
Nevertheless if S is the real physical surface of the earth, or a telluroid used to linearize the observation equations, the problem of knowing k_0 is almost impossible to be solved due to the enormous number of details of S in which geodesists are very much interested.
We all know what it might be the effort to solve twice the Dirihclet problem, in order to compute $k_0(P,Q)$ for each couple of points P,Q needed to apply collocation formulas.
Fortunately, we can greatly simplify the matter if we exploit a remark of §2 and accept to approximate $u \in H_{k_0}(S)$ by means of smoother functions $\tilde{u}$ belonging to a restricted space H_k . In physical geodesy this is done by taking $k(P,Q)$ and whence $\tilde{u}$ harmonic down to an internal sphere $\Sigma \subset B$.
Obviously this is meaningful only if the set of functions $\tilde{u} \in H_k$, restricted to A , is a dense subset of H_{k_0} , so that u can be approximated at will by such functions.
But this is exactly stated by a famous theorem which in the Hilbert space setting is known to geodesists as the <u>Runge-Krarup theorem</u>.
This theorem has been proved in several ways, here we can say that it holds if B is a simply connected domain, S satisfies a cone condition (*), the topology of H_{k_0} is regular in the sense that it is weaker than some Sobolev topology on S and H_k is a RKHS of functions harmonic down to any fixed internal sphere Σ .
What we do therefore is that the set of observations $\{q_i\}$, modelled as in §3, contains the values of the functionals $L_i u$ we attribute them to the same functionals applied to a $\tilde{u} \in H_k$, $L_i u = L_i \tilde{u}$ and we proceed as if $\tilde{u}$ were our unknown function.
Since $\tilde{u}$ is harmonic down to a sphere, the rotation group formalism becomes available and we can apply it to $\tilde{u}$.
In particular a covariance function can be computed, either by discretizing (6.34), i.e.

$$k(\theta_{PQ}) \sim \frac{1}{N} \sum_{|\theta_{P_iP_j} - \theta_{PQ}|<\delta} u(P_i)\, u(P_j) \tag{7.5}$$

or by applying (6.36) with empirically observed values of u_{nm}.
The first approach is usually emploied to estimate local covariance functions; after

---

(*) We can define a cone c_δ of fixed semiaperture δ and height h , such that at each point P of S we can attach c_δ with vertex in P and all contained in B.

empirical values of $k(\theta_{PQ})$ are computed by (7.5) a theoretical model of some definite positive function is fitted to them: the second approach is usually applied to global problems: a number of degree variances k_n $(n=1...N)$ is empirically estimated and then some simple analytic form $k_n = f(n)$ is fitted to them in such a way that the series (6.36) can be summed.

Remark 7.1: as it is known the solution of the outer Dirichlet problem for a sphere Σ with radius R is given by

$$\begin{cases} u(P) = \sum_{nm} (\frac{R}{r})^{n+1} u_{nm} Y_{nm}(P) = \mathcal{D}(u|_{\Sigma}) \\ u_{nm} = \int_{\sigma} u(R,\sigma) Y_{nm}(\sigma) d\sigma \quad ; \end{cases} \tag{7.6}$$

it follows that moving from Σ outside, on concentric spheres, $u(r,\sigma)$ is continuously smoothed since the higher is the degree n (frequency) the smaller is the smoothing factor $(R/r)^{n+1}$.

Therefore the inverse operation, i.e. the backward harmonic continuation, is a highly unstable operation. It follows that in order to reproduce small variations of u , $\tilde{u}$ might be obliged to make strong oscillations (see Fig. 7.1).

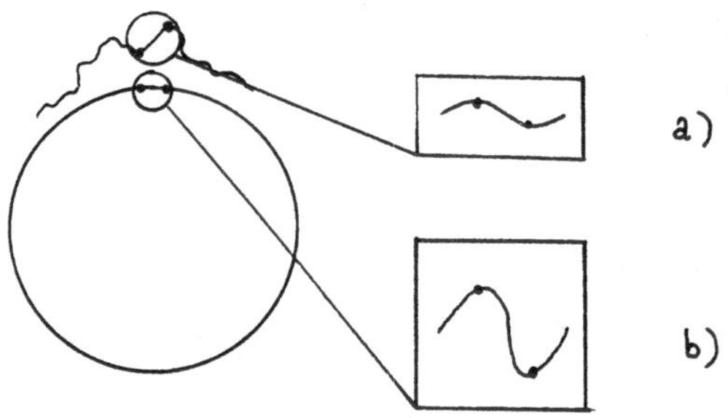

FIG. 7.1 - a) variations of u on S ; b) variations of $\tilde{u}$ on Σ .

In order to reduce this phenomenon, is is preferred nowadays to subtract from u the influence of the sorrounding topographic masses: this operation in fact has the effect to smooth u and hence also $\tilde{u}$.

Still, if we transport without any change the value of $\tilde{u}(P)$ from the surface S to Σ in order to compute then the spherical estimates of the covariance function $k(P,Q)$, we commit an error which might be relevant in mountainous areas. This could be avoided either by reducing our data in free air, or by correcting in $k(P,Q)$ the part of the signal which comes from the covariance of the terrain heights, computed with the same invariance group (cfr. /1/).

We note that these reductions or corrections should be applied only to compute k, while afterwards the estimation formulas are applied with observational functionals referring to the true measure points.

Exercise 7.1: verify that the extension of the covariance function to all of A has the form

$$k(P,Q) = \Sigma\, k_n \left(\frac{R^2}{r_P r_Q}\right)^{n+1} Y_{nm}(P)\, Y_{nm}(Q) \tag{7.7}$$

(Hint: apply twice the Dirichlet resolvent (7.6) to (6.36)).

Exercise 7.2: show that $k(P,Q)$, as spatial function, is invariant under rotations G_ω,

$$G_\omega\, k(P,Q) = k(g_\omega P, g_\omega Q) = k(P,Q) \quad , \tag{7.8}$$

both by using directly (7.7) and by observing from (7.6) the Dirichlet resolvent $\mathcal{D}$ satisfies the commutation relation

$$G_\omega \mathcal{D} = \mathcal{D} G_\omega \quad . \tag{7.9}$$

Remark 7.2: we must underline that it is exactly the property (7.8) which makes the use of the rotation group so attractive, at least for an almost spherical body like the earth. In fact (7.8) and the related shape (7.7) of the covariance function permits to find closed expressions for $k(P,Q)$ also in space, what is essential if we want to leave in computations the measure points in their correct positions, maybe at very different heights, as in the case of space geodesy observations.

Ex. 7.1 - Assume to have data $u(P_i)$ on the earth's ellipsoid, instead than on the earth's sphere.
Introduce on and outside it the ellipsoidal coordinate system (t,θ,λ) related to cartesian coordinates by

$$x = \sqrt{t^2 + E^2} \sin\theta \cos\lambda$$

$$y = \sqrt{t^2 + E^2} \sin\theta \sin\lambda \tag{7.10}$$

$$z = t \cos\theta \qquad (E^2 = a^2 - b^2)$$

(cfr. Heiskanen-Moritz /6/).
The earth's ellipsoid corresponds to

$$t = b \qquad \text{(semiminor axis)} \quad .$$

To any point P on this ellipsoid we can associate uniquely a point P_o on the unit sphere σ, i.e.

$$P_o = \Gamma P = \begin{vmatrix} x/b \\ y/b \\ z/a \end{vmatrix} \rightarrow \begin{cases} \theta_{P_o} = \theta_P \\ \lambda_{P_o} = \lambda_P \end{cases} \tag{7.11}$$

Correspondingly the values of the anomalous potential $u(P)$ can also be viewed as defining a function on the unit sphere

$$u_o(P_o) = u(P) = u(\Gamma^{-1}P_o) \quad .$$

On $u_o(P_o)$ we can construct a stochastic process by using the rotation group, namely

$$u_o(P_o,\omega) = G_\omega\, u_o(P_o) = u_o(g_\omega P_o) = u\,(\Gamma^{-1} g_\omega P_o)$$

Finally we can go back to the ellipsoid by using again (7.11) and get

$$u(P,\omega) = u_o(P_o,\omega) = u\,(\Gamma^{-1} g_\omega \Gamma P) \quad . \tag{7.12}$$

The covariance function of this process is easily found: in fact it is

$$k(P,Q) = E\,\{u(P,\omega)u(Q,\omega)\} = E\,\{u_o(P_o,\omega)u_o(Q,\omega)\} \quad . \tag{7.13}$$

If we develop $u_o(P_o)$, i.e. $u(P)$ in a series of spherical harmonics

$$u_o(P_o) = u(P) = \Sigma\, u_{nm}\, Y_{nm}(\theta,\lambda)$$

we see that (7.13) is simply

$$\begin{cases} k(P,Q) = \Sigma\, k_n\, Y_{nm}(\theta_P,\lambda_P)\, Y_{nm}(\theta_Q,\lambda_Q) \\ k_n = \dfrac{1}{2n+1}\, \Sigma_m\, u_{nm}^2 \end{cases} \tag{7.14}$$

This at least is the covariance function on the ellipsoid. If we want $k(P,Q)$ in space, we must solve twice the corresponding Dirichlet problem, what can be done with the help of ellipsoidal harmonics (cfr. /6/, 1-20): we find

$$k(P,Q) = \Sigma_{nm}\, k_n\, \frac{Q_{nm}(i\frac{t_P}{E})\, Q_{nm}(i\frac{t_Q}{E})}{Q_{nm}^2(i\frac{b}{E})}\, Y_{nm}\,(\theta_P,\lambda_P)\, Y_{nm}(\theta_Q,\lambda_Q) \quad . \tag{7.15}$$

It is remarkable to note that if we transport naturally (i.e. through the coordinates θ,λ) the transformation group G , defined by (7.12) on the ellipsoid $t=b$, to the external ellipsoids $t>b$, we see that $k(P,Q)$ is not anymore invariant under G in space; in fact its coefficients depend also on the order m and therefore k cannot be a function of $\theta_{P_oQ_o}$ only.

We make the reasonable conjecture that the spherical coordinates system and the rotation group are the only ones such that the Δ operator is separated into a radial and a Laplace-Beltrami part and that the group

G , naturally propagated in space, commutes with the resolvent of the Dirichlet operator, so that $k(P,Q)$ is invariant under G in space too.

Remark 7.3: if instead of $u(P_i)$ we know the values of functionals of u (cfr. the last note in §6) we can first compute their covariance function and then derive the covariance function of u . For instance if

$$L_P\, u = \left(-\frac{\partial}{\partial r_P} - \frac{2}{r}\right) u = \Delta g(P)$$

from (7.7) we see that

$$C_{\Delta g \Delta g}(P,Q) = L_P\, L_Q\, k(P,Q) = \frac{1}{R^2} \Sigma\, (n-1)^2\, k_n\, Y_{nm}(P)\, Y_{nm}(Q) \quad ,$$

so that from the degree variances c_n of Δg , computed on the sphere with the rotationally invariant formulas, we can find

$$k_n = \frac{R^2}{(n-1)^2}\, c_n \quad .$$

Exercise 7.3: find the relation between the covariance function $k(P,Q)$ of u and that one of Lu when

1) $Lu = -\frac{\partial}{\partial r} u = \delta g$ (true anomaly)

2) $Lu = -\frac{\partial^2 u}{\partial r^2} = \frac{\partial}{\partial r} \delta g$ (gradiometry)

Exercise 7.4: prove that, if

$$\xi(P) = \underline{e}_{\phi P} \cdot \nabla_P\, u \quad , \quad \eta(P) = \underline{e}_{\lambda P} \cdot \Delta_P\, u$$

are the deflections of the vertical, then for a spherical invariant process

$$E\,\{\xi(P)\, \eta(P)\} = 0$$

(Hint: note that $k(P,Q) = F\,((\underline{r}_P/r_P).(\underline{r}_Q/r_Q))$ and that

$$(\underline{e}_{\phi P}\cdot\nabla_P)(\underline{e}_{\lambda Q}\cdot\nabla_Q)F = (r_P r_Q)^{-1}\ \ (\underline{e}_{\phi P}\cdot\underline{e}_{rQ})(\underline{e}_{\lambda Q}\cdot\underline{e}_{rP})F'' - (\underline{e}_{\lambda Q}\cdot\underline{e}_{\phi P})F'$$

and let $Q \to P$).

Remark 7.4: when we estimate empirically a covariance function for instance by (7.5) as an approximation of (6.34) we always commit a discretization error. This is reflected in that there might be a jump in the origin of the covariance function.

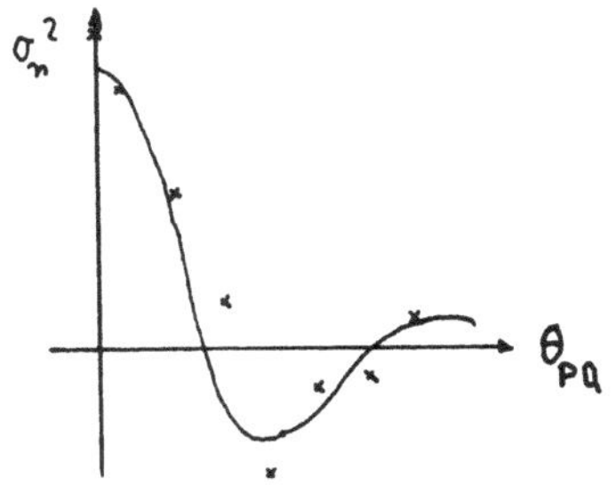

FIG. 7.2

Usually we model this fact by considering the residual σ_n^2 as the variance of a noise n : it is important to stress that this noise has nothing to do with the measurement error ν but it rather expresses a model error due to the coarseness with which we have computed the averages (6.34). As a matter of fact if we knew exactly the continuous function $u(P)$ we would derive from (6.34) a necessarily everywhere continuous covariance function $k(P,Q)$.

In any way it is always better to smooth as in Fig. 7.2 the covariance at the origin, so that the numerical computation of the collocation solution is stabilized (cfr. Ex. 4.2): in this way the long wavelength part of the signal can be estimated maybe using only a part of the data, conveniently spaced. Once this has been done, a second step of collocation can be performed to make the approximation on a more local basis. In this case, what appeared in the first step as a noise with respect to the long wavelength component, becomes a new signal when the former is subtracted: a new local

covariance function is estimated and a new solution found.
This stepwise approach is a fundamental tool to keep collocation formulas in the realm of computability.
It is interesting to notice that if it is very simple to go from a global to a local approach by the stochastic interpretation, the same is not tr true with a functional one; this is one of the important points where a certain greater flexibility of the stochastic approach is preceived.
We conclude by observing that it is also possible to find the asymptotic transition from a global (spherical) covariance function and a local one, in plane approximation.
In fact by applying the asymptotic relation

$$P_l(\cos\theta) = J_o(l\theta) \quad 1 + O(l^{-1}) \tag{7.16}$$

we see that, for Q ranging in a small neighborhood of P, setting ρ_{PQ}= distance $\overline{PQ}$ projected on the tangent plane at P, we find

$$\begin{aligned} k(P,Q) &= \Sigma_l \; k_l(2l+1) \; P_l(\cos\theta_{PQ}) \sim \\ &\sim \Sigma_l \; k_l(2l+1) \; J_o(l\,\theta_{PQ}) = \\ &= \Sigma_l \; k_l(2l+1) \; J_o(l\,\frac{\rho_{PQ}}{R}) \quad . \end{aligned} \tag{7.17}$$

From (7.17) we deduce that $k(P,Q)$ as seen on the tangent plane, is naturally expressed in Fourier-Bessel series and that the process $u(P,\omega)$ on the tangent plane is seen as rototranslational invariant, since its covariance function depends only on ρ_{PQ}.

APPENDIX 1

In this appendix we review some of the basic properties of Hilbert spaces: the selection of the examples is done in order to provide useful material to understand the rest of the lectures.

Def. 1: a real pre-Hilbert space H is a linear space on which a scalar product $\langle , \rangle$ has been defined, satisfying the following axioms:

$$\begin{aligned} &a)\ \langle x,y\rangle = \langle y,x\rangle \\ &b)\ \langle yx_1 + \mu x_2\ ,\ y\rangle = \lambda\langle x_1,y\rangle + \mu\langle x_2,y\rangle \\ &c)\ \langle x,x\rangle > 0 \quad \forall x \neq 0 \end{aligned} \tag{1}$$

Remark 1: a pre-Hilbert space is normed under the assignement

$$||x||^2 = \langle x,x\rangle \qquad , \tag{2}$$

i.e. the (non-linear) functional defined by (2) satisfies the norm axioms

$$\begin{aligned} &||x + y|| \leq ||x|| + ||y|| \\ &||\lambda x|| = |a|\ ||x|| \qquad (\ \lambda \ \text{real}) \\ &||\ x|| > 0 \qquad \forall x \neq 0 \end{aligned} \tag{3}$$

Def. 2: a pre-Hilbert space H is a Hilbert space if it is complete. i.e. if all Cauchy sequences $\{x_n\}$ $(\lim\limits_{nm\to\infty} ||x_n - x_m|| = 0\ )$ admit a limit x in H ,

$$\lim_{n\to\infty} ||x_n - x|| = 0 \tag{4}$$

Theorem 1 (Schwarz inequality): the following holds

$$\begin{aligned} &\forall x,y \in H \qquad |\langle x,y\rangle| \leq ||\ x||\ ||y|| \quad , \\ &|\langle x,y\rangle| = ||x||.||y|| \leftrightarrow x = \lambda y \end{aligned} \tag{5}$$

Remark 2: a pre-Hilbert space H, when it is not closed, can always be completed as follows

Define $H_c = \{(x_n);(x_n) \text{ is Cauchy in } H\}$

Define equivalence classes in H_c according to the rule

$$(x_n) \sim (y_n) \leftrightarrow \lim_{n\to\infty} \|x_n - y_n\|_H = 0$$

and let $\bar{H}$ be the space of such equivalent classes (one single sequence (x_n) can be taken as representer of its equivalen class).

Define a scalar product in $\bar{H}$ as

$$(x_n),(y_n) \in \bar{H} \; ; \quad \langle (x_n),(y_n)\rangle_{\bar{H}} = \lim_{n\to\infty} \langle x_n,y_n\rangle_H$$

(it is proved that the limit exists and that it is the same whatever are the choices of $(x_n),(y_n)$ in their equivalence classes).
Then $\bar{H}$ is a Hilbert space: if to any element x in H we associate the Cauchy sequence $(x_n;\ x_n=x)$ we state an isometric isomorphism between H and a subspace of $\bar{H}$

$$\langle (x_n,(y_n)\rangle_{\bar{H}} = \lim_{n\to\infty} \langle x_n,y_n\rangle_H = \langle x,y\rangle_H \qquad ;$$

this subspace of $\bar{H}$ is dense in $\bar{H}$, whence $\bar{H}$ is unique, apart from isomorphic representations.

Examples of Hilbert spaces.

Ex. 1: let $\bar{\Omega}$ be a closed domain in R^n, $C(\bar{\Omega})$ denotes the linear space of functions, continuous in $\bar{\Omega}$; $C(\bar{\Omega})$ is a pre-Hilbert space under the scalar product

$$\langle u,v\rangle = \int_\Omega u(\omega)\ v(\omega)\ d\Omega \quad . \tag{6}$$

The completion of this space is called $L^2(\bar{\Omega})$; its elements are equivalent classes of square integrable, measurable functions; two functions coinciding almost everywhere are in the same class and are considered as the same element in $L^2(\bar{\Omega})$ (whence $L^2(\bar{\Omega}) = L^2(\Omega)$).

Ex. 2: if Ω is an abstract set, $\mathcal{A}$ a σ-algebra of subsets of Ω (which we agree to call measurable sets) and $P(A)$ a probability measure on $(\Omega,\mathcal{A})$ (i.e. $P(A) \geq 0$, $P(\cup A_n) = \Sigma P(A_n)$ if A_n are disjoint , $P(\Omega) = 1$), we say that $(\Omega,\mathcal{A}, P)$ is a probability space. A real random variable on $(\Omega,\mathcal{A},P)$ is a function $\Omega \to R$ $(x = x(\omega))$ such that every set of the form $\{\omega;\ x(\omega) \leq \bar{x}\}$ is in $\mathcal{A}$. For any r.r.v. we can define the distribution function

$$F_x(\bar{x}) = P\ \{\omega;\ x(\omega) \leq \bar{x}\}$$

and, if they exist, the mean or expectation

$$\mu_x = E\ \{x\} = \int x\ dF_x(x) = \int_\Omega x(\omega)\ dP(\omega) \tag{7}$$

and the variance

$$\sigma^2(x) = E\ \{(x - \mu_x)^2\} \tag{8}$$

The space of r.r.v. on $(\Omega,\mathcal{A},P)$ with zero mean and finite variance is a Hilbert space with scalar product

$$\langle y,y\rangle = E\ \{xy\} = \int_\Omega x(\omega)\ y(\omega)\ dP(\omega) \tag{9}$$

This space is called $\mathcal{L}^2(\Omega,\mathcal{A},P)$.
The convergence in $\mathcal{L}^2(\Omega,\mathcal{A},P)$ is called the quadratic mean convergence(*).

Ex. 3: a second order stochastic process is a function mapping a set T into some $\mathcal{L}^2(\Omega,\mathcal{A},P)$, and we shall assume this mapping to be continuous.

(*) Incidentally $x_n \to x$ in $\mathcal{L}^2$ implies convergence in law, i.e. $F_{x_n}(x) \to F_x(x)$ at any continuity point.

$$t \in T \rightarrow x(t,\cdot) \in {}^2(\Omega,\mathcal{A},P) \quad : \tag{10}$$

the set T could be a finite interval, the whole real line, all of R^n or a manifold (for instance a sphere in R^3), etc.
We can consider the smallest closed subspace of $\mathcal{L}^2$ containing all $x(t,\cdot)(t \in T)$: one can show that this subspace is the same obtained by closing in $\mathcal{L}^2$ the linear manifold

$$\begin{aligned} V &= \{\sum_1^N{}_i \lambda_i \, x(t_i,\cdot) \ , \ \forall \lambda_i \ , \ N = 1,2...\} = \\ &= \mathrm{Span}\,\{x(t_i,\cdot)\} \quad . \end{aligned} \tag{11}$$

This subspace (being closed) is again a Hilbert space with scalar product (9) and we shall call it H_x , the Hilbert space spanned by the process $x(t,\cdot)$.

Ex. 4: the (Sobolev) space $H^{1,2}(0,1)$ of functions u which are in $L^2(0,1)$ and have also (generalized) derivatives in $L^2(0,1)$, with scalar product

$$\langle u,v\rangle_{H^{1,2}} = \int_0^{-1} [u(t)\, v(t) + u'(t)\, v'(t)]\, dt \tag{12}$$

is a Hilbert space.
The only difficult point is to prove the completeness: this however is more easily seen if we consider $H^{1,2}(0,1)$ as a subspace of $L^2(0,1) \oplus L^2(0,1)$, namely that of the couples (f,g) $(f \in L^2, g \in L^2)$ satisfying the linear relation

$$f(t) = f(0) + \int_0^t g(\tau)\, d\tau \quad .$$

It is not difficult to see that this condition is preserved under L^2 convergence of a sequence (f_n,g_n) .

Ex. 5: the (Sobolev) space $H_0^{1,2}(0,1)$, which is the subspace of $H^{1,2}$, of functions which go to zero at the ends of the interval:

$$u \in H^{1,2} \ , \ u(1) = u(0) = 0 \quad , \tag{13}$$

is a Hilbert space with the same scalar product (12). It is also possible to prove that $H_o^{1,2}$ is Hilbert also with the equivalent scalar product

$$\langle u,v\rangle_{H_o^{1,2}} = \int_o^{1} u'(t)\ v'(t)\ dt \ ^{(*)}. \tag{14}$$

Here equivalent means that the norm generated by (14) is equivalent to that one generated by (12), i.e. there is a constant A such that

$$A^{-1}\,\|u\|_{H^{1,2}} \leq \|u\|_{H_o^{1,2}} \leq A\,\|u\|_{H^{1,2}} \quad .$$

Ex. 6: the Sobolev space $H^{1,2}$ of functions defined on the unit sphere is a Hilbert space with scalar product

$$\langle u,v\rangle_{H^{1,2}} = \int_\sigma \left[u(\sigma)\ v(\sigma) + \nabla_\sigma u(\sigma)\ .\ \nabla_\sigma v(\sigma)\right]\ d\sigma \tag{15}$$

where $\nabla_\sigma = \underline{e}_\phi \frac{\partial}{\partial\phi} + \underline{e}_\lambda \frac{1}{\cos\phi}\frac{\partial}{\partial\lambda}$ is the tangential gradient.

Def. 3: a Hilbert space H is separable if there is a sequence $\{u_n\}$ which is everywhere dense in H .

Theorem 2: a Hilbert space is separable if there is in it a complete orthonormal system (CONS), i.e. a sequence $\{h_n\}$ such that

$$\langle h_n,h_m\rangle = \delta_{nm} \qquad n = 1,2\ldots$$

$$\langle u,h_n\rangle = 0 \quad , \quad \forall n \leftrightarrow u = 0$$

In this case the Fourier expansion $\forall u \in H$,

$$u = \sum_{1}^{+\infty}{}_n \langle u,h_n\rangle\ h_n \tag{16}$$

(*) In $H_o^{1,2}$ this product is definite positive since $\|u\|^2_{H_o^{1,2}} = 0 \rightarrow u' = 0$ on $|0,1|$, i.e. u = const; but since u satisfies (13), we have also $u = 0$.

and the Parseval's identity

$$\|u\|^2 = \sum_{1}^{+\infty} {}_n <u,h_n>^2 \qquad (17)$$

hold true.

Remark 3: once a CONS is fixed in H it is created a congruence (isometric isomorphism) between H itself and the space l^2 of square summable sequences of real numbers. More precisely

$$\begin{aligned} u \in H &\xrightarrow[\Psi]{} \{u_n\} = \{<u,h_n>\} \in l^2 \\ \{u_n\} \in l^2 &\xrightarrow[\Psi^{-1}]{} u = \sum_{n}^{+\infty} u_n h_n \in H \quad : \end{aligned} \qquad (18)$$

the isometry is a consequence of Parseval's identity (17).

Remark 4: if a sequence of elements of H , $\{h_n\}$ is orthonormal, one can always consider the linear manifold

$$V = \text{Span } \{h_n\} = \{\sum_{1}^{N} {}_n \lambda_n h_n, \text{ N variable but finite}\}$$

and close it in the H-norm: in this way one gets a closed subspace $\bar{V}$ of H . If $\bar{V}$ coincides with H , $\{h_n\}$ is also complete, otherwise for any u in H one can take its projection Pu on $\bar{V}$; it turns out that

$$Pu = \sum_{1}^{+\infty} {}_n <u,h_n> u_n \qquad (19)$$

and that

$$\|Pu\|^2 = \sum_{1}^{+\infty} {}_n <u,h_n>^2 \le \|u\|^2 \quad . \qquad (20)$$

From this remark it follows that in order to ascertain the completeness of $\{h_n\}$ it is sufficient to prove that $u = \Sigma_{1}^{+\infty}{}_n <u,h_n> h_n$ holds for a set D of u , densely contained in H .

In this case in fact it is obvious that $D \subset \bar{V} \subset H$, but since D is densely embedded in H , we have also $\bar{V} = H$.

Ex. 7: we assume to know that any C^∞ function on the circle $0 \leq \theta < 2\pi$ (i.e. any function continuous with all the derivatives of any order) can be developed into a Fourier series

$$u(\theta) = \frac{a_0}{2} + \sum_1^{+\infty}{}_n (a_n \cos n\theta + b_n \sin n\theta) \tag{21}$$

$$a_n = \frac{1}{\pi} \int_0^{2\pi} u(\theta) \cos n\theta \, d\theta$$

$$b_n = \frac{1}{\pi} \int_0^{2\pi} u(\theta) \sin n\theta \, d\theta \quad ,$$

which is uniformly convergent together with all the derivatives of any order. Since uniform convergence on a bounded set implies L^2 convergence, we have also that (21) holds for all C^∞ in L^2 and the former is a dense subset of the latter.

Since

$$\int_0^{2\pi} \cos n\theta \cos m\theta \, d\theta = \int_0^{2\pi} \sin n\theta \sin m\theta \, d\theta = 0 \qquad n \neq m$$

$$\int_0^{2\pi} \cos n\theta \sin m\theta \, d\theta = 0 \tag{22}$$

$$\int_0^{2\pi} \cos^2 n\theta \, d\theta = \int_0^{2\pi} \sin^2 n\theta \, d\theta = \pi \qquad n = 1,2\ldots$$

we can deduce that the sequence

$$\left\{ \frac{\sin n\theta}{\sqrt{\pi}} , \frac{1}{\sqrt{2\pi}} , \frac{\cos n\theta}{\sqrt{\pi}} \qquad n = 1,2\ldots \right\} \tag{23}$$

is a CONS in $L^2(0,2\pi)$.

Ex. 8: for the same reason mentioned in the Ex. 7, the sequence (23) is a COS in $H^{1,2}(0,2\pi)$, however the normalization constants are to be changed. In fact

$$\|\cos n\theta\|^2_{H^{1,2}} = \int_0^{2\pi} [\cos^2 n\theta + n^2 \sin^2 n\theta]\, d\theta = \pi(1+n^2) \quad ; \tag{24}$$

the same formula holds for $\sin n\theta$.

Ex. 9: take any function $u \in L^2(0,1)$, then if we continue u in $(-1,0)$ either by setting $u(t) = u(-t)$, or $u(t) = -u(-t)$, we obtain two functions, respectively u_s, u_a, which are both again in $L^2(-1,1)$. Then, also taking symmetries into account, we have

$$\begin{aligned} u_s(t) &= \frac{a_0}{2} + \sum_1^{+\infty}{}_n\, a_n \cos n\pi t \\ u_a(t) &= \sum_1^{+\infty}{}_n\, b_n \sin n\pi t \end{aligned} \tag{25}$$

The two series are to converge in $L^2(-1,1)$, whence also in $L^2(0,1)$ and

$$\{1, \sqrt{2} \cos n\pi t\} \quad , \quad \{\sqrt{2} \sin n\pi t\} \tag{26}$$

are two different CONS in $L^2(0,1)$.
Also, directly from (23), by transforming θ in $\theta = 2t\pi$, we see that

$$\{\sqrt{2} \sin 2n\pi t \ , \ 1 \ , \ \sqrt{2} \cos 2n\pi t\} \tag{27}$$

is still another CONS in $L^2(0,1)$.

Ex. 10: it is easy to verify that (27) is a COS in $H^{1,2}(0,1)$ although it should be normalized by proper constants.
We want to look now for a CONS in $H_0^{1,2}$; it is obvious that any $u \in H_0^{1,2}$ can also be developed on the base (27) since $H_0^{1,2} \subset H^{1,2}$, however (27) is not a natural base for $H_0^{1,2}$ since $\cos 2n\pi t \neq 0$ for $t=0$ and $t=1$, so that these functions are not in $H_0^{1,2}$.
On the other hand, since u' is in $L^2(0,1)$ too we have available also the development

$$u'(t) = \frac{a_0}{2} + \sum_1^{+\infty}{}_n\, a_n \cos n\pi t \quad : \tag{28}$$

since

$$\int_0^1 u'(t)\, dt = u(1) - u(0) = 0 \tag{29}$$

we find that $a_0 = 0$, while

$$a_n = 2 \int_0^1 u'(t) \cos n\pi t\, dt = 2 \int_0^1 u'(t) \left(\frac{\sin n\pi t}{n\pi}\right)' dt \quad . \tag{30}$$

Integrating (28) from 0 to t and taking into account that $u(0) = 0$, that the scalar product in $H_0^{1,2}$ can be written as in (14), that the functions sin n t are in $H_0^{1,2}$ (i.e. they go to zero at the ends of the interval (0,1)), we find

$$u(t) = \sum_{1}^{+\infty}{}_n \langle u, \frac{\sqrt{2} \sin n\pi t}{n\pi} \rangle_{H_0^{1,2}} \sqrt{2} \frac{\sin n\pi t}{n\pi} \quad ,$$

what proves that the sought CONS is $\{ \frac{\sqrt{2} \sin n\pi t}{n\pi} \}$.

Ex. 11: let us take any C^∞ function defined on the circle $f(\theta)$ $(0 \leq \theta \leq 2\pi)$; now define a r.r.v. $\{\eta\}$ as a uniform r.v. on the circle itself, so that the probability low of η is

$$P(\eta \in d\eta) = \frac{d\eta}{2\pi} \quad ; \tag{31}$$

define a stochastic process according to the law

$$x(\theta,\eta) = f(\theta + \eta) \quad . \tag{32}$$

From its very definition we get

$$E\{x(\theta,\eta)\} = \frac{1}{2}\int_0^{2\pi} f(\theta + \eta)\, d\eta = \frac{1}{2\pi}\int_0^{2\pi} f(\theta)\, d\theta = f_0 \quad ; \tag{33}$$

for the sake of simplicity we assume

$f_0 = 0$.

Moreover the covariance function is given by

$$C(\theta,\theta') = E\{x(\theta,\eta)\, x(\theta',\eta)\} = \frac{1}{2\pi}\int_0^{2\pi} f(\theta+\eta)\, f(\theta'+\eta)\, d\eta = \tag{34}$$
$$= \frac{1}{2\pi}\int_0^{2\pi} f(\eta)\, f(\theta'-\theta+\eta)\, d\eta = C(\theta'-\theta)\ .$$

Let us consider the complex r.v.'s

$$x_n(\eta) = \int_0^{2\pi} \frac{e^{jn\theta}}{\sqrt{2\pi}}\, x(\theta,\eta)\, d\theta \quad (*) \quad : \tag{35}$$

it is not difficult to prove that $\{u_n\} \in H_x$, i.e. the Hilbert space spanned by the process $x(\theta,\eta)$. Since

$$x_n(\eta) = \int_0^{2\pi} \frac{e^{jn\theta}}{\sqrt{2\pi}}\, f(\theta+\eta)\, d\theta = e^{-jn\eta}\int_0^{2\pi} \frac{e^{jn\theta}}{\sqrt{2\pi}}\, f(\theta)\, d\theta = f_n\, e^{-jn\eta} \quad ,$$

we have also

$$\langle x_n, x_m\rangle_{H_x} = E\{x_n\, x_m^*\} = f_n\, f_m^* \cdot \int_0^{2\pi} e^{-jn\eta}\, e^{jm\eta}\, \frac{d\eta}{2\pi} = \tag{36}$$
$$= |f_n|^2\, \delta_{nm} \quad .$$

Whence $\{\frac{x_n}{|f_n|}\}$ is an ON system in H_x .
We want to prove that it is complete to. To this aim it is enough to prove that for any $\theta \in [0,2]$ we have the convergent expansion

$$x(\theta,\eta) = \sum_{-\infty}^{+\infty}{}_n \langle x(\theta,\cdot)\ ,\ \frac{x_n(\cdot)}{|f_n|}\rangle\, \frac{x_n}{|f_n|}(\eta) \quad , \tag{37}$$

because in this case the same is true for all the finite combinations of the type $\Sigma\, \lambda_i\, x(\theta_i,\eta)$ which make up a linear manifold dense in H_x by definition; but then, applying the argument of Rem. 4, we see that $\{x_n\}$ is complete in H_x .
As a matter of fact we can compute

(*) In this example it is simpler to use the complex Fourier series; the same definitions and theorems hold in complex Hilbert spaces as in the real ones with the only change that $\langle u,v\rangle = \overline{\langle v,u\rangle}$ and $\langle u,\lambda v\rangle = \bar{\lambda}\langle u,v\rangle$.

$$\langle x(\theta,\cdot),x_n(\cdot)\rangle = f_n^* \int_0^{2\pi} f(\theta+\eta)\, e^{jn\eta} \frac{d\eta}{2\pi} =$$

$$= |f_n|^2 \frac{e^{-jn\theta}}{\sqrt{2\pi}} \quad ,$$

so that

$$\sum_{-\infty}^{+\infty}{}_n \frac{\langle x(\theta,\cdot),\, x_n(\cdot)\rangle}{|f_n|^2}\, x_n(\eta) = \sum_{-\infty}^{+\infty}{}_n \frac{e^{-jn\theta}}{\sqrt{2\pi}}\, f_n\, e^{-jn\eta} = \qquad (38)$$

$$= f(\theta+\eta) \quad ;$$

the series converges uniformly since $f \in C^{\infty}$ and hence also in quadratic mean with respect to η, i.e. also in H_x and (38) is identical with (37).

Remark 5: indeed it is not by chance that $\{\frac{x_n}{|f_n|}\}$ defined as in (35) is a CONS in H_x ; this is related to the fact that $\{\frac{e^{jn\theta}}{\sqrt{2\pi}}\}$ is the complete orthonormal system of eigenfunctions of the integral operator with kernel (34), so that

$$C(\theta' - \theta) = \sum_{-\infty}^{+\infty}{}_n |f_n|^2 \frac{e^{jn\theta'}}{\sqrt{2\pi}} \frac{e^{-jn\theta'}}{\sqrt{2\pi}} \quad .$$

This fact is general and it can be proved that if

$$C(t,t') = \Sigma\, c_n\, e_n(t)\, e_n(t')$$

is the covariance function of the u process, almost all of whose realizations are in L^2 , then defining

$$x_n(\omega) = \int x(t,\omega)\, e_n(t)\, dt$$

we obtain a complete orthonormal system normalized by the relations

$$\|x_n\|^2 = E\,\{x_n^2\} = c_n \quad ,$$

so that the expansion

$$x(t,\omega) = \sum_{-\infty}^{+\infty}{}_n \, x_n(\omega)\, e_n(t) \tag{39}$$

holds: the convergence of (39) takes place in H_x for every t.

Ex. 12: we assume that the reader is familiar with the elementary construction of the spherical harmonics $\{Y_{1m}(\theta,\lambda)\} = \{e^{jm\lambda}\, \bar{P}_{1|m|}(\cos\theta); m=-1,..1\}$ as a complete set of the eigenfunctions of the Laplace-Beltrami operator

$$\Delta_\sigma = \frac{1}{\sin\theta}\, \partial_\theta(\sin\theta\, \partial_\theta \cdot) + \frac{1}{\sin^2\theta}\, \partial_\lambda^2 \cdot \quad ,$$

among functions square integrable on the unit sphere (cfr. Heiskanen-Moritz /6/).

More precisely we assume to know that if

$$\Delta_\sigma u = \mu u \tag{40}$$

then $\mu = -1(1+1)$ for some $1=0,1,...$ and

$$u = \sum_{-1}^{1}{}_m \, u_{1m}\, Y_{1m}(\theta,\lambda) \quad ; \tag{41}$$

so that $Y_{1m}(\theta,\lambda)$ is characterized as the unique set of joint eigenfunctions of the two (commuting) selfadjoint operators Δ_σ and $j\frac{\partial}{\partial\phi}$ (*).

The multiplicative constant in the definition of $Y_{1m}(\theta,\lambda)$ is chosen so that they are $L^2(\sigma)$ normalized

$$\int_\sigma |Y_{1m}(\theta,\lambda)|^2\, d\sigma = 1 \quad ;$$

furthermore it is easy to verify directly that

$$\int_\sigma Y_{1m}(\theta,\lambda)\, Y^*_{ik}(\theta,\lambda)\, d\sigma = \delta_{1j}\, \delta_{mk} \tag{42}$$

(*) The fact that Δ_σ and $j\partial_\lambda$ commute tells us that these two operators must have at least one set of common eigenfunctions: since $j\partial_\phi u = \gamma u \rightarrow u(\theta,\lambda) = A(\theta)\, e^{jm\lambda}$, the classical choice os separating the variables becomes obligatory and not just one pos-

so that $\{Y_{lm}\}$ is an ONS in $L^2(\sigma)$.
It is also easy to ascertain that $\{Y_{lm}\}$ is an orthogonal system in $H^{1,2}(\sigma)$ (cfr. (15)) since, by applying the identity $\int \nabla u \cdot \nabla v \, d\sigma = - \int u \, \Delta_\sigma v \, d\sigma$, we get

$$\langle Y_{lm}, Y_{jk}\rangle_{H^{1,2}} = \int \{Y_{lm} Y^*_{jk} + \nabla_\sigma Y_{lm} \cdot \nabla_\sigma Y^*_{jk}\} \, d\sigma = \int [1 + l(l+1)] Y_{lm} Y^*_{jk} \, d\sigma = [1 + l(l+1)]\langle Y_{lm}, Y_{jk}\rangle_{L^2} \tag{43}$$

We want to see first that $\{Y_{lm}\}$ is a COS in $H^{1,2}(\sigma)$ and after we shall deduce that it is also a CONS in $L^2(\sigma)$.
To this aim we reorder the sequence $\{Y_{lm}\}$ according to a unique numeration

$$\{e_n\} = \{Y_{00}, Y_{1-1}, Y_{10}, Y_{11}, Y_{2-2}, \ldots\} \qquad n = 1,2\ldots \tag{44}$$

so that for each l,m we fix a certain $n = n(l,m)$; we put also

$$\mu_n = l(l+1) \qquad n = n(l,m) \tag{45}$$

and we see that in this way μ_n is a non decreasing sequence tending to infinity

$$\mu_0 \leqq \mu_1 \leqq \mu_2 \leqq \mu_3 \leqq \mu_4 \ldots \to \infty \qquad (\mu_1 \neq \mu_2 = \mu_3 = \mu_4 \neq \mu_5 = \mu_6 = \mu_7 = \mu_8 \;\; \ldots)$$

Now consider the following minimum problem

$$u \in H^{1,2} : \quad \|u\|^2_{L^2} = 1 \quad , \quad u \perp \text{Span}\,\{e_k, \, k = 1,\ldots n\} \;\text{ in }\; L^2(\sigma) \qquad \|u\|^2_{L^2} = \min \; ; \tag{46}$$

if such a function exists we see that, by using the Lagrange multiplier

---

sibility, so that the classical construction, showing that $A(\theta) = P_{n|m|}(\cos\theta)$, yields the complete set of eigenfunctions.

method, it should minimize

$$\int (u^2 + |\nabla_\sigma u|^2)\, d\sigma - \mu \int u^2 d\sigma + 2 \sum_{1k}^{n} \gamma_k \langle e_k, u\rangle_{L^2}$$

so that we get

$$(1 - \Delta_\sigma)u - \mu u + \Sigma \gamma_k e_k = 0 \tag{47}$$

If we take the L^2 product of (47) with e_j $(j = 1,2...n)$ we immediately find for all $\gamma_u^{'}s$, $\gamma_k = 0$, and (47) yields

$$(1 - \Delta_\sigma)u = \mu u \quad .$$

Since u has to be orthogonal to $\{e_k ; k = 1 - u\}$ u can only be one of the eigenfunctions $\{e_{n+1}, e_{n+2} ...\}$; among these one with minimum $H^{1,2}$ norm is certainly e_{n+1} since (cfr. (43))

$$\|e_{n+1}\|^2_{H^{1,2}} = 1 + \mu_{n+1} \qquad (\leqq 1 + \mu_{n+2} \dots) \quad .$$

Now assume that $\{e_k\}$ is not a COS in $H^{1,2}(\sigma)$: then there is one $v \in H^{1,2}$, $v \neq 0$ such that

$$\langle v, e_k\rangle_{H^{1,2}} = (1 + \mu_k) \langle v, e_k\rangle_{L^2} = 0 \qquad k = 1,2... \quad ;$$

since $v \neq 0$ we can take it with $\|v\|_{L^2} = 1$.
On the other hand choose n such that

$$1 + \mu_n \leqq \|v\|^2_{H^{1,2}} < 1 + \mu_{n+1} \quad ;$$

but in this case it is not true that the problem (46), for the mentioned value of n , has the solution e_{n+1} and there should be another eigenvalue μ of $-\Delta_\sigma$ between μ_n and μ_{n+1} , what is absurd. Whence $\{Y_{1m}\}$ is a COS in $H^{1,2}(\sigma)$.
In this case we have also, according to (43), that $\{\frac{Y_{1m}}{\sqrt{1+1(1+1)}}\}$ is a CONS in $H^{1,2}$ and that the series expansion

$$u = \Sigma \langle u, Y_{lm}\rangle_{H^{1,2}} \frac{Y_{lm}}{\sqrt{1+l(l+1)}} \tag{48}$$

is valid and convergent in $H^{1,2}$.

On the other hand we know that

$$\langle u, Y_{lm}\rangle_{H^{1,2}} = \int (u\, Y^*_{lm} + \nabla_\sigma u . \nabla_\sigma Y^*_{lm})\, d\sigma = [1 + l(l+1)] \int u\, Y^*_{lm}\, d\sigma \quad ,$$

so that (48) becomes identically

$$u = \Sigma \langle u, Y_{lm}\rangle_{L^2}\, Y_{lm} \quad . \tag{49}$$

Now (49) converges in $H^{1,2}$ (since it is the same as (48)) and then also in L^2 : whence $\{Y_{lm}\}$ is an orthonormal system in $L^2(\sigma)$ which is complete for $H^{1,2}$, i.e. for a subset dense in $L^2(\sigma)$, so that according to the Rem. 3, $\{Y_{lm}\}$ is a CONS in $L^2(\sigma)$ too.

Remark 6: if, instead of the complex $\{Y_{lm}\}$, we use the real spherical harmonics, defined for instance as

$$Y_{lm} = \begin{cases} \cos m\lambda\, P_{lm}(\cos\theta) & m \geq 0 \\ \\ \sin |m|\lambda\, P_{l|m|}(\cos\theta) & m < 0 \end{cases} \tag{50}$$

we get again a CONS, which might be more suited to the expansion of real functions.

Remark 7: it is easy to verify that if Sobolev spaces of higher order $H^{s,2}(\sigma)$, are conveniently defined, $\{Y_{lm}\}$ is an orthogonal system for any s. For instance if

$$\langle u,v\rangle_{H^{2,2}} = \int_\sigma \{uv + (\nabla_\sigma \nabla_\sigma u \,.\, \nabla_\sigma \nabla_\sigma v)\}\, d\sigma \quad ,$$

then by applying twice the Green formula on the sphere we find

$$\langle Y_{lm}, Y_{jk} \rangle_{H^{2,2}} = [1 + l^2(l+1)^2]\ \delta_{lj}\ \delta_{mk} \quad .$$

But then $\{Y_{lm}\}$ is also a CONS in $H^{s,2}(\sigma)$ since otherwise by repeating the same reasoning of (48),(49) we would be able to find a $u \in H^{s,2} \subset L^2$ for which the Fourier series representation (49) does not hold, which is absurde.

Def. 4: a linear functional on a Hilbert space H is a function $F : H \to R$ which is linear, i.e.

$$F\ (\lambda x + \mu y) = \lambda F(x) + \mu F(y) \tag{51}$$

Def. 5: a linear functional F is bounded if

$$\forall x \in H\ ,\quad |F(x)| < +\infty \quad ; \tag{52}$$

F is continuous if

$$|F(x)| \leq A\ .\ \|x\| \quad . \tag{53}$$

In this case it is defined a norm of the functional F as

$$\|F\| = \text{Min}\ \{A;\ |F(x)| \leq A\|x\|\ \} \tag{54}$$

It is possible to see that this is equivalent to the other definition

$$\|F\| = \text{Sup}\ \{|F(x)|\ ;\ \|x\| = 1\ \} \quad . \tag{55}$$

The following remarkable theorem holds.

Theorem 3: every bounded linear functional is continuous. In other words if we can compute $F(x)$ for all x in H, then F is continuous, i.e. it has a bounded norm (54).

Def. 6: the set of bounded linear functionals on H is a complete Banach space with norm (54); it is called H^* the dual space of H.

Ex. 13: for any $y \in H$ we can define a functional $F_y(x)$ as

$$F_y(x) = \langle y,x\rangle_H \quad . \tag{56}$$

F_y is linear and bounded: according to Theorem 1

$$|\langle y,x\rangle| \leq \|y\|_H \cdot \|x\|_H$$

so that by (54) $\|F_y\|_{H^*} \leq \|y\|_H$. On the other hand

$F(\frac{y}{\|y\|}) = \|y\|$ so that $\|F_y\|_{H^*} \geq \|y\|_H$ by (55): whence we have

$$\|F_y\|_{H^*} = \|y\|_H$$

Theorem 4 (Riesz): for any linear bounded functional F in H^* we can find an element y_F in H such that the representation

$$F(x) = \langle y_F,x\rangle \tag{57}$$

holds; moreover y_F depends linearly on F and the identity

$$\|F\|_{H^*} = \|y_F\|_H \tag{58}$$

is fulfilled.

Remark 8: with Riesz theorem we say that the two isometries (56) and (57) are one the inverse of the other and that there is an isomorphism between H and H^* . In this way we see also that H^* is a Hilbert space too.

Def. 7: a non linear functional $F ; H \to R$ is bounded on a set $M \subset H$ if $F(x) < +\infty$, $\forall x \in M$; it is continuous on M if $\lim_{x\to\bar{x}} F(x) = F(\bar{x})$, $\forall x \in M$.

Def. 8: $F(x)$ is called Frechet differentiable on an open set $M \subset H$ if the following relation holds for some $f(\bar{x}) \in H$;

$$\forall \bar{x} \in M \ , \ \forall h \ \{\bar{x} + h \in M\} \ , \ F(\bar{x} + h) = F(\bar{x}) + \langle f(\bar{x}),h\rangle + o(\|h\|) \quad : \tag{59}$$

in this case $f(x)$ is called the Frechet derivative of F, or sometimes its gradient, and indicated as

$$f(\bar{x}) = F_x(\bar{x}) = \partial_x F(\bar{x}) \quad . \tag{60}$$

Since f in general depends on x, we can think of differentiating it once more: we say that $f(x)$ has as Frechet derivative a linear (bounded) operator $G(x)$ if

$$f(\bar{x} + h) = G(\bar{x})h + o(\|h\|) \quad ; \tag{61}$$

this time $o(\|h\|)$ is a vector in H such that $\|o\|/\|h\| \to 0$.

Since it is not always easy to compute directly a Frechet differential, we can take advantage of a different definition.

Def. 9: we say that $F(x)$ is Gateaux differentiable if there is $f(x) \in H$

$$\frac{d}{dt} F(x + th)\Big|_{t=0} = \langle f(x),h\rangle \tag{62}$$

The relation between Frechet and Gateaux differential is given by the following theorem 5.

Theorem 5: a Gateaux differentiable functional is also Frechet differentiable on $M \subset H$, if $f(x)$ as defined by (62), is continuous at every x in M .

Whence to compute a Frechet differential we can always compute, if it exists, the Gateaux differential by applying (62) and then verify whether $F_x(\bar{x})$ is continuous on M .

APPENDIX 2

In this appendix we want to study the distributions of all orders of a process of the type $u(P,\omega) = u(g_\omega P)$ where $u(P)$ is defined on the unit sphere and g_ω is a random rotation, uniformly distributed on the rotation group: our aim is also to construct some examples of processes of the above type, with a normal distribution of the first order.

Let us first of all define a set σ_i of the points P_i on the unit sphere such that for a given value u_i, fulfil the relation

$$u(P_i) \leqq u_i \quad . \tag{1}$$

Then we define the distribution function of order N of the process $u(P,\omega)$ as

$$\begin{aligned} F(u_1,\dots u_N;\, P_1\dots P_N) &= P\,(u(P_1,\omega) \leqq u_1,\dots\, u(P_N,\omega) \leqq u_N) = \\ &= P\,(g_\omega P_1 \in \sigma_1,\dots\, g_\omega P_N \in \sigma_N) \quad . \end{aligned} \tag{2}$$

Moreover for any given configuration $(P_1, P_2 \dots P_N)$ we define its base configuration $(Q_1, Q_2 \dots Q_N)$ as the one which is obtained by rotating the former in such a way that

$$\begin{aligned} &\bar{g}P_1 = Q_1 = N \qquad \text{(the north pole)} \\ &\bar{g}P_2 = Q_2 \qquad \text{(on the main meridian, } \lambda_{Q_2} = 0\text{)} \end{aligned} \tag{3}$$

We note that all the configurations $(P'_1,\dots P'_N)$ which can be obtained from $(P_1,\dots P_N)$ by a rotation, have the same base configuration $(Q_1,\dots Q_N)$.

We say that a distribution function (2) is invariant under G, if for any $g \in G$

$$F\,(u_1,\dots u_N;\, gP_1,\dots gP_N) = F\,(u_1,\dots u_N;\, P_1,\dots P_N) \quad . \tag{4}$$

<u>Theorem 1:</u> let $u(P,\omega) = u(g_\omega P)$, with g_ω uniformly distributed on G, then the distribution functions of any order of $u(P,\omega)$ are invariant under G. In fact, fix N and let $(P_1,\dots P_N)$ be any configuration and $(Q_1,\dots Q_N)$

its base configuration; let us fix also the values $u_1 \dots u_N$ and define $G_{1,..N} \subset G$ as the set of the rotations satisfying $g_\omega Q_1 \in \sigma_1, .. g_\omega Q_N \in \sigma_N$. We have

$$\{g_\omega ; g_\omega P_1 \in \sigma_1 \dots g_\omega P_N \in \sigma_N\} = \{g_\omega ; g_\omega g^{-1} Q_1 \in \sigma_1 \dots g_\omega \bar{g}^{-1} Q_N \in \sigma_N\} =$$

$$= \{g_\omega ; g_\omega \bar{g}^{-1} \in G_{1\dots N}\} = G_{1\dots N} \cdot \bar{g} \qquad :$$

therefore, recalling (2),

$$F_N(u_1 \dots u_N ; P_1 \dots P_N) = P\,(g_\omega \in G_{1\dots N}\bar{g}) = P\,(g_\omega \in G_{1\dots N}) \tag{5}$$

by dint of the uniformity of our probability distribution.
From (5) it follows that F_N depends only on the configuration, so that it is invariant under G .

As a consequence of this theorem we can find a simple expression for the distribution of order 1. In fact,

$$F(u_1) = P(g_\omega \in G_1) = P(g_\omega N \in \sigma_1) \quad :$$

on the other hand a volume element of the set G_1 has probability

$$P(dG_1) = \frac{1}{8\pi^2}\, d\,\sigma_1(P_1)\, d\alpha\, (\bar{\alpha})$$

with $d\sigma_1$ a neighborhood of $P_1 \in \sigma_1$ and $d\alpha$ an interval of azimuths around an azimuth $\bar{\alpha}$. Since the relation

$$u\,(P_1) \leq u_1$$

is verified irrespectively of the azimuth α we see that we can put

$$P\,(G_1) = \frac{1}{8\pi^2} \int_{\sigma_1} d\sigma_1 \int_0^{2\pi} d\alpha = \frac{\mu(\sigma_1)}{4\pi} \quad . \tag{6}$$

With the rule (6) in mind we try to find examples where

$$F(u) = \text{erf}\ (u) = \int_{-\infty}^{u} \frac{e^{-\frac{t^2}{2}}}{\sqrt{2\pi}}\ dt \quad : \tag{7}$$

the case that the distribution $F(u)$ has a variance $\sigma^2 \neq 1$ follows.

Ex. 1: let us consider the function

$$u(P) = \text{erf}^{(-1)}\ (\frac{1+\cos\ \theta}{2}) \quad :$$

the process $u(g_\omega P)$ has a normal distribution. According to (7) we have only to compute $\mu(\sigma_1)$. Take $\sigma_1^c = \sigma \backslash \sigma_1$,

$$u(P) \geqq u_1 \qquad P \in \sigma_1^c \quad : \tag{8}$$

we have for $P \in \sigma_1^c$

$$\cos\ \theta \geqq 2\ \text{erf}\ (u_1) - 1 \quad ,$$

i.e.

$$\theta \leqq \theta_1 \ , \quad (\cos\ \theta_1 = 2\ \text{erf}\ (u_1) - 1) \quad .$$

Whence

$$\mu(\sigma_1^c) = \int_0^{\theta_1} \sin\ \theta\ d\theta\ .\ 2\pi = 2\pi\ (1 - \cos\ \theta_1) = 4\pi\ (1 - \text{erf}(u_1))$$

so that

$$\frac{\mu(\sigma_1)}{4\pi} = \text{erf}\ (u_1) \quad .$$

Ex. 2: obviously the function in Ex. 1 is a very simple one, depending only on the θ coordinate. However consider any area-preserving transformation of the unit sphere in itself;

$$\begin{cases} \theta = a(\theta',\lambda') \\ \lambda = b(\theta',\lambda') \end{cases} \quad , \tag{9}$$

and define the new function

$$u(P') = \mathrm{erf}^{-1}\left(\frac{1+\cos a(\theta',\lambda')}{2}\right) \quad . \tag{10}$$

It is obvious that the set σ_1' , such that

$$u(\theta',\lambda') \leqq u_s$$

corresponds to the set

$$\theta_1 \leqq a(\theta',\lambda') \leqq \pi \quad , \quad \forall b \quad ,$$

i.e. to σ_1 ; but then

$$\mu(\sigma_1') = \mu(\sigma_1)$$

and so the process

$$u(P',\omega) = u(g_\omega P')$$

has a normal distribution of the first order.
It would be interesting to enquire how large is the family (10) when $a(\theta',\lambda')$ is left to vary on all the components of area-preserving transformations.

References

/ 1/ Barzaghi R, La stima del geoide in Italia, Doctoral Dissertation, Manuscript, 1986.

/ 2/ Barzaghi R, Sansò F, New results on convergence problems in collocation theory, I Hotine-Marussi Symposium on Math. Geodesy (Rome, 1985), Proceedings, Ist. di Topografia, Politecnico di Milano, 1986.

/ 3/ Grafarend E W, Geodetic stochastic processes in International Summer School (Ramsau, 1973), B.I. Wissenschaftsverlag, 1975.

/ 4/ Grafarend E W, Operational geodesy, in Approximation methods in geodesy, by H Moritz-H Sünkel, H Wichmann Verlag, Karlsruhe, 1978.

/ 5/ Grafarend E W, The gaussian structure of the gravity field, Studia geoph. et geod., 14, 159-167, 1970.

/ 6/ Heiskanen W, Moritz H, Physical Geodesy, W H Freeman, 1967.

/ 7/ Ito K, Foundations of stochastic differential equations in infinite dimensional spaces, CBMS 47 - SIAM, Philadelphia, 1984.

/ 8/ Kallianpur G, Stochastic filtering theory, Springer Verlag, 1980.

/ 9/ Krarup T, A contribution to the mathematical foundation of physical geodesy, Rub. n 44, Geodetic Institute, Copenhagen, 1969.

/10/ Krarup T, On potential theory, in Int. Summer School (Ramsau, 1973), B.I. Wissenschaftsverlag, 1975.

/11/ Lauritzen S L, The probabilistic background of some statistical methods in physical geodesy, Pub. n 48, Geodetic Institute, Copenhagen, 1973.

/12/ Miller W Jr, Symmetry groups and their applications, Academic Press, 1972.

/13/ Moritz H, Advanced physical geodesy, H Wichmann Verlag, Karlsruhe, 1980.

/14/ Moritz H, Sansò F, Dialogue on collocation, Boll. di Geod. e Sc. Affini, 1980.

/15/ Petrov V V, Sums of independent random variables, Springer Verlag, 1975.

/16/ Romanov Y A, Markov random fields, Springer Verlag, 1982.

/17/ Sansò F, The minimum mean square estimation error in physical geodesy, Boll. di Geod. e Sc. Affini, 1980.

/18/ Sansò F, A note on density problems and the Runge-Krarup theorem, Boll. di Geod. e Sc. Affini, Proceed. VIII Symp. on Math. Geodesy (Como, 1981), Firenze, 1983.

/19/ Sansò F, Tscherning C C, On convergence problems in collocation theory, Boll. di Geod. e Sc. Affini, 1980.

/20/ Tscherning C C, A note on the choice of norm when using collocation for the computation of the approximations to the anomalous potential, Bull. Géod., vol 51, n 2, 1977.

/21/ Weyl H, The theory of groups and quantum mechanics, Dover.

INTRODUCTION TO SPECTRAL ANALYSIS

by

B. Hofmann-Wellenhof and H. Moritz

Institute of Theoretical Geodesy
Division of Physical Geodesy
Technical University Graz
Rechbauerstraße 12
A-8010 Graz, Austria

Lecture Notes in Earth Sciences, Vol. 7
Mathematical and Numerical Techniques in Physical Geodesy
Edited by H. Sünkel

1. INTRODUCTION

1.1. Diagonal matrices

Consider the linear equation system

$$\underline{y} = \underline{A}\,\underline{x} \qquad (1\text{-}1)$$

or if $\underline{A}$ is a square 3x3 matrix:

$$\begin{aligned} y_1 &= a_{11}x_1 + a_{12}x_2 + a_{13}x_3 \\ y_2 &= a_{21}x_1 + a_{22}x_2 + a_{23}x_3 \\ y_3 &= a_{31}x_1 + a_{32}x_2 + a_{33}x_3 \end{aligned} \qquad (1\text{-}2)$$

which can briefly be written as

$$y_i = \sum_{j=1}^{3} a_{ij}x_j \qquad i = 1, 2, 3 \qquad (1\text{-}3)$$

We see that y_i is obtained from x_j by a "discrete convolution", that is multiplication followed by summation.

In this context "convolution" means simply multiplication followed by summation. Later we shall get convolutions consisting of multiplication followed by integration. The goal of the spectral analysis is the elimination of the second part of the convolutions, i.e., in our two examples either summation or integration.

Matters are much simpler if we consider a diagonal matrix:

$$\underline{B} = \begin{bmatrix} b_1 & 0 & 0 \\ 0 & b_2 & 0 \\ 0 & 0 & b_3 \end{bmatrix} \qquad (1\text{-}4)$$

Instead of (1-1) we now have the linear equation system

$$\underline{y} = \underline{B}\,\underline{x} \qquad (1\text{-}5)$$

which yields the following three equations:

$$\begin{aligned} y_1 &= b_1x_1 \\ y_2 &= b_2x_2 \\ y_3 &= b_3x_3 \end{aligned} \qquad (1\text{-}6)$$

Eqs. (1-6) are the analogue to (1-2) but now for a diagonal matrix. The simplification is obvious: the original convolution, multiplication followed by summation, is replaced by a pure multiplication.

The spectrum of a matrix are the diagonal elements of the diagonal form of the matrix. The spectral domain comprises the frequencies $i = 1,2,3$.

Using indices we may replace (1-6) by

$$y_i = b_i x_i \tag{1-7}$$

or we can use still another form if we regard i as a discrete variable:

$$\boxed{y(i) = b(i) \cdot x(i)} \tag{1-8}$$

This is the simplest example of an equation in the spectral domain.

Now we consider the transformation inverse to (1-5):

$$\underline{z} = \underline{B}^{-1} \underline{x} \tag{1-9}$$

The simplification is particularly striking if we consider the inverse of a diagonal matrix, since then (1-9) reduces to

$$z(i) = \frac{x(i)}{b(i)} \tag{1-10}$$

that is, inversion reduces to a simple division.

Clearly, $n = 3$ was chosen only for simplicity and the procedure is valid for any n, but finding such a linear transformation, if actually possible, is in general certainly no less difficult than solving the original system of linear equations.

1.2. *Stokes'* formula in the spectral domain

We consider *Stokes'* formula for the computation of the geoid height N, that is

$$N = \frac{R}{4\pi G} \iint_\sigma \Delta g \, S(\psi) \, d\sigma \tag{1-11}$$

where

$$R = 6371 \text{ km}$$

$$G = 9.81 \text{ m s}^{-2} \qquad (1\text{-}12)$$

σ ... unit sphere

In addition, Δg is the gravity anomaly and $S(\psi)$ is the *Stokes* function. More precisely, we may replace (1-11) by

$$N(P) = \frac{R}{4\pi G} \iint_\sigma S(\psi_{PQ}) \, \Delta g(Q) \, d\sigma_Q \qquad (1\text{-}13)$$

where P is the computation point and Q is the variable point moving over the whole unit sphere. Now we introduce the abbreviations

$$\frac{R}{4\pi G} S(\psi) = a(P,Q) \qquad (1\text{-}14)$$

and

$$\iint_\sigma \mathrel{\hat{=}} \int_\sigma \qquad (1\text{-}15)$$

which leads to

$$N(P) = \int_\sigma a(P,Q) \, \Delta g(Q) \, d\sigma \qquad (1\text{-}16)$$

Eq. (1-16) is analogous to

$$y(i) = \sum_{j=1}^{3} a(i,j) \, x(j) \qquad (1\text{-}17)$$

where the sum corresponds to the integral, x plays the role of Δg, and matrix $a(i,j)$ corresponds to $a(P,Q)$.

This is a typical example of a convolution on the sphere! Convolution is now multiplication

$$a(P,Q) \cdot \Delta g(Q) \qquad (1\text{-}18)$$

followed by the integration

$$\int_\sigma \qquad (1\text{-}19)$$

where $a(P,Q)$ is the kernel of the convolution.

Now, it would be very convenient if we could eliminate the integration since then we are only to perform the remaining multiplication.

The kernel is isotropic because $S(\psi)$ only depends on the distance but not on the direction.

Now we write the equation in spherical harmonics. For this reason we introduce the base functions $R_{nm}(\vartheta,\lambda)$, $S_{nm}(\vartheta,\lambda)$ where ϑ, λ are the independent variables. We get

$$\begin{aligned} R_{nm}(\vartheta,\lambda) &= P_{nm}(\cos\vartheta)\cos m\lambda \qquad n = 0, 1, \ldots, \infty \\ S_{nm}(\vartheta,\lambda) &= P_{nm}(\cos\vartheta)\sin m\lambda \qquad m = 0, 1, \ldots, n \end{aligned} \tag{1-20}$$

As it is indicated, n and m are integer numbers. This causes an <u>infinite</u> integer spectrum. Of course you know that the $P_{nm}(\cos\vartheta)$ are *Legendre* functions. The appropriate expansion for N reads:

$$N(\vartheta,\lambda) = \sum_{n=0}^{\infty}\sum_{m=0}^{n}[N_{nm}R_{nm}(\vartheta,\lambda) + \bar{N}_{nm}S_{nm}(\vartheta,\lambda)] \tag{1-21}$$

and also one often writes

$$N(\vartheta,\lambda) = \sum_{n=0}^{\infty} N_n(\vartheta,\lambda) \tag{1-22}$$

where

$$N_n(\vartheta,\lambda) = \sum_{m=0}^{n}[N_{nm}R_{nm}(\vartheta,\lambda) + \bar{N}_{nm}S_{nm}(\vartheta,\lambda)] \tag{1-23}$$

In the same manner we proceed with Δg, therefore we get on the one hand

$$\Delta g(\vartheta,\lambda) = \sum_{n=0}^{\infty}\sum_{m=0}^{n}[G_{nm}R_{nm}(\vartheta,\lambda) + \bar{G}_{nm}S_{nm}(\vartheta,\lambda)] \tag{1-24}$$

and on the other hand

$$\Delta g_n(\vartheta,\lambda) = \sum_{m=0}^{n}[G_{nm}R_{nm}(\vartheta,\lambda) + \bar{G}_{nm}S_{nm}(\vartheta,\lambda)] \tag{1-25}$$

As a summary, the formulae for the expansions of N and of Δg on the sphere read:

$$\begin{aligned} N(\vartheta,\lambda) &= \sum_{n=0}^{\infty} N_n(\vartheta,\lambda) \\ \Delta g(\vartheta,\lambda) &= \sum_{n=0}^{\infty} \Delta g_n(\vartheta,\lambda) \end{aligned} \tag{1-26}$$

The *Stokes* formula is equivalent to

$$\boxed{N_n(\vartheta,\lambda) = \frac{R}{G(n-1)}\,\Delta g_n(\vartheta,\lambda) \qquad \text{for} \quad n \geq 2} \tag{1-27}$$

This equation only contains a multiplication. But this is not exactly yet the equation in the frequency domain since N_n still depends on ϑ, λ. Using (1-23) and (1-25) we find, however,

$$N_{nm} = \frac{R}{G(n-1)} G_{nm}$$
$$\text{for} \quad n \geq 2 \qquad (1\text{-}28)$$
$$\bar{N}_{nm} = \frac{R}{G(n-1)} \bar{G}_{nm}$$

where the dependence on ϑ, λ has disappeared. This can be written in a more general way, e.g. for the first of the two equations (1-28):

$$N(n,m) = H(n,m) \cdot G(n,m) \qquad (1\text{-}29)$$

where

$$H(n,m) = \frac{R}{G(n-1)} \qquad (1\text{-}30)$$

Eq. (1-29) is a typical equation in the spectral domain. (It is of secondary importance that $H(n,m)$ in (1-30) does not depend on m, it only depends on n.) This is the simplest case of a transfer function or system function. The principle convolution is replaced by a pure multiplication in the spectral domain.

2. *FOURIER* SERIES

2.1. Onedimensional

Periodic functions with the period 2π, that is

$$f(x+2\pi) = f(x) \qquad (2\text{-}1)$$

can be considered as functions defined on the unit circle and can be expanded into *Fourier* series. A function $f(x)$ is expanded as:

$$f(x) = a_0 + a_1 \cos x + a_2 \cos 2x + \ldots + b_1 \sin x + b_2 \sin 2x + \ldots \qquad (2\text{-}2)$$

This can be written in the form of a sum:

$$f(x) = \sum_{n=0}^{\infty} (a_n \cos nx + b_n \sin nx)$$

Formulae for the coefficients are: (2-3)

$$a_n = \frac{1}{\pi}\int_0^{2\pi} f(x) \cos nx \, dx \qquad \text{for} \quad n \neq 0$$

$$b_n = \frac{1}{\pi}\int_0^{2\pi} f(x) \sin nx \, dx \qquad \text{for} \quad n \neq 0$$

For $n = 0$ we have the corresponding coefficients

$$a_0 = \frac{1}{2\pi}\int_0^{2\pi} f(x) \, dx$$

$$b_0 = 0 \tag{2-4}$$

Be aware of the symmetry sum <--> integral. This situation sum <--> integral can be generalized to integral <--> integral, this is the *Fourier* integral. Another generalization is sum <--> sum, this is the discrete *Fourier* transform (which is nothing else but the FFT, i.e., the fast *Fourier* transform).

We can write the *Fourier* series in complex form:

$$f(x) = \sum_{n=-\infty}^{\infty} c_n \, e^{inx}$$

$$c_n = \frac{1}{2\pi}\int_0^{2\pi} f(x) \, e^{-inx} \, dx \tag{2-5}$$

This form of representation of the *Fourier* series is better suited for theoretical investigations because then one deals only with one kind of functions and one kind of coefficients. The identity between (2-5) and (2-3) can be proved using the relations

$$e^{inx} = \cos nx + i \cdot \sin nx$$

$$c_n = \alpha_n + i \cdot \beta_n \tag{2-6}$$

Now we consider an arbitrary period $\lambda \neq 2\pi$, cf. Fig. 2.1. In contrast to the previous case of a period 2π we now have a period λ, which means a change of scale of the x-axis which transforms the interval $[0,2\pi]$ into the interval $[0,\lambda]$. This means that x must be replaced by $\frac{2\pi}{\lambda} x$. Then we obtain

$$\begin{aligned} f(x) &= \sum_{n=-\infty}^{\infty} c_n \, e^{i\frac{2\pi}{\lambda}nx} \\ c_n &= \frac{1}{\lambda}\int_0^{\lambda} f(x)\, e^{-i\frac{2\pi}{\lambda}nx}\, dx \end{aligned} \qquad (2\text{-}7)$$

Be aware of the symmetry sum <--> integral (apart from the factor $\frac{1}{\lambda}$).

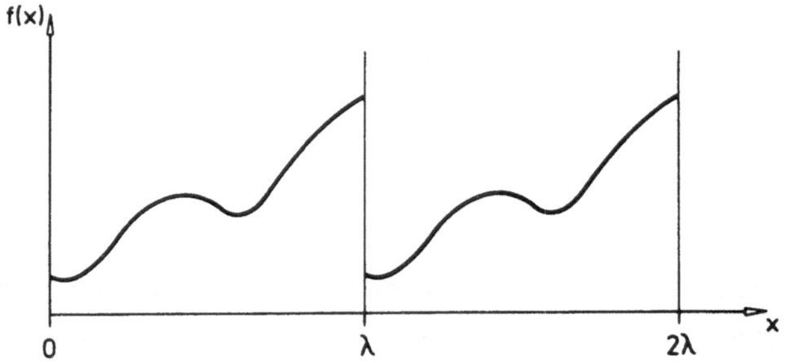

Figure 2.1. Periodic function with an arbitrary period.

2.2. Twodimensional

In Fig. 2.2 we show a twodimensional base domain.

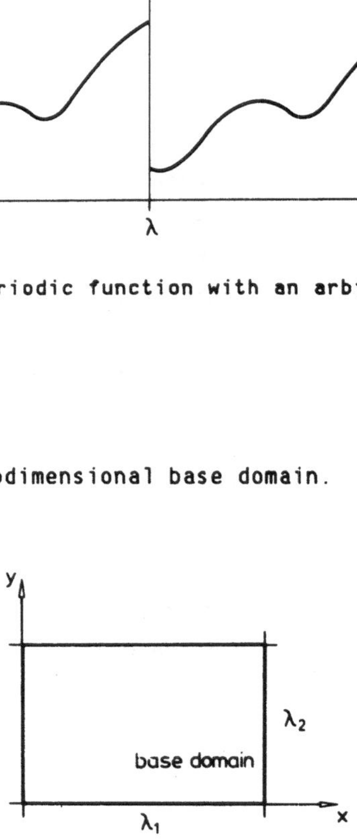

Figure 2.2. Twodimensional base domain.

Again we prefer the complex form. For the periods $\lambda_1 = \lambda_2 = 2\pi$ we obtain:

$$\begin{aligned} f(x,y) &= \sum_{m=-\infty}^{\infty} \sum_{n=-\infty}^{\infty} c_{mn} \, e^{i(mx+ny)} \\ c_{mn} &= \frac{1}{4\pi^2}\int_0^{2\pi}\int_0^{2\pi} f(x,y)\, e^{-i(mx+ny)}\, dxdy \end{aligned} \qquad (2\text{-}8)$$

For arbitrary periods $\lambda_1, \lambda_2 \neq 2\pi$ we get the following result:

$$f(x,y) = \sum_{m=-\infty}^{\infty} \sum_{n=-\infty}^{\infty} c_{mn} \, e^{i\,2\pi[(m/\lambda_1)x + (n/\lambda_2)y]}$$

$$c_{mn} = \frac{1}{\lambda_1 \lambda_2} \int_0^{\lambda_1}\!\!\int_0^{\lambda_2} f(x,y) \, e^{-i2\pi[(m/\lambda_1)x + (n/\lambda_2)y]} dxdy \tag{2-9}$$

We notice an almost perfect symmetry sum ←→ integral that is only disturbed by the factor $1/\lambda_1\lambda_2$ where $\lambda_1\lambda_2$ is the area of the rectangle of the base domain. We see that in (2-9) the expression

$$f(x,y) \, e^{-i\,2\pi[(m/\lambda_1)x + (n/\lambda_2)y]} \tag{2-10}$$

is averaged over the base domain, since the average is the integral divided by the area (the calculation of the *Fourier* coefficients is always an average!). The rectangle is the topologic product of the two intervals λ_1 and λ_2. (The plane is the topologic product of two straight lines).

The spectrum consists of the total set of integer numbers from $-\infty$ through $+\infty$. The spectral domain contains the total set of integer numbers, cf. Fig. 2.3.

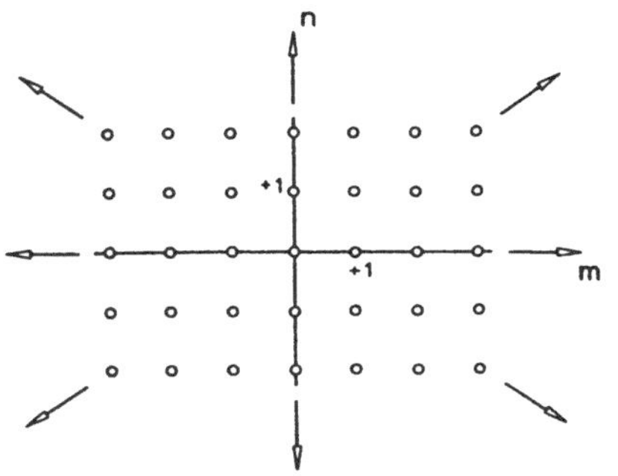

Figure 2.3. Spectral domain for the twodimensional *Fourier* expansions.

Note: the base domain is the rectangle λ_1, λ_2. The entire space domain is obtained if the base domain is continued periodically outside the rectangle λ_1, λ_2, cf. Fig. 2.2. The "basic" space domain, however, remains the rectangle.

The spectral domain is the entire integer point grid in the plane.

Geometrically: the space domain is a torus (i.e., a rectangle in which the two pairs of opposite sides are identified). The torus is the "topologic product" of two circles.

2.3. Spherical harmonics

The base domain is a sphere (which is obtained if half a circle rotates around an axis). A function represented by spherical harmonics reads

$$f(\vartheta,\lambda) = \sum_{n=0}^{\infty} \sum_{m=0}^{n} [a_{nm} R_{nm}(\vartheta,\lambda) + b_{nm} S_{nm}(\vartheta,\lambda)] \tag{2-11}$$

where

$$a_{nm} = \text{const} \int_{\lambda=0}^{2\pi} \int_{\vartheta=0}^{\pi} f(\vartheta,\lambda)\, R_{nm}(\vartheta,\lambda)\, d\sigma$$

$$b_{nm} = \text{const} \int_{\lambda=0}^{2\pi} \int_{\vartheta=0}^{\pi} f(\vartheta,\lambda)\, S_{nm}(\vartheta,\lambda)\, d\sigma \tag{2-12}$$

and the spectrum (cf. also Fig. 2.4) is obviously

$$n = 0, 1, \ldots, \infty$$

$$m = 0, 1, \ldots, n \tag{2-13}$$

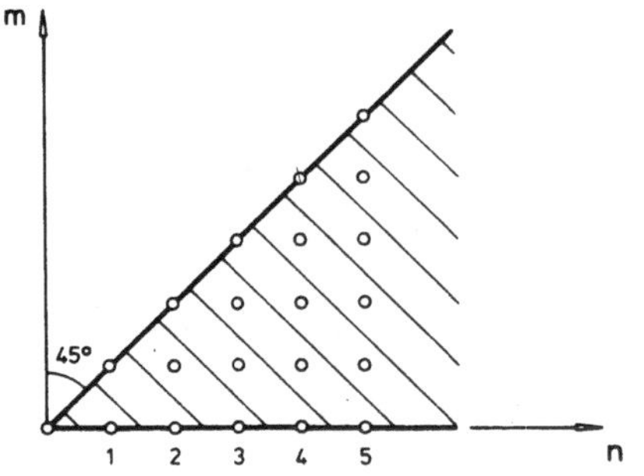

Figure 2.4. Spectrum of a function represented by spherical harmonics.

Remark: Additional information related to sec. 2 can be found in the appendix.

3. *FOURIER* INTEGRAL AND *FOURIER* TRANSFORM

We apply the following principle. We set up a *Fourier* series and let $\lambda \to \infty$, as always heuristically and with total disregard of mathematical rigor.

The formula for the onedimensional *Fourier* series with period λ reads:

$$f(x) = \sum_{n=-\infty}^{\infty} c_n \, e^{i\frac{2\pi}{\lambda}nx}$$

$$c_n = \frac{1}{\lambda}\int_{-\lambda/2}^{+\lambda/2} f(x)\, e^{-i\frac{2\pi}{\lambda}nx}\, dx \qquad (3\text{-}1)$$

Now we cannot simply let $\lambda \to \infty$ because this would give no result. But we can introduce an appropriate scale in the frequency domain (that is, the spectral domain). Note that a point grid finally becomes continuous if the distance between the points is taken smaller and smaller. We choose the new scale by

$$2\pi\, n = \lambda\, u \qquad (3\text{-}2)$$

and setting

$$\frac{2\pi}{\lambda} = \Delta u \qquad (3\text{-}3)$$

which implies $\Delta n = 1$, then the influence of the scale on the coefficients is

$$c_n = c(u)\cdot\Delta u \qquad \text{i.e.,} \qquad c(u) = \frac{c_n}{\Delta u} \qquad (3\text{-}4)$$

and for the onedimensional *Fourier* series we now get

$$f(x) = \sum_{u=-\infty}^{+\infty} c(u)\, e^{iux}\, \Delta u \qquad (3\text{-}5)$$

instead of the first equation in (3-1). For the coefficients we have from (3-4) the equation

$$c_n = c(u)\cdot\Delta u \qquad (3\text{-}6)$$

this leads for the coefficients c_n in (3-1) to the new expression

$$c(u)\cdot\Delta u = \frac{\Delta u}{2\pi}\int_{x=-\lambda/2}^{+\lambda/2} f(x)\, e^{-iux}\, dx \qquad (3\text{-}7)$$

As we can see, we may cancel by Δu and now we are able to perform the transition to the limit. Letting

$$\lambda \to \infty \qquad \text{and} \qquad \Sigma\,\Delta u \to \int du \qquad (3\text{-}8)$$

results in

$$\boxed{\begin{aligned} f(x) &= \int_{-\infty}^{+\infty} c(u)\, e^{iux}\, du \\ c(u) &= \frac{1}{2\pi}\int_{-\infty}^{+\infty} f(x)\, e^{-iux}\, dx \end{aligned}} \tag{3-9}$$

The spectrum *c(u)* is now continuous. The formulae are completely symmetric, apart from the disturbing factor $\frac{1}{2\pi}$. The space domain expands from $-\infty < x < +\infty$ and the spectral domain from $-\infty < u < \infty$.

Due to reasons of convention we set

$$F(u) = 2\pi c(u) \tag{3-10}$$

and *F(u)* is called *Fourier* transform or spectrum of *f(x)*. Taking into account eq. (3-10) we obtain

$$\boxed{\begin{aligned} F(u) &= \int_{-\infty}^{+\infty} f(x)\, e^{-iux}\, dx \\ f(x) &= \frac{1}{2\pi}\int_{-\infty}^{+\infty} F(u)\, e^{iux}\, du \end{aligned}} \tag{3-11}$$

as transformation equations.

<u>Remark</u>: on comparing this representation with others, an eye should be kept on two aspects: where is the factor 2π and which of the two equations contains $+i$ and which $-i$; this is largely conventional.

In symbolic form we can write the transformation equations (3-11) as

$$\begin{aligned} F(u) &= \mathcal{F}\{f(x)\} \\ f(x) &= \mathcal{F}^{-1}\{F(u)\} \end{aligned} \tag{3-12}$$

where $\mathcal{F}$ is the *Fourier* operator (which is a linear operator). The frequency is often denoted by ω therefore

$$u = \omega \tag{3-13}$$

and accordingly

$$\begin{aligned} F(\omega) &= \mathcal{F}\{f(x)\} \\ f(x) &= \mathcal{F}^{-1}\{F(\omega)\} \end{aligned} \tag{3-14}$$

How can we avoid the factor 2π ? We set

$$u = \omega = 2\pi\nu \tag{3-15}$$

This has a physical background, note that

$$\begin{aligned} &T \text{ ... period} \\ &\nu = T^{-1} \text{ ... frequency} \\ &\omega = 2\pi\nu \text{ ... circular frequency} \end{aligned} \tag{3-16}$$

In many cases ω is much more important than ν and therefore ω is often called "frequency" instead of "circular frequency". Now we have

$$F(u) = F(\nu) \tag{3-17}$$

and from (3-15) we immediately get

$$\frac{1}{2\pi}\, du = d\nu \tag{3-18}$$

The corresponding transformation equations are:

$$\boxed{\begin{aligned} F(\nu) &= \int_{-\infty}^{+\infty} f(x)\, e^{-2\pi i\nu x}\, dx \\ f(x) &= \int_{-\infty}^{+\infty} F(\nu)\, e^{+2\pi i\nu x}\, d\nu \end{aligned}} \tag{3-19}$$

What we have gained is the complete symmetry of the two equations in (3-19), but the factor 2π appears now in the exponent.

$F(u)$ or $F(\nu)$ is real if $f(x)$ is symmetric. In general $f(x)$ is not symmetric. But sometimes it is possible to take it symmetric (e.g., the covariance function).

An expression equivalent to (3-19) can be obtained by using the circular frequency $\omega = 2\pi\nu$ and following from this $d\omega = 2\pi d\nu$. We obtain

$$\boxed{\begin{aligned} F(\omega) &= \int_{-\infty}^{+\infty} f(x)\, e^{-i\omega x}\, dx \\ f(x) &= \frac{1}{2\pi}\int_{-\infty}^{+\infty} F(\omega)\, e^{+i\omega x}\, d\omega \end{aligned}} \tag{3-20}$$

<u>Example 3.1</u>

We consider the rectangular impulse, cf. eq. (3-21) and Fig. 3.1.

$$f(x) = \begin{cases} \frac{1}{2} & -1 < x < +1 \\ 0 & \text{elsewhere} \end{cases} \tag{3-21}$$

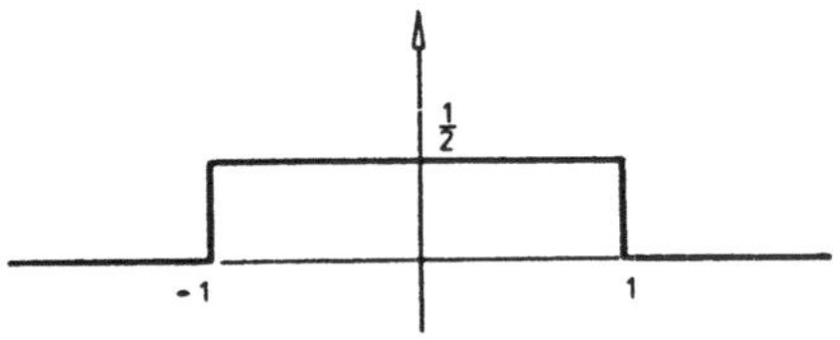

Figure 3.1. Rectangular impulse.

Using the formula containing the frequency ω, i.e. (3-20), and substituting eq. (3-21) into $f(x)$, we obtain

$$F(\omega) = \int_{-1}^{+1} \frac{1}{2} e^{-i\omega x} dx \tag{3-22}$$

The integration yields

$$F(\omega) = \frac{1}{2} (-i\omega)^{-1} e^{-i\omega x} \Big|_{-1}^{+1} \tag{3-23}$$

so that

$$F(\omega) = -\frac{1}{\omega} \frac{1}{2i} (e^{-i\omega} - e^{+i\omega}) \tag{3-24}$$

which we can slightly modify to obtain

$$F(\omega) = \frac{1}{\omega} \frac{1}{2i} (e^{i\omega} - e^{-i\omega}) = \frac{\sin\omega}{\omega} \tag{3-25}$$

The result is shown in Fig. 3.2. It is remarkable that this apparently simple function has a rather complicated spectrum.

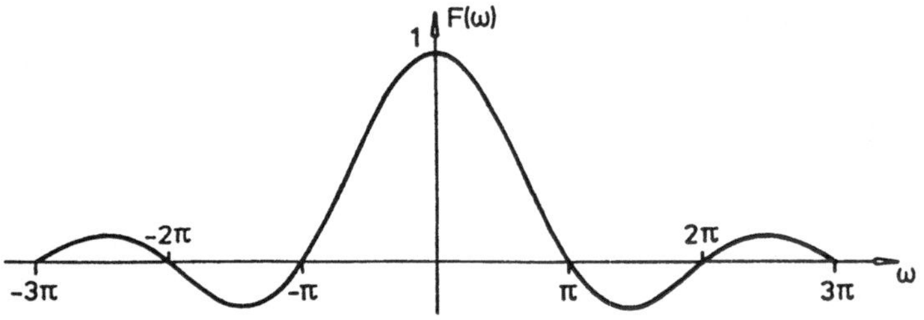

Figure 3.2. Graphic representation of the *Fourier* transform of the rectangular impulse.

Example 3.2, cf. Fig. 3.3 and Fig. 3.4.

$$f(x) = e^{-\alpha|x|}, \quad \alpha > 0 \qquad F(\omega) = \frac{2\alpha}{\omega^2 + \alpha^2} \tag{3-26}$$

"Markoff function" "Hirvonen function"

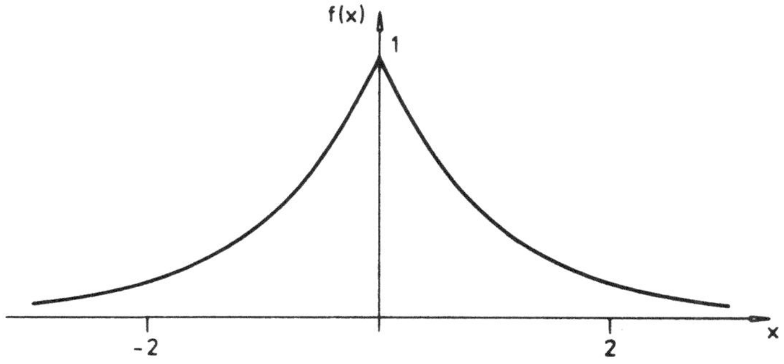

Figure 3.3. *Markoff* function (we have chosen $\alpha = 1$ for the figure).

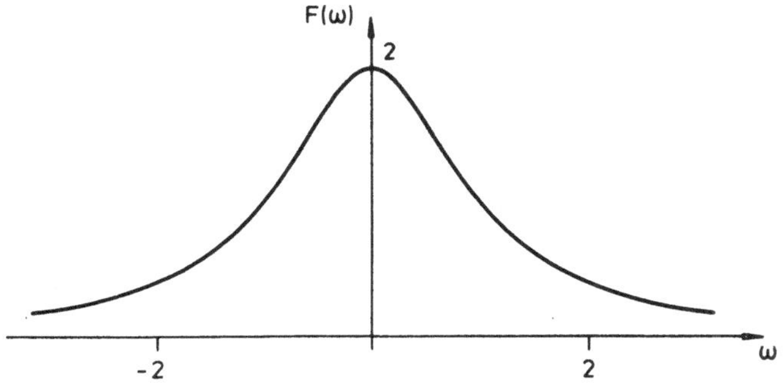

Figure 3.4. *Hirvonen* function (we have chosen $\alpha = 1$ for the figure).

If we exchange the whole matter by taking *Hirvonen* as $f(x)$ then the *Hirvonen* covariance function has a *Markoff* spectrum.

Example 3.3.

We consider the *Gauss* function and its transform:

$$f(x) = e^{-\alpha x^2}, \quad \alpha > 0$$

$$F(\omega) = \sqrt{\frac{\pi}{\alpha}}\, e^{-\omega^2/4\alpha} \tag{3-27}$$

Neglecting a factor we may say symbolically:

$$\text{Fourier}\{\text{Gauss}\} = \text{Gauss} \tag{3-28}$$

The *Gauss* function is therefore (apart from a factor) invariant with respect to a *Fourier* transform. As a consequence, the *Gauss* function is almost an eigenfunction of the *Fourier* transform. Note that

$$\alpha \xrightarrow{\text{by transformation}} \frac{1}{4\alpha} \tag{3-29}$$

If we now choose $\alpha = \frac{1}{2}$ then we obtain

$$\mathcal{F}\{e^{-x^2/2}\} = \sqrt{2\pi}\, e^{-\omega^2/2} \tag{3-30}$$

which means that

$$e^{-x^2/2} \tag{3-31}$$

is an eigenfunction of the *Fourier* operator where the eigenvalue is $\sqrt{2\pi}$.

Operators may be considered some kind of infinite matrices. The eigenfunction of an operator corresponds to the eigenvector of a finite matrix.

Remark: Additional information related to sec. 3 can be found in the appendix.

4. CONVOLUTION

The convolution is the reason why the *Fourier* transform is used. Consider a linear transformation of vectors $\underline{f}$, $\underline{g}$ in R^3. The corresponding formula is

$$\underline{g} = \underline{H}\,\underline{f} \tag{4-1}$$

The matrix $\underline{H}$ describes the linear transformation. Note: in sec. 1 we had $\underline{y} = \underline{A}\underline{x}$. Using indices, we can rewrite eq. (4-1) as

$$g_i = \sum_{j=1}^{3} h_{ij} f_j \tag{4-2}$$

or we can regard indices as independent variables:

$$g(i) = \sum_{j=1}^{3} h(i,j) \cdot f(j) \tag{4-3}$$

This equation is a convolution in R^3 in the form: multiplication followed by summation.

Now we try to find the continuous analogue. Then the index i changes to a variable x and the index j changes to a variable ξ accordingly. Naturally, the sum must be replaced by an integral. Thus the continuous analogue to (4-3) reads

$$g(x) = \int_a^b h(x,\xi)\, f(\xi)\, d\xi \tag{4-4}$$

Now we see that (symbolically!)

$$\boxed{\text{convolution} = \text{multiplication} + \text{integration}} \tag{4-5}$$

Especially important is the case

$$h(x,\xi) = h(x-\xi) \tag{4-6}$$

which is a function of the difference $x-\xi$ only (this is the case of homogeneity, cf. Fig. 4.1).

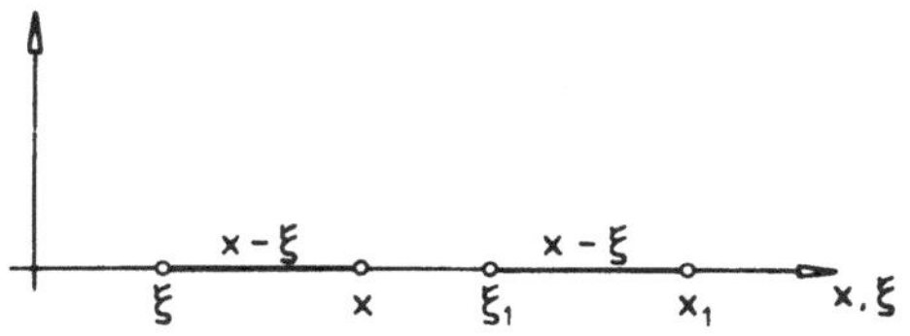

Figure 4.1. Homogeneous function. A translation along the x-axis does not change anything (e.g. time series).

Considering another generalization by letting $a, b \to \infty$, we get

$$\boxed{g(x) = \int_{-\infty}^{+\infty} h(x-\xi)\, f(\xi)\, d\xi} \tag{4-7}$$

This is the "classic" convolution formula. Instead of (4-7) we often write briefly (but deceptively simply!)

$$g(x) = h(x)*f(x) \tag{4-8}$$

The *Fourier* transform replaces a convolution by a simple multiplication.

Example: "running mean", i.e., the average from $x-a$ to $x+a$. The appropriate function is

$$h(x) = \begin{cases} \frac{1}{2a} & -a < x < a \\ 0 & \text{elsewhere} \end{cases} \tag{4-9}$$

A graphic representation of the function is given in Fig. 4.2.

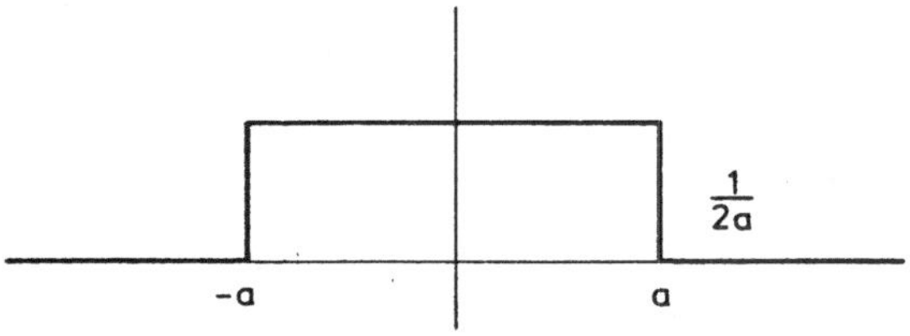

Figure 4.2. Graphic representation of h(x).

By applying (4-7) we obtain

$$g(x) = \frac{1}{2a} \int_{x-a}^{x+a} f(\xi)\, d\xi \tag{4-10}$$

The result is a slightly smoothed function, cf. Fig. 4.3.

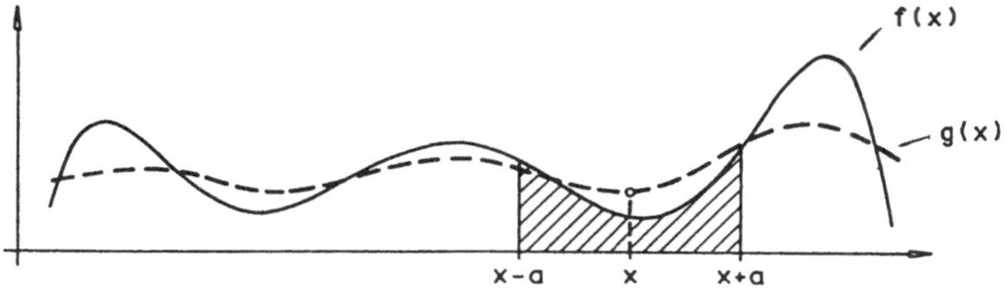

Figure 4.3. Graphic representation of f(x) and g(x).

-Convolution theorem

The situation of the space domain with the functions *f(x)*, *g(x)*, *h(x)* and the convolution formula are illustrated on the left side of (4-11) and in the frequency domain the corresponding functions *F(ω)*, *G(ω)*, *H(ω)* and the multiplication resulting from the convolution are shown on the right side of (4-11).

space domain	$\xrightarrow{\mathcal{F}}$ $\xleftarrow{\mathcal{F}^{-1}}$	frequency domain	
f(x)		F(ω)	
g(x)		G(ω)	(4-11)
h(x)		H(ω)	

$g(x) = h(x)*f(x)$	$G(\omega) = H(\omega)\cdot F(\omega)$
convolution	multiplication

The convolution in the space domain reduces to a simple multiplication in the frequency (spectral) domain.

We may say that the *Fourier* transform corresponds to the "diagonalization" of operators in *Hilbert* space (which is the infinite dimensional analogy of n-dimensional space R^n). Using slightly other words: the *Fourier* transform performs in *Hilbert* space what in R^3 is accomplished by matrices. The key of the whole matter lies in the homogeneity!

Shortly summarizing we can start with the convolution formula

$$g(x) = h(x)*f(x) \tag{4-12}$$

in detail written as

$$g(x) = \int_{-\infty}^{+\infty} h(x-\xi)\ f(\xi)\ d\xi \tag{4-13}$$

and this convolution is replaced by the multiplication

$$G(\omega) = H(\omega)\cdot F(\omega) \tag{4-14}$$

in the spectral domain. The function $h(x)$ in (4-12) is often denoted as weighting function since the convolution can be interpreted as a weighted mean. The function $H(\omega)$ which is $\mathcal{F}\{h(x)\}$, i.e. the *Fourier* transform of $h(x)$, is denoted as system function or transfer function.

Sometimes one speaks of a linear system $\mathcal{L}$, cf. (4-15).

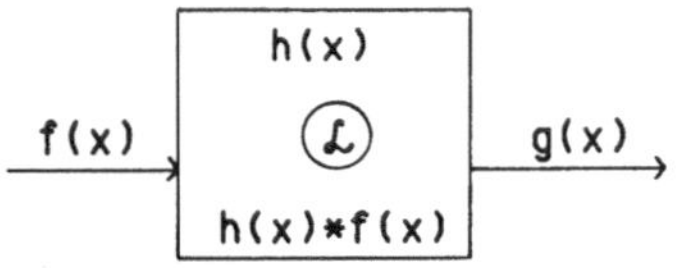

(4-15)

The "black box" indicated in (4-15) contains $h(x)$. In the same way we can see the situation in the spectral domain:

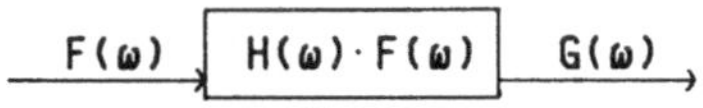

(4-16)

Any record player works in this way: from the tone-arm comes a signal, this signal is amplified in the system ("black box"), and there results music (hopefully linear, i.e., without distortion).

Therefore we have two basic physical interpretations:

1) "Weighted mean"

2) Linear system.

To find an empiric interpretation, we can proceed in the following way. We consider a special case: $f(x) = e^{i\omega x}$ as harmonic entry or test signal. Naturally there holds

$$e^{i\omega x} = \cos\omega x + i \sin\omega x \tag{4-17}$$

Now we ask what happens if we apply the system ("black box") to this test signal (4-17). According to (4-13) we get

$$g(x) = \int_{-\infty}^{+\infty} h(x-\xi)\, e^{i\omega\xi}\, d\xi \tag{4-18}$$

Introducing a new variable

$$x' = x - \xi \tag{4-19}$$

and differentiating this equation yields

$$dx' = -d\xi \tag{4-20}$$

because the integration is performed over ξ so that x is constant for the integration. Inserting (4-19) and (4-20) into (4-18) leads to

$$g(x) = -\int_{+\infty}^{-\infty} h(x')\, e^{i\omega x}\, e^{-i\omega x'}\, dx' \tag{4-21}$$

which can be modified to

$$g(x) = e^{i\omega x} \int_{-\infty}^{+\infty} h(x')\, e^{-i\omega x'}\, dx' \tag{4-22}$$

Now we define for the integral expression in (4-22) the function

$$H(\omega) = \int_{-\infty}^{+\infty} h(x')\, e^{-i\omega x'}\, dx' \tag{4-23}$$

and therefore we get the simple equation

$$g(x) = H(\omega)\, e^{i\omega x} \tag{4-24}$$

According to our starting assumption we have

$$f(x) = e^{i\omega x} \tag{4-25}$$

and thereby finally

$$g(x) = H(\omega)\cdot f(x) \tag{4-26}$$

This means that the system ("black box") multiplies the test signal by the factor $H(\omega)$. For a harmonic test signal we therefore may write

$$\frac{g(x)}{f(x)} = H(\omega) \tag{4-27}$$

Now, for $\omega = \omega_1$ we get $H(\omega_1)$, for $\omega = \omega_2$ results $H(\omega_2)$, etc. In this way we can determine the function pointwise (between two points we can interpolate). This is an empiric determination of $H(\omega)$.

Mathematical interpretation of $H(\omega)$.

We denote the influence of the linear system by $\mathcal{L}$ and we write

$$\mathcal{L}\{f(x)\} = g(x) \tag{4-28}$$

where $\mathcal{L}$ is the convolution operator. With respect to the example considered above we have

$$\mathcal{L}\{e^{i\omega x}\} = H(\omega)\, e^{i\omega x} \tag{4-29}$$

this means that the function $e^{i\omega x}$ is only multiplied by a constant factor, therefore $e^{i\omega x}$ is an eigenfunction of $\mathcal{L}$ with the eigenvalue $H(\omega)$.

Analogy to eigenvalues and eigenvectors of matrices: the linear operator is the ∞-dimensional analogue of a matrix. The eigenvalues of matrices are always related to the diagonal form of the matrices. This diagonalizing corresponds to the *Fourier* transform. The diagonal matrices correspond to the spectral (frequency) domain.

A linear operator corresponds to an ∞-dimensional matrix, and a function of one variable corresponds to an ∞-dimensional vector.

Example: running mean.

The function representation is

$$h(x) = \begin{cases} \frac{1}{2a} & -a < x < a \\ 0 & \text{elsewhere} \end{cases} \tag{4-30}$$

and we already know the spectral transform

$$H(\omega) = \frac{\sin a\omega}{a\omega} \tag{4-31}$$

from sec. 3 where we considered the special case $a = 1$, cf. (3-20), (3-25). The appropriate convolution reads

$$g(x) = \frac{1}{2a}\int_{x-a}^{x+a} f(\xi)\, d\xi \tag{4-32}$$

and the running mean has the simple form

$$G(\omega) = \frac{\sin a\omega}{a\omega} F(\omega) \tag{4-33}$$

in the spectral domain.

Geodetic applications: here we need the twodimensional case (i.e., if we approximate the earth's surface by a twodimensional plane). Then

$$g(x,y) = h(x,y)*f(x,y) \tag{4-34}$$

describes the twodimensional convolution symbolically, which corresponds to

$$g(x,y) = \int_{-\infty}^{+\infty}\int_{-\infty}^{+\infty} h(x-\xi,y-\eta)\, f(\xi,\eta)\, d\xi\, d\eta \tag{4-35}$$

in detailed form. Now we need two spectral variables which we denote as u, v (previously we had ω as spectral variable). Therefore

$$F(u,v) = \int_{-\infty}^{+\infty}\int_{-\infty}^{+\infty} f(x,y)\, e^{-i(ux+vy)}\, dx\, dy = \mathcal{F}\{f(x,y)\} \tag{4-36}$$

is the *Fourier* transform of the twodimensional function $f(x,y)$. And in an analogous way:

$$\begin{aligned} G(u,v) &= \mathcal{F}\{g(x,y)\} \\ H(u,v) &= \mathcal{F}\{h(x,y)\} \end{aligned} \tag{4-37}$$

Finally:

$$f(x,y) = \mathcal{F}^{-1}\{F(u,v)\} = \frac{1}{4\pi^2}\int_{-\infty}^{+\infty}\int_{-\infty}^{+\infty} F(u,v)\, e^{i(ux+vy)}\, du\, dv \quad \ldots \tag{4-38}$$

The convolution theorem in two dimensions is therefore given by

$$g(x,y) = h(x,y)*f(x,y) \tag{4-39}$$

and in the frequency (spectral) domain we have

$$G(u,v) = H(u,v)\cdot F(u,v) \tag{4-40}$$

This shows that a convolution in the space domain is a simple multiplication in the frequency domain.

Remark: Additional information related to sec. 4 can be found in the appendix.

5. GEODETIC APPLICATIONS

5.1. Analytic continuation of gravity anomalies upward and downward

The downward continuation is practically much more important and much more difficult to calculate. An application of downward continuation to gravity occurs in geophysical prospecting.

Another example of an upward and downward continuation of gravity anomalies arises in aerial gravimetry, cf. Fig. 5.1.

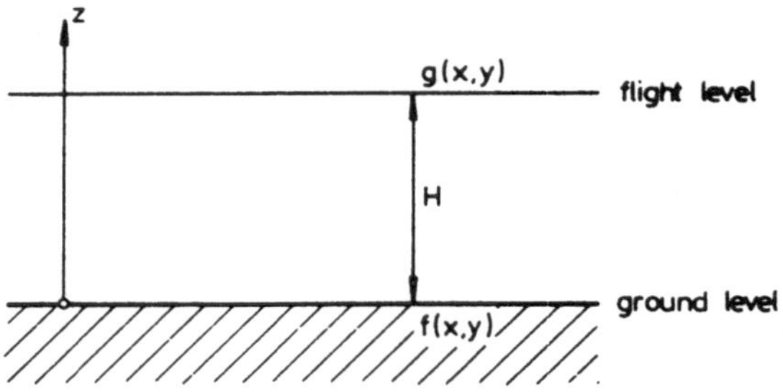

Figure 5.1. Aerial gravimetry.

Gravity values on the earth must be known in order to be able to calculate comparison values along the flight path by upward continuation. More important is the downward continuation of the measured values from flight elevation to ground.

The upward continuation can be solved with closed formulae; this is not possible with downward continuation.

Upward continuation

We consider the *Poisson* integral in the plane. Let the gravity anomaly $\Delta g(x,y,z)$ be a function in R^3. Then we have

$$\begin{aligned} f(x,y) &= \Delta g(x,y,0) \quad \text{(ground level)} \\ g(x,y) &= \Delta g(x,y,H) \quad \text{(flight level)} \end{aligned} \tag{5-1}$$

cf. Fig. 5.1. These functions are related by a convolution:

$$g(x,y) = \int_{-\infty}^{+\infty}\int_{-\infty}^{+\infty} h(x-\xi,y-\eta)\, f(\xi,\eta)\, d\xi\, d\eta \tag{5-2}$$

where $h(x,y)$ is the kernel or the weighting function. The formula for the

Poisson kernel in the plane is

$$h(x,y) = \frac{1}{2\pi} \frac{H}{(x^2+y^2+H^2)^{3/2}} \tag{5-3}$$

This formula describes upward continuation. For further information cf. *Heiskanen* and *Moritz (1967)*, page 35, eq. (1-89).

Symbolically we denote *Fourier* transforms by capital letters:

$$F,\ G,\ H = \mathcal{F}\{f,\ g,\ h\} \tag{5-4}$$

From

$$g = h*f \tag{5-5}$$

we get in the spectral domain

$$G(u,v) = H(u,v)\cdot F(u,v) \tag{5-6}$$

The function $F(u,v)$ must be computed numerically, e.g., by a Fast *Fourier* Transform, but the system function $H(u,v)$ (not to be confused with the elevation H!) is obtained analytically:

$$H(u,v) = \frac{H}{2\pi} \int_{-\infty}^{+\infty} \int_{-\infty}^{+\infty} \frac{e^{-i(ux+vy)}}{(H^2+x^2+y^2)^{3/2}}\, dx\, dy \tag{5-7}$$

The evaluation leads to

$$H(u,v) = e^{-H\sqrt{u^2+v^2}} \tag{5-8}$$

with the positive sign taken for the square root. We can define the so-called total frequency $\omega = \sqrt{u^2+v^2}$, where $\omega > 0$, and we get

$$\boxed{G(u,v) = e^{-H\omega}\cdot F(u,v)} \tag{5-9}$$

This holds for upward continuation. The spectrum of the gravity anomalies $F(u,v)$ is multiplied by a factor. The factor decreases exponentially with height as well as with frequency. One speaks of "attenuation" of the gravity anomalies in upward direction. We have $e^{-H\omega} < 1$. Note the tendency $e^{-H\omega} \rightarrow 0$ for $H \rightarrow \infty$ or $\omega \rightarrow \infty$. This means that the gravity anomalies field is much smoother at flight altitude than at ground level.

Downward continuation

Immediately from the upward continuation follows the downward continuation by

$$F(u,v) = e^{H\omega} G(u,v) \tag{5-10}$$

The problem lies in $e^{H\omega}$, cf. Fig. 5.2.

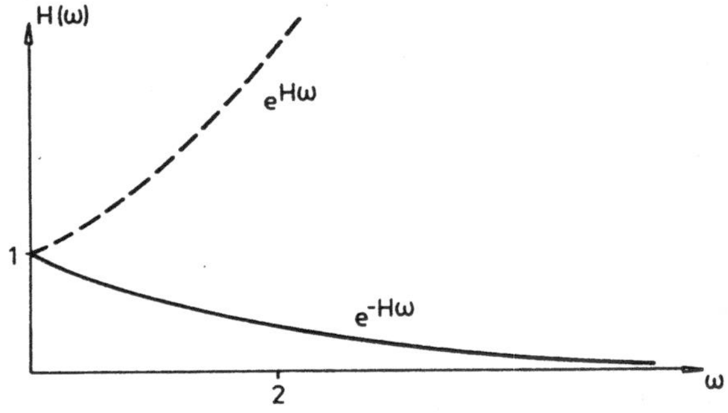

Figure 5.2. Graph of $e^{H\omega}$ and $e^{-H\omega}$.

The factor $e^{H\omega}$ implies an exponential amplification of the higher frequencies (i.e., larger ω), cf. Fig. 5.3.

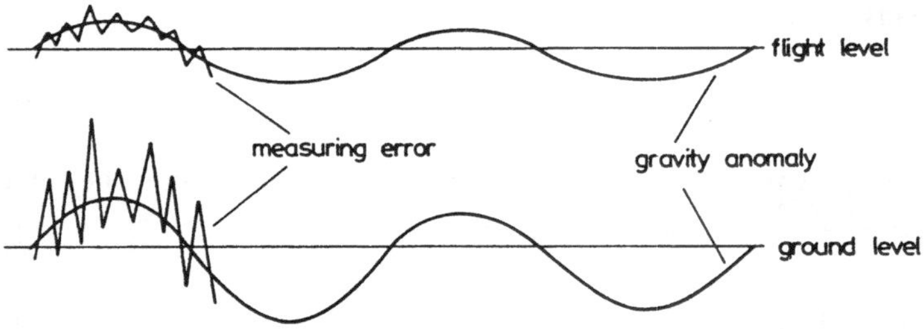

Figure 5.3. Amplifying effect for the downward continuation.

Continuing upward, the high frequencies are attenuated: the functions become smoother. Continuing downward, the functions become rougher. The transition from the flight level to the level of the earth causes an amplification of the high-frequencies. This implies various consequences. The high frequencies must be smoothed before downward continuation.

For the downward continuation there is no *Poisson* type integral. Here the spectral domain provides a considerable advantage since here an analytical representation is known, cf. formula (5-10).

Fig. 5.4 approximately illustrates the smoothing effect.

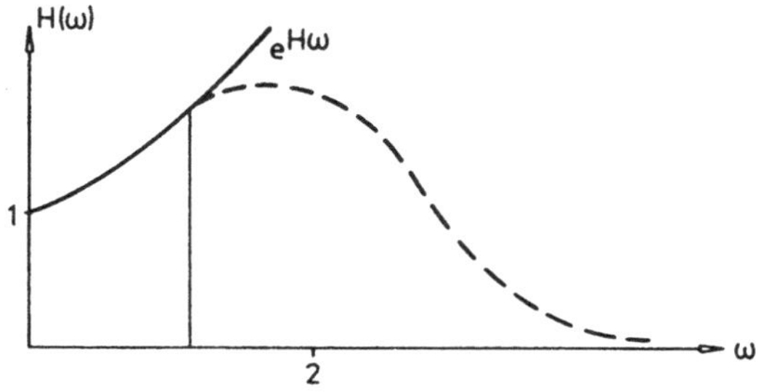

Figure 5.4. Smoothing effect.

In general the signal frequencies up to a certain limiting value are of interest only. By using a "low-pass-filter", only low frequencies can pass and we get again a *Poisson* integral:

$$f(x,y) \;=\; \bar{h}(x,y) * \Delta g(x,y) \qquad (5\text{-}11)$$

How can we represent the analytical upward continuation in spherical harmonics? We assume the situation of Fig. 5.5.

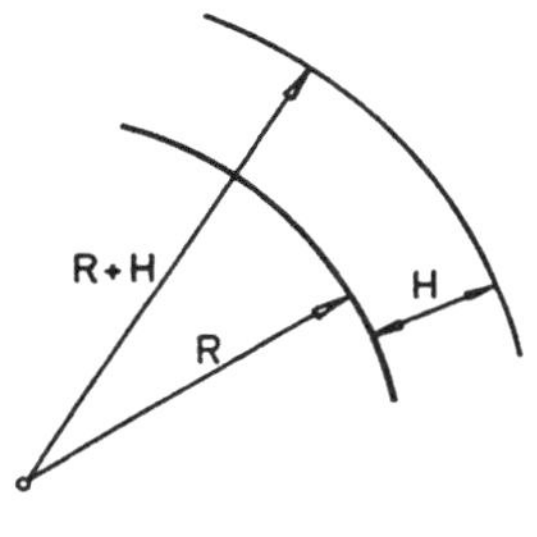

Figure 5.5. Ground level and flight level for the upward continuation.

The ground level is given by the mean terrestrial sphere:

$$r = R = 6371 \text{ km} \tag{5-12}$$

and the flight level consequently by

$$r = R + H \tag{5-13}$$

Therefore we have

$$f(\vartheta,\lambda) = \Delta g(R,\vartheta,\lambda) \tag{5-14}$$

and

$$g(\vartheta,\lambda) = \Delta g(R+H,\vartheta,\lambda) \tag{5-15}$$

According to (5-13) we can write this as

$$g(\vartheta,\lambda) = \Delta g(r,\vartheta,\lambda) \tag{5-16}$$

Now we perform an expansion into spherical harmonics:

$$f(\vartheta,\lambda) = \sum_{n=0}^{\infty} \Delta g_n(\vartheta,\lambda) \tag{5-17}$$

or equivalently:

$$f(\vartheta,\lambda) = \sum_{n=0}^{\infty} \sum_{m=0}^{n} \{F_{nm}^{(1)} R_{nm}(\vartheta,\lambda) + F_{nm}^{(2)} S_{nm}(\vartheta,\lambda)\} \tag{5-18}$$

Correspondingly we have:

$$g(\vartheta,\lambda) = \sum_{n=0}^{\infty} \left\{\frac{R}{r}\right\}^{n+2} \Delta g_n(\vartheta,\lambda) \tag{5-19}$$

or in detail:

$$g(\vartheta,\lambda) = \sum_{n=0}^{\infty} \sum_{m=0}^{n} [G_{nm}^{(1)} R_{nm}(\vartheta,\lambda) + G_{nm}^{(2)} S_{nm}(\vartheta,\lambda)] \tag{5-20}$$

Omitting the superscripts *(1)* and *(2)* we find by comparing (5-18) and (5-20), using (5-19):

$$\boxed{G(n,m) = \left\{\frac{R}{R+H}\right\}^{n+2} F(n,m)} \tag{5-21}$$

For reasons of a better understanding we set up the following comparison.

plane	sphere
u, v	n, m
$H(u,v) = e^{-H\sqrt{u^2+v^2}}$ $= a^{\sqrt{u^2+v^2}}$ $a < 1$	$H(n,m) = \left\{\frac{R}{R+H}\right\}^{n+2}$ $= b^{n+2}$ $b < 1$
u, v are equally relevant and $\sqrt{u^2+v^2}$ plays the role of frequency.	*m* does not explicitly occur and *n* plays the role of frequency.

5.2. Deflections of the vertical in the spectral domain

Again we consider the earth as a plane and *N(x,y)* shall denote the geoidal height in this plane. By

$$\xi = -\frac{\partial N}{\partial x}$$
$$\eta = -\frac{\partial N}{\partial y} \tag{5-22}$$

the deflections of the vertical are defined. We introduce the notation $N^*(u,v)$ for the spectrum of *N(x,y)*. Then the relation

$$N^*(u,v) = \mathcal{F}\{N(x,y)\} \tag{5-23}$$

holds. If we now want the inverse form, that is the representation of *N(x,y)* by its spectrum, then this may be symbolized by

$$N(x,y) = \mathcal{F}^{-1}\{N^*(u,v)\} \tag{5-24}$$

which is equivalent to

$$N(x,y) = \frac{1}{4\pi^2}\int_{-\infty}^{+\infty}\int_{-\infty}^{+\infty} N^*(u,v)\, e^{i(ux+vy)}\, du\, dv \tag{5-25}$$

According to (5-22) we must differentiate (5-25):

$$\frac{\partial N}{\partial x} = \frac{1}{4\pi^2}\int_{-\infty}^{+\infty}\int_{-\infty}^{+\infty} N^*(u,v)\,(iu)\, e^{i(ux+vy)}\, du\, dv \tag{5-26}$$

where we have interchanged differentiation and integration. Note that the spectrum depends on u, v only and is therefore not affected by the differentiation. The left-hand side of (5-26) is $\xi(x,y)$ and the product $N^*(u,v)\cdot(iu)$ can be denoted as $\xi^*(u,v)$ which leads to

$$\xi(x,y) = \frac{1}{4\pi^2}\int_{-\infty}^{+\infty}\int_{-\infty}^{+\infty}\xi^*(u,v)\, e^{i(ux+vy)}\, du\, dv \qquad (5\text{-}27)$$

In analogous way we can proceed for η. The situation in the space domain and in the spectral domain is shown by

$$\boxed{\begin{array}{lllll} \xi = -\frac{\partial N}{\partial x} & \longleftrightarrow & \xi^*(u,v) & = & -iu\, N^*(u,v) \\ \eta = -\frac{\partial N}{\partial y} & \longleftrightarrow & \eta^*(u,v) & = & -iv\, N^*(u,v) \end{array}} \qquad (5\text{-}28)$$

which again conforms to the pattern (5-6):

$$\begin{aligned} H(u,v) &= -iu \quad \text{for } \xi \\ H(u,v) &= -iv \quad \text{for } \eta \end{aligned} \qquad (5\text{-}29)$$

Now the question arises: what is simpler? A differentiation in the space domain corresponds to a multiplication by iu in the spectral domain. Linear differential equations in the space domain correspond to linear algebraic equations in the spectral domain. This answers our question in favour of the spectral domain.

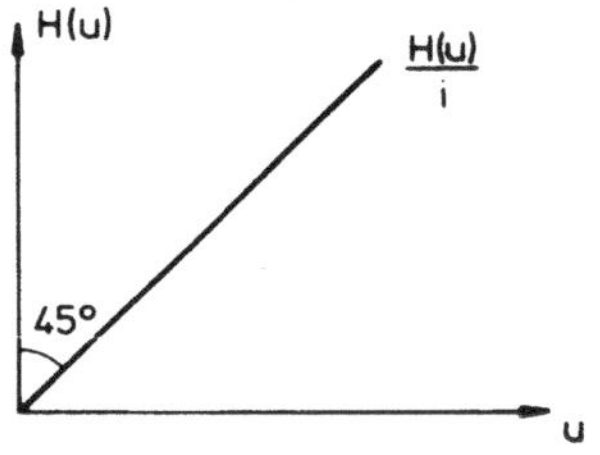

Figure 5.6. Amplification factor for the frequencies of the deflections of the vertical.

In Fig. 5.6 the amplification factor for the frequencies is shown. Frequency zero means a constant function. The higher the frequency the higher is the amplification (high-pass-filter). Note that differentiation is a "classic" example of a "roughening" operator ("roughening" means that high frequencies are amplified).

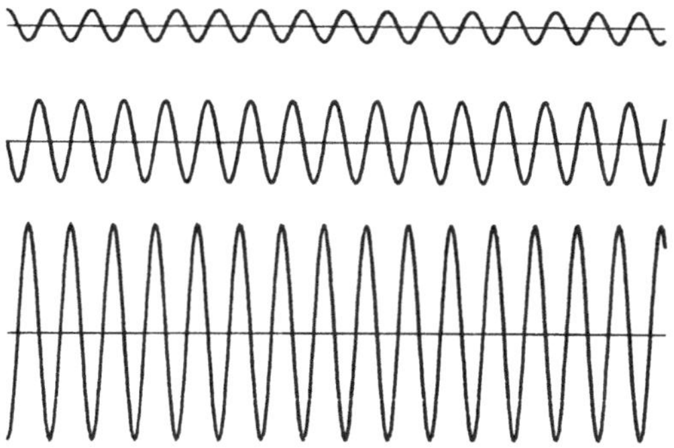

Figure 5.7. The effect of roughening.

To understand the effect of roughening, cf. Fig. 5.7. Assume as example $\varepsilon(x) = \varepsilon_0 \cos\omega x$ where ε_0 be very small and ω be very large. The differentiation yields $\varepsilon'(x) = -\varepsilon_0\omega \cdot \sin\omega x$ where the product $\varepsilon_0\omega$ is now large. What is the consequence? A small error $\varepsilon(x)$ generates arbitrarily large amplitudes in $\varepsilon'(x)$ if the small error has high frequencies.

Comparing deflections of the vertical and the analytical continuation, then we have for the deflections of the vertical "only" an amplification increasing linearly with frequency, whereas for downward continuation the amplification increases exponentially.

Remark: Additional information on geodetic applications can be found in the appendix.

6. *HANKEL* TRANSFORM

In case of isotropy, the *Fourier* transform becomes a *Hankel* transform. Consider again example 5.2: the *Poisson* kernel

$$h(x,y) = \frac{1}{2\pi} \frac{H}{(x^2+y^2+H^2)^{3/2}} = \bar{h}(r) \tag{6-1}$$

is homogeneous and isotropic. We consider this function as $\bar{h}(r)$ where $r = \sqrt{x^2+y^2}$; in this way a twodimensional function is reduced to a onedimensional function. This is possible in case of rotational symmetry.

Consider its *Fourier* transform

$$H(u,v) \quad = \quad e^{-H\sqrt{u^2+v^2}} \quad = \quad e^{-H\omega} \quad = \quad \bar{H}(\omega) \tag{6-2}$$

Also for this function it has been possible to reduce it to the single variable

$$\omega \quad = \quad \sqrt{u^2 + v^2} \tag{6-3}$$

Note that a rotationally symmetric function has a rotationally symmetric *Fourier* transform.

We add a few remarks on isotropy and homogeneity. In case of isotropy we have a dependence on the distance only but not on the azimuth (therefore "isotropy" is equivalent to "rotational symmetry"). Homogeneity depends (in the notation of the previous sections) on $x-\xi$, $y-\eta$ only (therefore "homogeneity" is equivalent to "translational symmetry" or "translational invariance"). Homogeneity always exists for convolutions by definition.

Considering a function of two variables, then in case of isotropy this can be regarded as a function of only one variable (if homogeneity is presupposed). In this way, a twodimensional *Fourier* transform is equivalent to a onedimensional transform, which is called *Hankel* transform. This advantage of the *Hankel* transform is lessened by the following disadvantage: the trigonometric functions *sin* and *cos* are replaced by *Bessel* functions. However, only J_0, the *Bessel* function of degree zero, is needed. The function J_0 behaves similar to a *cos* function (but amplitude and period are varying). It is a transcendental function, which does not belong to the elementary functions. The definition reads

$$J_0(x) \quad = \quad \frac{1}{2\pi}\int_{-\pi}^{+\pi} e^{ix\cos\vartheta}\, d\vartheta \tag{6-4}$$

or equivalently

$$J_0(x) \quad = \quad \frac{1}{\pi}\int_{0}^{\pi} \cos(x\cos\vartheta)\, d\vartheta \tag{6-5}$$

Moreover, J_0 can be represented by the following series expansion:

$$J_0(x) \quad = \quad 1 - \frac{x^2}{2^2} + \frac{x^4}{2^2 4^2} - \frac{x^6}{2^2 4^2 6^2} + \ldots \tag{6-6}$$

which converges for all x.

Now let us derive the details, starting with the function *H(u,v)*. For rotational symmetry we have

$$H(u,v) = \int_{-\infty}^{+\infty}\int_{-\infty}^{+\infty} h(x,y)\, e^{-i(ux+vy)}\, dx\, dy \qquad (6\text{-}7)$$

where we can substitute $h(x,y) = \bar{h}(r)$ in analogy to (6-1). We introduce the polar coordinates

$$\begin{aligned} x &= r\cos\alpha & \quad u &= \omega\cos\varphi \\ y &= r\sin\alpha & \quad v &= \omega\sin\varphi \end{aligned} \qquad (6\text{-}8)$$

and we get for the differential surface element

$$dx\, dy = r\, dr\, d\alpha \qquad (6\text{-}9)$$

Regarding the integrals we may put

$$\int_{-\infty}^{+\infty}\int_{-\infty}^{+\infty} dx\, dy = \int_{\alpha=0}^{2\pi}\int_{r=0}^{\infty} r\, dr\, d\alpha \qquad (6\text{-}10)$$

because the double integral on both sides covers the whole plane. In addition we have

$$ux + vy = r\omega\cos(\alpha-\varphi) \qquad (6\text{-}11)$$

which can easily be verified. The result of the transition to polar coordinates is:

$$H(u,v) = \int_{\alpha=0}^{2\pi}\int_{r=0}^{\infty} \bar{h}(r)\, e^{-ir\omega\cos(\alpha-\varphi)}\, r\, dr\, d\alpha \qquad (6\text{-}12)$$

We interchange the integrations and try to solve the integration over α:

$$\int_{\alpha=0}^{2\pi} e^{-ir\omega\cos(\alpha-\varphi)}\, d\alpha \qquad (6\text{-}13)$$

By the substitution

$$\alpha - \varphi = \vartheta \qquad (6\text{-}14)$$

we get

$$d\alpha = d\vartheta \qquad (6\text{-}15)$$

since φ is constant for the integration. Applying the substitution to (6-13) yields

$$\int_{\vartheta=-\pi}^{\vartheta=+\pi} e^{-ir\omega \cos\vartheta} \, d\vartheta \tag{6-16}$$

It is integrated over the whole circle. Note that the limits of integration are not transformed directly, but we only took care that the whole circle is covered. By putting

$$x = -r\omega \tag{6-17}$$

and considering the definition (6-4) we see that

$$\int_{\vartheta=-\pi}^{\vartheta=+\pi} e^{ix \cos\vartheta} \, d\vartheta = 2\pi \, J_0(-\omega r) = 2\pi \, J_0(\omega r) \tag{6-18}$$

Note the symmetry property of J_0. Thus (6-12) becomes

$$H(u,v) = 2\pi \int_{r=0}^{\infty} \bar{h}(r) \, J_0(\omega r) \, dr \tag{6-19}$$

which is a function of ω only which we denote by $\bar{H}(\omega)$. This gives the basic formulae for the *Hankel* transform:

$$\bar{H}(\omega) = 2\pi \int_{r=0}^{\infty} \bar{h}(r) \, J_0(\omega r) \, r \, dr \tag{6-20}$$

$$\bar{h}(r) = \frac{1}{2\pi} \int_{0}^{\infty} \bar{H}(\omega) \, J_0(\omega r) \, \omega \, d\omega \tag{6-21}$$

where the inverse relation (6-21) can be derived in an analogous way. These formulae replace the twodimensional *Fourier* integral by a onedimensional *Hankel* integral. The only disadvantage: the *sin*, *cos* functions are replaced by *Bessel* functions. But the advantage of one dimension prevails.

It is essential to state that the *Hankel* transform has all properties of the *Fourier* transform since the *Hankel* transform <u>is</u> a *Fourier* transform, only in another form.

Symbolically

$$\tilde{H}(\omega) = \mathcal{H}\{\bar{h}(r)\} \tag{6-22}$$

denotes the *Hankel* transform, so that for an arbitrary function we have

$$\tilde{f}(\omega) = \mathcal{H}\{f(r)\} \tag{6-23}$$

where

$$\tilde{f}(\omega) = \int_0^\infty f(r)\, J_0(\omega r)\, r\, dr$$

$$f(r) = \int_0^\infty \tilde{f}(\omega)\, J_0(\omega r)\, \omega\, d\omega \qquad (6\text{-}24)$$

Note the symmetry between $\tilde{f}(\omega)$ and $f(r)$ which has been made possible by using

$$\tilde{H}(\omega) = \frac{1}{2\pi} \bar{H}(\omega) \qquad (6\text{-}25)$$

instead of $\bar{H}(\omega)$. This symmetry can symbolically be expressed by writing

$$\mathscr{H}^{-1} = \mathscr{H} \qquad (6\text{-}26)$$

Example 6.1.

$$f(r) = \frac{1}{r} \qquad \tilde{f}(\omega) = \frac{1}{\omega} \qquad (6\text{-}27)$$

Example 6.2.

$$f(r) = e^{-\alpha r^2} \qquad \tilde{f}(\omega) = \frac{1}{2\alpha} e^{-\omega^2/4\alpha} \qquad (6\text{-}28)$$

From this example we see that, very symbolically,

$$\mathscr{H}\{\text{Gauß}\} = \text{Gauß} \qquad (6\text{-}29)$$

Example 6.3.

$$f(r) = e^{-\alpha r} \qquad \tilde{f}(\omega) = \frac{\alpha}{(\omega^2+\alpha^2)^{3/2}}$$

Markoff type covariance function — Hirvonen type covariance function (6-30)

We can interchange r and ω without any difficulty:

Example 6.4.

$$f(r) = \frac{\alpha}{(r^2+\alpha^2)^{3/2}} \qquad \tilde{f}(\omega) = e^{-\alpha\omega} \qquad (6\text{-}31)$$

7. COVARIANCE FUNCTION AND SPECTRUM

The term "spectrum" will now be considered in the sense of "power spectrum", i.e., as the *Fourier* transform of a covariance function. Let s denote a stochastic process on the sphere (or in plane or on a straight line). The notation s stems from signal. Stochastic roughly means non regular or random, cf. Fig. 7.1.

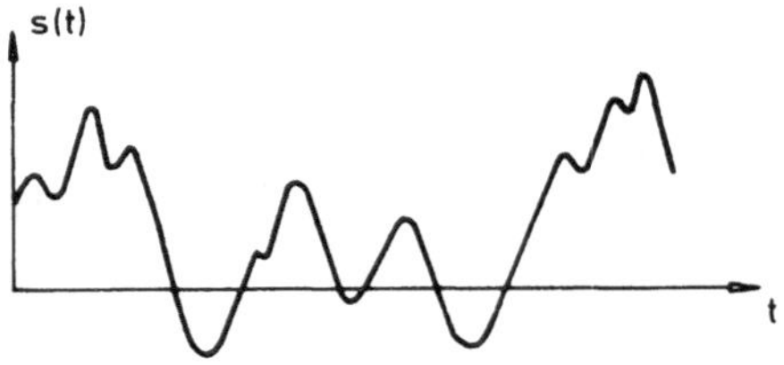

Figure 7.1. Time series as stochastic process.

In physical geodesy we say that the gravity anomalies Δg are a stochastic process (if certain systematic parts are split off and if all the discussions on this topic are disregarded).

Another example is the disturbing potential. The disturbing potential T is an especially convenient example since *Laplace*'s equation $\Delta T = 0$ holds in the outer space of the sphere because of harmonicity.

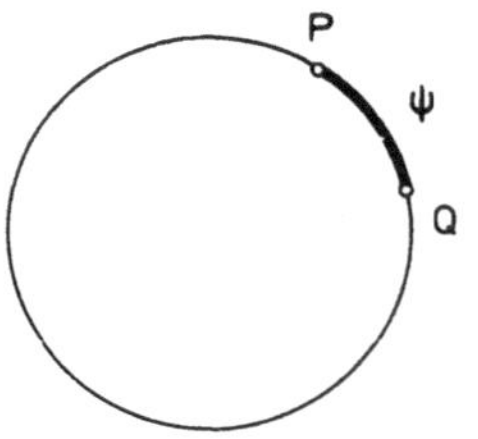

Figure 7.2. Forming the average over the sphere.

The covariance function is defined by

$$K(P,Q) = M\{s(P)s(Q)\} \qquad (7\text{-}1)$$

if

$$M\{s\} = 0 \qquad (7\text{-}2)$$

M can be e.g. the average over the sphere so that P and Q are points on the sphere. How can the average be determined? We let vary P and Q (cf. Fig. 7.2) over the whole sphere in such a way that their spherical distance ψ is constant (thereby isotropy is achieved too).

Considering the points P and Q outside the sphere, then we have the situation illustrated in Fig. 7.3.

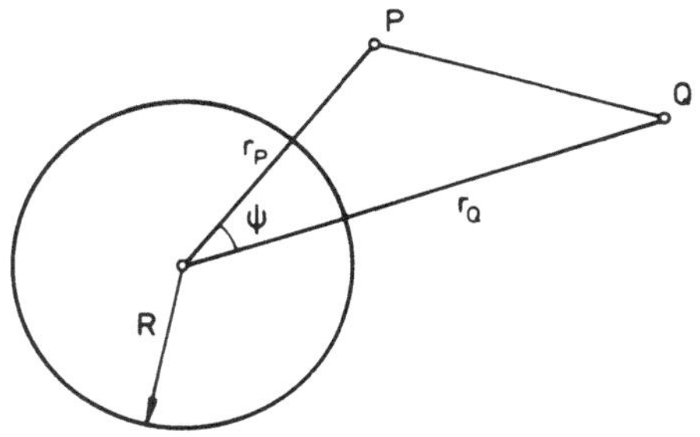

Figure 7.3. Definition of the covariance function if P and Q are outside the sphere.

Then

$$K(P,Q) = \sum_{n=0}^{\infty} \left\{\frac{R^2}{r_P r_Q}\right\}^{n+1} k_n P_n(\cos\psi) = K(r_P, r_Q, \psi) \tag{7-3}$$

is an appropriate analytical representation of the covariance function since the harmonicity in the outer space always implies a dependence on $1/r^{n+1}$. The function $K(P,Q)$ must be harmonic at P as well as at Q.

For an isotropic function a dependence on the azimuth must be absent, therefore the spherical harmonics

$$\begin{aligned} R_{nm} &= P_{nm}(\cos\psi)\cos m\alpha \\ S_{nm} &= P_{nm}(\cos\psi)\sin m\alpha \end{aligned} \tag{7-4}$$

cannot depend on the azimuth α. This is possible only for $m = 0$ which means that of all P_{nm} only P_n will occur.

On the sphere we have

$$r_P = r_Q = R, \qquad 0 \leq \psi \leq \pi \tag{7-5}$$

whence

$$K(P,Q) = K(\psi) = \sum_{n=0}^{\infty} k_n P_n(\cos\psi) \tag{7-6}$$

The extension into the outer space is very simply achieved by multiplying with the factor $\frac{R^2}{r_P r_Q}$. Therefore it is sufficient to study the situation on the sphere. We denote by

$$k_n = k(n) \tag{7-7}$$

the "spectrum" of the covariance function. The spectrum must now be seen in the sense of power spectrum of the stochastic process s, defined as the *Fourier* transform of the covariance function.

Slightly more precisely, the spectrum of a "stationary" stochastic process is the *Fourier* transform of the covariance function. In our sense, "stationary" means homogeneous and isotropic.

Note the difference between the conventional spectrum and the power spectrum. A spectrum in the former sense cannot be formed at all for the time series. On the sphere: if T has the conventional spectrum a_{nm}, b_{nm} (consisting of the spherical harmonic coefficients of T) then

$$k_n = \frac{1}{4n+2}\left\{2a_{n0}^2 + \sum_{m=1}^{n} \frac{(n+m)!}{(n-m)!}(a_{nm}^2 + b_{nm}^2)\right\}$$

is the power spectrum, cf. *Heiskanen* and *Moritz (1967)*, eqs. (7-28), (1-78).

Situation in the plane.

The transition from the sphere to the plane can be performed by letting $R \to \infty$ or by replacing the sphere by the tangent plane for a certain region. The distance between the two points P and Q is now denoted by r. Isotropy and homogeneity are assumed again (not necessary in the plain). Then we have to give the connection to the previous section.

$$\begin{array}{lll} C(r) & \text{instead of} & K(\psi) \\ \text{plane} & & \text{sphere} \end{array} \tag{7-8}$$

Then

$$S(\omega) = \mathcal{F}\{C(r)\} \tag{7-9}$$

is the spectrum, in detail written as

$$S(\omega) = S(u,v) = \int_{-\infty}^{+\infty}\int_{-\infty}^{+\infty} C(r)\, e^{-i(ux+vy)}\, dx\, dy \tag{7-10}$$

Now we can apply the *Hankel* transform:

$$S(\omega) = 2\pi \int_0^\infty C(r)\, J_0(\omega r)\, r\, dr \qquad (7\text{-}11)$$

where we have introduced the factor 2π and where

$$\omega = \sqrt{u^2 + v^2} \qquad (7\text{-}12)$$

Finally,

$$C(r) = \frac{1}{2\pi} \int_0^\infty S(\omega)\, J_0(\omega r)\, \omega\, d\omega \qquad (7\text{-}13)$$

is the inverse transform.

For a good understanding we show a comparison between the sphere and the plane.

	sphere	plane
distance	ψ	r
covariance function	$K(\psi)$	$C(r)$
spectrum	$k_n = k(n)$	$S(\omega)$
kind of spectrum	discrete	continuous
spectral transform	expansion into spherical harmonics	*Fourier* transform, 2-dimensional; or: *Hankel* transform in case of isotropy and homogeneity, 1-dimensional
formula	$K(\psi) = \sum_{n=0}^{\infty} k(n)\, P_n(\cos\psi)$	$C(r) = \frac{1}{2\pi} \int_0^\infty S(\omega)\, J_0(\omega r)\, \omega d\omega$
spectral variable	n	ω
	sum $\sum_0^\infty$	integral $\int_0^\infty$
base function	$P_n(\cos\psi) = f_1(\psi, n)$	$J_0(\omega r) = f_2(r, \omega)$

Considering a onedimensional variable (e.g. time t) then we have the covariance function

$$C(t) = M\{s(\tau)s(\tau+t)\} \qquad M\{s(\tau)\} = 0 \tag{7-14}$$

and the average could be defined as

$$C(t) = \lim_{T\to\infty} \frac{1}{T} \int_{-T/2}^{+T/2} s(\tau)\, s(\tau+t)\, d\tau \tag{7-15}$$

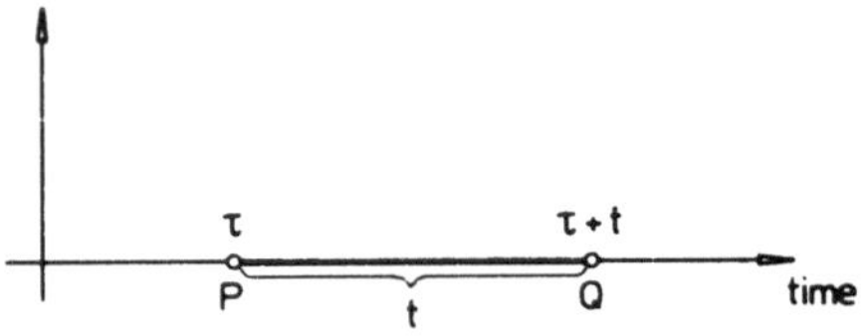

Figure 7.4. Onedimensional variable t.

To understand formula (7-15) imagine that a constant distance t is moved along the whole straight line, cf. Fig. 7.4.

Neglecting factors we have

$$S(\omega) = \mathcal{F}\{C(t)\} \tag{7-16}$$

as spectrum and

$$C(t) = \frac{1}{2\pi} \int_{-\infty}^{+\infty} S(\omega)\, e^{i\omega t}\, d\omega \tag{7-17}$$

is the corresponding inverse formula which can also be written as

$$C(t) = \frac{1}{\pi} \int_{0}^{\infty} S(\omega) \cos\omega t\, d\omega \tag{7-18}$$

8. COLLOCATION IN THE FREQUENCY DOMAIN

8.1. Filtering of an aerial gravimeter profile

Here the noise is much stronger than the signal. We show the corresponding situation in Fig. 8.1.

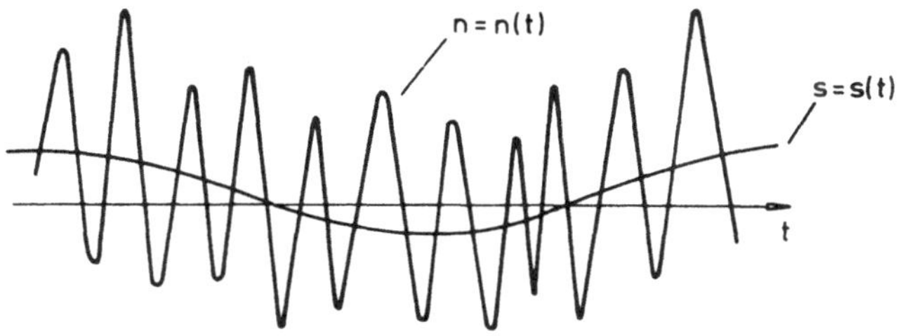

Figure 8.1. Signal and noise.

Considering gravity anomalies along a profile, we have

$$s(t) = \Delta g(t) \tag{8-1}$$

as signal and

$$n(t) \gg s(t) \tag{8-2}$$

indicates the noise whose impact can be seen in Fig. 8.1. The measured (registered) function is

$$x(t) = s(t) + n(t) \tag{8-3}$$

Now we try to derive this case from collocation. The collocation model

$$\underline{x} = \underline{A}\underline{X} + \underline{s} + \underline{n} \tag{8-4}$$

has the parameter vector $\underline{X}$ and the design matrix $\underline{A}$ and $\underline{s}$ and $\underline{n}$ denote signal and noise as before. The collocation can be seen as a generalization of prediction and filtering or as generalization of the adjustment by parameters.

adjustment by parameters	prediction, filtering
$\underline{s} = 0$	$\underline{X} = 0$
(Gauss-Markoff model)	(Kolmogorov-Wiener model)

For filtering we set $\underline{X} = 0$, getting the vector equation

$$\underline{x} = \underline{s} + \underline{n} \tag{8-5}$$

Introducing index notation and putting $x_i = x(i)$, we can write (8-5) as

$$x(i) = s(i) + n(i) \tag{8-6}$$

Replacing the index by a continuous variable t leads to

$$x(t) = s(t) + n(t) \tag{8-7}$$

We shall now generalize collocation from discrete to continuous observations.

<u>Prediction formula:</u>

$$\boxed{\underline{s} = \underline{C}_{sx}\underline{C}_{xx}^{-1}\underline{x}} \tag{8-8}$$

where $\underline{s}$ are the N filtered signals and $\underline{x}$ are the N unfiltered signals (discrete case).

Now we try to find the corresponding continuous formula. We set

$$\boxed{\underline{s} = \underline{H}\,\underline{x}} \tag{8-9}$$

where

$$\underline{H} = \underline{C}_{sx}\underline{C}_{xx}^{-1} \tag{8-10}$$

$\underline{H}$ can be seen as filter operator (as in the case of a record player, where $\underline{x}$ comes from the tone-arm and $\underline{s}$ is the filtered music which is heard). From (8-10) we immediately get

$$\boxed{\underline{H}\,\underline{C}_{xx} = \underline{C}_{sx}} \tag{8-11}$$

This is still the discrete case:

$\underline{C}_{xx}$,	elements	$(C_{xx})_{ij}$,	square matrix NxN
$\underline{C}_{sx}$,	elements	$(C_{sx})_{ij}$,	square matrix NxN
$\underline{H}$,	elements	h_{ij},	square matrix NxN

Thus we get the index notation of (8-9) as

$$s_i = \sum_{j=1}^{N} h_{ij}x_j \tag{8-12}$$

which is a discrete convolution, and in analogous way we obtain the index notation of (8-11) as

$$\sum_{j=1}^{N} h_{ij}(C_{xx})_{jk} = (C_{sx})_{ik} \tag{8-13}$$

which is also a discrete convolution. Formally, we can perform the transition to the continuous formula:

$$\boxed{s(t) = \int_{-\infty}^{+\infty} h(t-\tau)\, x(\tau)\, d\tau} \tag{8-14}$$

where i corresponds to t and j to τ, moreover we have assumed a homogeneous kernel $h(t-\tau)$.

In analogous way we proceed with (8-13) and

$$\boxed{\int_{-\infty}^{+\infty} h(t-\tau)\, C_{xx}(\tau)\, d\tau = C_{sx}(t)} \tag{8-15}$$

is the corresponding continuous analogue which is known as *Wiener Hopf* equation. Because of homogeneity C_{xx} depends on one variable only. We have got an integral equation. Because of the infinite limits of integration it is a singular integral equation which, in general, cannot be solved directly.

Both formulae (8-14) and (8-15) are convolutions. Therefore the two formulae can be transformed into the frequency domain. The convolutions are homogeneous since the kernel $h(t-\tau)$ is homogeneous. For the spectra we use the following notations:

$$\begin{aligned} \mathcal{F}\{h(t)\} &= H(\omega) \\ \mathcal{F}\{C_{xx}(t)\} &= S_{xx}(\omega) \\ \mathcal{F}\{C_{sx}(t)\} &= S_{sx}(\omega) \\ \mathcal{F}\{s(t)\} &= s(\omega) \\ \mathcal{F}\{x(t)\} &= x(\omega) \end{aligned} \tag{8-16}$$

Performing the *Fourier* transform for eq. (8-14) and eq. (8-15) yields

$$\begin{aligned} \mathcal{F}\{\text{eq.}(8\text{-}14)\} &\implies s(\omega) = H(\omega)\cdot x(\omega) \\ \mathcal{F}\{\text{eq.}(8\text{-}15)\} &\implies H(\omega)\cdot S_{xx}(\omega) = S_{sx}(\omega) \end{aligned} \tag{8-17}$$

From the second equation in (8-17) we get

$$H(\omega) = \frac{S_{sx}(\omega)}{S_{xx}(\omega)} \tag{8-18}$$

which can be inserted into the first equation:

$$\boxed{s(\omega) = \frac{S_{sx}(\omega)}{S_{xx}(\omega)}\, x(\omega)} \tag{8-19}$$

We can write this as

$$s(\omega) = H(\omega)\, x(\omega) \tag{8-20}$$

which is the first equation in (8-17). We see that $H(\omega)$ is a factor which depends on the frequency. It could be physically realized by an electronic filter or electric network.

Comparing (8-8) and (8-19) we see the advantage of the transform into the spectral domain: a very simple solution is possible for continuous collocation.

Now we return to

$$x = s + n \tag{8-21}$$

where signal and noise are not correlated. Therefore

$$C_{xx} = C_{ss} + C_{nn} \qquad C_{xs} = C_{ss} + \underbrace{C_{ns}}_{0} \tag{8-22}$$

are the appropriate covariances (no cross correlations!) and in an analogous way

$$S_{xx} = S_{ss} + S_{nn} \qquad S_{xs} = S_{ss} + 0 \tag{8-23}$$

This leads to

$$\boxed{s(\omega) = \frac{S_{ss}(\omega)}{S_{ss}(\omega) + S_{nn}(\omega)}\, x(\omega)} \tag{8-24}$$

We have

$$H(\omega) = \frac{S_{ss}(\omega)}{S_{ss}(\omega) + S_{nn}(\omega)} \leq 1 \tag{8-25}$$

If we have no noise then $S_{nn}(\omega) = 0$, the factor $H(\omega)$ becomes 1 and nothing happens. The stronger the noise the stronger must be the filtering (the denominator becomes larger, the factor $H(\omega)$ becomes smaller which is a strong filtering).

8.2. *Stokes' formula for erroneous* Δg

In 1975 *Pellinen* and *Moritz* answered the question how the *Stokes* formula must be modified if Δg is affected by measuring errors. Let

$$s = \Delta g \tag{8-26}$$

be the signal,

$$C_{ss}(\psi) = C(\psi) \tag{8-27}$$

the signal covariance function and

$$C_{nn}(\psi) = D(\psi) \tag{8-28}$$

the error covariance function where these two functions can be represented by

$$C(\psi) = \sum_{n=0}^{\infty} c_n P_n(\cos\psi) \tag{8-29}$$

and

$$D(\psi) = \sum_{n=0}^{\infty} d_n P_n(\cos\psi) \tag{8-30}$$

The *Stokes* formula

$$N = \frac{R}{4\pi\gamma} \iint_{\sigma} \Delta g \; S(\psi) \; d\sigma \tag{8-31}$$

enables the computation of the geoid height N where the factor $\frac{R}{4\pi\gamma}$ is constant and

$$S(\psi) = \sum_{n=2}^{\infty} \frac{2n+1}{n-1} P_n(\cos\psi) \tag{8-32}$$

is the *Stokes* function. In analogy to the filtering factor $H(\omega)$ in (8-25) we now have c_n which corresponds to $S_{ss}(\omega)$ and d_n $(n = 1,2,\ldots)$ which corresponds to $S_{nn}(\omega)$ (where n denotes noise). Therefore the filtering by the factor

$$\frac{S_{ss}(\omega)}{S_{ss}(\omega) + S_{nn}(\omega)} \tag{8-33}$$

dependent on frequency ω corresponds now to the factor

$$\frac{c_n}{c_n + d_n} \tag{8-34}$$

dependent on frequency n. Obviously, we must take into account this factor just there where the frequency appears explicitly, that is in (8-32). By the filtering the *Stokes* function is affected in the following way:

$$\bar{S}(\psi) = \sum_{n=2}^{\infty} \frac{c_n}{c_n + d_n} \frac{2n+1}{n-1} P_n(\cos\psi) \tag{8-35}$$

Every frequency n is treated separately by multiplication with a frequency dependent factor. For more details cf. *Moritz (1976)*, p.1.

9. SAMPLING THEOREM

We consider band-limited functions. If the function $f(x)$ has no frequencies higher than a limit frequency N then it is determined uniquely by its values for positive x at equidistant sampling points

$$x_n = \frac{n}{2N} \qquad n = 0, 1, 2, \ldots \tag{9-1}$$

The distance between the equidistant sampling points is therefore given by

$$\Delta x = \frac{1}{2N} \tag{9-2}$$

cf. Fig. 9.1.

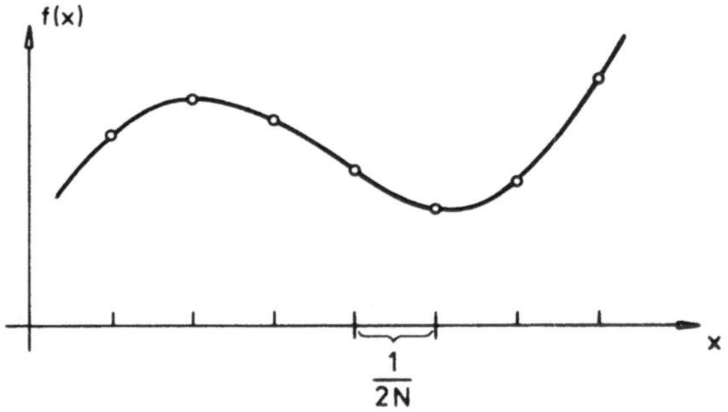

Figure 9.1. Space domain. The function is uniquely determined by its sampling points.

The frequency is

$$\nu = \frac{\omega}{2\pi} \leqslant N \tag{9-3}$$

where N is the so-called *Nyquist* frequency. In Fig. 9.2 band-limited functions are shown.

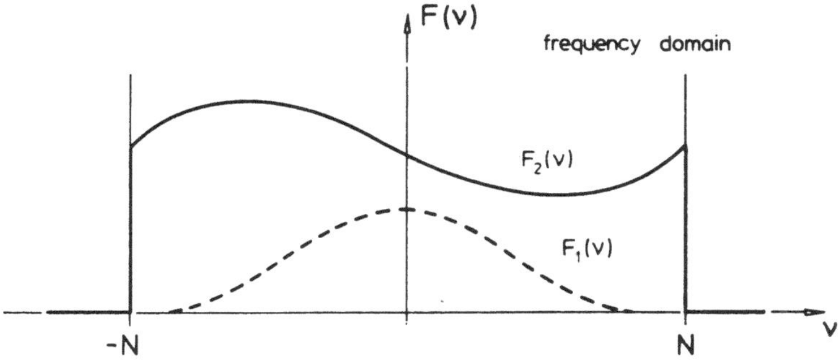

Figure 9.2. Frequency domain. $F_1(\nu)$ and $F_2(\nu)$ are band-limited functions.

We assume for $f(x)$ an even function, that is $f(-x) = f(x)$ where $f(-x)$ can be seen as the symmetric continuation of $f(x)$, cf. Fig. 9.3.

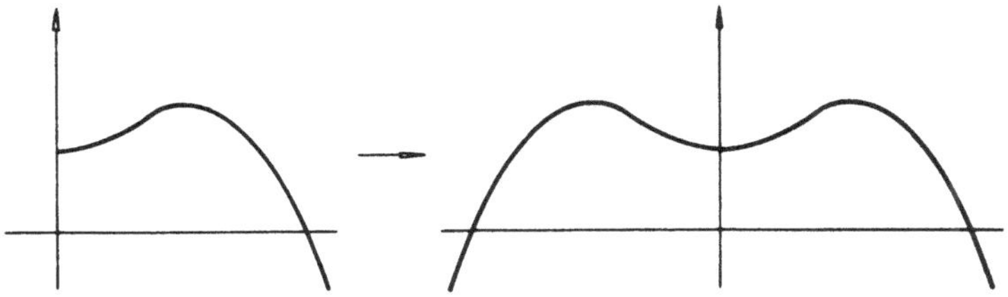

Figure 9.3. Even function composed of the symmetric parts $f(x)$ and $f(-x)$.

For symmetric functions the spectrum is <u>real</u>. But the spectrum is not necessarily symmetric.

We again use the frequency ω where $\omega = 2\pi\nu$ and

$$f(x) = \frac{1}{2\pi} \int_{-\infty}^{+\infty} F(\omega)\, e^{i\omega x}\, d\omega \tag{9-4}$$

is the *Fourier* transform. This can be simplified because of

$$F(\omega) = 0 \quad \text{if} \quad |\omega| > 2\pi N \tag{9-5}$$

which means that frequencies beyond of the *Nyquist* frequency are zero. Therefore:

$$f(x) = \frac{1}{2\pi} \int_{-2\pi N}^{+2\pi N} F(\omega)\, e^{i\omega x}\, d\omega \tag{9-6}$$

We have a situation as in the case of *Fourier* series but space domain and frequency domain are interchanged. As a consequence, (9-6) must be related to periodic functions. In order to be able to apply the theory of *Fourier* series we use a mathematical trick: we continue the spectrum periodically, cf. Fig. 9.4.

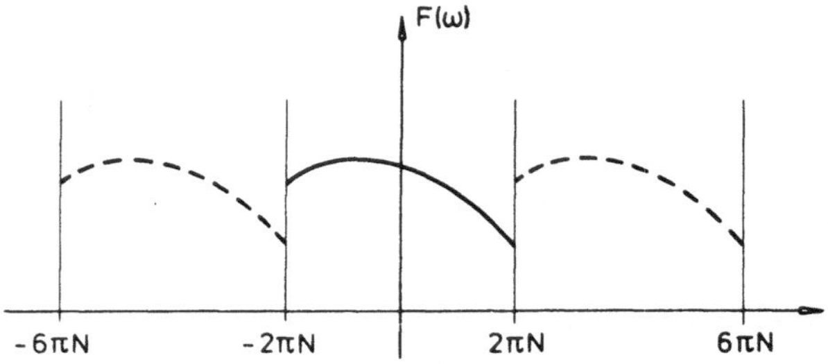

Figure 9.4. Periodic continuation of $F(\omega)$ with the period $4\pi N$.

Assuming this periodic continuation, then $F(\omega)$ can be expanded into a *Fourier* series with the period $\lambda = 4\pi N$. This results in

$$F(\omega) = \sum_{n=-\infty}^{n=+\infty} c_n \, e^{i\frac{2\pi}{\lambda}n\omega} \tag{9-7}$$

(This is an already well-known formula, cf. eq. (2-7) where we had x instead of ω). Because of $\lambda = 4\pi N$, (9-7) reduces to

$$F(\omega) = \sum_{n=-\infty}^{n=+\infty} c_n \, e^{in\omega/2N} \tag{9-8}$$

where

$$c_n = \frac{1}{4\pi N} \int_{-2\pi N}^{+2\pi N} F(\omega) \, e^{-in\omega/2N} \, d\omega \tag{9-9}$$

are the appropriate coefficients. Taking over (9-6) unchanged

$$f(x) = \frac{1}{2\pi} \int_{-2\pi N}^{+2\pi N} F(\omega) \, e^{i\omega x} \, d\omega \tag{9-10}$$

we insert (9-8) into this equation:

$$f(x) = \frac{1}{2\pi} \int_{-2\pi N}^{+2\pi N} \sum_{n=-\infty}^{\infty} c_n \, e^{in\omega/2N} \, e^{i\omega x} \, d\omega \tag{9-11}$$

and interchange integral and sum

$$f(x) = \frac{1}{2\pi} \sum_{n=-\infty}^{\infty} c_n \int_{-2\pi N}^{+2\pi N} e^{i\omega(x + n/2N)} \, d\omega \tag{9-12}$$

Note that interchange of integral and sum is allowed in this case although we have an infinite sum! We take out the integral

$$I = \int_{-2\pi N}^{+2\pi N} e^{i\omega(x + n/2N)} \, d\omega \tag{9-13}$$

whose solution is

$$I = \frac{1}{i(x + n/2N)} e^{i\omega(x + n/2N)} \Big|_{-2\pi N}^{+2\pi N} \tag{9-14}$$

The evaluation yields

$$I = \frac{4N}{2Nx + n} \sin\pi(2Nx + n) \tag{9-15}$$

where we have used the relation

$$\frac{1}{2i} (e^{i\alpha} - e^{-i\alpha}) = \sin\alpha \tag{9-16}$$

Now (9-15) is substituted into (9-12) with the result

$$f(x) = 2N \sum_{n=-\infty}^{\infty} c_n \frac{\sin\pi(2Nx + n)}{\pi(2Nx + n)} \tag{9-17}$$

Further we need the coefficients c_n. We consider the sampling points

$$x_k = \frac{k}{2N}, \quad k = 0, \pm 1, \pm 2, \ldots \tag{9-18}$$

and

$$\Delta x = x_k - x_{k-1} = \frac{1}{2N} \tag{9-19}$$

Note that we have written $\pm 1, \pm 2, \ldots$ in (9-18) which is now allowed since we have continued the function symmetrically. Using (9-18) we get from (9-6):

$$f(x_{-k}) = \frac{1}{2\pi} \int_{-2\pi N}^{+2\pi N} F(\omega) \, e^{-i\omega k/2N} \, d\omega \tag{9-20}$$

and from (9-9) follows

$$c_k = \frac{1}{4\pi N} \int_{-2\pi N}^{+2\pi N} F(\omega) \, e^{-i\omega k/2N} \, d\omega \tag{9-21}$$

We see that the two integrals of (9-20) and of (9-21) are the same with different factors only. From these two equations we therefore obtain

$$\boxed{c_k = \frac{1}{2N} f(x_{-k}) = \frac{1}{2N} f(x_k)} \tag{9-22}$$

where the second equality sign stems from the symmetry $x_{-k} = x_k$. Finally, substituting (9-22) into (9-17) gives

$$f(x) = \sum_{n=-\infty}^{\infty} \frac{\sin\pi(2Nx + n)}{\pi(2Nx + n)} f(x_n) \qquad (9\text{-}23)$$

This is the interpolation which reproduces the sampling points exactly in case of band-limited functions ("*sinx/x* interpolation", cf. Fig. 9.5).

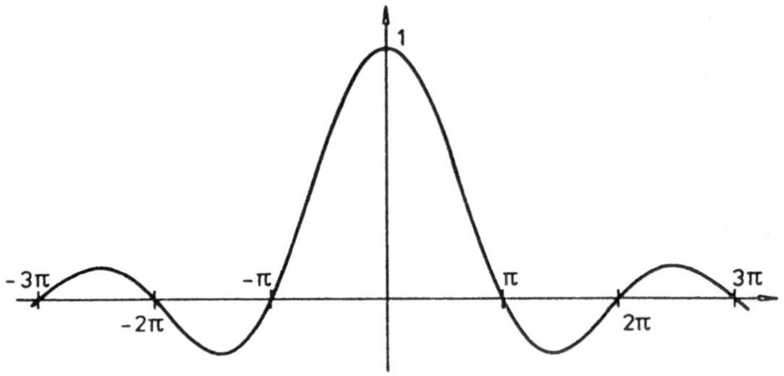

Figure 9.5. Function *sinx/x*.

The reproduction at the sampling points is exact since $f(x_n)$ at the sampling point considered is multiplied by *1* and all others are multiplied by *0*.

What happens if the *sinx/x* interpolation is applied to non-band-limited functions? The interpolation is applicable and yields a smooth function, cf. Fig. 9.6.

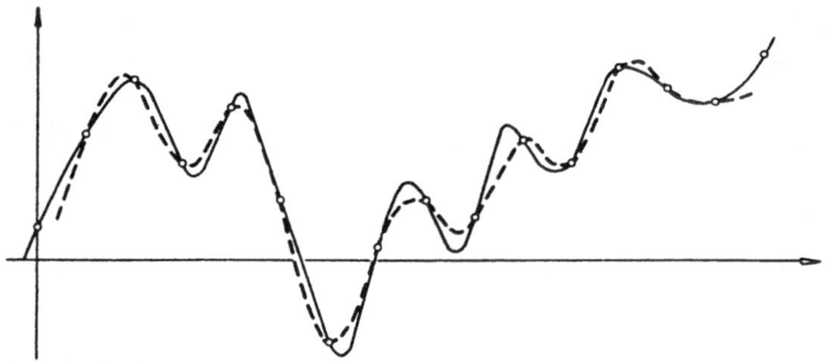

Figure 9.6. Application of the *sinx/x* interpolation to non-band-limited functions.

The original function obviously has frequencies $n > N$ which cannot be reproduced by the *sinx/x* interpolation if the distance between the sampling points is only *1/2N*.

Now we consider a function *f(x)* and its spectrum be *S(ω)*, symbolically

$$f(x) \longrightarrow S(\omega) \tag{9-24}$$

where we assume that the spectrum is $\neq 0$ for $|\omega| > 2\pi N$. Then the function $\bar{f}(x)$, which shall be the function interpolated by *sinx/x*, has a spectrum $\bar{S}(\omega)$, symbolically

$$\bar{f}(x) \longrightarrow \bar{S}(\omega) \tag{9-25}$$

where the spectrum is obtained by

$$\bar{S}(\omega) = \sum_{k=-\infty}^{\infty} S(\omega - k \cdot 4\pi N) \tag{9-26}$$

This is the so-called spectrum folding, cf. Fig. 9.7 and Fig. 9.8.

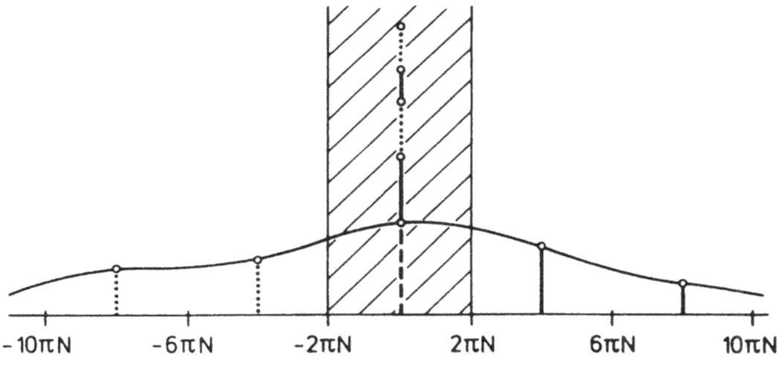

Figure 9.7. Example for spectrum folding.

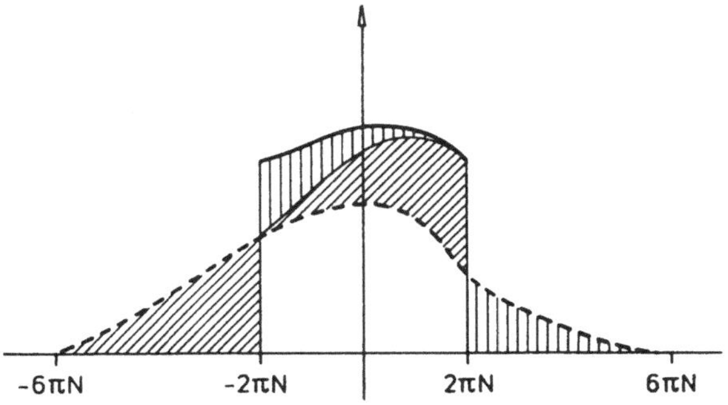

Figure 9.8. Example for spectrum folding.

Its effect in the space domain is called "aliasing"; it represents the fact that $\bar{f}(x)$ is slightly different from $f(x)$, although both functions coincide at the sampling points.

As an example think of the expansion into spherical harmonics. The series is cut off at a certain limit. The spectrum is affected by this cut off, and we again have an aliasing effect. The result is a function which is artificially band-limited.

We may also consider an analogous situation in the space domain, cf. Fig. 9.9.

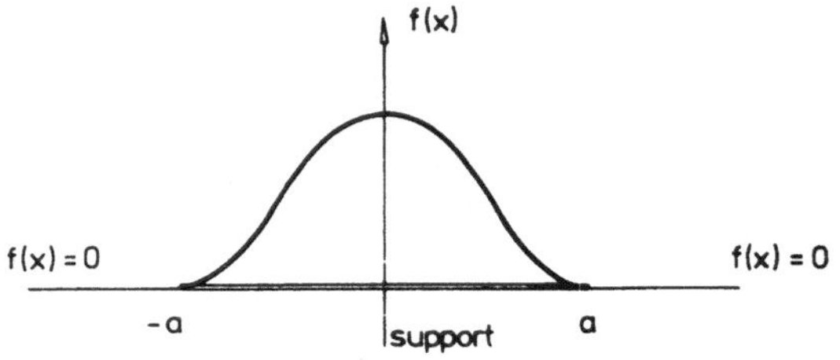

Figure 9.9. A finite function.

We have $f(x) = 0$ outside the interval ("support") $[-a,+a]$. These functions are denoted as functions with a finite support (or finite functions).

10. DISCRETE *FOURIER* TRANSFORM, FAST *FOURIER* TRANSFORM

The following abbreviations are usual: FFT for the Fast Fourier Transform and DFT for the Discrete Fourier Transform. The goal of this section is a numerical algorithm for the *Fourier* transform. The integrals must be discretized in order to get sums instead of integrals. The sums are evaluated by a computer.

Motivation

1) Band-limited functions have the following property: the spectrum is continuous but of finite support, i.e., the spectrum does not extend to infinity but is band-limited. The function is given at infinitely many discrete sampling points. (The function itself, of course, is not of finite support.)

2) Function with a finite support: the function is continuous and finite,

i.e., the function is zero outside a certain interval. The spectrum has the property to be given by infinitely many discrete sampling points.

What happens if we have a band-limited <u>and</u> finite function? Those functions, cf. Fig. 10.1, must somehow combine the two properties. There must exist finitely many discrete equidistant sampling points which are sufficient to determine the function <u>and</u> the spectrum. This is precisely the case we now will deal with.

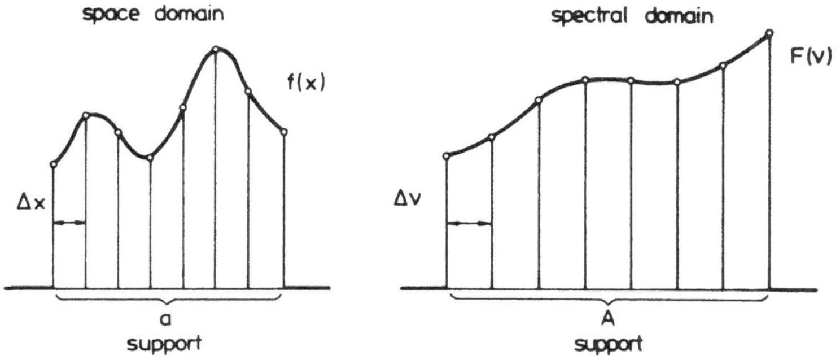

Figure 10.1. Function with a finite support and the corresponding band-limited spectrum.

For the space domain we have the function $f(x)$ with the support a. Outside the support the function is zero. For the spectrum $F(\nu)$ we have the support A and outside the support the spectrum is zero. Now there must exist an interpolation procedure which reproduces the function $f(x)$ exactly.

We assume the same number of sampling points for the spectrum as for the function. These discrete sampling points will be of interest only, the continuous function is now of secondary relevance. For the frequency we take

$$\nu = \frac{\omega}{2\pi} \tag{10-1}$$

instead of ω. Now we perform a heuristic transition from the *Fourier* integral to the DFT (Discrete *Fourier* Transform). The *Fourier* integral has an infinite support, whereas the DFT has a finite one. Our task is to discretize the formulae (3-19)

$$\begin{aligned} F(\nu) &= \int_{-\infty}^{+\infty} f(x)\, e^{-2\pi i \nu x}\, dx \\ f(x) &= \int_{-\infty}^{+\infty} F(\nu)\, e^{2\pi i \nu x}\, d\nu \end{aligned} \tag{10-2}$$

Basically, we want to get a certain number n of sampling points for x (that is the space domain) as well as for ν (that is the spectral domain). The distances between the sampling points are, cf. Fig. 10.1, on the one hand Δx and on the other hand $\Delta\nu$ as given by

$$\Delta x = \frac{a}{n} \qquad \Delta\nu = \frac{A}{n} \tag{10-3}$$

The sampling points of the two domains are

$$\begin{array}{lll} x = j\,\Delta x & \qquad & \nu = p\,\Delta\nu \\ \text{where } j = 1, 2, \ldots, n & & \text{where } p = 1, 2, \ldots, n \\ \text{or } j = 0, 1, 2, \ldots, n-1 & & \text{or } p = 0, 1, 2, \ldots, n-1 \end{array} \tag{10-4}$$

We additionally need a relation between Δx and $\Delta\nu$. In principle, this relation may be chosen arbitrarily. We choose

$$\boxed{\Delta\nu = \frac{1}{n\,\Delta x}} \tag{10-5}$$

where we get from

$$n\,\Delta x\,\Delta\nu = 1 \tag{10-6}$$

From (10-4) we can verify

$$\nu\,x = j\,p\,\Delta x\,\Delta\nu \tag{10-7}$$

and by

$$\Delta x\,\Delta\nu = \frac{1}{n} \tag{10-8}$$

from (10-6) we therefore obtain

$$\nu\,x = \frac{j\,p}{n} \tag{10-9}$$

By the choice (10-5) the formulae become very simple since the product $\nu\,x$ does not contain either Δx or $\Delta\nu$. This simplifies the following formulae essentially. The discretization leads to

$$\boxed{\begin{aligned} F(p\Delta\nu) &= \sum_{j=0}^{n-1} f(j\Delta x)\, e^{-2\pi i jp/n} \\ f(j\Delta x) &= \frac{1}{n}\sum_{p=0}^{n-1} F(p\Delta\nu)\, e^{+2\pi i jp/n} \end{aligned}} \tag{10-10}$$

These are the formulae for the DFT (Discrete *Fourier* Transform). The factor

1/n cannot be explained heuristically. For those who are not completely satisfied: the formulae can be deduced exactly rather simply as we shall show later. In (10-10) note that

$$i = \sqrt{-1} \qquad j = 0, 1, \ldots, n-1 \tag{10-11}$$

What must be known for an application of the DFT? We assume the support *a* of the function *f(x)* and *n* the number of the sampling points. This is sufficient, all other quantities are determined, since

$$\Delta x = \frac{a}{n} \tag{10-12}$$

and the frequency distance $\Delta\nu$ is given by

$$\Delta\nu = \frac{1}{n\ \Delta x} \tag{10-13}$$

In addition, we have

$$A = \frac{1}{\Delta x} \qquad \left(a = \frac{1}{\Delta\nu}\right) \tag{10-14}$$

The highest frequency is the *Nyquist* frequency and it is again denoted by *N*. Assuming symmetry, we have

$$\boxed{N = \frac{A}{2}} \tag{10-15}$$

or

$$N = \frac{1}{2\Delta x} \tag{10-16}$$

Now we demonstrate that the function *f(x)* has the period *a*. By (10-12) we have

$$f(j\Delta x + a) = f[(j+n)\Delta x] \tag{10-17}$$

By the second equation in (10-10) we can obtain from (10-17)

$$f(j\Delta x + a) = \frac{1}{n} \sum_{p=0}^{n-1} F(p\Delta\nu)\ e^{2\pi i(j+n)p/n} \tag{10-18}$$

whence

$$f(j\Delta x + a) = \frac{1}{n} \sum_{p=0}^{n-1} F(p\Delta\nu)\ e^{2\pi ijp/n}\ e^{2\pi ip} = f(j\Delta x) \tag{10-19}$$

proving the periodicity since $e^{2\pi ip} = 1$. To see this use the *Moivre* formula $e^{i\alpha} = \cos\alpha + i\sin\alpha$ and set $\alpha = 2\pi p$.

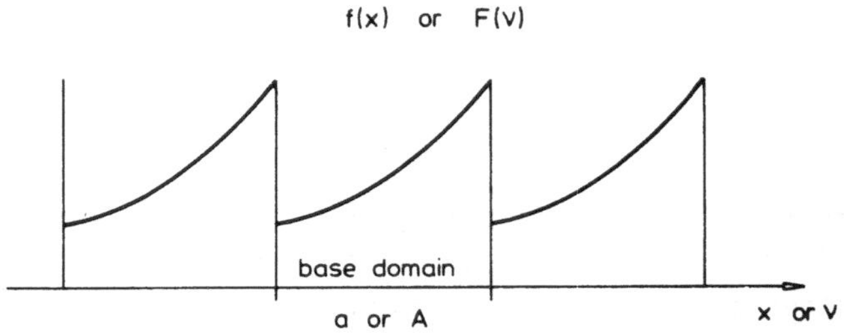

Figure 10.2. Periodic function, see formulae (10-17) to (10-19).

The values of the base domain, cf. Fig. 10.2, are not affected by such a periodic continuation.

Now we prove formulae (10-10) of the DFT (Discrete *Fourier* Transform) by insertion. Our starting point is the following argument: if the pair of the two formulae (10-10) is right then by inserting one into the other an identity must be obtained. We take from (10-10) the second formula

$$f(j\Delta x) = \frac{1}{n}\sum_{p=0}^{n-1} F(p\Delta\nu)\, e^{2\pi ijp/n} \tag{10-20}$$

and insert for $F(p\Delta\nu)$ the first formula of (10-10) with k instead of j:

$$f(j\Delta x) = \frac{1}{n}\sum_{p=0}^{n-1}\sum_{k=0}^{n-1} f(k\Delta x)\, e^{-2\pi ikp/n}\, e^{2\pi ijp/n} \tag{10-21}$$

This can be reduced to

$$f(j\Delta x) = \frac{1}{n}\sum_{p=0}^{n-1}\sum_{k=0}^{n-1} f(k\Delta x)\, e^{2\pi i(j-k)p/n} \tag{10-22}$$

Now we consider the auxiliary formula

$$\frac{1}{n}\sum_{p=0}^{n-1} e^{2\pi imp/n} = \begin{cases} 1 & \text{for } m = 0 \\ 0 & \text{for } m \neq 0 \end{cases} \tag{10-23}$$

which is a sum orthogonality quite analogous to

$$\frac{1}{2\pi}\int_0^{2\pi} e^{imx}\, dx = \begin{cases} 1 & \text{for } m = 0 \\ 0 & \text{for } m \neq 0 \end{cases} \tag{10-24}$$

Introducing the auxiliary formula (where we take $m = j-k$) yields

$$\frac{1}{n}\sum_{p=0}^{n-1} e^{2\pi i(j-k)p/n} = \delta_{jk} = \begin{cases} 1 & \text{for } j = k \\ 0 & \text{for } j \neq k \end{cases} \tag{10-25}$$

We see that δ_{jk} is the *Kronecker* delta. The substitution of (10-25) into (10-22) leads to

$$f(j\Delta x) = \sum_{k=0}^{n-1} f(k\Delta x)\,\delta_{jk} = f(j\Delta x) \tag{10-26}$$

Since δ_{jk} are the elements of the unit matrix I, we have

$$\delta_{jk} x_k = \sum_k \delta_{jk} x_k = x_j \tag{10-27}$$

(because of $Ix = x$). Thereby the proof of the DFT formulae is shown.

Space domain and spectral domain for the DFT

We have a discrete function which is continued periodically. The geometric situation is illustrated in Fig. 10.3.

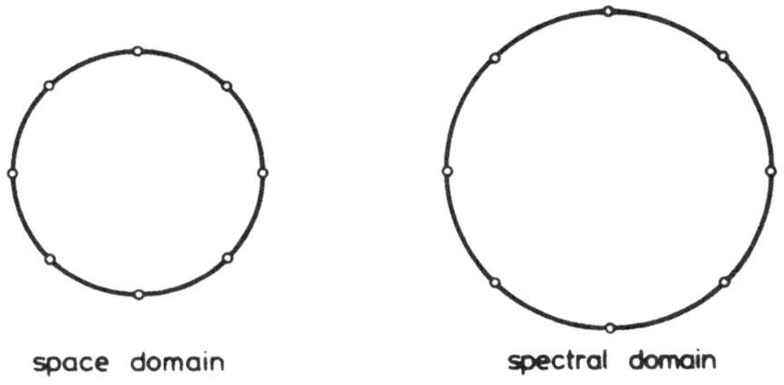

Figure 10.3. Space domain and spectral domain for the DFT.

The space domain only consists of equidistant points situated on a circle (like the hour marks on the dial of a clock) and the same holds for the spectral domain. The straight line is folded into a circle, corresponding to periodicity (think of linear time measured by a periodic clock!).

10.1. Twodimensional Discrete *Fourier* Transform

For the twodimensional DFT we shall use the plane. A discrete *Fourier* transform on the sphere is not possible, at least not in an equally simple and symmetric way, cf. *Colombo (1979)*.

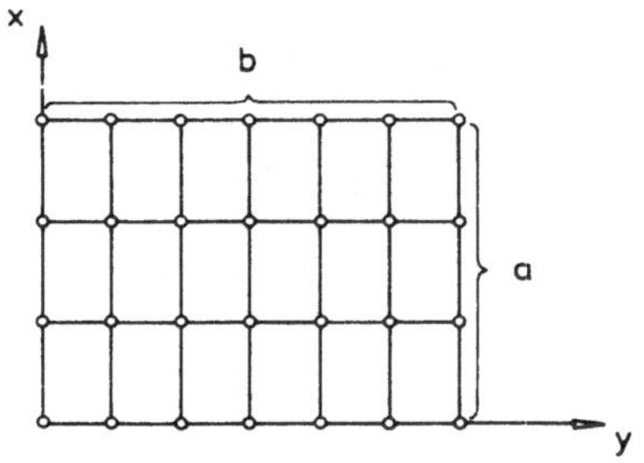

Figure 10.4. Twodimensional discrete *Fourier* transform.

We assume m equidistant sampling points in the x-direction and n equidistant sampling points in the y-direction, cf. Fig. 10.4. If n and m are given, all other necessary quantities can be computed. We get

$$\Delta x = \frac{a}{m} \qquad \Delta y = \frac{b}{n}$$
$$\Delta\mu = \frac{1}{m\ \Delta x} \qquad \Delta\nu = \frac{1}{n\ \Delta y} \tag{10-28}$$

The function $f(x,y)$ shall correspond to the spectrum $F(\mu,\nu)$. The formulae are analogous to the onedimensional case. The plane is the topologic product of two straight lines. We therefore have

$$F(p\Delta\mu,q\Delta\nu) = \sum_{j=0}^{m-1}\sum_{k=0}^{n-1} f(j\Delta x,k\Delta y)\ e^{-2\pi i(jp/m\ +\ kq/n)} \tag{10-29}$$

and

$$f(j\Delta x,k\Delta y) = \frac{1}{mn}\sum_{p=0}^{m-1}\sum_{q=0}^{n-1} F(p\Delta\mu,q\Delta\nu)\ e^{2\pi i(jp/m\ +\ kq/n)} \tag{10-30}$$

The *Nyquist* frequencies read

$$N_x = \frac{1}{2}\ m\ \Delta\mu = \frac{1}{2\Delta x}$$
$$N_y = \frac{1}{2}\ n\ \Delta\nu = \frac{1}{2\Delta y} \tag{10-31}$$

The *Nyquist* frequencies are the highest one which can be obtained from the data grid. The larger m and n are chosen, the better is the result as far as theory is concerned, since in this way the *Nyquist* frequencies become larger and the band of usable data will be wider.

Practical problems.

1) Choice of a, b. The rectangle must be taken so large that the outside region is of no interest for the area considered.

2) Choice of m, n (or Δx, Δy). The denser the grid is, the higher are the achievable frequencies (but also the higher the computation effort).

Space domain and the spectral domain for the twodimensional DFT

For the onedimensional DFT we had a circle. For the twodimensional DFT we form the topologic product of two circles and we get a torus.

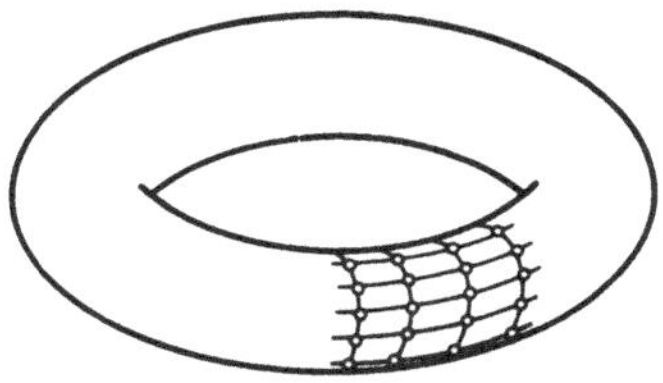

Figure 10.5. Space domain as well as spectral domain for the twodimensional DFT.

On the torus we have a discrete grid, cf. Fig. 10.5. The points indicated are the data points. Both the space domain and the frequency domain are regular grids of points on a torus.

For the numeric computation there exists a very economic algorithm for the *Fourier* transform, this is the Fast *Fourier* Transform (FFT). It is essentially based on the fact that the quotient of two exponential functions is again an exponential function, e.g., $e^a/e^b = e^{a-b}$. We shall purposely stop here, referring the reader to the extensive literature.

Instead, we shall try to give a general overview, emphasizing the basic common situation and perhaps clarifying differences between the various cases.

10.2. Overview of the transformations into the spectral domain

A transformation always has the form

$$\begin{aligned} f(x) &= \sum_{\omega} F(\omega)\, \varphi(\omega,x) \\ & \qquad\qquad \text{duality} \\ F(\omega) &= \sum_{x} f(x)\, \varphi(\omega,x) \end{aligned} \qquad (10\text{-}32)$$

if discrete variables x, ω are considered. Now

instead of $\sum_{\omega,x}$ we can also have $\int_{\omega,x}$

if one of the variables is continuous. Moreover, constant factors (2π, ...) may appear before the sum (or before the integral). The constant factors may be disregarded. The duality mentioned means a symmetric pair of formulae.

Notations

x ... space variable which covers the usual space or a part of it (straight line, plane, etc.). x is given; it is not necessarily onedimensional, it can e.g. be a point on the sphere; it may be continuous or discrete.

ω ... corresponding spectral variable, onedimensional or of higher dimension, discrete or continuous.

$f(x)$ ... given function (e.g. gravity anomalies of the whole earth or along a profile).

$F(\omega)$ ... spectrum (*Fourier* transform).

If the spectral variable ω is discrete, one speaks of a discrete spectrum, e.g.:

$$\omega = n = 0, 1, 2, \ldots, \infty \qquad (10\text{-}33)$$

A discrete spectral variable can also be twodimensional:

$$\omega = (m,n) \qquad (10\text{-}34)$$

It is important that the dimensions of the space variable and of the spectral variable are the same!

The spectrum can also be continuous:

$$\begin{aligned} \omega &= u && \text{one-dimensional} \\ \omega &= (u,v) && \text{two-dimensional} \end{aligned} \qquad (10\text{-}35)$$

$\varphi(\omega,x)$ ... performs the transition from the space domain into the spectral domain. This function is called base function.

Example (not serious). Space domain: daily life. Spectral domain: sleep. Base functions: sleeping-tablets.

The classic base functions are

$$\cos nx, \quad \sin nx, \quad e^{\pm inx} \qquad (10\text{-}36)$$

all other base functions are deduced from these base functions, e.g.

$$\cos(mx + ny) \qquad (10\text{-}37)$$

or

$$e^{\pm i(mx + ny)} \qquad (10\text{-}38)$$

Also the spherical harmonics

$$R_{nm}(\vartheta,\lambda), \quad S_{nm}(\vartheta,\lambda) \qquad (10\text{-}39)$$

as base functions on the sphere are generalizations of the classic base functions. The fact that the spherical harmonics are indeed base functions $\varphi(\omega,x)$ can be seen from the following scheme:

$$\underbrace{R_{\underbrace{nm}_{\omega}}(\underbrace{\vartheta,\lambda}_{x})}_{\varphi(\omega,x)} \qquad (10\text{-}40)$$

Now we give a comparative overview of the most important cases.

space domain	base functions	spectrum	spectral domain
(1) circle (periodic function on straight line)	$\cos nx$, $\sin nx$ or $e^{\pm inx}$	discrete a_n, b_n real c_n complex	 $n = 0,1,2,\ldots$ $n = \ldots,-1,0,1,\ldots$
(2) torus (periodic function in plane)	$\cos(mx+ny)$, $\sin(mx+ny)$ $e^{\pm i(mx+ny)}$	discrete a_{mn}, b_{mn} c_{mn}	$m,n = 0,1,2,\ldots$ (point grid) $m,n=\ldots,-1,0,1,\ldots$
(3) sphere	$R_{nm}(\vartheta,\lambda), S_{nm}(\vartheta,\lambda)$	discrete a_{nm}, b_{nm}	$n = 0,1,2,\ldots,\infty$ $m = 0,1,2,\ldots,n$
(4) straight line (circle with $R \to \infty$). Fourier integral	$e^{\pm iux}$ ($\cos ux$, $\sin ux$)	continuous $c(u)$	$-\infty < u < +\infty$ (straight line, like space domain)
(5) plane Fourier integral	$e^{\pm i(ux+vy)}$	continuous $c(u,v)$	$-\infty < u < \infty$ $-\infty < v < \infty$ (whole plane, like space domain)
(6) infinite integer point grid in plane $x,y=\ldots,-1,0,1,\ldots$	$e^{\pm i(ux+vy)}$	continuous $c(u,v)$	torus (follows from duality)
(7) periodic point grid (FFT, point grid on torus)	$e^{\pm i(mx+ny)}$	c_{mn}	point grid on torus (like space domain)
(8) one-dimensional point grid on circle (one-dimensional FFT)	$e^{\pm inx}$	c_n	spectral domain is again like space domain

The examples (4), (5), (7), and (8) are completely symmetric cases.

FFT is the Fast *Fourier* Transform. This is a discrete point transformation. If the whole infinite plane shall be treated, we get an integral for $f(x)$ and $F(\omega)$. By a discretization we get a sum. The geometric interpretation is a point grid.

If the space domain is an infinite straight line, case (4), then the spectral domain is also an infinite straight line (analogous for the plane).

Considerations with respect to the spectra of a torus, i.e., case (2), cf. Fig. 10.6 and Fig. 10.7. Compare the difference between the two spectra! The different spectra are generated by different base functions.

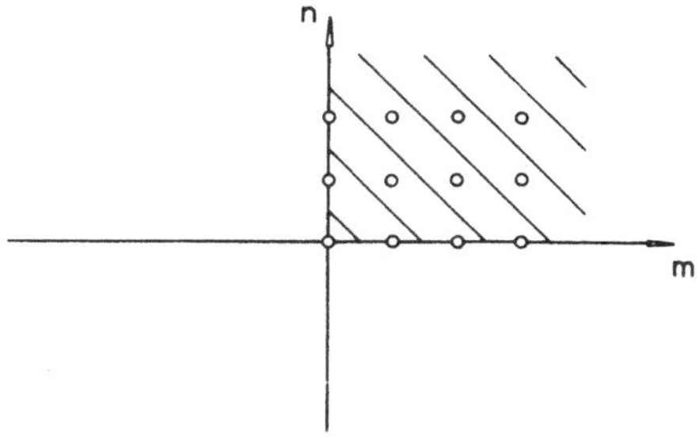

Figure 10.6. Torus. Spectral domain $m,n = 0,1,2,\ldots$

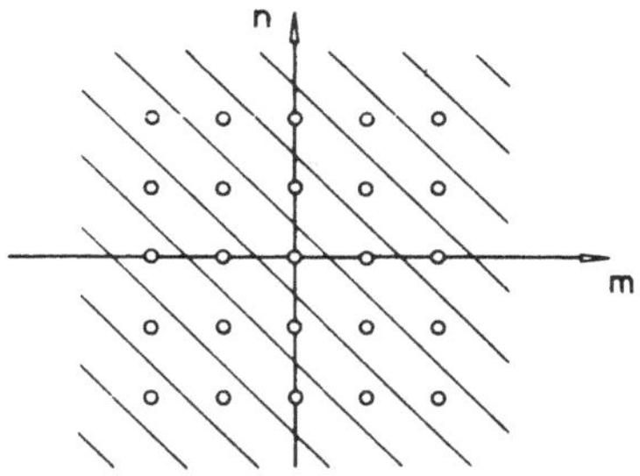

Figure 10.7. Torus. Spectral domain $m,n = \ldots,-2,-1,0,1,2,\ldots$

Spectrum of the sphere, cf. Fig. 10.8.

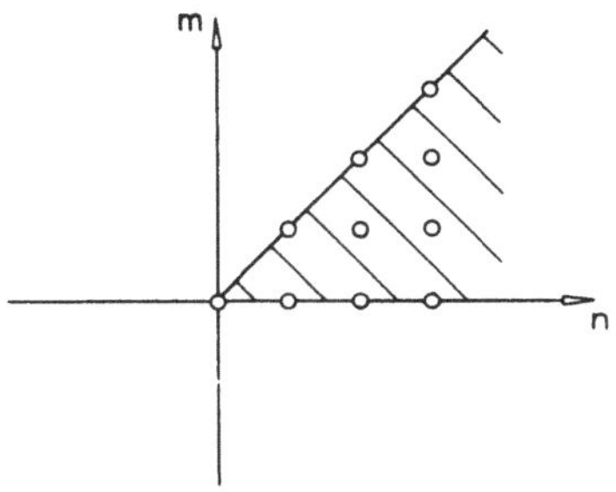

Figure 10.8. Spectrum of the sphere $n = 0,1,\ldots,\infty;\ m = 0,1,\ldots,n$

The variables x and ω can be interchanged in the pair of formulae for $f(x)$ and $F(\omega)$ (duality). In this way, the examples (2) and (6) are dual to each other, and any of the "symmetric cases" (4), (5), (7), or (8) is "self-dual". Take example (6), in which the space domain is an infinite regular point grid in the plane: the spectral domain is then a torus.

The onedimensional analogue is the following. Given are equidistant measurement values on the straight line from $-\infty$ to $+\infty$. This means that we have $f(x)$ where $x = \ldots,-2,-1,0,1,2,\ldots$ The spectrum is periodic, the spectral domain is a circle (this is the dual case to example (1)). Therefore we have a *Fourier* series <u>for the spectrum</u>:

$$F(u) = \sum_{x=-\infty}^{x=+\infty} f(x)\, e^{-iux} \tag{10-41}$$

and

$$f(x) = \frac{1}{2\pi} \int_{u=0}^{2\pi} F(u)\, e^{iux}\, du \tag{10-42}$$

are the *Fourier* coefficients (discrete values at the discrete sampling points x). Note that now $f(x)$ takes the place of c_n. A function at discrete sampling points thus has a spectrum from 0 to 2π, as we know already from sec. 9, cf. Fig. 10.9.

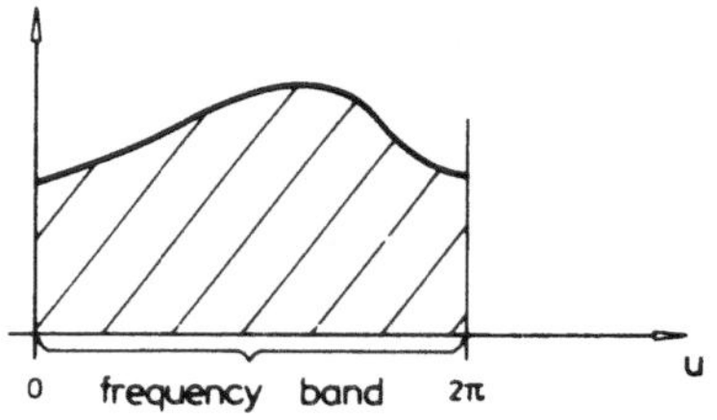

Figure 10.9. A function at discrete sampling points has a band-limited spectrum.

The function whose spectrum is shown in Fig. 10.9 is "band-limited", where band limited says that the frequency band is limited.

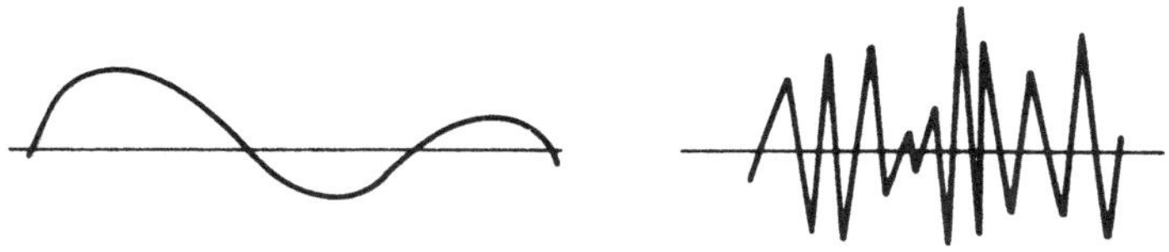

smooth, small bandwidth	higher frequent (more sampling points are necessary) large bandwidth
simple musical instrument	symphonic orchestra

Figure 10.10. Example of different bandwidths.

If the bandwidth is small, then the number of sampling points can be small, cf. Fig. 10.10. The necessary density of the sampling points depends on the bandwidth.

<u>Remark:</u> Additional information related to sec. 10 can be found in the appendix, e.g. an algorithm for the DFT.

APPENDIX

The objective of this appendix is to assist readers who are interested in additional examples, in another development of some formulae, in geodetic applications, etc. The following sections are not directed to very versatile readers who have already understood the whole theory in the preceding ten sections. Therefore one should not expect to be confronted with new theory.

The subsequent sections are indicated by the following notation. The capital letter *A* (for Appendix) is followed by a digit referring to one of the ten previous sections. Additional digits are for the segmentation of the appendix section. Since we have written supplements for the sections *2, 3, 4, 5,* and *10,* gaps occur in the enumeration of the appendix.

A.2.1. NUMERICAL EXAMPLES FOR ONEDIMENSIONAL *FOURIER* SERIES

A.2.1.1. Example 1

Consider the periodic function

$$f(x) = \begin{cases} 2x & \text{for} \quad 0 \leq x < 3 \\ 0 & \text{for} \quad -3 \leq x < 0 \end{cases} \qquad f(x \pm 6) = f(x) \qquad \text{(A2-1)}$$

and find the *Fourier* series for this function. The graphic representation of this function is illustrated in Fig. A.2.1.

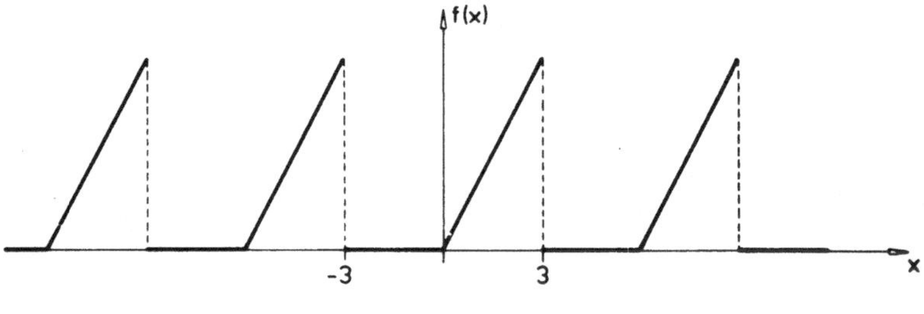

Figure A.2.1. Graphic representation of the periodic function of example 1.

In (2-3) and (2-4) we have given the coefficients for a periodic function with the period $\lambda = 2\pi$. For this example we need slightly different formulae since we have a period of $\lambda = 6$. For an arbitrary period $\lambda = 2L$ we have for the coefficients

$$a_n = \frac{1}{L}\int_c^{c+2L} f(x)\cos\frac{n\pi x}{L}\,dx$$

$$n = 0,\ 1,\ 2,\ \ldots \qquad \text{(A2-2)}$$

$$b_n = \frac{1}{L}\int_c^{c+2L} f(x)\sin\frac{n\pi x}{L}\,dx$$

where c is an arbitrary real number. These formulae are the real counterpart of (2-7). For the function of (A2-1) we choose

$$c = -3 \qquad \text{(A2-3)}$$

and we are able to compute the first coefficient by

$$a_0 = \frac{1}{3}\int_{-3}^{+3} f(x)\,dx \qquad \text{(A2-4)}$$

where we have taken into account $\lambda = 2L = 6$ and therefore $L = 3$. Because of (A2-1) we must split up the integral into two parts

$$a_0 = \frac{1}{3}\left\{\int_{-3}^{0} 0\,dx + \int_0^{+3} 2x\,dx\right\} \qquad \text{(A2-5)}$$

and we obtain for a_0 the result

$$a_0 = 3 \qquad \text{(A2-6)}$$

In order to get the other coefficients, we start with

$$a_n = \frac{1}{3}\int_0^3 2x\cos\frac{n\pi x}{3}\,dx \qquad \text{(A2-7)}$$

where we have already omitted the integral from -3 to 0 since its contribution is again zero because $f(x) = 0$ holds for this interval. We introduce an auxiliary variable

$$u = \frac{n\pi}{3} \qquad \text{(A2-8)}$$

and we can rewrite the coefficients by

$$a_n = \frac{2}{3}\int_0^3 x\cos ux\,dx \qquad \text{(A2-9)}$$

The integral can be evaluated directly using the formula

$$\int x\cos ux\,dx = \frac{\cos ux}{u^2} + \frac{x\sin ux}{u} \qquad \text{(A2-10)}$$

We obtain

$$a_n = \frac{2}{3}\left\{\frac{\cos ux}{u^2} + \frac{x \sin ux}{u}\right\}\Bigg|_{x=0}^{x=3} \tag{A2-11}$$

or

$$a_n = \frac{2}{3}\left\{\frac{\cos 3u}{u^2} + \frac{3 \sin 3u}{u} - \frac{1}{u^2} - 0\right\} \tag{A2-12}$$

or in equivalent form

$$a_n = \frac{2}{3u^2}\left[\cos 3u + 3u \sin 3u - 1\right] \tag{A2-13}$$

By resubstituting for u using (A2-8), we see that for every $n = 1, 2, \ldots$ the second term between brackets vanishes. The result reads

$$a_n = \frac{6}{n^2\pi^2}\left[\cos n\pi - 1\right] \tag{A2-14}$$

In the same way we proceed for b_k. We can be brief. From (A2-2) we have

$$b_n = \frac{1}{3}\int_{-3}^{+3} f(x) \sin\frac{n\pi x}{3}\, dx \tag{A2-15}$$

For the specific function in (A2-1) we get

$$b_n = \frac{2}{3}\int_{0}^{3} x \sin\frac{n\pi x}{3}\, dx \tag{A2-16}$$

and, using (A2-8), we obtain

$$b_n = \frac{2}{3}\int_{0}^{3} x \sin ux\, dx \tag{A2-17}$$

Now, for the integral we use the formula

$$\int \sin ux\, dx = \frac{\sin ux}{u^2} - \frac{x \cos ux}{u} \tag{A2-18}$$

which leads to

$$b_n = \frac{2}{3}\left\{\frac{\sin ux}{u^2} - \frac{x \cos ux}{u}\right\}\Bigg|_{x=0}^{x=3} \tag{A2-19}$$

The evaluation of the integration limits yields

$$b_n = \frac{2}{3}\left\{\frac{\sin 3u}{u^2} - \frac{3 \cos 3u}{u} - 0 + 0\right\} \tag{A2-20}$$

By substituting for u eq. (A2-8), we see that the first term between brackets vanishes for $n = 1, 2, \ldots$ and we therefore obtain

$$b_n = -\frac{6}{n\pi}\cos n\pi \qquad \text{(A2-21)}$$

Now, the *Fourier* series for a function $f(x)$ with period $\lambda = 2L$ is given by

$$f(x) = \frac{a_0}{2} + \sum_{n=1}^{\infty}\left\{a_n\cos\frac{n\pi x}{L} + b_n\sin\frac{n\pi x}{L}\right\} \qquad \text{(A2-22)}$$

This formula is the generalization of (2-3) and (2-4) to an arbitrary period λ (whereas in (2-3) and (2-4) the period must be $\lambda = 2\pi$).

The remaining work to do is very simple. We insert (A2-6), (A2-14) and (A2-21) into (A2-22):

$$\boxed{f(x) = \frac{3}{2} + \sum_{n=1}^{\infty}\left\{\frac{6(\cos n\pi - 1)}{n^2\pi^2}\cos\frac{n\pi x}{3} - \frac{6\cos n\pi}{n\pi}\sin\frac{n\pi x}{3}\right\}} \qquad \text{(A2-23)}$$

This is the *Fourier* series representation for the function $f(x)$ in (A2-1).

As we have seen, the expansion of a very simple function into a *Fourier* series is already a rather tedious work. However, sometimes it is possible to profit by specific properties of the function $f(x)$. Let us assume we have an <u>even function</u> $f(x)$, i.e., $f(x) = f(-x)$, then we can easily find

$$a_n = \frac{2}{L}\int_0^L f(x)\cos\frac{n\pi x}{L}\,dx \qquad n = 0, 1, 2, \ldots \qquad \text{(A2-24)}$$
$$b_n = 0$$

and the corresponding *Fourier* series reads

$$f(x) = \frac{a_0}{2} + \sum_{n=0}^{\infty} a_n\cos\frac{n\pi x}{L} \qquad \text{(A2-25)}$$

This kind of series is also called a *cosine* series since only *cos*-terms appear.

If we consider an <u>odd function</u> $f(x)$, i.e., $f(-x) = -f(x)$, then we get

$$a_n = 0$$
$$b_n = \frac{2}{L}\int_0^L f(x)\sin\frac{n\pi x}{L}\,dx \qquad \text{(A2-26)}$$

and the corresponding *Fourier* series reads

$$f(x) = \sum_{n=1}^{\infty} b_n \sin\frac{n\pi x}{L} \tag{A2-27}$$

which is also called a *sine* series.

Naturally, it is much more convenient to expand an even or an odd function into a series since only one type of coefficients must be calculated whereas the other are zero. Sometimes it is possible to perform a transformation of a periodic function in such a way that the function becomes an even or an odd function. This can e.g. sometimes simply be achieved by shifting the coordinate system which does not change the behaviour of the function or by artificially making a nonperiodic function periodic (periodic continuation). We must only endeavour to integrate over the full period. In sec. A.2.1.2 we demonstrate a corresponding example.

A.2.1.2. Example 2

Consider the periodic function

$$f(x) = \begin{cases} 1 & \text{for} \quad 0 < x < \frac{\pi}{2} \\ -1 & \text{for} \quad \frac{\pi}{2} < x < \frac{3\pi}{2} \\ 1 & \text{for} \quad \frac{3\pi}{2} < x < 2\pi \end{cases} \tag{A2-28}$$

The graphic representation of the function is given in Fig. A.2.2 where the function (A2-28) is the heavy line and the periodic continuation is the remaining part of the figure.

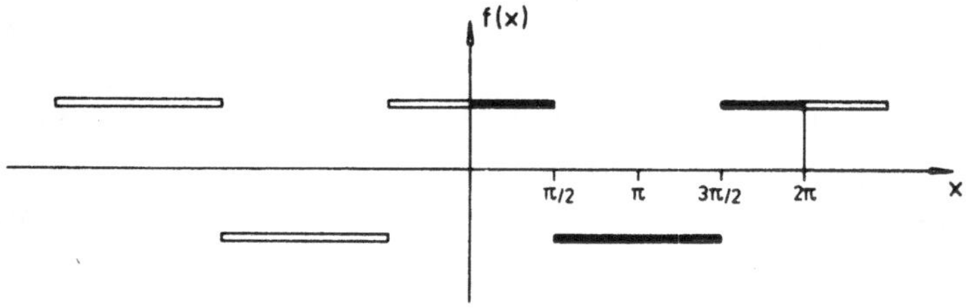

Figure A.2.2. Graphic representation of the function of example 2 and periodic continuation.

We immediately see from Fig. A.2.2 that (A2-28) is an even function, i.e.,

$f(x) = f(-x)$. Therefore we get $b_k = 0$; $k = 1, 2, \ldots$ and the *Fourier* series becomes a *cosine* series. We must only calculate the coefficients

$$a_k = \frac{1}{L}\int_c^{c+2L} f(x) \cos\frac{k\pi x}{L}\, dx \tag{A2-29}$$

We choose $c = -\pi/2$, and for the given period $T = 2L = 2\pi$ we obtain

$$a_k = \frac{1}{\pi}\int_{-\pi/2}^{3\pi/2} f(x) \cos kx\, dx \tag{A2-30}$$

According to (A2-28) we must insert the given values for the three defined pieces of the function. Considering the function periodically continued as shown in Fig. A.2.2, we can rewrite the function for the interval 2π in the form

$$f(x) = \begin{cases} 1 & \text{for } \frac{-\pi}{2} < x < \frac{\pi}{2} \\ -1 & \text{for } \frac{\pi}{2} < x < \frac{3\pi}{2} \end{cases} \tag{A2-31}$$

We only must always endeavour to cover the full period of the periodic function. This also corresponds to (A2-29) where c can be chosen arbitrarily because the difference of the two integration limits yields the full period $T = 2L$.

Thus we are now able to compute the coefficients a_k for (A2-31) by splitting up the integral into two parts, whereas for the original representation of the function (A2-28) three parts were necessary. We get

$$a_k = \frac{1}{\pi}\left\{\int_{-\pi/2}^{+\pi/2} 1 \cos kx\, dx + \int_{\pi/2}^{3\pi/2} (-1) \cos kx\, dx\right\} \tag{A2-32}$$

We obtain

$$a_k = \frac{1}{\pi k}\left\{\sin kx \,\Big|_{-\pi/2}^{+\pi/2} - \sin kx \,\Big|_{\pi/2}^{3\pi/2}\right\} \tag{A2-33}$$

and the evaluation leads to

$$a_k = \frac{1}{\pi k}\left\{3 \sin\frac{k\pi}{2} - \sin\frac{3k\pi}{2}\right\} \qquad k = 1, 2, \ldots \tag{A2-34}$$

which finally results in

$$a_2 = a_4 = a_6 = \ldots = 0 \tag{A2-35}$$

and

$$a_1 = \frac{4}{\pi}, \quad a_3 = \frac{-4}{3\pi}, \quad a_5 = \frac{4}{5\pi}, \ldots \tag{A2-36}$$

Furthermore we need a_0 which can be calculated from (A2-32):

$$a_0 = \frac{1}{\pi}\left\{\int_{-\pi/2}^{+\pi/2} 1 \, dx + \int_{\pi/2}^{3\pi/2} (-1) \, dx\right\} \qquad \text{(A2-37)}$$

The result is

$$a_0 = 0 \qquad \text{(A2-38)}$$

and we can write the *Fourier* series for $f(x)$ in the form

$$\boxed{f(x) = \frac{4}{\pi}\left\{\cos x - \frac{1}{3}\cos 3x + \frac{1}{5}\cos 5x - \frac{1}{7}\cos 7x + \ldots\right\}} \qquad \ldots \text{(A2-39)}$$

As expected we have obtained a *cosine* series.

Now it is very interesting to state that function $f(x)$ in (A2-28), which we have represented by the *cosine* function (A2-39), can also be expanded into a *sine* series. How can this be done? As we have already seen, we need an odd function in order to get a *sine* series. We can easily find an odd representation of function $f(x)$ in (A2-28). We only must shift the coordinate system as shown in Fig. A.2.3.

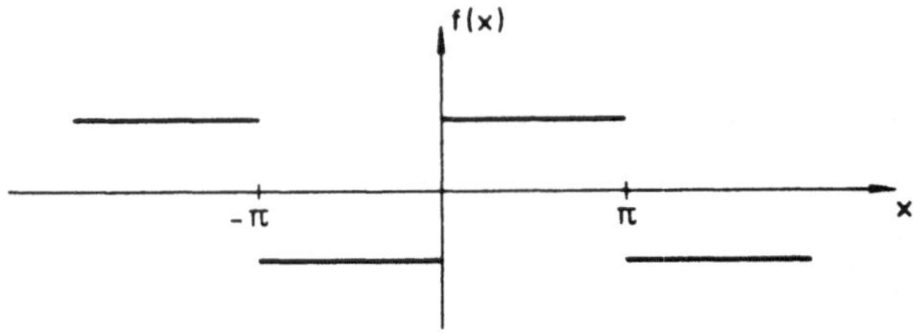

Figure A.2.3. Graphic representation of the function of example 2. Compared to Fig. A.2.3 we have only shifted the vertical axis by $-\pi/2$.

Now we have found for $f(x)$ an odd function $f(x) = -f(-x)$ without changing the function itself. For the *Fourier* representation of the function in Fig. A.2.3 we have $a_k = 0$; $k = 0, 1, 2, \ldots$ and the coefficients b_k may be calculated from

$$b_k = \frac{1}{\pi}\int_{-\pi}^{0} (-1) \sin kx \, dx + \int_{0}^{\pi} 1 \sin kx \, dx \qquad \text{(A2-40)}$$

where we immediately get

$$b_k = \frac{1}{\pi k} \left\{ \cos kx \Big|_{-\pi}^{0} - \cos kx \Big|_{0}^{\pi} \right\} \qquad \text{(A2-41)}$$

and evaluating this leads to

$$b_k = \frac{1}{\pi k} [1 - \cos(-\pi k) - \cos k\pi + 1] \qquad \text{(A2-42)}$$

which gives

$$b_k = \frac{2}{\pi k} [1 - \cos k\pi] \qquad \text{(A2-43)}$$

Therefore we obtain on the one hand

$$b_1 = \frac{4}{\pi}, \quad b_3 = \frac{4}{3\pi}, \quad b_5 = \frac{4}{5\pi}, \ \ldots \qquad \text{(A2-44)}$$

and on the other hand

$$b_2 = b_4 = b_6 = \ldots = 0 \qquad \text{(A2-45)}$$

and the *sine* series for the function $f(x)$ has the form

$$\boxed{f(x) = \frac{4}{\pi} \left\{ \sin x + \frac{1}{3} \sin 3x + \frac{1}{5} \sin 5x + \ldots \right\}} \qquad \text{(A2-46)}$$

Note that (A2-39) and (A2-46) are *Fourier* series for the <u>same</u> function $f(x)$ in (A2-28).

A.2.2. The equivalence of the real and the complex form of the *Fourier* series

In sec. 2.1 we have formulated the *Fourier* series in real form by (2-3) as

$$\begin{aligned} f(x) &= \sum_{n=0}^{\infty} (a_n \cos nx + b_n \sin nx) \\ a_n &= \frac{1}{\pi} \int_0^{2\pi} f(x) \cos nx \, dx \\ b_n &= \frac{1}{\pi} \int_0^{2\pi} f(x) \sin nx \, dx \end{aligned} \qquad \text{(A2-47)}$$

and in complex form by (2-5) as

$$\begin{aligned} f(x) &= \sum_{n=-\infty}^{+\infty} c_n e^{inx} \\ c_n &= \frac{1}{2\pi} \int_0^{2\pi} f(x) e^{-inx} \, dx \end{aligned} \qquad \text{(A2-48)}$$

We shall prove the equivalence of the two representations following *Baule (1966)*, pp. 63-64. Those who say this proof is completely simple should stop reading here and perform their own derivation. We start with the following consideration. Note that in (A2-48) the sum ranges from $-\infty$ to $+\infty$ (whereas in (A2-47) the range is 0 to $+\infty$). Therefore, apart from the case $n = 0$, for every $+n$ in the series there also exists the corresponding $-n$. We take out an arbitrary n and compute the sum of the term for $+n$ and of the term for $-n$. We obtain

$$c_{+n}e^{+inx} + c_{-n}e^{-inx} = \frac{e^{+inx}}{2\pi}\int_0^{2\pi} f(\xi)\, e^{-in\xi}\, d\xi + \frac{e^{-inx}}{2\pi}\int_0^{2\pi} f(\xi)\, e^{+in\xi}\, d\xi \tag{A2-49}$$

where we have changed the integration variable to ξ in order to avoid misunderstandings in the subsequent formulae. The two terms can now be written as

$$c_{+n}e^{+inx} + c_{-n}e^{-inx} = \frac{1}{2\pi}\int_0^{2\pi} f(\xi)\, e^{+in(x-\xi)}\, d\xi + \frac{1}{2\pi}\int_0^{2\pi} f(\xi)\, e^{-in(x-\xi)}\, d\xi \tag{A2-50}$$

or

$$c_{+n}e^{+inx} + c_{-n}e^{-inx} = \frac{1}{2\pi}\int_0^{2\pi} f(\xi)\, [e^{in(x-\xi)} + e^{-in(x-\xi)}]\, d\xi \quad \ldots \tag{A2-51}$$

Now we use the identity

$$\cos z = \frac{1}{2}\,(e^{iz} + e^{-iz}) \tag{A2-52}$$

and we obtain

$$c_{+n}e^{+inx} + c_{-n}e^{-inx} = \frac{1}{\pi}\int_0^{2\pi} f(\xi)\, \cos n(x-\xi)\, d\xi \tag{A2-53}$$

which yields

$$c_{+n}e^{+inx} + c_{-n}e^{-inx} = \frac{1}{\pi}\int_0^{2\pi} f(\xi)\, [\cos nx \cos n\xi + \sin nx \sin n\xi]\, d\xi \quad \ldots \tag{A2-54}$$

By introducing

$$a_n = \frac{1}{\pi}\int_0^{2\pi} f(\xi)\cos n\xi \, d\xi$$

$$b_n = \frac{1}{\pi}\int_0^{2\pi} f(\xi)\sin n\xi \, d\xi \qquad \text{(A2-55)}$$

we can rewrite (A2-54) as

$$c_{+n}e^{+inx} + c_{-n}e^{-inx} = a_n\cos nx + b_n\sin nx \qquad \text{(A2-56)}$$

This holds for any $n = 1, 2, \ldots$ and corresponds therefore to the sum in (A2-47). The only thing left to treat is the case $n = 0$. Inserting $n = 0$ into the second formula of (A2-48) yields

$$c_0 = \frac{1}{2\pi}\int_0^{2\pi} f(x)\, dx \qquad \text{(A2-57)}$$

Comparing c_0 with eq. (2-4) shows that $c_0 = a_0$ which completes the proof for the equivalence of the real and complex form of the *Fourier* series.

A.3.1. *FOURIER* INTEGRAL AND *FOURIER* TRANSFORM

By

$$F(\nu) = \int_{-\infty}^{+\infty} f(x)\, e^{-2\pi i \nu x}\, dx$$
$$f(x) = \int_{-\infty}^{+\infty} F(\nu)\, e^{+2\pi i \nu x}\, d\nu \qquad (A3\text{-}1)$$

we have found in (3-19) one of the possible representations of the *Fourier* integral where *F(ν)* is denoted as *Fourier* transform. Sometimes the two formulae of (A3-1) are combined to only one by inserting the *Fourier* transform *F(ν)* into the formula for the function *f(x)*:

$$f(x) = \int_{\nu=-\infty}^{+\infty} \int_{\xi=-\infty}^{+\infty} f(\xi)\, e^{2\pi i \nu (x-\xi)}\, d\xi\, d\nu \qquad (A3\text{-}2)$$

In order to avoid confusions, we have changed the integration variable x of the *Fourier* transform to ξ. Note that the complex form (A3-2) can also be replaced by a real form. To see this, we use the *Moivre* identity

$$e^{iz} = \cos z + i \sin z \qquad (A3\text{-}3)$$

and we get

$$f(x) = \int_{\nu=-\infty}^{+\infty} \int_{\xi=-\infty}^{+\infty} f(\xi) \cos 2\pi\nu(x-\xi)\, d\xi\, d\nu \; +$$
$$+ \; i \int_{\nu=-\infty}^{+\infty} \int_{\xi=-\infty}^{+\infty} f(\xi) \sin 2\pi\nu(x-\xi)\, d\xi\, d\nu \qquad (A3\text{-}4)$$

Considering in the second term the integral over ν, we can easily see that we have an odd function with respect to ν for which the integral from $\nu = -\infty$ to $\nu = +\infty$ vanishes, cf. *Baule (1966)*. Therefore (A3-4) reduces to

$$f(x) = \int_{\nu=-\infty}^{+\infty} \int_{\xi=-\infty}^{+\infty} f(\xi) \cos 2\pi\nu(x-\xi)\, d\xi\, d\nu \qquad (A3\text{-}5)$$

This equation is the *Fourier* integral in real form. The two representations (A3-2) and (A3-5) are equivalent. As it is the case for the complex form, we can also split up the real form into

$$f(x) = \int_{\nu=-\infty}^{+\infty} \int_{\xi=-\infty}^{+\infty} f(\xi) \cos 2\pi\nu x \cos 2\pi\nu\xi \, d\xi \, d\nu + \int_{\nu=-\infty}^{+\infty} \int_{\xi=-\infty}^{+\infty} f(\xi) \sin 2\pi\nu x \sin 2\pi\nu\xi \, d\xi \, d\nu \qquad \text{(A3-6)}$$

or

$$f(x) = \int_{\nu=-\infty}^{+\infty} [a(\nu) \cos 2\pi\nu x + b(\nu) \sin 2\pi\nu x] \, d\nu$$

where

$$a(\nu) = \int_{x=-\infty}^{+\infty} f(x) \cos 2\pi\nu x \, dx \qquad \text{(A3-7)}$$

$$b(\nu) = \int_{x=-\infty}^{+\infty} f(x) \sin 2\pi\nu x \, dx$$

This is the real analogue to the complex form (A3-1). Note that we have changed the integration variable of the coefficients $a(\nu)$ and $b(\nu)$ from ξ to x for "esthetic" reasons.

Simplifications occur for even or odd functions. If $f(x)$ is an even function so that $f(x) = f(-x)$, then $b(\nu)$ becomes zero and the *Fourier* integral reads

$$f(x) = \int_{\nu=-\infty}^{+\infty} a(\nu) \cos 2\pi\nu x \, d\nu$$

$$a(\nu) = \int_{x=-\infty}^{+\infty} f(x) \cos 2\pi\nu x \, dx \qquad \text{(A3-8)}$$

If $f(x)$ is an odd function so that $f(-x) = -f(x)$, then $a(\nu)$ becomes zero and the *Fourier* integral reads

$$f(x) = \int_{\nu=-\infty}^{+\infty} b(\nu) \sin 2\pi\nu x \, d\nu$$

$$b(\nu) = \int_{x=-\infty}^{+\infty} f(x) \sin 2\pi\nu x \, dx \qquad \text{(A3-9)}$$

A.3.2. Numerical example for the *Fourier* integral and the *Fourier* transform

Since we have already shown three examples in sec. 3, we may be brief here. We consider again the rectangular impulse given by (3-20). We have

$$f(x) = \begin{cases} \frac{1}{2} & -1 < x < +1 \\ 0 & \text{elsewhere} \end{cases} \qquad \text{(A3-10)}$$

We compute now <u>three times</u> the *Fourier* integral for this function with different formulae. For the <u>first computation</u> we use (A3-1). Note that we have already found the result for the *Fourier* transform in (3-21) to (3-25):

$$F(\omega) = \frac{\sin\omega}{\omega} \qquad \text{(A3-11)}$$

Compared to eq. (A3-1) we have substituted

$$\omega = 2\pi\nu \, , \qquad d\omega = 2\pi d\nu \qquad \text{(A3-12)}$$

Substituting this into (A3-1), cf. also eq. (3-20), we can compute the *Fourier* integral

$$f(x) = \frac{1}{2\pi}\int_{-\infty}^{+\infty} F(\omega)\, e^{i\omega x}\, d\omega \qquad \text{(A3-13)}$$

Inserting (A3-11) and applying the *Moivre* identity (A3-3) leads to

$$f(x) = \frac{1}{2\pi}\int_{-\infty}^{+\infty} \frac{\sin\omega}{\omega}\,(\cos\omega x + i\,\sin\omega x)\, d\omega \qquad \text{(A3-14)}$$

and because the imaginary term contains an odd function which vanishes for this integral, we finally obtain

$$\boxed{f(x) = \frac{1}{2\pi}\int_{-\infty}^{+\infty} \frac{\sin\omega}{\omega}\cos\omega x\, d\omega} \qquad \text{(A3-15)}$$

What we have gained is an analytic expression for the function $f(x)$!

Now we perform the <u>second computation</u> using (A3-5). Inserting for the function $f(\xi) = \frac{1}{2}$ and appropriate limits, and using again the substitution (A3-12), we obtain

$$f(x) = \frac{1}{2\pi}\int_{\omega=-\infty}^{+\infty}\int_{\xi=-1}^{+1} \frac{1}{2}\cos\omega(x-\xi)\, d\xi\, d\omega \qquad \text{(A3-16)}$$

or

$$f(x) = \frac{1}{2\pi}\int_{\omega=-\infty}^{+\infty} \frac{1}{2}\left\{\cos\omega x \int_{\xi=-1}^{+1} \cos\omega\xi \, d\xi \;+\; \sin\omega x \int_{\xi=-1}^{+1} \sin\omega\xi \, d\xi\right\} d\omega \qquad \ldots \text{(A3-17)}$$

The second term becomes zero and because of

$$\int_{-1}^{+1} \cos\omega\xi \, d\xi = \frac{2\sin\omega}{\omega} \qquad \text{(A3-18)}$$

we get

$$\boxed{f(x) = \frac{1}{2\pi}\int_{-\infty}^{+\infty} \frac{\sin\omega}{\omega}\cos\omega x \, d\omega} \qquad \text{(A3-19)}$$

which is the same as (A3-15).

For the <u>third computation</u> of the *Fourier* integral for the rectangular impulse we note that the given function is an even function. Therefore we may apply (A3-8) where we again substitute (A3-12). Therefore:

$$a(\omega) = \int_{-1}^{+1} \frac{1}{2}\cos\omega x \, dx \qquad \text{(A3-20)}$$

The integration yields

$$a(\omega) = \frac{1}{2\omega}\sin\omega x \,\Big|_{x=-1}^{x=+1} \qquad \text{(A3-21)}$$

and

$$a(\omega) = \frac{\sin\omega}{\omega} \qquad \text{(A3-22)}$$

after evaluating the limits of the integration. Eq. (A3-22) must be inserted into (A3-8), and, using the substitution (A3-12), we immediately get

$$\boxed{f(x) = \frac{1}{2\pi}\int_{-\infty}^{+\infty} \frac{\sin\omega}{\omega}\cos\omega x \, d\omega} \qquad \text{(A3-23)}$$

Why did we compute three times the *Fourier* integral for the same function? Our goal was to make the reader familiar with the different formulations of the *Fourier* integral. Naturally, all formulations are equivalent. It depends on the function *f(x)* and on the reader itself which of the formulae should be preferred.

A.4.1. DETAILED DERIVATION OF THE CONVOLUTION THEOREM

In sec. 4 we have seen that the transition from the space domain to the frequency domain is achieved by the *Fourier* transform, symbolically written as $\mathcal{F}$. Our task is to show that the convolution

$$\boxed{g(x) = h(x)*f(x)} \tag{A4-1}$$

in the space domain becomes the simple multiplication

$$\boxed{G(\omega) = H(\omega)\cdot F(\omega)} \tag{A4-2}$$

in the frequency domain where

$$\begin{aligned} G(\omega) &= \mathcal{F}\{g(x)\} \\ H(\omega) &= \mathcal{F}\{h(x)\} \\ F(\omega) &= \mathcal{F}\{f(x)\} \end{aligned} \tag{A4-3}$$

are the *Fourier* transforms for the functions of the space domain, that is,

$$\begin{aligned} G(\omega) &= \mathcal{F}\{g(x)\} = \int_{-\infty}^{+\infty} g(x)\, e^{-i\omega x}\, dx \\ H(\omega) &= \mathcal{F}\{h(x)\} = \int_{-\infty}^{+\infty} h(x)\, e^{-i\omega x}\, dx \\ F(\omega) &= \mathcal{F}\{f(x)\} = \int_{-\infty}^{+\infty} f(x)\, e^{-i\omega x}\, dx \end{aligned} \tag{A4-4}$$

In order to transform eq. (A4-1) from the space domain into the frequency domain we therefore apply the *Fourier* transform

$$\mathcal{F}\{g(x)\} = \mathcal{F}\{h(x)*f(x)\} \tag{A4-5}$$

The left-hand side of this equation may immediately be written as $G(\omega)$ according to (A4-3). We now have

$$G(\omega) = \mathcal{F}\{h(x)*f(x)\} \tag{A4-6}$$

For computing the *Fourier* transform of the convolution on the right-hand side of (A4-6), we write the convolution $h(x)*f(x)$ in a more detailed form, cf. eqs. (4-7), (4-8):

$$G(\omega) = \mathcal{F}\left\{ \int_{-\infty}^{+\infty} h(x-\xi)\, f(\xi)\, d\xi \right\} \tag{A4-7}$$

and according to (A4-4) we use instead of the symbol $\mathcal{F}$ the explicit form for the *Fourier* transform:

$$G(\omega) = \int_{x=-\infty}^{+\infty} \left[\int_{\xi=-\infty}^{+\infty} h(x-\xi)\, f(\xi)\, d\xi \right] e^{-i\omega x}\, dx \tag{A4-8}$$

Hoping that mathematicians allow an interchange of the integrals, we get

$$G(\omega) = \int_{\xi=-\infty}^{+\infty} f(\xi) \left[\int_{x=-\infty}^{+\infty} h(x-\xi)\, e^{-i\omega x}\, dx \right] d\xi \tag{A4-9}$$

The integral in the brackets has almost the form of the *Fourier* transform for h. In order to get an expression of type (A4-4), we use

$$x' = x - \xi \qquad \text{and} \qquad dx' = dx \tag{A4-10}$$

as substitution for (A4-9) and obtain

$$G(\omega) = \int_{\xi=-\infty}^{+\infty} f(\xi) \left[\int_{x'=-\infty}^{+\infty} h(x')\, e^{-i\omega x'}\, e^{-i\omega\xi}\, dx' \right] d\xi \tag{A4-11}$$

or

$$G(\omega) = \int_{\xi=-\infty}^{+\infty} f(\xi)\, e^{-i\omega\xi} \left[\int_{x'=-\infty}^{+\infty} h(x')\, e^{-i\omega x'}\, dx' \right] d\xi \tag{A4-12}$$

and now because of (A4-4) we obtain

$$G(\omega) = \int_{\xi=-\infty}^{+\infty} f(\xi)\, e^{-i\omega\xi}\, H(\omega)\, d\xi \tag{A4-13}$$

where $H(\omega)$ is constant with respect to the integration over ξ so that

$$G(\omega) = H(\omega) \cdot \int_{\xi=-\infty}^{+\infty} f(\xi)\, e^{-i\omega\xi}\, d\xi \tag{A4-14}$$

Finally we use again (A4-4), and the result is

$$G(\omega) = H(\omega) \cdot F(\omega) \tag{A4-15}$$

which finishes the proof. Therefore, expressing verbally (A4-1) and (A4-2), i.e., the convolution theorem, we may say: if the functions $h(x)$ and $f(x)$ in the space domain have the *Fourier* transforms $H(\omega)$ and $F(\omega)$ in the frequency domain, then the convolution $h(x)*f(x)$ in the space domain becomes the multiplication $H(\omega) \cdot F(\omega)$ of their transforms in the frequency domain.

The convolution theorem can also be written in a different way. Consider (A4-1) and (A4-2) and keep in mind that because of (4-11)

$$g(x) = \mathcal{F}^{-1}\{G(\omega)\} \tag{A4-16}$$

holds, then we get on the one hand by (A4-1):

$$h(x)*f(x) = \mathcal{F}^{-1}\{G(\omega)\} \tag{A4-17}$$

and on the other hand by (A4-2):

$$h(x)*f(x) = \mathcal{F}^{-1}\{H(\omega)\cdot F(\omega)\} \tag{A4-18}$$

Then the <u>convolution theorem</u> becomes

$$\boxed{\int_{-\infty}^{+\infty} h(x-\xi)\ f(\xi)\ d\xi = \frac{1}{2\pi}\int_{-\infty}^{+\infty} H(\omega)\ F(\omega)\ e^{i\omega x}\ dx} \tag{A4-19}$$

where we used (4-7), (4-8) for the left-hand side and (3-14), (3-20) for the right-hand side.

<u>A.4.2. Using the convolution theorem for the solution of an integral equation</u>

Given are *g(x)* and *h(x)* and the solution of the integral equation

$$f(x) = g(x) + \int_{-\infty}^{+\infty} f(\xi)\ h(x-\xi)\ d\xi \tag{A4-20}$$

is wanted. Assuming that the *Fourier* transforms

$$\begin{aligned} F(\omega) &= \mathcal{F}\{f(x)\} \\ G(\omega) &= \mathcal{F}\{g(x)\} \\ H(\omega) &= \mathcal{F}\{h(x)\} \end{aligned} \tag{A4-21}$$

exist and taking into account that the integral in (A4-20) is nothing else but a convolution *f(x)*h(x)*, then we can transform eq. (A4-20) into the frequency domain by

$$\mathcal{F}\{f(x)\} = \mathcal{F}\{g(x)\} + \mathcal{F}\{f(x)*h(x)\} \tag{A4-22}$$

which leads to

$$F(\omega) = G(\omega) + F(\omega)\cdot H(\omega) \tag{A4-23}$$

where we have used (A4-21) and the convolution theorem for $\mathcal{F}(f(x)*h(x))$, cf. eq. (A4-1), (A4-2). From (A4-23) we can now easily compute

$$F(\omega) = \frac{G(\omega)}{1 - H(\omega)} \tag{A4-24}$$

which is already the solution of the integral equation - but in the frequency domain. In order to obtain the solution in the space domain, we apply the inverse *Fourier* transform:

$$f(x) = \mathcal{F}^{-1}\left\{\frac{G(\omega)}{1 - H(\omega)}\right\} \tag{A4-25}$$

explicitly written as

$$\boxed{f(x) = \frac{1}{2\pi}\int_{-\infty}^{+\infty} \frac{G(\omega)}{1 - H(\omega)}\, e^{-i\omega x}\, d\omega} \tag{A4-26}$$

This is the solution of the integral equation (A4-20). Being a little more precise, we must presuppose that the integral exists.

A.4.3. Verifying the convolution theorem by a numerical example

Given are the functions

$$h(x) = f(x) = \begin{cases} 1 & \text{for} \quad |x| < 1 \\ 0 & \text{for} \quad |x| > 1 \end{cases} \tag{A4-27}$$

and the convolution theorem is to be verified. This means that we must show that the convolution

$$g(x) = h(x)*f(x) \tag{A4-28}$$

in the space domain becomes the multiplication

$$\boxed{G(\omega) = H(\omega)\cdot F(\omega)} \tag{A4-29}$$

in the frequency domain, where

$$G(\omega) = \mathcal{F}\{g(x)\} = \mathcal{F}\{h(x)*f(x)\} \tag{A4-30}$$

and

$$H(\omega) = \mathcal{F}\{h(x)\}$$
$$F(\omega) = \mathcal{F}\{f(x)\} \qquad \text{(A4-31)}$$

holds. Starting with the simpler part, we compute the *Fourier* transforms for $h(x)$ and $f(x)$. We can be brief since we have already shown the calculation of a very similar function in sec. 3. In order to avoid much redundancy we at least show a slightly different derivation of the *Fourier* transform. For the *Fourier* transform we take eq. (3-20):

$$\mathcal{F}\{h(x)\} = H(\omega) = \int_{-1}^{+1} 1 \cdot e^{-i\omega x} \, dx \qquad \text{(A4-32)}$$

We have changed the limits of integration to -1 and $+1$, since outside this interval the function $h(x)$ is zero. Now we split up into

$$H(\omega) = \int_{-1}^{+1} \cos\omega x \, dx - i \int_{-1}^{+1} \sin\omega x \, dx \qquad \text{(A4-33)}$$

using *Moivre*'s formula. The second integral vanishes because of its odd integrand. By integrating the first integral we get

$$H(\omega) = \left. \frac{\sin\omega x}{\omega} \right|_{-1}^{+1} \qquad \text{(A4-34)}$$

and

$$\boxed{H(\omega) = \frac{2\sin\omega}{\omega}} \qquad \text{(A4-35)}$$

after evaluating the limits of integration. Because of (A4-27) we immediately have

$$\boxed{F(\omega) = \frac{2\sin\omega}{\omega}} \qquad \text{(A4-36)}$$

and the multiplication of the two *Fourier* transforms reads

$$\boxed{H(\omega)\, F(\omega) = \frac{4\sin^2\omega}{\omega^2}} \qquad \text{(A4-37)}$$

Since $G(\omega) = H(\omega) \cdot F(\omega)$ holds, we must obtain the same result if we compute $G(\omega)$ according to (A4-30). The convolution $h(x)*f(x)$ is in (4-7), (4-8) defined by

$$g(x) = h(x)*f(x) = \int_{-\infty}^{+\infty} h(x-\xi)\ f(\xi)\ d\xi \qquad \text{(A4-38)}$$

and

$$g(x) = \int 1 \cdot 1\ d\xi \qquad \text{(A4-39)}$$

is obtained for the functions (A4-27) where the limits for the integration must still be determined. This can be done best by a graphic representation of the two functions, cf. *Brigham (1974)*, Ch. 4.2; cf. also Fig. A.4.1.

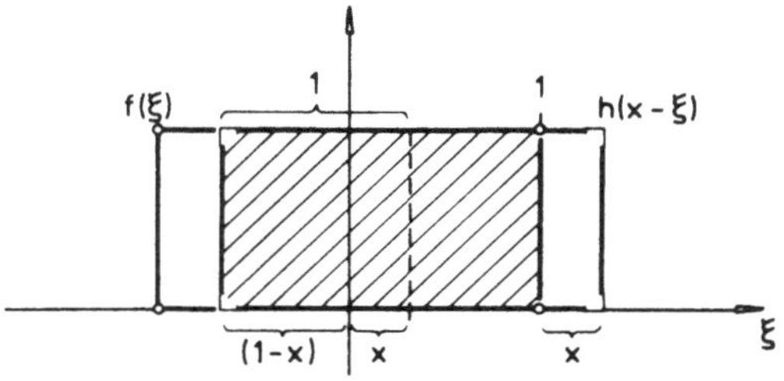

Figure A.4.1. Determination of the limits for the convolution. The shaded area represents the value of the integral of the product $g(x) \cdot h(x-\xi)$ for the indicated x. As x varies from 0 to $+2$, the shaded area changes. For $x > 2$ there is no shaded area, this means that the product $g(x) \cdot h(x-\xi)$ becomes zero.

From Fig. A.4.1 we can immediately see the limits for $0 < x < 2$, namely

$$g(x) = \int_{-(1-x)}^{1} 1 \cdot 1\ d\xi = 2 - x \quad \text{for} \quad 0 < x < 2 \qquad \text{(A4-40)}$$

In an analogous way we treat the case $-2 < x < 0$, cf. Fig. A.4.2.

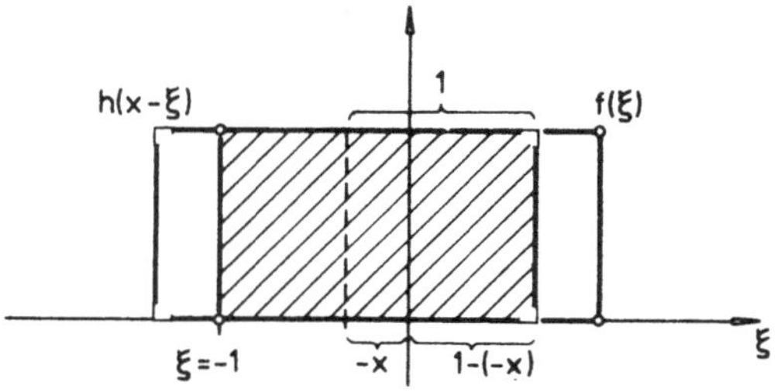

Figure A.4.2. Determination of the limits for the convolution. See also Fig. A.4.1.

From Fig. A.4.2 we get

$$g(x) = \int_{\xi=-1}^{1+x} d\xi = 2 + x \quad \text{for} \quad -2 < x < 0 \tag{A4-41}$$

and in summary we have

$$g(x) = \begin{cases} 2 + x & \text{for} \quad -2 < x < 0 \\ 2 - x & \text{for} \quad 0 < x < 2 \\ 0 & \text{elsewhere} \end{cases} \tag{A4-42}$$

as function for the convolution $g(x) = h(x)*f(x)$. The graphic representation of this function is given in Fig. A.4.3.

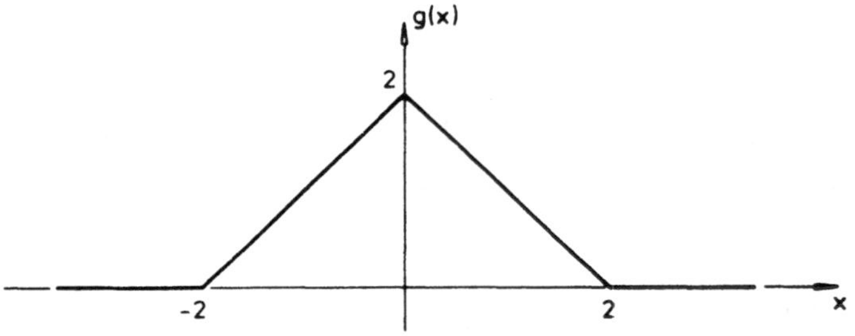

Figure A.4.3. Graphic representation of the convolution $g(x) = h(x)*f(x)$ for the functions (A4-27).

Now we need $G(\omega)$, this can be obtained by transforming $g(x)$ from the space domain to the frequency domain:

$$G(\omega) = \mathcal{F}\{g(x)\} = \int_{-\infty}^{+\infty} g(x)\, e^{-i\omega x}\, dx \tag{A4-43}$$

Substituting (A4-42) into this equation yields

$$G(\omega) = \int_{-2}^{0} (2 + x)\, e^{-i\omega x}\, dx + \int_{0}^{+2} (2 - x)\, e^{-i\omega x}\, dx \tag{A4-44}$$

or by splitting up into real and imaginary part:

$$\begin{aligned} G(\omega) &= \int_{-2}^{0} (2 + x) \cos\omega x\, dx + \int_{0}^{+2} (2 - x) \cos\omega x\, dx - \\ &- i \left[\int_{-2}^{0} (2 + x) \sin\omega x\, dx + \int_{0}^{+2} (2 - x) \sin\omega x\, dx \right] \end{aligned} \tag{A4-45}$$

The imaginary part of the formula vanishes. The exercise of this proof is left to the reader. Thus we only must calculate

$$G(\omega) = I_1 + I_2 \tag{A4-46}$$

where

$$I_1 = 2\int_{-2}^{0} \cos\omega x \, dx + \int_{-2}^{0} x \cos\omega x \, dx$$

$$I_2 = 2\int_{0}^{2} \cos\omega x \, dx - \int_{0}^{2} x \cos\omega x \, dx \tag{A4-47}$$

We obtain

$$I_1 = \frac{2}{\omega} \sin\omega x \Big|_{-2}^{0} + \left\{\frac{\cos\omega x}{\omega^2} + \frac{x \sin\omega x}{\omega}\right\} \Big|_{-2}^{0} \tag{A4-48}$$

or

$$I_1 = -\frac{2}{\omega} \sin(-2\omega) + \frac{1}{\omega^2} - \frac{\cos(-2\omega)}{\omega^2} + \frac{2 \sin(-2\omega)}{\omega} \tag{A4-49}$$

which reduces to

$$I_1 = \frac{1}{\omega^2}(1 - \cos 2\omega) \tag{A4-50}$$

and finally

$$I_1 = \frac{2 \sin^2\omega}{\omega^2} \tag{A4-51}$$

In the same way we have

$$I_2 = \frac{2 \sin^2\omega}{\omega^2} \tag{A4-52}$$

where the verification of this result can easily be done by the reader. Because of (A4-46) we form the sum of (A4-51) and (A4-52) getting

$$\boxed{G(\omega) = \frac{4 \sin^2\omega}{\omega^2}} \tag{A4-53}$$

Comparing this result with (A4-37) finishes our verification of

$$G(\omega) = H(\omega) \cdot F(\omega).$$

A.5. FURTHER GEODETIC APPLICATIONS

A.5.1. Solving *Molodensky*'s series by fast *Fourier* transform

The title of sec. A.5.1 stems from *Sideris* and *Schwarz (1985)*. We do not treat the whole contents of this publication but we only show one aspect where the principle of the *Fourier* transform can be seen very well.

The formulation of *Molodensky*'s problem is deceptively simple. Assume the gravity potential W and the gravity vector $\underline{g}$ at every point on the surface S are given, symbolically

$$\underline{g} \quad = \quad \text{function}(S,\ W) \tag{A5-1}$$

From this equation the physical surface S is to be determined, that is

$$S \quad = \quad \text{function}(W,\ \underline{g}) \tag{A5-2}$$

The solution leads to a series of double integrals. The zero-order term is the *Stokes* or *Vening Meinesz* integral. The first-order term is denoted as G_1 and can be approximated by

$$G_1(x_p, y_p) \quad = \quad \frac{1}{2\pi} \iint_E \frac{h(x,y) - h(x_p,y_p)}{[(x_p-x)^2 + (y_p-y)^2]^{3/2}} \, \Delta g(x,y) \, dx \, dy \tag{A5-3}$$

for a plane E. We have taken over this formula from *Sideris* and *Schwarz (1985)*, eq. (1.1). Some explanations of the notations: x_p, y_p is the computation point where G_1 is calculated for. The height of the computation point is $h(x_p,y_p)$. In addition, x, y and the height $h(x,y)$ denote the variable point which varies over the whole plane because of the double integral, and $\Delta g(x,y)$ are the gravity anomalies at those points.

We carry out a slight modification of eq. (A5-3) by the abbreviation

$$r(x,y) \quad = \quad (x^2 + y^2)^{-3/2} \tag{A5-4}$$

and we obtain

$$\begin{aligned} G_1(x_p, y_p) \quad = \quad & \frac{1}{2\pi} \iint_E \Big\{ [h(x,y)\ \Delta g(x,y)]\ r(x_p-x, y_p-y) \quad - \\ & - \ h(x_p,y_p)\ \Delta g(x,y)\ r(x_p-x, y_p-y) \Big\} \, dx \, dy \end{aligned} \tag{A5-5}$$

Using the symbolic form of convolution, we get

$$G_1(x,y) = \frac{1}{2\pi}\{[h(x,y)\Delta g(x,y)]*r(x,y) - h(x,y)\cdot[\Delta g(x,y)*r(x,y)]\} \tag{A5-6}$$

As we know, the convolution becomes a simple multiplication in the frequency domain. Thus we transform the convolutions into the frequency domain and after the evaluation in the frequency domain we apply the inverse transformation to the space domain. The final result for G_1 using the *Fourier* transform can therefore be written as

$$\boxed{G_1(x,y) = \frac{1}{2\pi}\left[\mathcal{F}^{-1}\{\mathcal{F}\{h(x,y)\Delta g(x,y)\}\cdot\mathcal{F}\{r(x,y)\}\} - h(x,y)\,\mathcal{F}^{-1}\{\mathcal{F}\{\Delta g(x,y)\}\cdot\mathcal{F}\{r(x,y)\}\}\right]} \tag{A5-7}$$

This result is given in *Sideris* and *Schwarz (1985)*, eq. (1.4). Five different *Fourier* transforms are necessary. The formulae shown here can also be found in *Sideris (1984)*.

We think that the principle of the application of the *Fourier* transform for the solution of *Molodensky*'s series should be clear. The reader is referred to the papers mentioned for more details and further developments.

A.5.2. Digital terrain model

The estimation of the accuracy of digital terrain models using spectral analysis is shown in *Tempfli (1982)*. In this section we restrict ourselves to a profile representation of the topography of the earth. We follow the paper by *Tempfli (1982)*, sec. 2.

Assume a function $h(\varphi,\lambda)$ where φ, λ are geographical coordinates and $h(\varphi,\lambda)$ is the corresponding height. If $h(\varphi,\lambda)$ were known at all points φ, λ, then many of the geodetic problems were solved. Fortunately (since otherwise many geodesists would be unemployed) this is not the case. Therefore we consider a profile of the earth's surface, that is, of the topography. This profile can be seen as a continuous signal $h(x)$ where x is the coordinate of a profile point and $h(x)$ can be discretized at, e.g., equidistant points $x = x_l$; $l = 1, 2, \ldots$ We denote the discrete signal by $h_l(x)$. The discrete signals $h_l(x)$ are measured, e.g. by photogrammetric methods, and are naturally affected by measuring errors. We use the notations $g_i(x)$ and $m_i(x)$ for the measurement values and the measuring errors, respectively. For a better understanding we illustrate this by a figure given in *Tempfli (1982)*, cf. Fig. A.5.1.

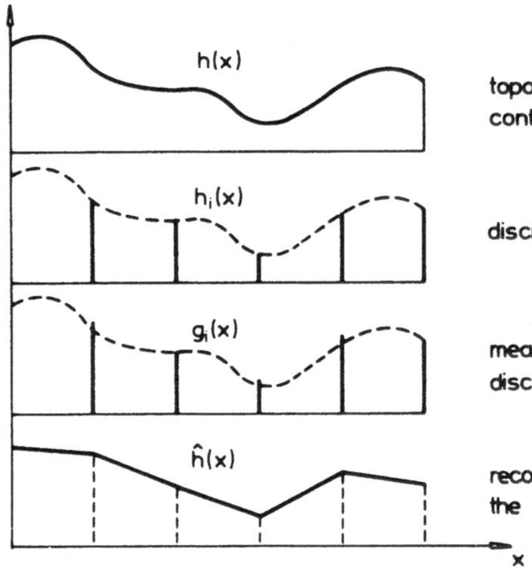

Figure A.5.1. Discretization of a continuous topography profile.

Assuming equidistant points, e.g.,

$$x_\ell = \ell \cdot \Delta x \quad \text{where} \quad \ell = \ldots, -2, -1, 0, 1, 2, \ldots \tag{A5-8}$$

where Δx is the distance between two neighboured discrete points, the discrete signal may be written (with a view to use in a convolution integral) as

$$h_i(x) = \sum_{\ell=-\infty}^{+\infty} h(\ell \cdot \Delta x)\, \delta(x - \ell \cdot \Delta x) \tag{A5-9}$$

where $\delta(\ldots)$ is the *Dirac* function which is defined as

$$\begin{aligned} \delta(t - t_0) &= 0 \quad \text{for} \quad t \neq t_0 \\ \delta(t - t_0) &= \infty \quad \text{for} \quad t = t_0 \\ \int_{-\infty}^{+\infty} \delta(t - t_0)\, dt &= 1 \end{aligned} \tag{A5-10}$$

Further details for the *Dirac* function can be found e.g. in *Brigham (1974)*, sec. A-1. Note that the product of the discrete function $h(\ell \cdot \Delta x)$ and

$\delta(x-\ell\cdot\Delta x)$ in (A4-9) reproduces the signal values $h_i(x)$ at $x = \ell\cdot\Delta x$. Considering measurements at the discrete points, we have

$$g_i(x) = h_i(x) + m_i(x) \tag{A5-11}$$

where $m_i(x)$ are the measuring errors. Now it is important to know that it is possible to use the spectral theory. The reconstruction of a continuous signal from the discrete signal $g_i(x)$ may be formulated by

$$\boxed{\hat{h}(x) = \int_{-\infty}^{+\infty} a(t)\, g_i(x - t)\, dt} \tag{A5-12}$$

where $a(t)$ is a weight function. Naturally, this formula is nothing else but a convolution, cf. eq. (4-7). As *Tempfli (1982)* shows, any procedure reconstructing the signal from a linear combination of the given reference values, i.e., the measurements, can be formulated as a convolution. In addition, the *Fourier* transform of the weight function

$$A(\omega) = \mathcal{F}\{a(t)\} \tag{A5-13}$$

is especially suited for theoretical investigations of the reconstruction procedure. This is the reason why eq. (A5-12) is transformed into the frequency domain. Knowing that a convolution in the space domain becomes a multiplication in the frequency domain, cf. eq. (4-11), we may write

$$H(\omega) = A(\omega)\cdot G_i(\omega) \tag{A5-14}$$

where

$$\begin{aligned} H(\omega) &= \mathcal{F}\{h(x)\} \\ A(\omega) &= \mathcal{F}\{a(t)\} \\ G_i(\omega) &= \mathcal{F}\{g_i(x-t)\} \end{aligned} \tag{A5-15}$$

Now we have shown how the theory of spectral analysis can be applied to digital terrain models. Readers interested in this topic are referred to *Tempfli (1982)* where special emphasis is put on accuracy estimates of the digital terrain models.

A.5.3. Geodetic networks

Sufficiently regular networks can be successfully studied by spectral methods, cf. *Meissl (1976)*. This presupposes a regular network structure, but gives extraordinary mathematical insight which can be used for understanding networks of a more irregular structure.

We shall try to illustrate this by means of a very simple example due to *Meissl (1982)*, unpublished. We follow *Sünkel (1985)*. Consider the regular leveling line of Fig. A.5.2 consisting of equidistant points.

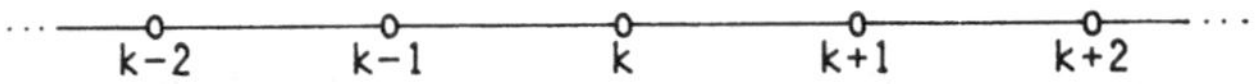

Figure A.5.2. Regular leveling line.

The leveling line is very long, so that it can be regarded as extending to infinity in both directions.

The observation equations are

$$h_{k+1} - h_k = \ell_{k,k+1} \tag{A5-16}$$

where h denotes the heights and ℓ the observed differences. Let us form the normal equations in the way familiar from adjustment computations:

$$-h_{k-1} + 2h_k - h_{k+1} = -\ell_{k,k+1} + \ell_{k-1,k} = r_k \tag{A5-17}$$

For an infinite leveling line, k may assume all positive or negative integer values.

The normal equations (A5-17) may be written in the matrix form

$$\underline{a}\,\underline{h} = \underline{r} \tag{A5-18}$$

or, more explicitly

$$\begin{bmatrix} \ddots & & & & & & & \ddots \\ \dots & 0 & -1 & 2 & -1 & 0 & \dots & \\ & \dots & 0 & -1 & 2 & -1 & 0 & \dots \\ & & \dots & 0 & -1 & 2 & -1 & 0 \quad \dots \\ & & & \ddots & & & & \ddots \end{bmatrix} \cdot \begin{bmatrix} \vdots \\ h_{-1} \\ h_0 \\ h_1 \\ \vdots \end{bmatrix} = \begin{bmatrix} \vdots \\ r_{-1} \\ r_0 \\ r_1 \\ \vdots \end{bmatrix} \tag{A5-19}$$

The matrix $\underline{a}$ is infinite but has a very regular band structure; it is a

Toeplitz matrix. Note that we have

$$a_{ij} = a_{i+1,j+1} = \ldots = a_{i+k,j+k} = \ldots \qquad \text{(A5-20)}$$

e.g. $a_{00} = a_{11} = a_{22} = \ldots = 2$, $a_{10} = a_{21} = a_{32} = \ldots = -1$, etc., reminiscent of (4-6). Because of this regularity, we may introduce the function

$$A(\lambda) = -e^{i\lambda} + 2 - e^{-i\lambda} = 2(1 - \cos\lambda) \qquad \text{(A5-21)}$$

to represent the *Fourier* transform of the matrix $\underline{a}$, the coefficients being the non-zero elements in any row (or column) of $\underline{a}$, and similarly

$$H(\lambda) = \sum_{-\infty}^{+\infty} h_k \, e^{-ik\lambda} \qquad \text{(A5-22)}$$

and

$$R(\lambda) = \sum_{-\infty}^{+\infty} r_k \, e^{-ik\lambda} \qquad \text{(A5-23)}$$

Since (A5-19) is a "discrete convolution", cf. (1-3), the *Fourier* transform of the infinite system (A5-19) is the simple equation

$$A(\lambda) \cdot H(\lambda) = R(\lambda) \qquad \text{(A5-24)}$$

corresponding to (4-11).

The equivalence of (A5-19) and (A5-24) can be shown in the following way. We substitute (A5-21), (A5-22) and (A5-23) into (A5-24) finding

$$(-e^{i\lambda} + 2 - e^{-i\lambda}) \sum_{-\infty}^{+\infty} h_k \, e^{-ik\lambda} = \sum_{-\infty}^{+\infty} r_k \, e^{-ik\lambda} \qquad \text{(A5-25)}$$

or

$$- \sum_{-\infty}^{+\infty} h_k e^{-i(k-1)\lambda} + 2 \sum_{-\infty}^{+\infty} h_k e^{-ik\lambda} - \sum_{-\infty}^{+\infty} h_k e^{-i(k+1)\lambda} = \sum_{-\infty}^{+\infty} r_k e^{-ik\lambda} \qquad \ldots \text{(A5-26)}$$

Now we change the summation index of the first sum by setting $k-1 \to k$ and for the third sum by setting $k+1 \to k$, then we obtain

$$- \sum_{-\infty}^{\infty} h_{k+1} e^{-ik\lambda} + 2 \sum_{-\infty}^{+\infty} h_k e^{-ik\lambda} - \sum_{-\infty}^{+\infty} h_{k-1} e^{-ik\lambda} = \sum_{-\infty}^{+\infty} r_k e^{-ik\lambda} \qquad \ldots \text{(A5-27)}$$

and

$$\sum_{-\infty}^{+\infty} \{-h_{k+1} + 2h_k - 2h_{k-1}\} e^{-ik\lambda} = \sum_{-\infty}^{+\infty} r_k e^{-ik\lambda} \tag{A5-28}$$

A comparison of the coefficients in this equation immediately gives

$$-h_{k+1} + 2h_k - h_{k-1} = r_k \tag{A5-29}$$

which is (A5-17) and thus the equivalence of (A5-19) and (A5-24) is shown.

Now we return to eq. (A5-24) where we get the solution by

$$H(\lambda) = \frac{R(\lambda)}{A(\lambda)} \tag{A5-30}$$

so that the unknowns h_k are given as the *Fourier* coefficients of $H(\lambda)$:

$$h_k = \frac{1}{2\pi}\int_0^{2\pi} H(\lambda)\, e^{ik\lambda}\, d\lambda = \frac{1}{2\pi}\int_0^{2\pi} \frac{R(\lambda)}{A(\lambda)}\, e^{ik\lambda}\, d\lambda \tag{A5-31}$$

by (2-5); note that the roles of "time domain" and "frequency domain" are now interchanged. The substitution of (A5-23), written with j as summation index:

$$R(\lambda) = \sum_{-\infty}^{+\infty} r_j\, e^{-ij\lambda} \tag{A5-32}$$

into (A5-26) finally gives h_k as a linear combination of the observations r_j defined by (A5-17):

$$h_k = \sum_{j=-\infty}^{+\infty} b_{k-j}\, r_j \tag{A5-33}$$

where

$$b_k = \frac{1}{2\pi}\int_0^{2\pi} \frac{e^{ik\lambda}}{A(\lambda)}\, d\lambda = \frac{1}{4\pi}\int_0^{2\pi} \frac{\cos k\lambda}{1-\cos\lambda}\, d\lambda \tag{A5-34}$$

by (A5-21) and by the symmetry of $A(\lambda)$. Using a trick to eliminate a singularity, the integral can be calculated to give

$$b_k = -\frac{1}{2}\, |k| \tag{A5-35}$$

so that (A5-33) becomes simply

$$h_k = -\frac{1}{2} \sum_{j=-\infty}^{+\infty} |k-j|\, r_j \tag{A5-36}$$

Similar mathematical structures occur in the spectral theory of spline interpolation, cf. *Sünkel (1984)*.

A.10.1. EXAMPLE FOR THE FAST *FOURIER* TRANSFORM

Our starting point is eq. (10-10) where we set $\Delta\nu = 1$ and $\Delta x = 1$ for simplicity:

$$F(p) = \sum_{j=0}^{n-1} f(j)\, e^{-2\pi ijp/n} \qquad p = 0, 1, 2, \ldots, n-1 \tag{A10-1}$$

Following *Brigham (1974)*, Ch. 10, we put

$$W^{jp} = e^{-2\pi ijp/n} \tag{A10-2}$$

where the superscripts j and p must be seen as a product $j \cdot p$! Substituting (A10-2) into (A10-1) leads to

$$F(p) = \sum_{j=0}^{n-1} f(j)\, W^{jp} \qquad p = 0, 1, 2, \ldots, n-1 \tag{A10-3}$$

Now we choose a certain case, e.g. we consider the example $n = 8$, then we obtain the following eight equations:

$$\begin{aligned}
F(0) &= f(0)W^0 + f(1)W^0 + f(2)W^0 + f(3)W^0 + f(4)W^0 + f(5)W^0 + f(6)W^0 + f(7)W^0 \\
F(1) &= f(0)W^0 + f(1)W^1 + f(2)W^2 + f(3)W^3 + f(4)W^4 + f(5)W^5 + f(6)W^6 + f(7)W^7 \\
F(2) &= f(0)W^0 + f(1)W^2 + f(2)W^4 + f(3)W^6 + f(4)W^8 + f(5)W^{10} + f(6)W^{12} + f(7)W^{14} \\
F(3) &= f(0)W^0 + f(1)W^3 + f(2)W^6 + f(3)W^9 + f(4)W^{12} + f(5)W^{15} + f(6)W^{18} + f(7)W^{21} \\
F(4) &= f(0)W^0 + f(1)W^4 + f(2)W^8 + f(3)W^{12} + f(4)W^{16} + f(5)W^{20} + f(6)W^{24} + f(7)W^{28} \\
F(5) &= f(0)W^0 + f(1)W^5 + f(2)W^{10} + f(3)W^{15} + f(4)W^{20} + f(5)W^{25} + f(6)W^{30} + f(7)W^{35} \\
F(6) &= f(0)W^0 + f(1)W^6 + f(2)W^{12} + f(3)W^{18} + f(4)W^{24} + f(5)W^{30} + f(6)W^{36} + f(7)W^{42} \\
F(7) &= f(0)W^0 + f(1)W^7 + f(2)W^{14} + f(3)W^{21} + f(4)W^{28} + f(5)W^{35} + f(6)W^{42} + f(7)W^{49}
\end{aligned}$$

... (A10-4)

or in matrix form

$$\begin{bmatrix} F(0) \\ F(1) \\ F(2) \\ F(3) \\ F(4) \\ F(5) \\ F(6) \\ F(7) \end{bmatrix} = \begin{bmatrix} W^0 & W^0 & W^0 & W^0 & W^0 & W^0 & W^0 & W^0 \\ W^0 & W^1 & W^2 & W^3 & W^4 & W^5 & W^6 & W^7 \\ W^0 & W^2 & W^4 & W^6 & W^8 & W^{10} & W^{12} & W^{14} \\ W^0 & W^3 & W^6 & W^9 & W^{12} & W^{15} & W^{18} & W^{21} \\ W^0 & W^4 & W^8 & W^{12} & W^{16} & W^{20} & W^{24} & W^{28} \\ W^0 & W^5 & W^{10} & W^{15} & W^{20} & W^{25} & W^{30} & W^{35} \\ W^0 & W^6 & W^{12} & W^{18} & W^{24} & W^{30} & W^{36} & W^{42} \\ W^0 & W^7 & W^{14} & W^{21} & W^{28} & W^{35} & W^{42} & W^{49} \end{bmatrix} \cdot \begin{bmatrix} f(0) \\ f(1) \\ f(2) \\ f(3) \\ f(4) \\ f(5) \\ f(6) \\ f(7) \end{bmatrix}$$

... (A10-5)

For (A10-4) or (A10-5) the computational effort can easily be estimated. We need n^2 (in general complex) multiplications and $n(n-1)$ (in general complex) additions. The only goal of the FFT (Fast *Fourier* Transform) is a reduction of the number of multiplications and additions. As we shall see, the reduction of those operations is very remarkable in case of FFT.

The principle of the FFT is the factorization of the matrix in (A10-5). Before doing this, a more accurate look at the coefficients helps much. Note that we have

$$W^{jp} = e^{-2\pi ijp/n} = \cos(2\pi\cdot\frac{jp}{n}) - i\sin(2\pi\cdot\frac{jp}{n}) \qquad \text{(A10-6)}$$

so that e.g. for $j = 0$ (and arbitrary p) we have

$$W^0 = 1 \qquad \text{(A10-7)}$$

and of course the same result for $p = 0$ and arbitrary j. The periodicity of the *cosine* and *sine* function in (A10-6) causes a nice property which can be formulated by

$$W^{jp} = W^{\mathrm{mod}(j\cdot p,\ n)} \qquad \text{(A10-8)}$$

where $mod(j\cdot p, n)$ must be understood in the sense of the intrinsic FORTRAN function, thus it is nothing else but the remainder upon division of $j\cdot p$ by n. To illustrate this property by an example, we set e.g. $j = 5$, $p = 7$ and we have $n = 8$, then we get

$$W^{35} = W^{\mathrm{mod}(5\cdot 7,\ 8)} = W^3 \qquad \text{(A10-9)}$$

since the remainder upon $5\cdot 7/8 = 3$. Not that (A10-9) only reflects the fact that $\cos(2\pi\cdot 35/8) = \cos(2\pi\cdot 3/8)$.

Taking into account (A10-8) in (A10-5), we obtain

$$\begin{bmatrix} F(0) \\ F(1) \\ F(2) \\ F(3) \\ F(4) \\ F(5) \\ F(6) \\ F(7) \end{bmatrix} = \begin{bmatrix} W^0 & W^0 & W^0 & W^0 & W^0 & W^0 & W^0 & W^0 \\ W^0 & W^1 & W^2 & W^3 & W^4 & W^5 & W^6 & W^7 \\ W^0 & W^2 & W^4 & W^6 & W^0 & W^2 & W^4 & W^6 \\ W^0 & W^3 & W^6 & W^1 & W^4 & W^7 & W^2 & W^5 \\ W^0 & W^4 & W^0 & W^4 & W^0 & W^4 & W^0 & W^4 \\ W^0 & W^5 & W^2 & W^7 & W^4 & W^1 & W^6 & W^3 \\ W^0 & W^6 & W^4 & W^2 & W^0 & W^6 & W^4 & W^2 \\ W^0 & W^7 & W^6 & W^5 & W^4 & W^3 & W^2 & W^1 \end{bmatrix} \cdot \begin{bmatrix} f(0) \\ f(1) \\ f(2) \\ f(3) \\ f(4) \\ f(5) \\ f(6) \\ f(7) \end{bmatrix} \qquad \text{(A10-10)}$$

Following *Bracewell (1985)*, p. 371, the factorization yields

$$\begin{bmatrix} F(0) \\ F(1) \\ F(2) \\ F(3) \\ F(4) \\ F(5) \\ F(6) \\ F(7) \end{bmatrix} = \begin{bmatrix} 1 & 0 & 0 & 0 & 0 & 0 & 0 & 0 \\ 0 & 0 & 0 & 0 & 1 & 0 & 0 & 0 \\ 0 & 0 & 1 & 0 & 0 & 0 & 0 & 0 \\ 0 & 0 & 0 & 0 & 0 & 0 & 1 & 0 \\ 0 & 1 & 0 & 0 & 0 & 0 & 0 & 0 \\ 0 & 0 & 0 & 0 & 0 & 1 & 0 & 0 \\ 0 & 0 & 0 & 1 & 0 & 0 & 0 & 0 \\ 0 & 0 & 0 & 0 & 0 & 0 & 0 & 1 \end{bmatrix} \cdot \begin{bmatrix} 1 & 1 & 0 & 0 & 0 & 0 & 0 & 0 \\ 1 & W^4 & 0 & 0 & 0 & 0 & 0 & 0 \\ 0 & 0 & 1 & 1 & 0 & 0 & 0 & 0 \\ 0 & 0 & 1 & W^4 & 0 & 0 & 0 & 0 \\ 0 & 0 & 0 & 0 & 1 & 1 & 0 & 0 \\ 0 & 0 & 0 & 0 & 1 & W^4 & 0 & 0 \\ 0 & 0 & 0 & 0 & 0 & 0 & 1 & 1 \\ 0 & 0 & 0 & 0 & 0 & 0 & 1 & W^4 \end{bmatrix} \cdot$$

$$\cdot \begin{bmatrix} 1 & 0 & 1 & 0 & 0 & 0 & 0 & 0 \\ 0 & 1 & 0 & W^2 & 0 & 0 & 0 & 0 \\ 1 & 0 & W^4 & 0 & 0 & 0 & 0 & 0 \\ 0 & 1 & 0 & W^6 & 0 & 0 & 0 & 0 \\ 0 & 0 & 0 & 0 & 1 & 0 & 1 & 0 \\ 0 & 0 & 0 & 0 & 0 & 1 & 0 & W^2 \\ 0 & 0 & 0 & 0 & 1 & 0 & W^4 & 0 \\ 0 & 0 & 0 & 0 & 0 & 1 & 0 & W^6 \end{bmatrix} \cdot \begin{bmatrix} 1 & 0 & 0 & 0 & 1 & 0 & 0 & 0 \\ 0 & 1 & 0 & 0 & 0 & W & 0 & 0 \\ 0 & 0 & 1 & 0 & 0 & 0 & W^2 & 0 \\ 0 & 0 & 0 & 1 & 0 & 0 & 0 & W^3 \\ 1 & 0 & 0 & 0 & W^4 & 0 & 0 & 0 \\ 0 & 1 & 0 & 0 & 0 & W^5 & 0 & 0 \\ 0 & 0 & 1 & 0 & 0 & 0 & W^6 & 0 \\ 0 & 0 & 0 & 1 & 0 & 0 & 0 & W^7 \end{bmatrix} \cdot \begin{bmatrix} f(0) \\ f(1) \\ f(2) \\ f(3) \\ f(4) \\ f(5) \\ f(6) \\ f(7) \end{bmatrix} \quad \ldots \text{(A10-11)}$$

As we can see, only two non-zero elements are left in each row (or column). This causes that we have only $2n$ multiplications per factor. The number of factors M is given by $2^M = n$ if we neglect the first factor (matrix multiplication) which is a mere rearrangement of the vector elements. Thus the total number of multiplications is $2n \log_2 n$ in contrast to n^2 for the system (A10-5). Now it is obvious why the fast *Fourier* transform is called "fast".

A.10.2. Development of an algorithm for the FFT

In sec. A.10.1 we have shown the concept and the principle of the FFT for the example $n = 8$. Now we try to develop an algorithm for the same example where we make extensive use of *Brigham (1974)*, Ch. 11. In the discrete

Fourier transform (A10-3):

$$F(p) \;=\; \sum_{j=0}^{n-1} f(j)\, W^{jp} \qquad p = 0,\ 1,\ 2,\ \ldots,\ n-1 \tag{A10-12}$$

we represent the integer numbers p and j as binary numbers. For $n = 8$ we need 3-bit binary numbers. Necessary are the numbers from *0* to *7* for j and p in the binary system. We have

$$\begin{array}{ll} \text{decimal system} & \text{binary system} \\ p = 0 & p_2p_1p_0 = 000 \\ p = 1 & p_2p_1p_0 = 001 \\ p = 2 & p_2p_1p_0 = 010 \\ p = 3 & p_2p_1p_0 = 011 \\ p = 4 & p_2p_1p_0 = 100 \\ p = 5 & p_2p_1p_0 = 101 \\ p = 6 & p_2p_1p_0 = 110 \\ p = 7 & p_2p_1p_0 = 111 \end{array} \tag{A10-13}$$

Denoting the three bits of the binary system for the representation of p with $p_0,\ p_1,\ p_2$ (from right to left!), then the relation to the decimal system is given by

$$p \;=\; 2^2 \cdot p_2 + 2^1 \cdot p_1 + 2^0 \cdot p_0 \tag{A10-14}$$

Consider e.g. in (A10-13) the sixth binary number, i.e., *101* where we have $p_0 = 1,\ p_1 = 0,\ p_2 = 1,$ then we get

$$p \;=\; 2^2 \cdot 1 + 2^1 \cdot 0 + 2^0 \cdot 1 \tag{A10-15}$$

and the result is $p = 5$ which corresponds to (A10-13).

Now, we thus may replace (for $n = 8$)

$$p \text{ (decimal system) } \longrightarrow p_2p_1p_0 \text{ (3-bit binary system)} \tag{A10-16}$$

Obviously, if $n > 8$ we need more bits, since in the binary system each bit may only assume the values *0* or *1*. Note that the binary representation $p_2p_1p_0$ is <u>not</u> a product of the three bits but a number consisting of *3* bits (exactly as in decimal system, compare e.g. *297* in the decimal system which is a three digit number and not the product of the three digits).

As for p, the same can be done for j. In the 3-bit binary system we have $j_2 j_1 j_0$ and the transition from the binary system to the decimal system is achieved by

$$j = 2^2 j_2 + 2^1 j_1 + 2^0 j_0 \qquad \text{(A10-17)}$$

Using the binary system we can represent (A10-3) by

$$F(p_2 p_1 p_0) = \sum_{j_0=0}^{1} \sum_{j_1=0}^{1} \sum_{j_2=0}^{1} f(j_2 j_1 j_0)\, W^{(4p_2+2p_1+p_0)(4j_2+2j_1+j_0)} \qquad \text{... (A10-18)}$$

According to the $\underline{3}$-bit binary representation we must perform 3 summations (over the three bits). From (A10-18) we take out

$$W^{(4p_2+2p_1+p_0)(4j_2+2j_1+j_0)} \qquad \text{(A10-19)}$$

and perform the factorization. Writing

$$W^{(4p_2+2p_1+p_0)(4j_2)}\, W^{(4p_2+2p_1+p_0)(2j_1)}\, W^{(4p_2+2p_1+p_0)(j_0)} \qquad \text{(A10-20)}$$

and taking into account that because of (A10-8) and (A10-7) holds: $W^{16} = W^{8} = W^{0} = 1$, we get e.g.:

$$W^{4p_2 \cdot 4j_2} = W^{16p_2 j_2} = W^0 = 1$$

since p_2 and j_2 may only assume the values 0 or 1. Instead of (A10-20) we therefore obtain

$$W^{4p_0 j_2}\, W^{(2p_1+p_0)(2j_1)}\, W^{(4p_2+2p_1+p_0)(j_0)} \qquad \text{(A10-21)}$$

This is the factorization of W. By (A10-21) we may write eq. (A10-18) as

$$F(p_2 p_1 p_0) = \sum_{j_0} \sum_{j_1} \sum_{j_2} f(j_2 j_1 j_0) \cdot W^{4p_0 j_2}\, W^{(2p_1+p_0)(2j_1)}\, W^{(4p_2+2p_1+p_0)(j_0)} \qquad \text{... (A10-22)}$$

We set

$$f_1(p_0 j_1 j_0) = \sum_{j_2} f(j_2 j_1 j_0)\, W^{4p_0 j_2} \qquad \text{(A10-23)}$$

and

$$f_2(p_0 p_1 j_0) = \sum_{j_1} f_1(p_0 j_1 j_0)\, W^{(2p_1+p_0)(2j_1)} \qquad \text{(A10-24)}$$

and

$$f_3(p_0p_1p_2) = \sum_{j_0} f_2(p_0p_1j_0)\, W^{(4p_2+2p_1+p_0)(j_0)} \tag{A10-25}$$

and obtain

$$F(p_2p_1p_0) = f_3(p_0p_1p_2) \tag{A10-26}$$

Formulae (A10-23) to (A10-26) are the desired factorization for the case $n = 8$. For an arbitrary n the development of the factorization can be done in the same way. We do not extend the formulae to this general case but refer the reader to *Brigham (1974)*, Ch. 11.3. In addition, in sec. A.10.3 the *Cooley-Tukey* algorithm is shown which treats the general case.

We add a short remark concerning the bits of the left-hand sides in (A10-23), (A10-24) and (A10-25). Consider e.g. (A10-23). On the right-hand side we have the summation index j_2 whereas p_0, j_1 and j_0 are "free" indices. Thus the bits on the left-hand side are the free indices. The same holds for (A10-24) and (A10-25) in a corresponding way.

A.10.3. The *Cooley-Tukey* algorithm

By kind permission of *Springer* Verlag we reproduce the following algorithm from the Mathematical Intelligencer, vol. 7, No. 2, 1985, pp. 62-64, which appeared as an appendix to the review by *L. Auslander* of the book "Fast Transforms Algorithms, Analysis, Applications" by *D.F. Elliott* and *R.R.Rao (1981)*.

The Cooley-Tukey Algorithm

Currently the most popular algorithm for computing the finite Fourier transforms is the FFT or Cooley-Tukey algorithm. The history of this algorithm has been set forth in an interesting article by Cooley et al. [2] and the original paper is Cooley and Tukey [1]. Because it is so important in numerical computations, and because it is so brief, we will begin by reproducing the few paragraphs in Cooley and Tukey that set forth the idea of their algorithm.

"Consider the problem of calculating the complex Fourier series

$$X(j) = \sum_{k=0}^{N-1} A(k)W^{jk}, \qquad j = 0, \ldots, N-1, \tag{1}$$

where the given Fourier coefficients $A(k)$ are complex and W is the principal Nth root of unity:

$$W = e^{2\pi i/N}. \tag{2}$$

A straightforward calculation using (1) would require N^2 operations where 'operation' means, as it will throughout this note, a complex multiplication followed by a complex addition.

The algorithm described here iterates on the array of given complex Fourier amplitudes and yields the result in less than $2N\log_2 N$ operations without requiring more data storage than is required for the given array A. To derive the algorithm, suppose N is composite, i.e., $N = r_1r_2$. Then let the indices in (1) be expressed

$$\begin{aligned} j &= j_1r_1 + j_0, & j_0 &= 0, 1, \ldots, r_1 - 1, \\ & & j_1 &= 0, 1, \ldots, r_2 - 1, \\ k &= k_1r_2 + k_0, & k_0 &= 0, 1, \ldots, r_2 - 1, \\ & & k_1 &= 0, 1, \ldots, r_1 - 1. \end{aligned} \tag{3}$$

Then, one can write

$$X(j_1,j_0) = \sum_{k_0}\sum_{k_1} A(k_1,k_0)W^{jk_1r_2}\, W^{jk_0} \tag{4}$$

since

$$W^{jk_1r_2} = W^{j_0k_1r_2}. \tag{5}$$

The inner sum, over k_1, depends only on j_0 and k_0 and can be defined as a new array,

$$A_1(j_0,k_0) = \sum_{k_1} A(k_1,k_0)W^{j_0k_1r_2}. \tag{6}$$

This result can then be written

$$X(j_1,j_0) = \sum_{k_0} A_1(j_0,k_0)W^{(j_1r_1+j_0)k_0}. \tag{7}$$

There are N elements in the array A_1, each requiring r_1 operations, giving a total of Nr_1 operations to obtain A_1. Similarly, it takes Nr_2 operations to calculate X from A_1. Therefore, this two-step algorith, given by (6) and (7), requires a total of

$$T = N(r_1 + r_2)$$

operations."

Of course, the Cooley-Tukey algorithm can be iterated. To see how, we will work out the details for $N = 8$.

Suppose we want to compute the discrete Fourier transform of $[A(0), A(1), \ldots, A(7)]$. Let $W = e^{2\pi i/8}$ and write

$$j = j_0 + j_12 + j_24, \quad j_t = 0, 1, t = 0, 1, 2$$
$$k = k_0 + k_12 + k_24, \quad k_t = 0, 1, t = 0, 1, 2.$$

Then

$$X(j) = \sum_{k=0}^{7} A(k)W^{jk} \qquad j = 0, 1, \ldots, 7$$

can be written

$$X(j_0 + j_12 + j_24) = \sum_{k_0,k_1} W^{(k_0+k_12)j} \sum_{k_2} A(k_0 + k_12 + k_24)W^{j_0k_24}$$

Step 1: For each k_0, k_1, j_0 compute

$$C_1(k_0, k_1, j_0) = \sum_{k_2} A(k_0 + k_12 + k_24)W^{j_0k_24}$$

(For each k_0,k_1 this is simply the Fourier transform on two points.)

Step 2: For each k_0, k_1, j_0 compute

$$M_1(k_0, k_1, j_0) = W^{2k_1j_0}C_1(k_0, k_1, j_0)$$

Step 3: For each k_0, j_0, j_1 compute

$$C_2(k_0, j_1, j_0) = \sum_{k_1} M_1(k_0, k_1, j_0)W^{k_1j_14}$$

(Again, this is the Fourier transform on two points.)

Step 4: For each k_0, j_1, j_0 compute

$$M_2(k_0, j_1, j_0) = W^{k_0(j_0+2j_1)}\, C_2(k_0, j_1, j_0).$$

Step 5: For each j_0, j_1, j_2 compute

$$C_3(j_0, j_1, j_2) = \sum_{k_0} M_2(k_0, j_1, j_0)W^{k_0j_24}.$$

Finally, we see that

$$X(j_0 + j_12 + j_24) = C_3(j_0,j_1,j_2).$$

The discrete Fourier transform can be thought of as a linear transformation. Since each step of the Cooley-Tukey algorithm is a linear transformation, it can be viewed as simply a factorization of the Fourier transform matrix into a product of sparse matrices. Indeed, it should be remarked that all known efficient Fourier transform algorithms amount to a factorization of the Fourier matrix.

In the case of $N = 8$ we can translate the Cooley-Tukey algorithm above into the following factorization. Let $F(8)$ denote the Fourier transform matrix (W^{ij}), $0 \leq i,j \leq 7$. Then

$$F(8) = P\begin{pmatrix} F(2) & & & 0 \\ & F(2) & & \\ & & F(2) & \\ 0 & & & F(2) \end{pmatrix} M_2 \begin{pmatrix} I_2 & I_2 & & 0 \\ I_2 & -I_2 & & \\ & & I_2 & I_2 \\ 0 & & I_2 & -I_2 \end{pmatrix} M_1 \begin{pmatrix} I_4 & I_4 \\ & \\ I_4 & I_4 \end{pmatrix}$$

where

$I_n = n \times n$ identity matrix

$$F(2) = \begin{pmatrix} 1 & 1 \\ 1 & -1 \end{pmatrix}$$

$$M_1 = \begin{pmatrix} 1 & & & & & & & 0 \\ & 1 & & & & & & \\ & & 1 & & & & & \\ & & & 1 & & & & \\ & & & & 1 & & & \\ & & & & & 1 & & \\ & & & & & & i & \\ 0 & & & & & & & i \end{pmatrix}$$

$$M_2 = \begin{pmatrix} 1 & & & & & & & 0 \\ & 1 & & & & & & \\ & & 1 & & & & & \\ & & & i & & & & \\ & & & & 1 & & & \\ & & & & & W & & \\ & & & & & & 1 & \\ & & & & & & & W^3 \end{pmatrix}$$

and

P is a permutation matrix.

The matrices in the Cooley-Tukey factorization have been given picturesque names by the engineers. The matrices M_1, M_2 are called "twiddle factors", the permutation matrix P is called the "bit reversal permutation", and the matrices formed by the $F(2)$'s are called "butterfly matrices". In the above factorization we are working on column vectors, so that the diagram below depicts the successive operations

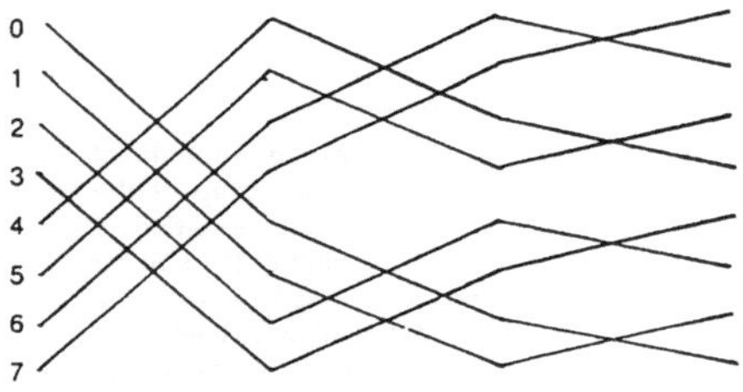

of the matrices formed out of the $F(2)$'s, where downward slanting lines denote addition and upward slanting lines denote subtraction.

References

1. J. Cooley and J. Tukey (1965) An algorithm for the machine calculation of complex Fourier series, *Math. Comput.* 19, (1967) 297–301.
2. J. W. Cooley, P. A. Lewis, and P. P. Welch (1967) Historical notes on the last Fourier transform, *Proc. IEEE*, 55, 1675–1677.

Acknowledgements

These lecture notes have been worked out by *B. Hofmann-Wellenhof* on the basis of a brief introductory course on spectral analysis given by *H. Moritz* in 1985. This course emphasizes intuitive and heuristic ("physical") aspects and basic logical structures, with almost total disregard of mathematical rigor. It is hoped that this course motivates the reader to get the desired mathematical details from the extensive literature.

Dipl.-Ing. *N. Kühtreiber* has kindly drawn the figures and he and Dipl.-Ing. *M. Wei* have checked the formulae.

The appendix A.10.3 on the *Cooley-Tukey* algorithm (following a book review by *L. Auslander*) has been reproduced from *The Mathematical Intelligencer*, vol. 7, No. 2, 1985 by kind permission of *Springer* Verlag.

References

BAULE, B. (1966): *Die Mathematik des Naturforschers und Ingenieurs. Band II. Ausgleichs- und Näherungsrechnung.* S. Hirzel Verlag Leipzig. 101 pp.

BRACEWELL, R.N. (1985): *The Fourier transform and its applications.* 2nd edition. McGraw-Hill International Book Company. xviii+444 pp.

BRIGHAM, E.O. (1974): *The fast Fourier transform.* Prentice-Hall, Inc. Englewood Cliffs, New Jersey. xiii+252 pp.

COLOMBO, O.L. (1979): *Optimal estimation from data regularly sampled on a sphere with applications in geodesy.* Department of Geodetic Science. Ohio State University, Columbus, Ohio. Report No. 297.

HEISKANEN, W.A.; H. MORITZ (1967): *Physical geodesy.* W.H. Freeman & Co. San Francisco, London. xi+364 pp.

MEISSL, P. (1976): *Strength analysis of twodimensional angular Anblock networks.* Manuscripta Geodaetica, Vol. 4(1976), pp. 293-334.

MORITZ, H. (1976): *Integral formulas and collocation.* Manuscripta Geodaetica, Vol. 1(1976), pp. 1-40.

SIDERIS, M. (1984): *Computation of gravimetric terrain corrections using fast Fourier techniques.* University of Calgary, Division of Surveying Engineering. UCSE Report No. 20007. 110 pp.

SIDERIS, M.G.; K.P. SCHWARZ (1985): *Solving Molodensky's series by fast Fourier transform.* Proceedings of the I. Hotine-Marussi Symposium on Mathematical Geodesy. Rome, June 3-6, 1985, pp. 493-511.

SÜNKEL, H. (1984): *Splines: their equivalence to collocation.* Department of Geodetic Science. Ohio State University, Columbus, Ohio. Report No. 353.

SÜNKEL, H. (1985): *Fourier analysis of geodetic networks.* In: *Optimization and design of geodetic networks (E.W. Grafarend* and *F. Sansò*, eds.), Springer, Berlin. pp. 257-300.

TEMPFLI, K. (1982): *Genauigkeitsschätzung digitaler Höhenmodelle mittels Spektralanalyse.* Veröffentlichungen des Instituts für Photogrammetrie der Technischen Universität Wien. Geowissenschaftliche Mitteilungen, Heft 22. 125 pp.

PART B

NOTES ON THE MAPPING OF THE GRAVITY FIELD USING SATELLITE DATA

by

O.L. Colombo

EG&G Washington Analytical Services Center, Inc.
5000 Philadelphia Way, Suite J
Lanham, Maryland 20706, U.S.A.

Lecture Notes in Earth Sciences, Vol. 7
Mathematical and Numerical Techniques in Physical Geodesy
Edited by H. Sünkel

INTRODUCTORY CHAPTER

The objective of these lectures is to explain some basic theoretical principles involved in the mapping of the gravity field using data from artificial satellites, tracking data in particular. This mapping of the field is one of three major problems in satellite geodesy, the other two being the precise calculation of spacecraft orbits, and the determination of the position of tracking stations. All three problems enter, to some extent, in any application of space techniques to geodesy and geophysics, and they are interrelated: to determine orbits it is necessary to know both the gravity field and the position of the tracking stations; to obtain the position of a station with tracking data, one must know the orbit and, thus, the field, and to map the field one needs an approximate orbit as a first estimate of where the satellite was when the measurements were taken.

In general, the observed quantities have a nonlinear mathematical representation, and, beguinning with approximate values for all parameters, one linearizes the observation equations about those values, proceeds to estimate corrections to them using some linear estimation technique (least squares adjustment, least squares collocation, etc.), recalculates the orbits on the basis of these corrections, and starts a new iteration using the corrected orbits and values as an improved guess. The usual algorithm is Newton-Gauss, or some variant of it. There is a deep similarity, both at the formal and at the conceptual levels, with the free boundary value problem of physical geodesy, which is also nonlinear, is also linearized and solved (in principle) iteratively, and nobody knows for sure that the iterations must always converge to a meaningful solution. Experience shows that things do work out more often than not, but there is no general proof that this should necessarily happen. In last analysis, one must assume that the problem is well-behaved, so it has a solution that can be approached further at each iteration, if one begins searching for it from a reasonable starting point, while enforcing a reasonable set of constraints.

This work shall concentrate on the determination of orbit and gravity field parameters, although there are many others (station positions, transponder delays, etc.) that must be estimated also before the data can be translated into a map of that field. Among the orbital parameters, the main ones are the six components of the initial state (three position and three velocity components, or else six Keplerian--or similar--elements), which are the initial conditions for integrating the orbit using the *known* values of the forces acting on the satellite. The initial position and velocity are those of the center of mass of the spacecraft, which is an extended object. Measurements such as radar ranging, etc., are made at or to some other point in an antenna, beacon, or sensor, and must be reduced to the center of mass in some way, to make the measurements compatible with the mathematical description of the orbit. This can be a complex and laborious task, of which no more shall be said here. In addition to solving for the parameters of the gravity force (point masses, spherical harmonic potential coefficients, mean anomalies, etc.), it is usually necessary to estimate other parameters associated with the forces of drag and solar radiation pressure.

First, some comments on notation:

$\underline{x}$	is a n-dimensional vector with components x_i,
$\underline{x} \cdot \underline{y}$	is the scalar product of $\underline{x}$ and $\underline{y}$,
$\underline{x} \times \underline{y}$	is the vector product of $\underline{x}$ and $\underline{y}$,
$\lvert\underline{x}\rvert$	is the modulus of $\underline{x}$,
$D_b a$	is the partial derivative of scalar a with respect to scalar b,
$D_{\underline{x}} a$	is the n-vector of the partial derivatives of the scalar a with respect to the n components x_i of $\underline{x}$,
$D_{\underline{x}} y$	is the matrix of the partial derivatives of all the components of $\underline{y}$ with respect to all the components of $\underline{x}$.

Let $\underline{s}$ be the 6-vector representing the state of the satellite's center of mass at any instant t: three components of $\underline{s}$ represent the position and three the velocity. Accordingly, $\underline{s}(0)$ is the initial state vector, corresponding to the beginning of the orbit, which here is chosen to coincide with the time-origin. Let $\underline{f}$ be a n-vector of parameters associated with the various forces acting on the spacecraft and shaping that orbit (gravity, solar radiation pressure, and drag), and let $\underline{q}$ be an m-vector of all other parameters of interest, such as station coordinates, etc. In general, an observation "w" (range, range-rate, etc.) is a nonlinear function of some, or all, the parameters, plus some noise n:

$$w = w(\underline{s}, \underline{f}, \underline{q}) + n \ . \tag{I.1}$$

Let also $\underline{s}_o$, $\underline{f}_o$, and $\underline{q}_o$ be the vectors of the known, or nominal, values of $\underline{s}$, $\underline{f}$, and $\underline{q}$, respectively. The idea here is to use many observations such as "w" to improve these initial estimates by calculating corrections to them. If the initial values are good, the corrections are proportionally small and it may be possible to obtain them by solving a first order approximation of the nonlinear estimation problem, based on linearized observation equations of the general form

$$\Delta w = D_{\underline{s}}\, w \cdot \Delta\underline{s} + D_{\underline{f}} w \cdot \Delta\underline{f} + D_{\underline{q}} w \cdot \Delta\underline{q} + n \ , \tag{I.2}$$

where all derivatives are taken along the nominal orbit $\underline{s}_o$ defined by $\underline{s}(0)_o$ and $\underline{f}_o$, with $\underline{q} = \underline{q}_o$. For its part, the state $\underline{s}$ of the satellite depends, non-linearly, on the initial state $\underline{s}(0)$ and on the forces $\underline{f}$ along the orbit:

$$\underline{s}(t) = \underline{s}(\underline{s}(0), \underline{f}) \ . \tag{I.3}$$

Small errors $\Delta\underline{s}(0)$ and $\Delta\underline{f}$ in $\underline{s}(0)$ and $\underline{f}$ result, usually, in small orbit errors (or perturbations) $\Delta\underline{s}$ which can be related to the former by a first order expression

$$\Delta\underline{s} = D_{\underline{s}(0)}\underline{s}\ \Delta\underline{s}(0) + D_{\underline{f}}\underline{s}\ \Delta\underline{f} \ . \tag{I.4}$$

Replacing (I.4) in (I.2):

$$\Delta\underline{w} = D_{\underline{s}}w \cdot D_{\underline{s}(0)}\underline{s}\ \Delta\underline{s}(0) + D_{\underline{s}}w \cdot D_{\underline{f}}\underline{s}\ \Delta\underline{f} + D_{\underline{f}}w \cdot \Delta\underline{f} + D_{\underline{q}}w \cdot \Delta\underline{q} + n, \tag{I.5}$$

where $\Delta\underline{f}$ appears multyplying not only the direct derivatives in $D_{\underline{f}}w$, but also the indirect derivatives in $D_{\underline{s}}w\, D_{\underline{f}}\underline{s}$.

The components of the matrices $D_{\underline{s}(0)}\underline{s}$ and $D_{\underline{f}}\underline{s}$ are the solutions of a special form of the linearized equations of motion, known as *variational equations*. More will be said about those at the end of this chapter. Formulating and solving these equations to a satisfactory degree of accuracy is a central problem in the theory of satellite geodesy.

One approach towards formulating and solving the linearized perturbation equations consists *in making* a number of careful simplifications to get equations that have *analytical solutions*, or integrals. Because of these approximations, this approach is used mostly to gain an understanding of the properties of orbits, the sensitivity of various types of observations to the information one is trying to extract from them, and, generally, to get a good grasp of the problem at hand. Sometimes, this analytical approach can be used for making quantitative determinations. Often, however, the helpful approximations made in order to obtain direct integrals are paid in a loss of accuracy in numerical results that cannot be tolerated, and a different method is then needed. This method consists in the *numerical integration* of the exact linearized equations. Both the *analytical* and the *numerical* techniques complement each other, and a good knowledge of both is needed to work effectively in satellite geodesy. Roughly put, the former gives insight, the latter, results. Here I shall concentrate on the analytical approach, because my purpose is to share whatever insight I may have in the basic problems of this discipline.

There are many ways of defining the coordinate system in which the orbit is integrated, either analytically or numerically. Each choice leads to a particular form of the equations of motion, always nonlinear in the coordinates. To describe the motion of the center of mass of a satellite one needs six coordinates, as already mentioned. In general, it is convenient to work with the six Keplerian elements, which allow an equally accurate study of approximately elliptical orbits of *both low and high mean eccentricity*, so most textbooks present the theory in terms of those elements. However, in the particular case where the orbit is *nearly circular* ($e < 0.005$), one can use other coordinates that are, perhaps, more appealing to intuition, as they correspond directly to the usual ideas of position and velocity. As it happens, most spacecraft of interest in satellite geodesy follow nearly circular orbits, so they are amenable to the more intuitive treatment, at least for developing an approximate, analytical theory. I have chosen such coordinates to present that type of theory in what follows, in the hope that the reader will find the formulation as convenient and helpful as I do.

CHAPTER ONE : THE LINEARIZED EQUATIONS OF MOTION

REFERENCE FRAMES AND EQUATIONS OF MOTION

Usually, the orbits of artificial satellites are integrated in a system of Cartesian coordinates with origin at the geocenter, and with axes whose directions are fixed with respect to the stars (Figure 1.1-a). Since the earth follows an orbit largely determined by the gravitation of the sun and, to a lesser extent, of the moon and the planets, that geocentric frame has the same acceleration as the earth's center of mass in inertial space. The Newtonian equations of motion in an accelerated frame are:

$$\ddot{\underline{r}} = \underline{a}(\underline{r}, \dot{\underline{r}}) + \underline{g}_b - \underline{b} , \qquad (1.1)$$

where $\underline{r}$, $\dot{\underline{r}}$, and $\ddot{\underline{r}}$ are the geocentric position, velocity and acceleration vectors of the center of mass of a spacecraft; $\underline{a}(r,\dot{\underline{r}})$ is the acceleration due to the sum of terrestrial gravitation, solar radiation pressure, and drag; $\underline{b}$ is the acceleration of the origin (geocenter) respect to a true inertial frame, and $\underline{g}_b$ is the sum of the gravitational accelerations of the satellite caused by the sun, the moon and other celestial bodies. To a close approximation,

$$\underline{b} = \underline{g}_b \text{ (at the geocenter), so}$$

$$\ddot{\underline{r}} = \underline{a} + \underline{g}_{tidal},$$

where

$$\underline{g}_{tidal} = \underline{g}_b - \underline{g}_b\text{(at the geocenter)} \qquad (1.2)$$

is known as the tidal acceleration. Clearly, expression (1.1) can be written

$$\ddot{\underline{r}} = \underline{a}_T(\underline{r},\dot{\underline{r}}) , \qquad (1.3)$$

where $\underline{a}_T = \underline{a} + \underline{g}_{tidal}$, so that the tidal efects are treated as ordinary accelerations. In this way, the equations of motion can be formulated as if the earth-centered frame were truly inertial. This <u>pseudo-inertial</u> treatment of the equations, with the gravitational effects of bodies other than the earth described in terms of tidal fields (with their corresponding accelerations and potentials), is used extensively--both in theory and in practice--in space geodesy.

In what follows, instead of working with the usual pseudo-inertial frame just described, most theoretical derivations will be made in a frame whose origin is also the geocenter, where the attractions of the moon, etc., still appear in the form of tidal accelerations, but where the axes rotate in space. This <u>rotating system</u> is very convenient for studying the orbital perturbations of satellites in nearly circular orbits, which are the orbits of most satellites of interest in space geodesy.

As shown in Figure 1.1-b, the system of coordinates has the y and z axes in the plane of the orbit, z pointing approximately to the spacecraft, because the system rotates uniformly, with the mean angular speed n_0 of the satellite about the x axis. This axis, in turn, being perpendicular to the orbit plane, rotates with this plane very slowly about the earth's axis. This movement is known as the precession of the line of nodes of the orbit (this line is the intersection of the orbital and the equatorial planes) and is caused mostly by the gravitational pull of the equatorial bulge. It amounts

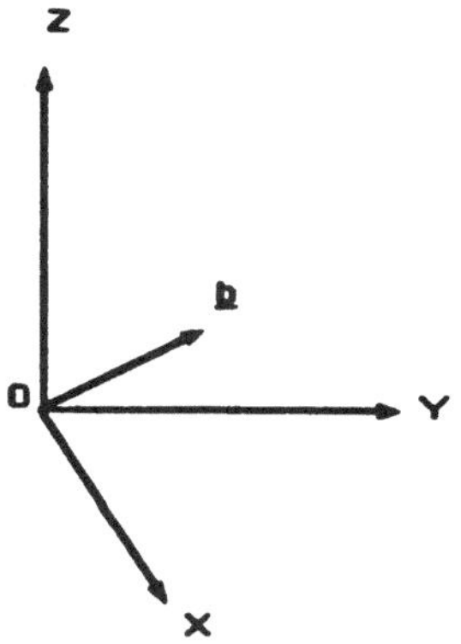

Figure 1.1(a)
Usual quasi-inertial cartesian frame of satellite geodesy. Axes have fixed orientation relative to stars.
<u>b</u>: acceleration of the frame
O: geocenter

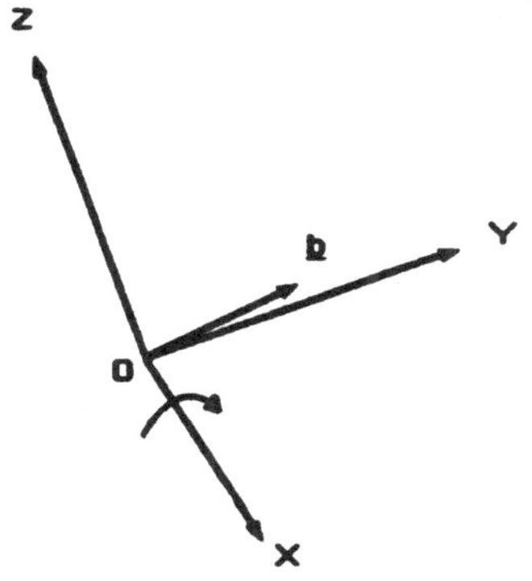

Figure 1.1(b)
Rotating cartesian frame.
X axis is normal to orbital plane. Z axis in radial direction
Same acceleration <u>b</u> and origin O as in 1.1(a).

to, at most, a few degrees per day, and it is zero for polar orbits. For nearly circular orbits, the x, y and z axes can be described as pointing in the across-track, along-track, and radial directions (some authors use "transverse" instead of "along-track").

The Newtonian equations of motion in the rotating, geocentric coordinates x, y, z are, in vector form:

$$\ddot{\underline{r}} = \underline{a}_T - n_o\underline{N} \times (n_o\underline{N} \times \underline{r}) - 2n_o\underline{N} \times \dot{\underline{r}} - n_o\dot{\underline{N}} \times \underline{r} \ , \qquad (1.4)$$

where $\underline{N}$ is a unit vector normal to the orbit plane (i.e., along the x axis) and n_o is the angular velocity of the rotating system. The last term can be ignored, because the main change in $\underline{N}$, for nearly circular orbits, is due, mostly, to the precession of the line of nodes, which is very slow, so $\dot{N}$ is small.

The relationship between the coordinates x, y, z of the rotating frame and those, x_{NR}, y_{NR}, z_{NR}, of a geocentric, Non-Rotating frame whose axes coincide with the rotating ones at any given time t, are trivial for position and acceleration, but more complex for velocity:

$$x_{NR} = x, \ y_{NR} = y, \ z_{NR} = z, \ \ddot{x}_{NR} = \ddot{x}, \ \ddot{y}_{NR} = \ddot{y}, \ \ddot{z}_{NR} = \ddot{z},$$

while

$$\dot{y}_{NR} = \dot{y} + n_o z$$
$$\dot{z}_{NR} = \dot{z} - n_o y \ . \qquad (1.5)$$

These expressions are needed to relate the results of the theory to be presented here to the more usual description of orbital perturbations in terms of across-track, along-track and radial components along the axes of the x_{NR}, y_{NR}, z_{NR} system.

LINEARIZED EQUATIONS OF MOTION (HILL'S EQUATIONS)

To develop an analytical theory of the effect on the orbit of small forces, such as those associated with the anomalous gravitational field of the earth, the first step is to define a nominal force and the corresponding nominal orbit, and to linearize the equations of motion about them. For the purposes of this study, it is enough to use the acceleration GM/r^2 of a central force field (i.e., a spherical approximation) and a circular orbit whose radius ρ_o equals the average radius of the actual, slightly elliptical orbit, and whose angular frequency corresponds to this radius (and is close to that of the actual orbit) in the central field:

$$n_o = (GM/\rho_o^3)^{1/2} , \quad (1.6)$$

while

$$\underline{a}_{To} = -(GM/\rho_o^2)\ \underline{R} \quad (1.7)$$

is the central field acceleration along a circular orbit of radius ρ_o, $\underline{R}$ being a unit vector that points from the geocenter to the spacecraft. Ignoring the small term in $\dot{\underline{N}}$, and taking first order differentials of both sides of (1.4),

$$\Delta\ddot{\underline{r}} = \Delta\underline{a}_{T_o} - n_o\underline{N} \times (n_o\underline{N} \times \Delta\underline{r}) - 2\ n_o\underline{N} \times \Delta\dot{\underline{r}} . \quad (1.8)$$

The term $\Delta\underline{a}_{T_o}$ is the sum of two effects: (a) the change in $\underline{a}_{T_o}$ due to the change $\Delta\underline{f}$ in force parameters, and (b) the effect of the change in position $\Delta\underline{r}$. Consequently, the previous expression can be written as

$$\Delta\ddot{\underline{r}} = D_{\underline{f}}\ \underline{a}_{T_o} \cdot \Delta\underline{f} + D_{\underline{r}}\ \underline{a}_{T_o} \cdot \Delta\underline{r} - n_o\underline{N} \times (n_o\underline{N} \times \Delta\underline{r}) - 2\ n_o\underline{N} \times \Delta\dot{\underline{r}} .$$

(1.9)

This vector formula can be replaced with three scalar expressions, one for each of the coordinates x, y, z of the rotating system. From (1.7) we get

$$D_{\underline{r}}\ \underline{a}_{T_o} = -GM/\rho_o^3 \begin{bmatrix} 1 & 0 & 0 \\ 0 & 1 & 0 \\ 0 & 0 & -2 \end{bmatrix} . \quad (1.10)$$

Calling Δa_x, Δa_y, Δa_z to the x, y, z components of $\Delta\underline{a}$,

$$\begin{aligned} \Delta\ddot{x} &= \Delta a_x - GM/\rho_o^3\ \Delta x \\ \Delta\ddot{y} &= \Delta a_y + (n_o^2 - GM/\rho_o^3)\Delta y - 2n\ \Delta\dot{z} \\ \Delta\ddot{z} &= \Delta a_z + (n_o^2 + 2\ GM/\rho_o^3)\Delta z + 2n_o\Delta\dot{y} , \end{aligned} \quad (1.11)$$

with Δx, Δy, Δz being the perturbations in x, y, z. Finally, taking (1.6) into account, this becomes, after grouping terms,

$$\Delta\ddot{x} = \Delta a_x - n_o^2\,\Delta x$$

$$\Delta\ddot{y} = \Delta a_y - 2n_o\Delta\dot{z} \qquad (1.12)$$

$$\Delta\ddot{z} = \Delta a_z + 3n_o^2\,\Delta z + 2n_o\Delta\dot{y}\ .$$

Equations (1.12) are known sometimes as Hill's equations, and are the basis of the first order perturbation theory to be presented here. Their main characteristics are: their simplicity; the fact that their solutions, given in terms of across-track, along-track and radial perturbations, are easy to visualize; the complete independence of the first equation (which is that of a simple harmonic oscillator) from the other two; the time-invariant nature of all three equations (they have constant coefficients). The last property is important because linear, time-invariant systems of differential equations are particularly easy to solve when the perturbing forces can be expressed as sums of sines and cosines whose angles increase secularly (i.e., at a constant rate) with time. Such sums appear quite naturally when trying to represent anomalous gravitational accelerations in the x, y, z rotating coordinates.

THE HOMOGENEOUS SOLUTION OF THE EQUATIONS

The initial conditions of the equations are the errors, or perturbations, of the initial state $\underline{s}(0)$: $\Delta x(0)$, $\Delta y(0)$, $\Delta z(0)$, $\Delta\dot{x}(0)$, $\Delta\dot{y}(0)$, $\Delta\dot{z}(0)$. The homogeneous solutions are time functions that, together with their first and second derivatives, satisfy the equations when the forcing terms Δa_x, Δa_y, Δa_z are all identically zero. Those solutions are:

$$\Delta x(t) = \Delta x(0)\ \cos n_o t + \frac{\Delta\dot{x}(0)}{n_o}\sin n_o t$$

$$\Delta y(t) = \frac{2}{n_o}\,\Delta\dot{z}(0)\ \cos n_o t + \left(\frac{4}{n_o}\,\Delta\dot{y}(0) + 6\Delta z(0)\right)\sin n_o t$$

$$+ \left(\Delta y(0) - \frac{2}{n_o}\,\Delta\dot{z}(0)\right) - \left(3\Delta\dot{y}(0) + 6n_o\,\Delta z(0)\right)t \qquad (1.13)$$

$$\Delta z(t) = \frac{\Delta\dot{z}(0)}{n_o} \sin n_o t - (\frac{2}{n_o} \Delta\dot{y}(0) + 3\Delta z(0)) \cos n_o t + (\frac{2}{n_o} \Delta\dot{y}(0) + 4\Delta z(0)) . \qquad (1.13) \text{ (cont.)}$$

The reader can see for himself that the perturbations in x, y, z given above do satisfy the equations and fulfill the initial conditions, by replacing them and their appropriate time derivatives on both sides of equations (1.12).

THE RESPONSE TO AN OSCILLATION

The accelerations caused by the disturbing gravitational potential can be represented, as shown later on, by sums of sinusoidal oscillations of different frequencies. It is easier to understand the effect of such a sum if one studies first the case where there is only one frequency.

Consider, then, small perturbing accelerations Δa_x, Δa_y, and Δa_z of the form

$$\begin{aligned} \Delta a_x &= A_x \cos \omega t + B_x \sin \omega t \\ \Delta a_y &= A_y \cos \omega t + B_y \sin \omega t \qquad (1.14) \\ \Delta a_z &= A_z \cos \omega t + B_z \sin \omega t , \end{aligned}$$

where A_x, B_x, etc., are the amplitudes, and ω is their common angular frequency. These accelerations change the orbit by amounts that, to first order, can be approximated by the corresponding forced solutions of equations (1.12). These solutions must satisfy the initial conditions; if these conditions are all zero, both for position and for velocity, then the solutions are:

$$\Delta x(t)=\frac{1}{(n_o^2-\omega^2)}\{A_x\cos\,\omega t + B_x\sin\,\omega t\}-\frac{1}{(n_o^2-\omega^2)}\{A_x\cos\,n_o t + \frac{\omega B_x}{n_o}\sin\,n_o t\}$$

$$\Delta y(t)=\frac{1}{\omega^2(n_o^2-\omega^2)}\{(\omega^2+3n_o^2)(A_y\cos\omega t+B_y\sin\omega t)+2n_o\omega(A_z\sin\omega t-B_z\cos\,\omega t)\}$$

$$+\frac{1}{(n_o^2-\omega^2)}\{-\frac{1}{n_o}(2\omega B_z+4n_oA_y\cos n_o t)\cos n_o t+[\frac{1}{\omega}(12n_oB_y-6\omega A_z)$$

$$-\frac{1}{\omega n_o}\;(4(\omega^2+3n_o^2)B_y+8n_o\omega A_z)]\;\sin\,n_o t$$

$$+\;[\frac{1}{\omega^2}\;((2n_o\omega B_z-(\omega^2+3n_o^2)A_y)+\frac{1}{n_o}\;(2\omega B_z+4n_oA_y)]$$

$$+\;\frac{1}{\omega}\;[(3(\omega^2+3n_o^2)B_y+6n_o\omega A_z)+6n_o\omega A_z-12n_o^2B_y]t\} \qquad (1.15)$$

$$\Delta z(t)=\frac{1}{\omega(n_o^2-\omega^2)}\;\{\omega(A_z\cos\omega t+B_z\sin\omega t)+2n_o(A_y\sin\omega t-B_y\cos\omega t)\}$$

$$\frac{1}{\omega(n_o^2-\omega^2)}\;\{-\;[\frac{\omega}{n_o}\;(\omega^2\;B_z\;+\;2n_o\;\omega\;A_y)]\;\sin\;n_o t$$

$$+\quad[\frac{2}{n_o}\;(\omega^2+3n_o^2)B_y+4\omega A_z+3\omega A_z-6n_oB_y]\;\cos\;n_o t$$

$$-\;[\frac{2}{n_o}\;(\omega^2+3n_o^2)B_y+4\omega A_z+4\omega A_z-8n_oB_y]\}\;.$$

Notice the presence of oscillatory terms of the same frequency ω as those in the forcing functions: this is an important property of the solution of linear differential equations with constant coefficients. In each formula, these terms have been grouped within curly brackets at the beginning: they are the respective special solutions, or integrals, of the equations. Likewise, the terms corresponding to the homogeneous part of the solution are grouped within the pairs of brackets at the end.

If Δa_x, Δa_y, and Δa_z are sums of oscillations of different frequencies, then the corresponding solution is the sum of the solutions for the individual forcing terms, each of the form shown above (this is a consequence of the linear nature of the equations).

If the initial conditions $\Delta x(0)$, $\Delta y(0)$, $\Delta z(0)$, $\Delta\dot{x}(0)$, $\Delta\dot{y}(0)$, $\Delta\dot{z}(0)$, are not zero, then the complete solution is the sum of the homogeneous solution (1.13), satisfying those conditions, and of the forced solutions for zero initial conditions, of the form (1.15).

RESONANCE

Expressions (1.15) show that the response to an oscillatory acceleration of frequency ω is larger the closer ω is to either the zero frequency or to n_0. At those precise frequencies, the amplitude of the response becomes, in fact, infinite. This indicates that expressions (1.15) are no longer valid at those two critical frequencies, and that special solutions have to be derived for both cases. This increase in the amplitude of the response as the frequency of the driving oscillation approaches some critical value is known as resonance. The infinite limit value for the amplitude arises, in this case, from having neglected the existence of damping forces (that absorb energy) when formulating the nominal force for the equations of motion according to (1.7). In reality, there is always some slight damping due to forces such as drag, but it is so small that it does not affect the main dynamic characteristics of the orbit at the level of approximation needed here.

The inmediate consequence of the resonant character of the system, i.e., a great exageration of the response to forcing oscillations whose frequencies are close to either zero or n_0, can be seen also as a relative attenuation of oscillations outside those two resonant bands. Accordingly, the Fourier spectra of the orbital perturbations Δx, Δy, Δz are likely to show two marked peaks, one centered at zero and the other at n_0; this has been found to be true in practice. Consequently, orbital perturbations tend to be simpler functions of time than the accelerations that produce them: the dynamics of the orbit exerts a strong filtering action outside the two "pass-bands" mentioned above.

When excited at a resonant frequency, an undamped physical system will respond with increasing oscillations of its own; if the critical frequency is zero, then the corresponding response may include linear, quadratic and other indefinitely increasing polynomial terms. The system under consideration here is described by linear differential equations such as (1.12) only as long as the perturbations Δx, Δy and Δz are small enough to grant a first order treatment. Therefore, indefinitely increasing solutions would be meaningful only over a period of time within which they remain sufficiently small. The orbital arcs whose perturbations are described satisfactorily by a first order theory may be several days (and even weeks) long, but a point will be reached beyond which, in the presence of resonant effects, the formulae are no longer valid; for this reason, the theory presented here is best suited for studying the perturbations in arcs about one week long (a length often adopted, in practice, when analyzing satellite data).

In the case at hand, disturbing accelerations of the form

$$\Delta a_x = A_x \cos n_o t + B_x \sin n_o t + C_x$$

$$\Delta a_y = A_y \cos n_o t + B_y \sin n_o t + C_y \qquad (1.16)$$

$$\Delta a_z = A_z \cos n_o t + B_z \sin n_o t + C_z ,$$

where A_x, B_z, C_x, etc., are constants, will excite the two resonances in the system simultaneously, because they include the two critical frequencies, 0 and n_o. The corresponding complete solution, when the initial conditions are $\Delta x(0)$, $\Delta y(0)$, $\Delta z(0)$, $\Delta\dot{x}(0)$, $\Delta\dot{y}(0)$, $\Delta\dot{z}(0)$, is given by the expressions:

$$\Delta x(t)=\frac{1}{2n_o} A_x\ t \sin n_o t - \frac{1}{2n_o} B_x\ t \cos n_o t + \frac{1}{n_o^2} (n_o^2\ \Delta x(0)-C_x) \cos n_o t$$

$$+ \frac{1}{n_o^2} (\tfrac{1}{2} B_x + n_o\ \Delta\dot{x}(0)) \sin n_o t + \frac{1}{n_o^2} C_x$$

$$\Delta y(t)=\frac{1}{n_o} (2A_y + B_z)t \sin n_o t + \frac{1}{n_o} (A_z - 2B_y)t \cos n_o t$$

$$+ \frac{1}{n_o^2} (3A_y + 2B_z - 4C_y + 2n_o\ \Delta\dot{z}(0)) \cos n_o t$$

$$+ \frac{1}{n_o^2} (5B_y - A_z + 2C_z + 4n_o\ \Delta\dot{y}(0) + 6n_o^2\ \Delta z(0)) \sin n_o t$$

$$- \frac{1}{n_o^2} (3A_y + 2B_z - 4C_y + 2n_o\ \Delta\dot{z}(0) - n_o^2\ \Delta y(0))$$

$$- \frac{1}{n_o} (2C_z + 3B_y + 3n_o\ \Delta\dot{y}(0) + 6n_o^2\ \Delta z(0))t - \frac{3}{2} C_y t^2 \qquad (1.17)$$

$$\Delta z(t) = \frac{1}{2n_o}(A_z - 2B_y)t \sin n_o t - \frac{1}{2n_o}(2A_y + B_z)t \cos n_o t$$

$$- \frac{1}{n_o^2}(2B_y + C_z + 2n_o \Delta\dot{y}(0) + 3n_o^2 \Delta z(0)) \cos n_o t \qquad (1.17)\ (\text{cont.})$$

$$+ \frac{1}{2n_o^2}(2A_y + B_z - 4C_y + 2n_o \Delta\dot{z}(0)) \sin n_o t$$

$$+ \frac{1}{n_o^2}(2B_y + C_z + 2n_o \Delta\dot{y}(0) + 4n_o^2 \Delta z(0)) + \frac{2}{n_o} C_y t ,$$

as the reader could verify (with enough patience) by replacing these formulae, their time derivatives, and the expressions for the forcing terms, in the differential equations (1.12).

The increasing oscillations are caused by the excitation terms of frequency n_o: the linear and quadratic terms $\Delta y(t)$ and $\Delta z(t)$ result from constant radial and along-track accelerations. This behaviour greatly resembles that of an undamped oscillator consisting of a mass that "accumulates" kinetic energy, and a spring that "gathers" potential energy as it stretches; the oscillations that occur when the mass is released can be described as exchanges of energy between mass and spring. In the case of a spacecraft, its mass plays the same role as in the oscillator, while the gravitational field acts as the spring. The analogy is particularly appropriate in the case of Δx, that (as (1.12) shows) is identical to the response of a simple harmonic oscillator of natural frequency n_o to a forcing acceleration Δa_x.

Resonance may occur under the influence of gravitational and non-gravitational forces (such as drag and solar radiation pressure). Certain features of the anomalous gravitational field, for example, may excert a pull on the satellite that is constant, or that repeats itself once in every revolution, almost exactly one orbital period apart. The same could be the case with drag, with a constant along-track effect, and also a repeating fluctuation, as the orbit crosses again and again the atmospheric bulge. Solar radiation pressure, resulting from the electromagnetic interaction between light from the sun and the surfaces of the spacecraft, may repeat itself at each revolution, if the orientation of the satellite towards the sun is also repeated (this happens in such diverse spacecraft as those of the GPS system and of the geosynchronous TDRS tracking system). In some cases (such as the repeat, "frozen" orbit that could be chosen for the GRM satellite-to-satellite tracking mission discussed at the end of this work), gravitational resonances can be very substantial.

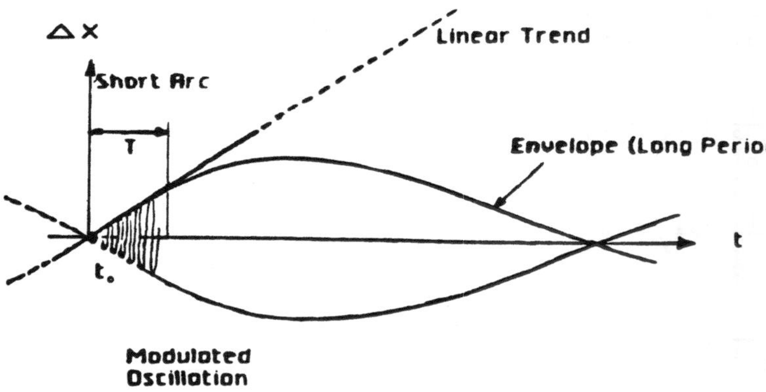

An oscillation modulated by a very long period envelope can be explained, over a relatively short interval T, as a linear resonance.

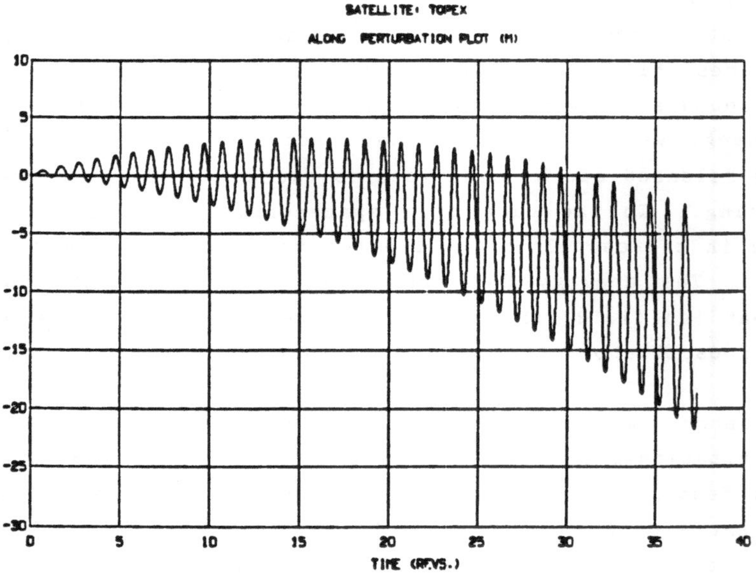

Resonant perturbation (notice the increasing oscillations of frequency n_e, and the parabolic "bend" due to linear and quadratic terms in (1.17) for Δy).
(Author's simulation of solar radiation pressure effects on TOPEX)

Orbit Resonances

Figure 1.2

Perfect resonance, where the forcing terms have exactly one or both critical frequencies, is most unlikely. But there are "deep" resonances, where those terms have frequencies very close to critical. If $\Delta\omega$ is the difference between the actual forcing frequency and either zero or n_o, then the solution of the equations is given by expressions (1.15), as we have a "non-resonant" oscillation. However, $\Delta x(t)$, $\Delta y(t)$, and $\Delta z(t)$ will resemble oscillations of frequency ω, with periodically modulated envelopes that have finite amplitudes and a beat frequency equal to $\Delta\omega$. This shape will result from the interaction of the special integral terms and the homogeneous response terms in (1.15). As $\Delta\omega$ is very small, the period of the envelope must be very long. Over a relatively short interval of time, the solution will seem to consist of oscillations of linearly varying amplitudes that strongly resemble the resonant solution (1.17). Only when the interval becomes comparable with the envelope period ($2\pi/\Delta\omega$) will the curvature and finite amplitude of the envelope become evident, as shown in Figure 1.2.

It is interesting to analyze expressions (1.17) to get some understanding of certain unusual characteristics of orbital perturbations associated with their resonant nature. For example, notice the efect of a constant along-track acceleration of amplitude $-C_y$: the quadratic term in Δy, which is also along-track, would be $\Delta y = 3/2\ C_y t^2$ for a constant force opposing the movement of the satellite, such as drag. This means that, under a breaking force, the satellite tends to travel further in a given period of time and, therefore, to go faster than if the force had not been applied. This conclusion is quite counterintuitive: in the familiar case of a boat coasting along the surface of a still lake, drag will be also constant (as long as the velocity does not change much) and opposed to the movement. For the boat, the effect is a loss of velocity, a shorter distance being travelled than if no drag were present. Since drag oposses movement, the net effect is always a loss of energy. But the satellite, according to the expression (1.17) for Δy, must gain velocity and, thus, kinetic energy. This gain must be offset by the loss of potential gravitational energy brought about by the corresponding decrease in height (i.e., in z) described by the term $-2n_o^{-1}\ C_y t$, in (1.17) for Δz (after replacing C_y with $-C_y$). A physical explanation for this peculiar behavior is that, for nearly circular orbits, the velocity of the spacecraft must be larger for a smaller radius (i.e., for a lower orbit), so the centrifugal force can balance the pull of gravity, which increases with decreasing height. Drag (or any other braking force) will bring the satellite gradually down, to a lower and faster orbit. This remains true as long as drag is much weaker than gravity. When this is no longer so, drag cannot be ignored in the equations of motion before they are linearized, the present theory ceases to be valid, and the movement of a reentering spacecraft begins to exhibit a "normal" behaviour: the satellite slows down.

Another characteristic of the resonant solutions worth noting is that the increasing along-track and radial oscillations are related in amplitude, the along-track ones growing always twice as fast as the radial ones, regardless of the actual sizes of the corresponding parameters A_y, B_y and A_z, B_z.

Finally, a constant radial acceleration C_z, such as may be caused by an error in GM, must result in a constant height offset Δz_c, in oscillations of frequency n_o (i.e., once per revolution), and a constant drift along-track equal to $2\, n_o\, \Delta z_c t$.

THE "BUTTERFLY" PATTERN

When an orbit is <u>estimated</u> by adjusting the initial state (so as to fit tracking data in the least squares sense), the radial, across-, and along-track differences Δx, Δy, Δz between the true and the estimated orbits are likely to have the general shape shown in Figure 1.3. This shape is known as a "butterfly" pattern for obvious reasons, and can be explained in terms of resonance. Plots of this type are routinely observed when comparing simulated "true" and "estimated" orbits, the latter fitted to "tracking" based on the former. The "estimated" orbits are integrated with slightly different force parameters than the "true" ones in order to see what effect such differences, or errors, may have on actual orbits obtained with real data. From expressions (1.13) one can see that the perturbations due to discrepancies between the initial conditions of the true and the estimated orbits consist of oscillations of frequency n_o, as well as a linear drift along-track, and offsets in x, y and z. Resonant perturbations are caused by errors in the force model, and the adjustment of the initial conditions will tend to reduce the efects of such errors by substracting oscillations, drift, and offsets from them. If one takes a linearly increasing (resonant) oscillation $Dt \sin n_o t$, and replaces the independent variable t with t-t', the result is a function with a butterfly envelope that pinches

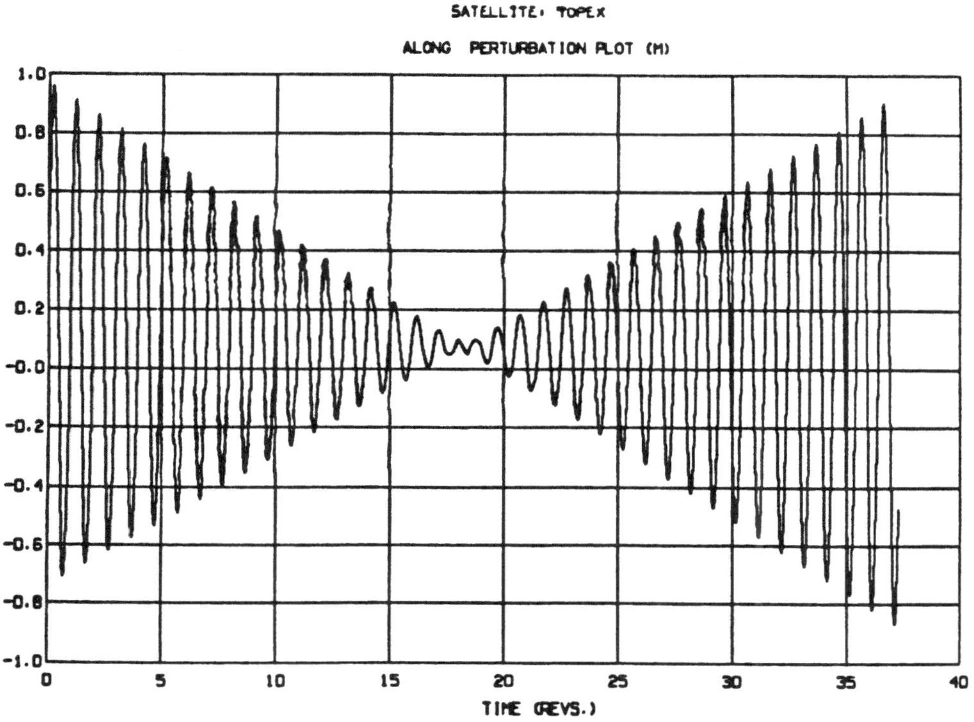

"Butterfly" pattern in the resonant perturbation of an orbit with adjusted initial state.

Figure 1.3

off at t=t' while retaining its rate of growth, or slope. Its mathematical expression is

$$D(t-t')\sin n_o(t-t') = [D\cos n_o t']t\sin n_o t - [D\sin n_o t']t\cos n_o t$$

$$- [D\,t'\cos n_o t']\sin n_o t + [D\,t'\sin n_o t']\cos n_o t, \qquad (1.18)$$

where the terms between square brackets are constants. So adding or substracting a steady oscillation of appropriate amplitude to an increasing oscillation results in a "butterfly". This is precisely what an adjustment of the initial conditions may do. If the tracking coverage is sufficiently dense, the minimization of the squares of the observation residuals tends to reduce also the r.m.s value, over the arc, of the perturbations themselves. As it happens, a "butterfly" pinching off at the mid-point of the arc has the smallest r.m.s. of all "butterflies" with the same rate of growth. Therefore, orbits with resonant perturbations, estimated with a good tracking coverage, tend to exhibit errors of the "butterfly" type that are smallest near the middle of the arc.

If an orbit is predicted, i.e., is integrated forwards from given initial conditions which are not adjusted, then the resonant oscillations grow steadily from the start (as in Figure 1.2), and there is no "butterfly" effect.

THE VARIATIONAL EQUATIONS

Consider the linearized equations of motion (1.12), and let Δp be an error, or disturbance, in the value of a parameter p, such as one of the components of the initial state (like Δx_o), or a potential coefficient, etc. Dividing both sides of (1.12) by Δp and taking the limit for $\Delta p \to 0$:

$$D_p\ddot{x}(t) = D_p a_x - n_o^2\, D_p\, x(t)$$

$$D_p\ddot{y}(t) = D_p a_y - 2\, n_o\, D_p\dot{z}(t) \tag{1.19}$$

$$D_p\ddot{z}(t) = D_p a_z + 3\, n_o^2\, D_p z(t) + 2\, n_o\, D_p\dot{y}(t)\ ,$$

because, as Δp is not a function of time (parameters are, of course, constants), it is valid to exchange the operations of limit and differentiation.

The solutions of equations (1.19) above are precisely the perturbations caused by a unit Δp; they represent the sensitivity of the orbit to small perturbations in the parameter p. As mentioned in the Introduction, these sensitivities have to be known before setting up the linearized observation equations needed to estimate parameters such as p from satellite data. Differential equations whose solutions are those sensitivities are called variational equations. Expressions (1.19) are, therefore, the variational equations corresponding to the choice of state variables $(x, y, z, \dot{x}, \dot{y}, \dot{z})$ made here. In fact, they are only approximations to the actual variationals, because they have been obtained after making a number of simplifying assumptions. Their main advantage is that they can be solved analytically, using tables of explicit integrals (or of Laplace transforms). As an example of Δp being a force parameter, consider $\Delta p = B_x$ in (1.15). In that case

$$D_{B_x} x(t) = \frac{1}{n_o^2-\omega^2} \sin \omega t - \frac{\omega}{n_o(n_o^2-\omega^2)} \sin n_o t$$

$$D_{B_x} y(t) = 0 \qquad (1.20)$$

$$D_{B_x} z(t) = 0$$

(Δy and Δz are independent from Δx and from the associated across-track accelerations). Alternatively, p could be one of the six components of the initial state. For example, if $\Delta p = \Delta \dot{y}_o$, the corresponding sensitivities would be:

$$D_{\dot{y}(o)} x(t) = 0$$

$$D_{\dot{y}(o)} y(t) = \frac{4}{n_o} \sin n_o t - 3t \qquad (1.21)$$

$$D_{\dot{y}(o)} z(t) = -\frac{2}{n_o} \cos n_o t + \frac{2}{n_o} ,$$

which are obtained from the homogeneous solutions (1.13) of the linearized equations, with $\Delta\dot{y}(0) = 1$ and $\Delta x(0)$, $\Delta y(0)$, $\Delta z(0)$, $\Delta\dot{x}(0)$, and $\Delta\dot{z}(0)$ all equal to zero.

While the analytical solutions of (1.19) are only approximations to the exact sensitivities, they are good enough for understanding the main characteristics of orbital perturbations. In some applications, they may be accurate enough for practical calculations, but often this is not the case. The usual thing to do is to obtain the rigurous form of the linearized equations of motion (i.e., without any simplifications). As these equations are, in general, not time-invariant, they are very difficult, or even impossible, to solve analytically, and have to be integrated numerically. This is done, for each parameter p, using as forcing terms disturbing accelerations where $\Delta p = 1$ and all the other parameter errors are zero. To obtain the sensitivities to a component of the initial state, the homogeneous, or unforced, equations are integrated with that state component equal to one, while all the others are zero.

CHAPTER 2. ORBITAL PERTURBATIONS OF GRAVITATIONAL ORIGIN

NOMINAL ORBITS

The gravity field can be described in terms of its potential $V(r, \phi, \lambda)$, where r, ϕ, and λ are the geocentric distance, the latitude and the longitude in an equatorial, earth-fixed system of spherical coordinates. The expansion of V in spherical harmonics is

$$V(r, \phi, \lambda) = \frac{GM}{R} \sum_{nm\alpha} C_{nm\alpha} P_{nm}(\cos \phi) \cos (m\lambda - \frac{\pi}{2} \alpha) , \qquad (2.1)$$

where R is the mean equatorial radius of the earth; G is the gravitational constant; M is the mass of the earth; P_{nm} is the asociated Legendre function of the first kind of integer degree n and order m; $C_{nm\alpha}$ is the corresponding potential coefficient (for zonal terms, the notation used here is C_{no}). The symbol " $\sum_{nm\alpha}$ " stands for " $\sum_{n=0}^{\infty} \sum_{m=0}^{n} \sum_{\alpha=0}^{1}$ ".

Because of the factors $(R/r)^{n+1}$ in (2.1), changes in r due to the lack of circularity of the orbit will cause changes in the values of the harmonics proportional to $n+1$ (to first order). Therefore, departures from a circular orbit are more important the higher the degree of the spherical harmonics, and have to be taken into account when studying the fine details of the field (with short spatial wavelengths) using satellite data. A very close approximation to the actual shape of the orbit may be hard (or impossible) to use to formulate an analytical theory of orbital perturbations. But a simple non-circular approximation (like the secularly precessing ellipse discussed below) is already much better than a circle. Notice that the non-circularity of the orbit is only a problem when trying to bring the non-spherical part of the gravity field into the theory, to get accurate expressions for the forcing accelerations Δa_x, Δa_y, and Δa_z in the linearized equations of motion (1.12). For the dynamic terms of those equations, which govern the unforced response, the circular orbit approximation is quite adequate as long as the average eccentricity of the actual orbit is small.

The field that corresponds to a potential $V = GM/r$, or central force field, is that of a homogeneous sphere with the same mass as the earth. In such a field, a body moving at less than the escape velocity and a mass negligible compared to the earth's (e.g., a spacecraft), will follow an elliptical orbit obeying the three Laws of Kepler (including expression (1.6) , the Third Law , for n_o). One focus of the ellipse will be at the geocenter, and the orbit itself will lie entirely on a plane with a fixed orientation in space. Because the earth is not a perfect sphere, a real orbit departs from the Keplerian ellipse in several ways. The major ones are slow and steadily increasing, or secular, changes in the orientation of both the orbital plane in space, and of the ellipse inside this plane. These changes are caused largely by the zonal features of the anomalous field, as it will be explained. Most prominent of all zonal features is the equatorial bulge, whose effect can be described mostly in terms of the second zonal C_{20}. The gravitation of this bulge "tries" to bring the orbit plane to rest on the equator, and gyroscopic reaction causes the plane to turn, instead, about the axis of symmetry of the bulge, which is very close to the spin axis of the earth. This turning is known as the precession of the nodes, and is measured by an increase in the right ascension Ω of the ascending node (where the orbit crosses the equator into the Northern hemisphere), at a constant rate $\dot{\Omega}$ (the origin of right ascensions is the projection on the equator of a line drawn from the geocenter towards the spring equinox). The rate $\dot{\Omega}$ due to C_{20} is

$$\dot{\Omega} = \frac{3}{2} n_o C_{20} \left(\frac{R}{a}\right)^2 \left(1-e^2\right)^{-2} \cos I , \qquad (2.2)$$

where n_o is given by (1.6). Notice $\dot{\Omega} = 0$ when $I = \pm \frac{\pi}{2}$ (polar orbits).

The definition of Ω and the other Keplerian elements: the semimajor axis a, the eccentricity e, the mean anomaly M, the inclination I, and the argument of perigee ω, are explained in Figure 2.1. Together, they determine the size, shape, and orientation of the orbit, as well as the position and velocity of the orbiting body, so the these six elements are completely equivalent to the six cartesian components of the state vector $\underline{s}$ (three for position and three for velocity) introduced in the previous chapter.

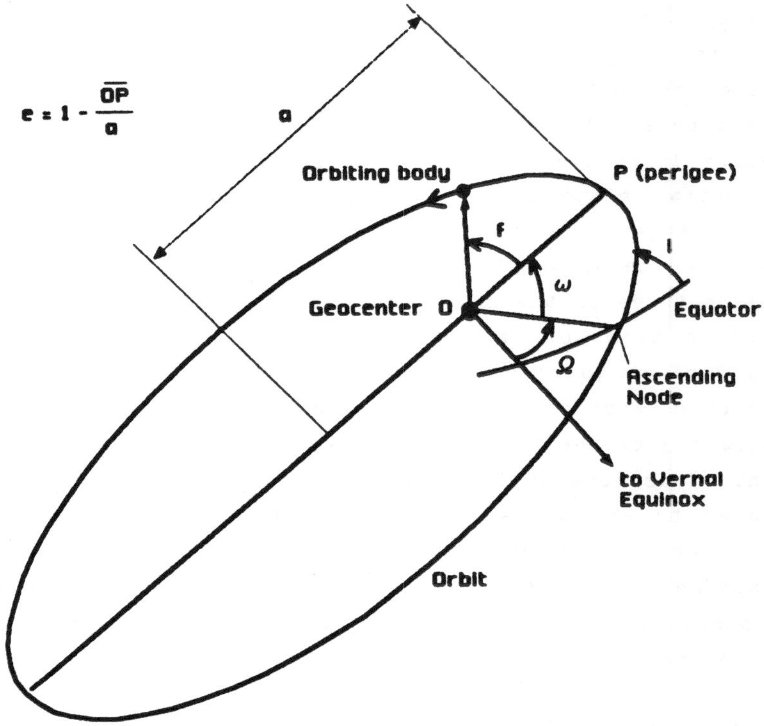

The Keplerian Elements
Figure 2.1

The second main effect of the bulge is a secular change in the angle ω between the major axis of the elliptical orbit and the equator. This angle is also known as the argument of perigee. This rotation of the ellipse in its plane occurs because the equatorial bulge makes the force of gravity increase faster as one approaches the earth than it would if the earth were a sphere. As a result, the orbiting body makes a tighter turn at perigee (its closest point) than it would make in a central field. The orbit is no longer a closed ellipse, but a rosetta: a pattern that can be obtained by rotating the Keplerian ellipse while the satellite is turning along it. The rate of this rotatiom, or precession of the perigee, is

$$\dot{\omega} = \frac{3}{4} n_o C_{20} \left(\frac{R}{a}\right)^2 (1-e^2)^{-2} (1-5 \cos^2 I). \qquad (2.3)$$

In an elliptical orbit, the instantaneous angular velocity is slowest at apogee and fastest at perigee. This angular velocity is the time derivative of the angle f, known as the true anomaly (see Figure 2.1). The quantity M, known as the mean anomaly, is defined as

$$M = \dot{M}(t-t_o) , \qquad (2.4)$$

where t_o is a time of passage through perigee, and $\dot{M} = 2\pi/\text{Orbit Period}$. (The mass of the earth "M" is always part of the product "GM", so it can not be confused with the mean anomaly.) For a spherical earth, $\dot{M}$ would be the same as n_o. Under the effect of the bulge, it changes to

$$\begin{aligned}\dot{M} &= n_o[1 - \frac{3}{4} C_{20} \left(\frac{R}{a}\right)^2 (1-e^2)^{-3/2} (3 \cos^2 I-1)] \\ &= n_o + O(C_{20}) .\end{aligned} \qquad (2.5)$$

It is easier, in practice, to work with M than with the more geometrically intuitive true anomaly f. Moreover, for the nearly circular orbits to be studied here, it is adequate to assume that $f = M$. The earth's bulge has the additional effect of introducing an up-and-down "wiggle" in the orbit twice per revolution, and much lesser oscillations at higher frequencies. These fluctuations are periodical and do not add up over time the way the secular changes in Ω, ω, and M do, so they are not included in the definition of the approximate, nominal orbit. The ellipse on which this orbit is based is known as the mean ellipse, because all variations faster than the secular ones in ω and Ω have been smoothed out; for the same reason, the Keplerian elements of this ellipse are known as mean elements. These are also the nominal elements of the orbit, a_o, e_o, I_o, Ω_o, ω_o, and M_o, where

$$\Omega_o(t) = \dot{\Omega}t + \Omega(t\ )$$

$$\omega_o(t) = \dot{\omega}t + \omega_o(t_o) \qquad (2.6)$$

$$M_o(t) = \dot{M}t + M_o(t_o)\ .$$

Here t_o is the time origin, usually the beginning of the orbit arc. The other three elements: a_o, e_o, and I_o, are constants in the nominal orbit. The rates $\dot{\Omega}$, $\dot{\omega}$, and $\dot{M}$ are those given by (2.2), (2.3), and (2.5).

$\dot{M} + \dot{\omega}$ is a closer approximation to the actual orbital frequency than n_o or $\dot{M}$; but the difference is small enough to ignore it in many cases.

To give some idea of the magnitudes of $\dot{\Omega}$, $\dot{\omega}$, and $\dot{M}$ for ordinary spacecraft, at an altitude of about 1000 km $\dot{\Omega}$ and $\dot{\omega}$ are of the order of a few degrees per day, with periods of two or three months, while the period for $\dot{M}$ is of about ninety minutes, or close to 14 revolutions per day.

For nearly circular orbits, the following approximations are valid with an error of the order of the square of the eccentricity:

$$f = M \qquad (2.7)$$

and

$$r = a\,(1-e\cos M)\ . \qquad (2.8)$$

"FROZEN" AND REPEATING ORBITS

In orbits of very small eccentricity, the secular effects on ω of the zonals C_{no} of degree n higher than 2 can become comparable to those of C_{20} itself, because the position of the perigee is ill-defined for such orbits. Since the effects of some zonals (notably C_{30}) are opposite in sign to those of C_{20}, in some cases their sum may cancel out, so $\dot{\omega} = 0$. An orbit without a precessing perigee (or one where precession is negligible over a long period of time) is known as a "frozen" orbit. To obtain a nearly "frozen" orbit, the Keplerian elements at the start must satisfy the (approximate) relationships given by Cook (see References at the end of this work):

$$\omega = \frac{\pi}{2} \qquad (2.9)$$

$$e = \frac{1}{3} \sum_{n=3}^{N(odd)} C_{no}\left(\frac{R}{a}\right)^{n} \frac{(n-1)}{n(n+1)} P_{n1}(0)\ P_{n1}(\cos I)\left[C_{20}\left(\frac{R}{a}\right)^{2}\left(1-\frac{5}{4}\sin^{2} I\right)\right]^{-1}$$

$$= 1.182 \times 10^{-3} \left(\frac{R}{a}\right) \sin I + O(C_{50}) , \qquad (2.9)\ (\text{cont.})$$

where N is a conveniently large number, and N(odd) indicates that the sum is over odd values of n. The values of eccentricity given by (2.9) are always quite small, so "frozen" orbits are also nearly circular. Notice that the perigee must stand on the Northern hemisphere. In terms of Cook's theory, this is due to the opposing signs of C_{20} and C_{30} (the "pear-shaped" zonal coefficient).

Another special case where the perigee is "frozen" and $\dot{\omega} = 0$ occurs, according to (2.3), when $I = \cos^{-1}(\sqrt{1/5})$. This happens for $I = 63.4°$ and $I = 116.6°$; both values of I are known as "critical inclinations". At them, the effect of the bulge on the perigee simply vanishes. To obtain a cancellation of all zonal secular effects (which, different from that of C_{20}, do not become zero at these inclinations), certain special conditions must be fulfilled by the orbit. They are similar to (2.9) (which are no longer valid in this case), but more complex, and have been derived by Hough (see References).

In general, no orbit is perfectly "frozen", but experiences some fluctuations in ω about the value of $\pi/2$ given by (2.9). These are periodical, and receive the name of librations: the perigee of a nearly "frozen" orbit librates instead of precessing. The period of libration is the same as that of precession. What matters here is not whether the orbit is perfectly "frozen", but whether the amplitude of the libration of its perigee is sufficiently small to be ignored.

One reason for "freezing" the perigee is to get an orbit that repeats itself almost exactly from the point of view of an earth-bound observer. Such an orbit can be useful in surveying both lands and oceans, for example with radar altimeters. They may be also important for the global mapping of the gravity field with satellite-to-satellite tracking data, as explained in the last Chapter. Another reason for choosing a "frozen" orbit is to ensure that a spacecraft attains the same heights at the same latitudes; this can be of some importance in altimeter missions.

For the orbit to repeat itself after a period of time T_r, the orbital frequencies $\dot{\Omega}$, $\dot{\omega}$, and $\dot{M}$ must be harmonics of the repeat frequency ω_r, where

$$\omega_r = 2\pi/T_r \,. \tag{2.10}$$

Usually, T_r is less than a month; as $\dot{\omega}$ cannot be smaller than ω_r unless it is zero (i.e., unless $\dot{\omega}$ is the zero harmonic of ω_r), the orbital eccentricity has to be very high, according to (2.3), except in the case $\dot{\omega} = 0$. This means that to have a repeating orbit with small eccentricity, one must choose a "frozen" orbit.

An important condition to be satisfied so that the orbit does repeat only once in each interval T_r (avoiding unintended "cycles-within-cycles"), is that the number N_R of orbital revolutions and the number N_D of turns of the earth respect the orbital plane must be relative prime integers: the only common factor they can have is the unity. If θ is the hour angle of Greenwich, so $\dot{\theta}$ is the spin rate of the earth, then the relative angular speed of the earth respect to the orbit plane is $\dot{\theta} - \dot{\Omega}$. Since $\dot{\theta} - \dot{\Omega} \approx \dot{\theta}$, the number N_D and the value of T_r, in days, are almost the same.

Since a number of forces acting on a spacecraft tend to modify its orbit gradually, it is necessary to execute trimming maneuvers from time to time, to keep it in a "frozen" repeating orbit. This means that satellites meant to follow this type of orbit must have rocket engines and carry fuel for such maneuvers.

Among spacecraft with geodetic applications, those with repeating, "frozen" orbits that have been (or are expected to be) in operation include: the altimeter satellites SEASAT, GEOSAT, ERS-1, TOPEX, and POSEIDON; the navigational satellites of the Global Positioning System, or GPS; and the two spacecraft of the projected Gravity Mapping Mission (GRM), at least in their "low-low" configuration.

THE GROUND-TRACK OF AN ORBIT

As a satellite orbits the earth, the point directly underneath it describes a curve on the surface of the planet that is known as the ground-track. In many studies involving spacecraft data, one often needs to calculate approximately where the satellite may be at a given time, in earth-fixed coordinates. The following expressions give the geocentric distance, the latitude, and the longitude of a satellite as functions of time t; their accuracy is of a few kilometers if the orbit is close to circular (one needs to know the initial values of the nominal Keplerian elements and of θ at $t = t_0$).

$$r = a(1 - \cos M)$$

$$\phi = \sin^{-1} [\sin I \sin(\omega + M)] \qquad (2.11)$$

$$\lambda = \sin^{-1} [\cos I \sin(\omega + M)/\cos\phi] + \Omega - \theta ,$$

where ω, M, and Ω are given by (2.6), while $\theta = \dot{\theta}(t-t_o)+\theta(t_o)$. Notice that the expression for r is the same as (2.8). There is more than one possible solution for λ. The correct one puts the angle $\lambda - \Omega + \theta$ in the same quadrant as $(\omega+M)$ sign $\lfloor\frac{\pi}{2} - I\rfloor$. If $I = \frac{\pi}{2}$, then $\lambda = \Omega - \theta$ if $\omega + M < \pi$, and $\lambda = \pi + \Omega - \theta$ if $\omega + M \geq \pi$.

THE EXPRESSION OF THE DISTURBING POTENTIAL T IN MEAN KEPLERIAN ELEMENTS

In what follows, using the precessing mean ellipse as the nominal orbit, the difference between it and the true orbit caused by the anomalous gravity field will be treated as a first order perturbation. This perturbation will be assumed to be equal to the algebraic sum of the effects of each term in the spherical harmonic expression of the potential (other than the central field and the secular effects of C_{20}). Let the $\tilde{C}_{nm\alpha}$ be the known values of the potential coefficients $C_{nm\alpha}$ in (2.1), and the $\Delta C_{nm\alpha}$, the errors in the $\tilde{C}_{nm\alpha}$. Then, to first order, the difference between the true orbit and that integrated numerically with the $\tilde{C}_{nm\alpha}$ will be also the difference between the perturbations of the nominal orbit caused by the $C_{nm\alpha}$ and those caused by the $\tilde{C}_{nm\alpha}$. Therefore, the same first order formulae that give the difference between the true and the nominal orbits, must give also that between the true and the numerically integrated orbits, or orbit errors, provided that one replaces in them $C_{nm\alpha}$ with $\Delta C_{nm\alpha}$.

The field that has a spherical harmonic expansion with the $\Delta C_{nm\alpha}$ as potential coefficients is, of course, the disturbing field of physical geodesy. The main purpose of what follows is to discuss the effect of that disturbing field on the orbits of artificial satellites.

The spherical harmonics expansion of the disturbing potential $T(r, \phi, \lambda)$ can be obtained by replacing in (2.1) $C_{nm\alpha}$ with $\Delta C_{nm\alpha}$. As the mean Keplerian elements give the nominal position of the spacecraft on the reference orbit at time t, it is possible to use them as coordinates instead of r, ϕ, and λ. When this is done, the spherical harmonic expansion of T^o (i.e., T on the nominal orbit) becomes

$$T^o = \frac{GM}{R} \sum_{nm\alpha} \left(\frac{R}{a_o}\right)^{n+1} \Delta C_{nm\alpha} \sum_{p=0}^{n} F_{nmp}(I_o) \sum_{q=-\infty}^{\infty} G_{npq}(e_o) \cos[(n-2p+q)(\omega_o+M_o)$$

$$-q\omega_o + m(\Omega_o-\theta) + \phi_{nm\alpha}] , \qquad (2.12)$$

where

$$\phi_{nm\alpha} = -\frac{\pi}{2} [\alpha + \frac{1}{2} (1-(-1)^{n-m})] .$$

The $F_{nmp}(I_o)$ are known as the <u>inclination functions</u>, and $G_{npq}(e_o)$, as the <u>eccentricity functions</u>. They are defined in detail in Kaula's book (see References), where one also finds the complete derivation of (2.12).

ECCENTRICITY AND INCLINATION FUNCTIONS

As a rule, the expansion in q can be truncated at $q=\pm 1$ for harmonics of degree $n \leq 10$, and $q=\pm 2$ for $10<n<30$. In general, the range of q needed to maintain a minimum of accuracy increases with n, and can be as high as $q=\pm 5$ for $n=300$.

The following approximations give $G_{npq}^{(e)}$ correct to the order of e^2, for $-1 \leq q \leq 1$; they can be useful to derive some rough approximations to the formulae introduced in the remainder of this Chapter, approximations that can be adequate for preliminary error analyses of satellite missions, and similar work:

$$G_{npo}^{(e)} = 1$$

$$G_{np\pm 1}^{(e)} = \frac{e}{2} [n \pm 2(n-2p)+1] . \qquad (2.13)$$

To calculate the inclination functions efficiently, the following procedure may be used. The expression for a spherical harmonic along a major circle on the unit sphere, when the plane of the circle is inclined by an angle I relative to the equator (the plane where $\phi=0$) is

$$P_{nm}(\cos \phi)\cos[m\lambda-\frac{\pi}{2} \alpha] = \sum_{p=0}^{n} F_{nmp}(I) \cos [(n-2p)F+m\lambda_o+\phi_{nm\alpha}] , \qquad (2.14)$$

where $\phi_{nm\alpha}$ is given in (2.12), λ_o is the longitude where the great circle intersects the equator and F=0, and F is the angle along the great circle from the equator to the point (ϕ,λ). Clearly, the $F_{nmp}(I)$ are the Fourier coefficients of the spherical harmonic along the circle, i.e., regarded as a function of F. Since cos(x)=cos(-x), sin(x)=-sin(-x), there are always two values of p: p_1, and $p_2 = n-p_1$, such that $\cos[(n-2p_1)F+\phi_{nm\alpha}]=\cos[(n-2p_2)F+\phi_{nm\alpha}](-1)^{\alpha+n-m}$. Therefore, the previous expression can be writter, for $\alpha = 0$ and $\lambda_o = 0$, as

$$P_{nm}(\cos\phi)\cos m\lambda = \sum_{p_1=0}^{n^*/2} [F_{nmp_1}(I)+(-1)^{n-m} F_{nm(n-p_1)}(I)]\cos[(n-2p_1)F+\phi_{nm\alpha=0}],$$

where n*/2 indicates the closest integer less or equal to n/2. If, on the other hand, one choses $\alpha=1$, the result is

$$P_{nm}(\cos\phi)\cos\left(m\lambda-\frac{\pi}{2}\right) = \sum_{p_1=0}^{n^*/2} [F_{nmp_1}(I)-(-1)^{n-m} F_{nmp(n-p_1)}(I)]\cos[(n-2p_1)F+\phi_{nm\alpha=1}].$$

Beginning by calculating $P_{nm}(\cos\phi)\cos\left(m\lambda-\frac{\pi}{2}\right)$ at evenly spaced points along the great circle (which can be done using methods that are well known to geodesists) first with $\alpha=0$, then with $\alpha=1$, and getting the Fourier coefficients for each case (preferably with a Fast Fourier Transform algorithm), one finishes by solving the equations in $F_{nmp_1}(I)$ and $F_{nmp_2}(I)$:

$$\begin{aligned} F_{nmp_1}(I) + (-1)^{n-m} F_{nmp_2}(I) &= C_{nmp_1 0} \\ F_{nmp_1}(I) - (-1)^{n-m} F_{nmp_2}(I) &= C_{nmp_1 1}(-1)^{n-m}, \end{aligned} \tag{2.15}$$

where $C_{nmp_1 0}$ and $C_{nmp_1 1}$ are the Fourier coefficients for $\alpha=0,1$. If one is working with normalized spherical harmonics, this procedure yields the values of the corresponding normalized inclination functions.

THE FORCING TERMS IN HILL'S EQUATIONS CORRESPONDING TO THE DISTURBING POTENTIAL T

Writing Ω_o, ω_o, and M_o as the time-functions given in (2.6), and remembering that

$$\theta = \dot{\theta}(t-t_o) + \theta(t_o) \ , \qquad (2.16)$$

the expansion of T can be given as a sum of cosine functions of time:

$$T^o(t) = \sum_{nmpq\alpha} T_{nmpq\alpha} \cos (f_{nmpq} t + \phi_{nmpq\alpha}) \ , \qquad (2.17)$$

where the amplitude, frequency, and phase are given by

$$T_{nmpq\alpha} = \frac{GM}{R} \left(\frac{R}{a_o}\right)^{n+1} \Delta C_{nm\alpha} F_{nmp}(I_o) G_{npq}(e_o) \ ,$$

$$f_{nmpq} = (n-2p+q)(\dot{\omega} + \dot{M}) - q\dot{\omega} + m(\dot{\Omega} - \dot{\theta}) \qquad (2.18)$$

$$\phi_{nmpq\alpha} = (n-2p+q)(\omega_o(t_o) + M_o(t_o)) - q\omega_o(t_o) + m(\Omega_o(t_o) - \theta(t_o)) + \phi_{nm\alpha}.$$

Here " $\sum_{nmpq\alpha}$ " represents the combination of the various sumations in (2.12).

The forcing terms Δa_x, Δa_y, and Δa_z in the linearized equations of motion are perturbing accelerations related to T_o as follows:

$$\Delta a_x = D_x T^o = \frac{1}{r_o \sin(f_o + \omega_o)} D_{I_o} T^o$$

$$= \frac{1}{a_o \sin(M_o + \omega_o)} D_{I_o} T^o \qquad \text{(all partial derivatives are taken on the nominal orbit)}$$

$$\Delta a_y = D_y T^o = \frac{1}{a_o} (D_{M_o} T^o + D_{\omega_o} T^o) \qquad (2.19)$$

$$\Delta a_z = D_z T^o = D_r T^o = D_{a_o} T^o \ ,$$

where use has been made of the approximations in (2.7) and (2.8) for f and r in terms of M and a. To simplify the mathematics of the along-track perturbations Δy, a term proportional to $D_{\Omega_o} T^o$ has been omitted in Δa; as a result, the theory for Δy in what follows is valid only for high inclination orbits. These are the orbits chosen for altimeter and other earth-surveying satellites, such as the GRM pair discussed in Chapter 3.

Taking the various partial derivatives of (2.12) acording to (2.19), one finds that the disturbing accelerations have the general form

$$\Delta a_x = \sum_{nmpq\alpha} \Delta \bar{C}_{nm\alpha} A_{x\ nmpq} \cos((f_{nmpq} \pm (\dot{\omega}+\dot{M}))t + \phi_{nmpq\alpha} + \frac{\pi}{2})$$

$$\Delta a_y = \sum_{nmpq\alpha} \Delta C_{nm\alpha} A_{y\ nmpq} \cos(f_{nmpq}t + \phi_{nmpq\alpha} + \frac{\pi}{2}) \qquad (2.20)$$

$$\Delta a_z = \sum_{nmpq\alpha} \Delta C_{nm\alpha} A_{z\ nmpq} \cos(f_{nmpq}t + \phi_{nmpq\alpha}) .$$

The derivations of the expressions for Δa_y and Δa_z are immediate, while the case of Δa_x is considerably more complicated, because the factor $(a_o \sin(M_o+\omega_o))^{-1}$ is singular whenever $M_o+\omega_o = 0, \pi, 2\pi, \ldots$. The expression for Δa_x in (2.18) can be obtained by working backwards from

$$\Delta x = r(\Delta I \sin(\omega+f) - \Delta\Omega \sin I)$$
$$= a_o(\Delta I \sin(\omega_o + M_o) - \Delta\Omega \sin I_o) , \qquad (2.21)$$

which is based on purely geometrical considerations. ΔI and $\Delta\Omega$ are perturbations of I_o and Ω_o, and the corresponding expressions (minus homogeneous terms) can be found in Kaula's book (see References). As a result, the expression for Δa_x is undefined when n, m, p and q are such that $f_{nmpq} = 0$.

The amplitudes of the cosines are:

$$A_{x\ nmpq} = \frac{GM}{R^2} \left(\frac{R}{a_o}\right)^2 \frac{(n_o^2 - f^2_{nmpq} \pm)}{n_o(1-e_o^2)^{1/2}} \frac{G_{npq}(e_o)}{f_{nmpq}\pm} \{ \mp [\frac{(n-2p)\cos I_o - m}{\sin I_o}] F_{nmp}(I_o)$$
$$\langle f_{nmpq} \pm (\dot{\omega}+\dot{M}) \rangle$$
$$+ \frac{\partial}{\partial I_o} F_{nmp}(I_o) \}$$
$$\langle f_{nmpq} \rangle \qquad (2.22)$$

$$A_{y\ nmpq} = \frac{GM}{R^2} \left(\frac{R}{a_o}\right)^{n+2} F_{nmp}(I_o) G_{npq}(e_o)(2n-4p+q)$$

and

$$A_{z\ nmpq} = -\frac{GM}{R^2} \left(\frac{R}{a_o}\right)^{n+2} (n+1) F_{nmp}(I_o) G_{npq}(e_o) .$$

The signs "$f_{nmpq}\pm$", "$<f_{nmpq} \pm (\dot{\omega}+\dot{M})>$" and "$<f_{nmpq}>$" indicate that the first term within the curly brackets in $A_{x\ nmpq}$ is associated with two frequencies (with "-" sign for the first and "+" sign for the second), while the other term corresponds to f_{nmpq} alone. Note that "α" is not present in any of these expressions: "cosine" and "sine" terms ($\alpha = 0$, $\alpha = 1$) have the same amplitudes.

Fortunately, Δy and Δz are both independent of Δx, and Δa_x, according to the linearized equations (1.12). Moreover, in the cases of satellite altimetry and of GRM satellite-to-satellite tracking (the two examples of the application of this theory given in Chapter Three) there is no need to use the expressions for Δx or Δa_x, because the across-track perturbations do not appear in a first order theory for either type of satellite data.

LUMPED COEFFICIENTS

According to (2.15), all terms of the same frequency f_{nmpq} in the cosine expansions of T, Δa_x, Δa_y, and Δa_z must satisfy the conditions that $(n-2p+q)$ = constant, that n is only even or only odd, and that both q and m are constants as well. Therefore, all coefficients $\Delta C_{nm\alpha}$ of the disturbing potential T of the same order m and of the same parity in n create perturbations at that frequency, and their joint contributions to the disturbing accelerations are:

$$\Delta a_{x\ f_{nmpq\ \alpha}} = \sum_{nmpq\alpha}^{(constr)} \Delta C_{nm\alpha}\ A_{x\ nmpq} \cos\left((f_{nmpq} \pm (\dot{\omega}+\dot{M}))t + \phi_{nmpq\alpha} + \frac{\pi}{2}\right)$$

$$\Delta a_{y\ f_{nmpq\ \alpha}} = \sum_{nmpq\alpha}^{(constr)} \Delta C_{nm\alpha}\ A_{y\ nmpq} \cos\left(f_{nmpq} t + \phi_{nmpq\alpha} + \frac{\pi}{2}\right) \qquad (2.23)$$

$$\Delta a_{z\ f_{nmpq\ \alpha}} = \sum_{nmpq\alpha}^{(constr)} \Delta C_{nm\alpha}\ A_{z\ nmpq} \cos\left(f_{nmpq} t + \phi_{nmpq\alpha}\right) .$$

where "(constr)" indicates that $n,m,p,$and q are subject to the constraints mentioned above, and "$\pm(\dot{\omega}+\dot{M})$" indicates that there are three frequencies involved in the case of Δa_x, as already explained.

The perturbations in x, y, z related to the disturbing accelerations (2.23) are the forced responses of equations (1.12) to sinwaves. According to (1.15), they are oscillations of the same frequency f_{nmpq}, of the form

$$\Delta x_{f_{nmpq\,\alpha}}(t) = X_{f_{nmpq\,\alpha}} \cos((f_{nmpq} \pm (\dot{\omega}+\dot{M}))t + \phi_{nmpq\alpha}+\frac{\pi}{2}) + \text{homogeneous terms}$$

$$\Delta y_{f_{nmpq\,\alpha}}(t) = Y_{f_{nmpq\,\alpha}} \cos(f_{nmpq}\, t + \phi_{nmpq\alpha} + \frac{\pi}{2}) + \text{homogeneous terms}$$

$$\Delta z_{f_{nmpq\,\alpha}}(t) = Z_{f_{nmpq\,\alpha}} \cos(f_{nmpq} t + \phi_{nmpq\alpha}) + \text{homogeneous terms} \;, \tag{2.24}$$

where

$$X_{f_{nmpq\,\alpha}} = [\sum_{nmpq}^{(constr)} \Delta C_{nm\alpha} A_{x\; nmpq}][n_o^2 - (f_{nmpq} \pm (\dot{M}+\dot{\omega}))^2]^{-1}$$

$$Y_{f_{nmpq\,\alpha}} = [(f_{nmpq}^2 + 3n_o^2)\sum_{nmpq}^{(constr)} \Delta C_{nm\alpha} A_{y\; nmpq} - 2n_o f_{nmpq} \sum_{nmpq}^{(constr)} \Delta C_{nm\alpha} A_{z\; nmpq}]$$

$$\times\, [f_{nmpq}^2\,(n_o^2 - f_{nmpq}^2)]^{-1}$$

$$Z_{f_{nmpq\,\alpha}} = [f_{nmpq} \sum_{nmpq}^{(constr)} \Delta C_{nm\alpha} A_{z\; nmpq} + 2n_o \sum_{nmpq}^{(constr)} \Delta C_{nm\alpha} A_{y\; nmpq}]$$

$$\times\, [f_{nmpq}(n_o^2 - f_{nmpq}^2)]^{-1} \;. \tag{2.25}$$

Clearly, $X_{f_{nmpq\,\alpha}}$, $Y_{f_{nmpq\,\alpha}}$, and $Z_{f_{nmpq\,\alpha}}$ are linear combinations of $\Delta C_{nm\alpha}$ of the same order m, and either even or odd degree n, all "lumped" together. For this reason, they are known as "lumped coefficients".

ORBITAL SENSITIVITIES TO ERRORS IN POTENTIAL COEFFICIENTS

The sensitivity of satellite data w to a coefficient $\Delta C_{nm\alpha}$ of the disturbing potential T has a mathematical expression of the general form

$$D_{C_{nm\alpha}} w = D_{C_{nm\alpha}} w^{(direct)} + D_{\underline{s}}\, w\, D_{C_{nm\alpha}} \underline{s} \tag{2.26}$$

which results from dividing (I.5) by $\Delta C_{nm\alpha}$ and taking the limit for $\Delta C_{nm\alpha} \to 0$. The partial derivative $D^{(direct)}_{C_{nm\alpha}} w$, if non-vanishing, as with gradiometer data, is related to the spherical harmonic $\frac{GM}{R} (\frac{R}{r}) P_{nm}(\cos \phi) \cos (m\lambda - \frac{\pi}{2} \alpha)$. This is the direct sensitivity of the observation w with respect to $\Delta C_{nm\alpha}$. As for the other term in (2.26), or indirect sensitivity, the partial derivatives in $D_{\underline{s}} w$ reflect the geometry of the obervation, and may include the direction cosines of the line from a tracking station to a satellite, etc. The partial derivatives in $D_{C_{nm\alpha}} \underline{s}$ on the other hand, are solutions of the variational equations (discussed at the end of Chapter I), and can be obtained from (2.24-25) after making $\Delta C_{nm\alpha} = 1$ and all the other potential coefficient errors zero in (2.22).

CHAPTER 3. TOPICS RELATED TO THE MAPPING OF THE GRAVITY FIELD

"M-DAILIES" AND SHALLOW RESONANCES

The amplitudes (2.25) of the gravitational perturbations (2.24) at frequency f_{nmpq} are proportional to the inverse of either this frequency, or its square, and to the inverse of $(n_o^2 - f^2_{nmpq})$. For certain combinations of values of n, m, p, and q , these quantities are going to be very small, and their inverses very large, the same as the amplitudes, or "lumped coefficients", in (2.25). Therefore, certain "lumped coefficients" will have a much larger effect than others on Δx, Δy, and Δz, and also a larger influence on the "observed minus computed" residuals Δw of the tracking data. As the Δw are the source of information on the disturbing potential and on its coefficients $\Delta C_{nm\alpha}$, it is clear that those coeficients that create the largest perturbations and contribute the most to the Δw are also the ones that can be estimated better. Moreover, the factors in (2.25) give different weights to the different $\Delta C_{nm\alpha}$ that make up a given "lumped coefficient". These uneven weights depend on the degree n, and also on the mean inclination I_o and the mean eccentricity e_o of the orbit, through the inclination and eccentricity functions in (2.22). Therefore, different orbits are most sensitive to different coefficients. One could also say that specific spacecraft "see" best specific parts of the gravity field; this is why people who work at getting general purpose satellite models prefer to use data from spacecraft in different orbits (by contrast, "single-satellite" field models are sometimes produced; they are said to be "tuned" or "tailored" to that spacecraft; and help to calculate its orbit accurately). As most satellites of geodetic interest have orbits with similar (small) e_o, the element that they most differ in is I_o, so the tendency is to use simultaneously data from many spacecraft whose orbital inclinations are as different as possible. At least, this has been the case until now, because a conjunction of relatively low accuracy measurements and poor data coverage has forced reliance on the largest perturbations as sources of information. With a dedicated satellite mission of the sort described at the end of this Chapter, a single orbit (with two spacecraft tracking each other along it) could be enough for mapping the field to a degree of accuracy and detail far in excess of what has been possible until now with conventional tracking.

From the formula $f_{nmpq} = (n-2p+q)(\dot{\omega}+\dot{M})-q\dot{\omega}+m(\dot{\Omega}-\dot{\theta})$, it follows that, as $\dot{M}$ is much larger than $\dot{\omega}$ or $\dot{\Omega}-\dot{\theta}$, the following combinations of n,m,p, and q will result in large along-track and radial perturbations (for the across-track, (n-2p+q+1) and (n-2p+q-1) should be considered, as well as (n-2p+q)):

(a) (n-2p+q) = k, where k = -1, 0, 1; m nonzero. Here the frequencies have the form

$$f_{nmpq} = -q\dot{\omega}+m(\dot{\Omega}-\dot{\theta})+k\dot{M} , \tag{3.1}$$

and the corresponding perturbations are usually known as "m-dailies", as they repeat about m times a day for k = 0 (remember that $\dot{\Omega}-\dot{\theta} \gg \dot{\omega}$); or else their envelopes do, for k = ±1.

(b) $(n-2p+q)(\dot{\omega}+M)+m(\dot{\Omega}-\dot{\theta})$ close to 0 or to $\pm\dot{M}$: this happens when m is a multiple of the approximate number of revolutions the satellite completes each day. This number is close to 13 or 14 for most spacecraft of interest (height ≈0.1 earth radius). The corresponding perturbations are called "shallow resonances", because the condition resembles that of deep resonance to be described later on, where f_{nmpq} is much closer to zero or to n_o. With approximately 13 revolutions per day, for example, the first shallow resonance involves the coefficients of order m = 13, the second shallow resonance, those of order m = 26, etc. The coefficients of those orders are usually referred to as "resonant" coefficients, although it would be more proper to call them "shallow resonant". The periods of these perturbations (or their envelopes) range from several days to several weeks.

(c) $m = 0$ (zonals), $(n-2p+q) = k$, $k = -1, 0, 1$. Here, for $k=0$, the significant perturbations have for frequencies $\dot{\omega}$ and its low harmonics $q\dot{\omega}$, with periods of the order of months or years (and envelopes with the same periods, for $k=\pm1$).

As already noted in expressions (2.23) and (2.25), all $\Delta C_{nm\alpha}$ in a "lumped coefficient" have the same order m, the same α, and degrees n of the same parity. Examination of the formula for the frequency f_{nmpq} shows that, for the same order m, there are pairs of frequencies f_{nmpq} corresponding to even and odd degree coefficients. The separation in frequency within each pair is $q\dot{\omega}(q=0, \pm1, \pm2, \ldots)$, and $q\dot{\omega}$ has a period up to 1000 times longer than that of $\dot{M}$ (for most spacecraft above 600 km, the expansions in (2.24) can be truncated at $q = \pm2$). This means that, in order to separate well the even from the odd degree coefficients, it is important to gather data for each spacecraft over a length of time comparable to the period of $\dot{\omega}$. This is not essential, but can be quite helpful, particularly when the data are taken at irregular intervals and in the vecinity of a few, unevenly distributed tracking stations, a situation still common, and one in which the normal equations of the estimation procedure are likely to be ill-conditioned.

DEEP ZONAL RESONANCES AND PRECESSION

For the even zonal harmonics of the potential, $m = 0$, and $(n-2p) = 0$. So, if $q = 0$, then $f_{nmpq} = 0$ as well, and the zero frequency resonance is excited both in Δy and Δz, while the resonance at $n_o \approx \dot{M}$ is also excited in Δx. Moreover, from (2.22) follows that only the disturbing acceleration Δa_z has a zero frequency term in this particular case (provided that the inclination is high). The resonant response, according to (1.17), must consist of "homogeneous" oscillations of frequency n_o, a constant offset, and a drift along-track equal to $[-2/n_o \sum_{npq} \Delta C_{no} A_{z\ nopq}]t$, where $A_{z\ nmpq}$ is given by (2.22). This drift represents a change in the combined rate $\dot{\omega}+\dot{M}$, as both $\dot{M}$ and $\dot{\omega}$ are modified by this resonance. Therefore, resonance with the even zonals affects the rate of precession of the perigee, and also the overall orbital frequency $\dot{\omega}+\dot{M}$. The second zonal has the greatest effect of all even ones, being so much larger than all the others. This effect is the main driver of the precession of perigee. Both expressions (2.3) and (2.5) can be deduced from the theory given here, provided that I_o is close to $\pm \frac{\pi}{2}$ (otherwise, expression (2.20) for Δa_y is no longer adequate, as explained in the previous Chapter).

For even zonals, and with p, and q as above, expressions (2.20) and (2.22) show that the forcing acceleration Δa_x must have a term of frequency $\pm\dot{M}$. The difference between $|\dot{M}|$ and n_o being very small, there is a virtually exact resonance in the across-track direction, consisting of steadily increasing oscillations at a rate R_x that can be obtained by replacing B_x with $-A_{x\ nmpq}$ in (1.17). These oscillations have a frequency of once per revolution, and attain their maximum value at the equator, because of the rotational symmetry of the zonal field. As a result, at each passage through the equator the node is rotated by a constant angle $2\pi R_x/(n_o a_o)$ from its position at the previous passage, while the amplitude of the perturbation increases by the amount $2\pi R_x/n_o$.

GRAVITATIONAL PERTURBATIONS IN A "FROZEN", REPEATING ORBIT

In this case, $\dot{\omega} = 0$, and both $\dot{M}$ and $\dot{\Omega}-\dot{\theta}$ are harmonics of the repeat frequency ω_r: $\dot{M} = N_R\omega_r$, $\dot{\Omega}-\dot{\theta} = N_D\omega_r$, where the integers N_R and N_D are the number of revolutions, and the approximate number of days in each repeat period, respectively. Accordingly, the frequency

$$
\begin{aligned}
f_{nmpq} &= (n-2p+q)\dot{M} + m\,(\dot{\Omega}-\dot{\theta}) \\
&= [(n-2p+q)N_R + mN_D]\omega_r \qquad (3.2) \\
&= i\ \omega_r,\quad i = 0, \pm 1, \pm 2, \ldots
\end{aligned}
$$

is also a harmonic of ω_r. This has two important consequences:

(a) The perturbations are periodical with period $T_r = 2\pi/\omega_r$. This can be seen by replacing f_{nmpq} according to (3.2) in (2.20) through (2.24). Except for a "homogeneous" along-track drift (see (1.15)), and for the secular effects of the resonance mentioned next, the expressions for Δa_x, Δa_y, and

Δa_z, and for Δx, Δy, Δz become Fourier series with fundamental frequency ω_r.

(b) there are deep resonances associated with coefficients $\Delta C_{nm\alpha}$ whose orders are multiples of the number of revolutions N_R:

$$m = kN_R, \; k = 0, 1, 2, 3 \; ... \tag{3.3}$$

The zonals (m = 0) are a special case of the rule just given. Since the degree n is never smaller than the order m, the zonals include the lowest degree resonant coefficients, and their effects are larger than those of the other resonant orders, as a consequence of the attenuation with height represented by the factors $(\frac{R}{a_o})^{n+2}$ in (2.22). The main effects of these resonances, according to (1.17), are a secular drift along-track, as well as continuously (or secularly) increasing oscillations in all three directions, corresponding to changes in orbital shape and rates of precession of the perigee and the node (the radial and along-track increasing oscillations reflect gradual changes in the shape of the mean orbit, which affect mostly the eccentricity; such changes are predicted by the theory of "frozen" orbits (see References), and are actually slow oscillations about the mean value e_o given by (2.9)).

The perturbations Δx, Δy, and Δz consist, therefore, of the homogeneous response, of increasing oscillations, of a secular along-track drift, and of periodical parts of period $T_r = 2\pi/\omega_r$. These periodical parts must be the same at times t, $t + T_r$, $t + 2T_r$, $t + 3T_r$, ..., which are also times when the spacecraft must pass over the same place on earth, with the same latitude and longitude, because this is a repeating orbit. This means that these periodical parts of Δx, Δy and Δz depend on the geographical position of the satellite, and in this sense can be regarded as "geographical". This relationship with geographical position is not quite one-to-one. At certain points on earth, known as crossover points, orbital passes running towards the north (ascending) cross south-bound ones (descending). These crossing passes occur at times in the orbital repeats that are not separated by an exact number of periods T_r. Therefore, the periodical perturbations, although they happen at the same locations, do not have the same values (but if one were to group separately the ones that occur along ascending passes, from those along descending passes, the perturbations in either set could be represented with truly geographical functions). The only periodical perturbations that are entirely geographical are those of zonal origin, because their expansions ((2.24) with m=0) depend only on M_o when the orbit is "frozen" [so $\dot{\omega} = 0$, $\omega_o = \frac{\pi}{2}$]. From the equations of the ground-track (2.11) one can see that latitude is a unique function of M, once ω is given, so these perturbations are the same at each pass over a crossover point, regardless of whether this is an ascending or a descending pass. The same is true of the periodical perturbations caused by coefficients of all resonant orders (i.e., those that satisfy ($m = kN_R$). Also the terms in cos $n_o t$ are purely geographical, because $n_o t = M_o$ + constant.

A NUMERICAL TECHNIQUE FOR FINDING THE FOURIER COEFFICIENTS OF THE SENSITIVITIES FOR A FROZEN, REPEATING ORBIT

In this type of orbit, the expressions (2.20) for the perturbing accelerations become true Fourier series, with all the terms being harmonics of the fundamental repeat frequency ω_r. Moreover, as their Fourier coefficients, according to (2.20) and (2.22), are not functions of ω_o, Ω_o, θ, or $\dot{\Omega}-\dot{\theta}$, one could imagine the earth slowing down its spin rate until it ceases turning relative to the orbit plane (i.e., $\dot{\Omega}-\dot{\theta} = 0$), while those coefficients remain constant. Calling $\Delta\tilde{a}_x$, $\Delta\tilde{a}_y$, and $\Delta\tilde{a}_z$ to the accelerations along the same orbit but for a "stationary earth," choosing $M_o(t_o) = 0$ and $\omega_o = \frac{\pi}{2}$, replacing $\dot{M}t$ with M_o, and choosing Ω_o and θ so that $\Omega_o - \theta = 0$, one can write, from (2.20):

$$\begin{aligned}
\Delta\tilde{a}_x &= \sum_{nmj\alpha} - \Delta C_{nm\alpha}\, A_{x\ nmj}\, \sin(jM_o + \phi_{nmj\alpha}) \\
\Delta\tilde{a}_y &= \sum_{nmj\alpha} - \Delta C_{nm\alpha}\, A_{y\ nmj}\, \sin(jM_o + \phi_{nmj\alpha}) \qquad (3.4) \\
\Delta\tilde{a}_z &= \sum_{nmj\alpha} \Delta C_{nm\alpha}\, A_{z\ nmj}\, \cos(jM_o + \phi_{nmj\alpha}) ,
\end{aligned}$$

where j = 0, ±1, ±2, ±3, ... stands for (n-2p+q) and also for the set of all comnbinations of n, p, and q that give the same value j = (n-2p+q).

While the coefficients remain the same as with a normally rotating earth, the expansions are Fourier series with arguments that depend only on the mean anomaly M_o. Considering the terms associated with a single $\Delta C_{nm\alpha} = 1$, one gets from (3.4) the corresponding accelerations $\Delta\tilde{a}_{x\ nm\alpha}$, $\Delta\tilde{a}_{y\ nm\alpha}$, and $\Delta\tilde{a}_{z\ nm\alpha}$. The solutions of the linearized equations of motion with these accelerations are the sensitivities of x, y, and z to $\Delta C_{nm\alpha}$. The interesting thing now is that expressions (3.4) for such accelerations are quite similar to (2.14) for a single spherical harmonic along a major circle, which is the basis of the efficient method for calculating inclination functions described in Chapter 2. But instead of being on a circle, here the values of the accelerations are given over the mean ellipse of the "frozen" orbit. The trick is to calculate these accelerations at regular intervals, and then get their Fourier coefficients by an efficient numerical method, such as the Fast Fourier Transform. Although the expansions have infinite terms, they can be truncated quite accurately at $j = \pm N_{max}$, where N_{max} is the highest degree in the spherical harmonic expansion that has an appreciable effect on the orbit of the spacecraft. The values of the accelerations should be calculated at least at $2N_{max}$ points equispaced in M_o along the orbit, with the perigee of the mean ellipse at $\omega_o = -\frac{\pi}{2}$ By doing these calculations twice, first for $\alpha = 0$, and then for $\alpha = 1$, as in Chapter 2, one gets all the coefficients. The following expressions can be used to compute these accelerations at their "sampling points", for a given n and m:

$$\Delta\tilde{a}_{x\ nm\alpha} = D_x\, Y_{nm\alpha}$$

$$= \frac{1}{r}\,[D_\phi\, Y_{nm\alpha} \sin\mu - D_\lambda\, Y_{nm\alpha}\, (\cos\phi)^{-1} \cos\mu]$$

$$\Delta\tilde{a}_{y\ nm\alpha} = D_y\, Y_{nm\alpha}$$

$$= \frac{1}{r}\,[D_\phi\, Y_{nm\alpha} \cos\mu + D_\lambda\, Y_{nm\alpha}(\cos\phi)^{-1}\sin\mu]$$

$$\Delta\tilde{a}_{z\ nm\alpha} = D_z\, Y_{nm\alpha}$$

$$= -\frac{1}{r}\, Y_{nm\alpha}\ , \qquad (3.5)$$

where

$$Y_{nm\alpha} = \frac{GM}{R}\left(\frac{R}{r}\right)^{n+1} P_{nm}(\cos\phi)\cos(m\lambda - \frac{\pi}{2}\,\alpha)$$

and

$$\mu = \cos^{-1}\,(-\cos\lambda\, \sin I)\ .$$

The origin for λ is the ascending node of the orbit. Once the Fourier coefficients of these accelerations for a single, unitary $\Delta C_{nm\alpha}$ have been found, they can be used according to (2.25) to find the $X_{f_{nmpq\alpha}}$, $Y_{f_{nmpq\alpha}}$, and $Z_{f_{nmpq\alpha}}$ for the sensitivities. Notice that these coefficients are the same for $\alpha = 0$ and $\alpha = 1$. These results are valid for all inclinations, can be obtained without the calculation of inclination and eccentricity functions, and take directly into account the slightly elliptical shape of the orbit. One could go one step further, and replace the mean ellipse with a numerically integrated orbit that includes in its shape the full effect of C_{20} and all other zonals C_{n0}, up to a conveniently high degree n, as long as this orbit is symmetrical respect to a meridian plane (as is the mean ellipse of a "frozen" orbit, where $\omega = \frac{\pi}{2}$). Doing this would result in a theory where the nominal orbit "wiggles" realistically about the mean ellipse. In the case of C_{20}, in particular, the "wiggles" it produces amount to several kilometers, so their inclusion can certainly improve the accuracy of the theory. For further details, see my work on satellite-to-satellite tracking, Chapter 2 and Appendix II, (1984), mentioned in the References.

SATELLITE ALTIMETRY

Calling: H to the height of the sea surface above the ellipsoid, N to the geoid height, S to the stationary mean sea surface topography (s.s.t), T to time-dependent changes of the sea surface (tides, mesoscale eddies, etc.), -n to the error in the altimeter measurement, and ΔH to the difference between the observed value of H and that calculated from the available satellite ephemeris, from the known part of the geoid, tidal models, etc., is

$$\Delta H = -\Delta z + \Delta N + \Delta S + \Delta T + n, \qquad (3.6)$$

where ΔN, ΔS and ΔT are the errors in the computed values of N, S, and T. To a large extent, the radial orbit error Δz shares a common origin with ΔN: An incomplete knowledge of the gravity field. From expressions (2.24) for Δz, (2.14) for a spherical harmonic on a circle, and (3.6) for ΔH, one gets that the sensitivity of ΔH to a particular error $\Delta C_{nm\alpha}$ in a potential coefficient (assuming that both orbit and geoid are calculated with the same field model) is

$$\Delta H_{nm\alpha} = \sum_{pq} \left(-Z_{nmpq\alpha} + RF_{nmp}(I_o)\right)\cos\left(f_{nmpq}t + \phi_{nmpq\alpha}\right) , \qquad (3.7)$$

where q is not zero only in operations involving the factor $-Z_{nmpq\alpha}$. This formula is arrived at after approximating the ellipsoid with a sphere of radius R. Clearly, altimetry can be regarded as a source of information on the gravity field, this information coming from both Δz and ΔN. Since the orbit is hundreds of kilometers above the sea surface, the frequency contents of Δz and ΔN are quite different. The orbit error has very long spatial wavelength features (of the order of thousands of kilometers), while the geoid is much richer in high frequency signal (down to a hundred kilometers or less in wavelength). That high frequency content corresponds to such pronounced details of the geoid as its wrinkles over ocean trenches and its dimples over seamounts. To use altimetry as data for mapping the anomalous field requires either filtering out in some way the short-wave features, or using a model with a very large number of parameters to represent the highly complex geoid surface. This latter approach would involve solving for the values of all those parameters in some form of adjustment. This can be done, and has been done many times in the past, in a piece-wise fashion, mapping the field over limited regions of the ocean, often after substracting the geoid calculated according to some low-resolution satellite model to eliminal global trends. These solutions usually filter out the orbit error Δz by the method of crossing arcs, after removing the known tidal and oceanographic effects from the data. For a global mapping of the field, one could conceivably solve for all the required parameters taking into account the whole gravitational signal in (3.7), and not just the part corresponding to ΔN. This approach is impractical, however, because the irregular shape of the oceans, which cover the earth only in part, precludes the use of efficient methods like that proposed for satellite-to-satellite tracking and discussed at the end of this Chapter. Therefore, at least for the present, a likely way of combining altimetry with conventional satellite tracking to obtain a global model in a simultaneous solution would be to eliminate all information in ΔN of higher spatial frequency than that in Δz; this could be done using some kind of "low-pass" filter. At present, while there are many ways in which this filtering could be done, the question of which would be adequate remains open, because nobody has yet tried this type of combined solution, except in a very rudimentary form (i.e., averaging the altimetry over 5° x 5° blocks, or simply ignoring the high frequencies in ΔN). An important source of error in any attempt to model the field with altimetry is the stationary mean sea surface topography, which cannot be separated from the geoid without making a priori assumptions that may or may not be correct (unless, of course, one already has a good model for N, but this begs the question in this context). Among various ideas of how both N and the s.s.t could be estimated simultaneously, an interesting one has been suggested by Wunsch and Gaposchkin (see References). In essence, theirs is a generalization of least squares collocation, involving the covariance functions of both geoid and s.s.t (or the degree variances of their spherical harmonic coefficients). This idea is now under investigation as part of the activities related to the preparation of an improved field model for calculating the orbit of the altimeter satellite TOPEX, which will be launched early in the 1990's (see the recent study by Wagner (1986) in the References).

In most planned altimeter missions (ERS-1, TOPEX, etc.) the spacecraft will follow "frozen", repeating orbits, as SEASAT did during its last active weeks. The purpose is to sample the same places of the sea surface at repeated intervals, in order to observe changes of oceanographic and geophysical interest. From what has been said about these orbits in previous sections, one would expect Δz to have both a component that repeats periodically along the groundtrack, and slowly increasing oscillations of one cycle per revolution due to resonance. From the point of view of studying the gravity field, repeating ground tracks present difficulties, because they give a coarser sampling of the geoid surface than can be obtained with a non-repeating orbit. Also some information in Δz could be lost through "frequency folding" similar to that encountered in Fourier analysis when undersampling, if the number of revolutions N_R per repeat period is smaller than twice the highest degree of any potential coefficient that has an appreciable effect on the orbit (this degree, at the usual altitude of 1000 km, is around n = 50).

If one substracts from each other successive values of ΔH obtained at the same places of a repeating groundtrack, their differences will not contain the periodical part of Δz, or any part of ΔN. What remains is the secularly increasing, resonant oscillations in the orbit error, and the changes in the ocean surface which the altimeter mission is meant to study. The resonant orbit error Δz_{res} can be modelled very simply (at least over arcs only a few days long, in which the curvature of the resonant envelope would not show), with expressions like

$$\Delta z_{res} = A + B \cos n_o t + C \sin n_o t + Dt \cos n_o t + Et \sin n_o t \; . \tag{3.8}$$

Notice that within the same arc, A and B are unobservable by this method, but that the changes ΔA and ΔB from arc to arc are not. In a least square adjustment, or similar estimation procedure, one could allocate a few "nuisance parameters" like A,B,C,D,E, to model the residual orbit error, and the rest to tides and other phenomena of interest. Of course, with this approach almost all gravitational information would be lost, but the oceanographers would find their analysis of the data greatly simplified.

"Frozen", repeating orbits present special problems (besides possible undersampling) to geodesists interested in mapping the ocean geoid, and to oceanographers who would like to do the same with the sea surface topography. What is needed here is a sea surface free of the effects of the orbit error, and this has been sought many times in the past by the method of crossing arcs. In this method, the orbit error Δz is frequently modelled as a low degree polynomial, and the diferences between altimetry profiles along ascending and descending passes at crossover points are used as observations of this error. As explained earlier , there is a periodical (and geographical) part in Δz that is most pronounced in "frozen", repeating orbits (for other orbits, only terms with $q = 0$ would have something like a geographical character). This periodicity limits the observability of the orbit error. In particular, terms in Δz associated with the zonal errors ΔC_{no}, and with the cosine part of the homogeneous response (partly due to the initial state error), are unobservable because they have the same values for ascending and descending passes at the same latitudes and cancel out in the crossover differences (although changes in the cosine part would show up if the crossing passes belonged to different ephemeris arcs). All this would tend to leave "zonal wrinkles" in a map of the ocean geoid obtained by the crossing arcs method. Of course, because of the complex way in which the orbit errors may propagate in this method, it is likely that the ultimate effect would not be entirely zonal, but might spread over the whole spectrum of spherical harmonics. The periodical errors in this kind of orbits have received much attention in recent years, because of the implications for the analysis of altimetry. Various workers that have contributed, among them Wagner, Rosborough, and this author, are mentioned in the References.

SATELLITE-TO-SATELLITE TRACKING

In tracking satellites from terrestrial stations, as has been mostly the case until now, the data gathered are confined to those periods and places where the spacecraft can be seen from the tracking sites. These sites are irregularly distributed over the earth, so the observation of the orbits is patchy. Moreover, because of atmospheric drag, satellites have to be quite high to remain aloft for considerable periods of time. This reduces their sensitivity to the short wavelengths of the gravity field, making their tracking data good only for mapping the broader features, less attenuated by height. To overcome these problems, several ways have been proposed. Among them is the idea, still under study, of constructing two drag-free satellites (where drag and other non-gravitational forces are sensed and compensated by the gentle action of special reaction engines), and placing them in essentially the same orbit, some hundreds of kilometers apart. Each satellite would track the other, and either their separation or their relative velocity would be measured every few seconds for several months. Being drag-free, the common orbit could be much lower than that of other spacecraft used until now, making the data much more sensitive to the fine details of the field. This, of course, limits the mission to last only for as long as there is fuel for the compensating engines to work. Thanks to the much shorter distance over which tracking is to take place, and to the development of a highly specialized Doppler radar, an extraordinary accuracy of measurement is expected: about one micron per second, or 10^{-6} m/sec. The combination of all these features could achieve the main goal of mapping the geoid over the whole world, for wavelengths longer than 100 km, with an accuracy of about 10 cm, and gravity to within 1 mgal. The mission, as described above, is being considered now by NASA in the USA, under the name of Geopotential Research Mission, or GRM for short (previously known as GRAVSAT). For further details, the reader is referred to the NASA Report by Keating et al., mentioned in the References. The main justification for this mission is that it will provide scientists with a homogeneous and global data set: One source of measurements, a polar orbit covering the whole earth over time. From this much could be learned about gravity on the continents, where information is still patchy and of irregular quality, and about the difference between the geoid on the oceans and the mean sea surface. In any event, GRM could well become the first large scale mission that is primarily dedicated to the detailed mapping of gravity worldwide.

There are two ways, as with altimetry, of trying to map the field with satellite-to-satellite tracking: One is to use data within limited regions and produce local maps, the other is to solve for the whole world at once, with all the data. Several approaches to the <u>local</u> solution have been suggested, some being rather straightforward applications of standard techniques in either satellite or physical geodesy, adapted to the nature of the mission. Problems related to the stability of the downward continuation and, therefore, of any procedure that solves for field quantities at the earth's surface, remain to be answered and are now under investigation. The global solution requires the estimation of a number of parameters approximately equal to the square of the degree N_{max} at which the spherical harmonic expansion can be safely truncated at satellite altitude. With a planed height of some 160 km, this degree is $N_{max} = 300$, so there are some 100000 unknowns to be solved for, including also orbit parameters such as the initial state of each successive weekly orbit arc, etc. There are also (for a six-months mission) about 5000000 observations to process.

In recent years, several authors have suggested practical ways of making such a huge global adjustment, in spite of the number of observations and unknowns, by using both analytical perturbation theory and a "frozen", repeating polar orbit with a repeat period sufficiently long to obtain a fine coverage of the world. I indicated the advantages of a "frozen", repeating orbit in an OSU Report (No. 317, 1981); later on, Wagner and Goad (AGU Meeting, December 1982) showed how to formulate the problem using analytical orbit perturbation theory; more recently, I proposed some refinements to their theory. Preliminary tests have shown that the theory is now accurate to 1% for perturbations caused by potential coefficients up to degree 300. The works mentioned are listed in the References, including a report of mine written while visiting at the Technical University of Delft, Holland, and what follows is a simplified description of the method explained in that report.

The scalar relative velocity V_{12} is

$$V_{12} = \underline{e}_{12} \cdot (\underline{v}_1 - \underline{v}_2) , \tag{3.9}$$

where $\underline{e}_{12}$ is a unit vector pointing from the trailing satellite (No. 2) to the leading one (No. 1), and $\underline{v}_1$ and $\underline{v}_2$ are the velocities of each satellite. Their relative velocity is invariant with respect to coordinate transformations, so the frame of coordinates chosen does not affect the value of V_{12} (here the frame is still the rotating one used throughout this work). Substracting the relative velocity calculated according to a known gravity field model, the residual ΔV_{12} is (with some simplifications not made in the original work, but brought in here to expedite the discussion)

$$\Delta V_{12} = (\Delta\dot{y}_1 - \Delta\dot{y}_2)\cos\frac{\psi}{2} + (\Delta\dot{z}_1 - \Delta\dot{z}_2)\sin\frac{\psi}{2} , \tag{3.10}$$

where the expressions for $\Delta\dot{x}$ and $\Delta\dot{y}$ can be obtained by taking the time derivative of Δx and Δy in (2.24). ψ is the geocentric angle defined by the positions of the two satellites. From the expressions for the orbital perturbations one gets the following sensitivity to a given $\Delta C_{nm\alpha}$:

$$\Delta V_{nm\alpha} = \sum_{j=-n}^{n} \{2\omega_{jm}[\cos(\tfrac{\psi}{2})Y_{nmj}\sin(\omega_{jm}C)+\sin(\tfrac{\psi}{2})Z_{nmj}\cos(\omega_{jm}C)]\}\sin(\omega_{jm}t+\phi_{nmj\alpha})$$

$$+ A_o + A_1 \cos n_o t + A_2 \sin n_o t \qquad (3.11)$$

$$+ A_3 t \cos n_o t + A_4 t \sin n_o t ,$$

where

$$C = \frac{\psi}{2n_o}$$

and

$$\omega_{jm} = j\,\dot{M}_o + m\,(\dot{\Omega}_o - \dot{\theta}) = j(N_R + m\,N_D)\,\omega_r .$$

with $j = (n-2p+q) = 0, \pm1, \pm2, \pm3, \ldots$. The Z_{nmj} and Y_{nmj} could be calculated along the lines of (2.25) with $\Delta C_{nm\alpha} = 1$ and the other coefficients set to zero, but also (and more precisely) by the method explained earlier in this Chapter, using as a nominal orbit not a mean ellipse, but a numerically integrated, stationary orbit. The terms with coefficients A_0, A_1, etc., correspond to the homogeneous and resonant responses. The later are also related to the gravity field and, therefore, to the $\Delta C_{nm\alpha}$, but they are treated here as entirely separate unknowns, to get a system of normal equations which, in spite of its enormous size, can be set up using expressions to calculate the elements of the normal matrix that are much faster to execute than the scalar products of columns of the design matrix, millions of elements long. Moreover, the normal equations obtained with this choice of parameters, ordered in a proper way, can be very sparse and solved in a reasonably short time with a special technique, without the risk of serious numerical problems due to accumulated roundoff errors. This is because, as shown in (3.11), the periodical part of the perturbations, now separated from the aperiodical part by the introduction of the additional unknowns A_i, is a Fourier series of fundamental frequency equal to the orbital repeat frequency ω_r. Over the length of the repeat period, and as long as the sampling rate of the data is high enough, sine functions whose frequencies are harmonics of ω_r satisfy the orthogonal conditions:

$$\sum_{i=1}^{M} \sin(\omega_{jm}\, t+\phi_{nmj\alpha})\sin(\omega_{j'm'}t+\phi_{n'm'j'\alpha'}) = \begin{cases} \frac{M}{2} & \text{if } m=m', j=j', \alpha=\alpha' \\ 0 & \text{otherwise} \end{cases} , \tag{3.12}$$

M being the number of equally spaced measurements. As a consequence, it is easy to show from (3.11) and (3.12) that the element of the normal matrix G corresponding to two potential coefficients, $\Delta C_{nm\alpha}$ and $\Delta C_{n'm'\alpha'}$ is

$$g_{nm\alpha}^{n'm'\alpha'} = \begin{cases} \frac{\sigma^{-2}}{2} M \sum_{j=-J}^{J} V_{nmj\alpha} V_{n'm'j'\alpha'} & \text{if } m = m' \text{ and } \alpha=\alpha' \\ 0 \quad \text{otherwise} & . \end{cases} \tag{3.13}$$

Here $J = \max(n,n')$, σ is the standard deviation of the data errors, and $V_{nmj\alpha}$ is the amplitude of $\sin(\omega_{mj}t + \phi_{nmj\alpha})$ in (3.11) The advantage of this formula over the usual scalar multiplication of columns of the design matrix, is due to the fact that the number of Fourier coefficients $V_{nmj\alpha}$ that are not negligibly small is much less than the number of observations. The elements of matrix G corresponding both to a potential coefficient and one of the A_i parameters can be calculated also analytically (although the formulation is more complex). The situation is somewhat different regarding the formation of the right hand sides of the equations. To calculate them one needs the Fourier coefficients of the data. Considering that there may be millions of measurements in one repeat period, this calls for the use of Fast Fourier Transform algorithms that can process amounts of data so large that they cannot be accomodated simultaneously in "real" memory. Fortunately, there is no lack of such methods. There are many other details, which the interested reader can find described in the Delft report. A large numerical experiment is now going on, based on the simulation of satellite-to-satellite data with a field provided by Rapp, complete to degree and order 180, plus terms of low order (up to m = 10) extending up to degree 300. The simultion has been carried out at the University of Texas at Austin by Schutz and colleagues, with starting elements for the repeating orbits of each satellite provided by me.

The simulated data have been distributed in magnetic tape to various investigators so they can test their methods to map the field (both locally and globally) with this data. This experiment is being carried out with support from NASA, and under the auspices of the Special Study Group No. 2.28 of the IUGG.

ACKNOWLEDGEMENTS

My thanks to the organizers of the Summer School for their cordial invitation to attend; to my colleagues at NASA's Goddard Space Flight Center and at EG&G for helping me find the time to write these notes; to John Robbins, who gave me a hand with the illustrations; to several friends, Dave Rowlands in particular, for helping me make them ready in time for publication, when I was struck by sudden illness; and to Nancy Gebicke, who prepared the final draft of these notes with such care and efficiency.

REFERENCES

Chapter 1: A description of the derivation of Hill's equations can be found in the book by Kaplan and in the first chapter of my Delft report on satellite-to-satellite tracking (see references for Chapter 3), which also introduces much of the theory presented throughout this work. The paper on the orbits of GPS gives a rather detailed account of various forces acting on spacecraft that are likely to excite orbital resonances.

Kaplan, M. H., Modern spacecrat dynamics and control, John Wiley & Sons, New York, 1976.
Colombo, O. L., Ephemeris errors of GPS satellites, Bull. Geod., Vol. 60, 64-84, 1986.

Chapter 2: The basic book on the Keplerian analytical approach to satellite geodesy remains Kaula's, which is now regrettably out of print, but is still being photocopied by most newcomers to the field. The theory of "frozen" orbits was first derived by Cook, for orbits at elevations of about 0.1 of the earth radius, and away from the critical inclination. The critical inclination/distant satellite gap has been filled more recently by Hough.

Cook, G.E.,Perturbations of near-circular orbits by the earth's gravitational potential, Planetary and Space Sciences, Vol. 14, 434-444, 1966.
Hough, M. E., Orbits near critical inclination, including luni-solar perturbations, Celestial Mechanics, Vol. 25, 111-136, 1981.
Kaula, W.M., Theory of satellite geodesy. Blaisdell Publishing Co., 1966.

Chapter 3: Work on the theory of the periodic orbit errors of "frozen", repeat orbits has been done mostly in connection with satellite altimetry and, therefore, it has concentrated on the radial errors. At this time, the main publications on this subject (that I know of) are listed below. Wunsch and Gaposchkin, and Wagner (1986) consider the simultaneous analysis of mean sea surface and geoid. The elegant idea of using both analytical orbital perturbations and a "frozen", repeating orbit to find a practical solution to the global mapping of gravity with satellite-to-satellite tracking is explained in Wagner's paper and also in my Delft report. GRAVSAT is a name used formerly for GRM. The NASA report by Keating et al., gives a comprehensive picture of the mission itself.

Colombo, O. L., The global mapping of gravity with two satellites, Report of the Geodetic Commission of the Netherlands, Vol. 7, No. 3, Delft, 1984.

Colombo, O.L., Altimetry, orbits and tides, TM 86180, Goddard Space Flight Center, NASA, Greenbelt, Maryland, 1984.

Keating, T., Taylor, P., Kahn, W., and F.J. Lerch, Geopotential Research Mission: Science, engineering, and program summary, NASA TM-86240, 1986.

Rosborough, G. W., Satellite orbit perturbations due to the geopotential, Report CSR-1, The University of Texas at Austin, Doctoral Dissertation, 1986.

Wagner, C. A., Direct determination of gravitational harmonics from low-low GRAVSAT data, J. Geophys. Res., Vol. 88, 10309-10322, 1983.

Wagner, C. A., Radial variations of a near circular satellite orbit due to gravitational errors: Implications for satellite altimetry, J. Geophys. Res., Vol. 90, 3027-3036, 1985.

Wagner, C. A., Accuracy estimate of geoid and ocean topography recovered jointly from satellite altimetry, J. Geophys. Res., Vol. 91, 453-461, 1986.

Wunsch, C., and Gaposchkin, E.M., On using satellite altimetry to determine the general circulation of the oceans with applications to geoid improvements, Rev. Geophys. Appl. Phys., Vol. 18, 725-745, 1980.

SATELLITE GRADIOMETRY

by

R. Rummel

Department of Geodesy
Delft University of Technology
Thijsseweg 11
NL-2600 GA Delft, The Netherlands

Lecture Notes in Earth Sciences, Vol. 7
Mathematical and Numerical Techniques in Physical Geodesy
Edited by H. Sünkel

1. INTRODUCTION.

Just when I started to write down the lecture notes I received a copy of the new Banff proceedings. There Prof. Moritz gave a keynote address entitled "Inertia and Gravitation in Geodesy" (1985). It contains on eight pages everything one could possibly ever like to know about gradiometry. It is concise, very general (in the language of space-time), and as usual very elegant. Hence I could preferably just refer you to this publication and for more details to "kinematical geodesy" (Moritz, 1968). The question is then, what is left to me. In the last paragraph of his Banff keynote address Prof. Moritz apologizes for presenting "only" the theory, for "the essential work, which is both extremely hard and extremely ingenious, is done by the constructing engineers". In this situation it would have been some comfort to belong to the latter group and to be able to lecture about the technology of gradiometry (see e.g. (Spaceborne Gravity Gradiometers, 1983)), but unfortunately I don't. After some days of frustration I was wondering whether there does not exist a third important terrain one should talk about. In Delft this terrain is called "inschakeling", the connection of the mathematical model with physical reality or more concrete with an experiment and the "uitschakeling", the physical interpretation of the result obtained inside the mathematical model, or loosely speaking: "given a certain problem, what are the proper experiments (instruments, measurements) to be chosen, what the mathematical and stochastic model, and what is the proper interpretation of the results in terms of the posed problem?". It seems as if this aspect has received some attention in land surveying and network computation, at least from some scientists, but very little in physical geodesy. It would therefore be interesting to work out the "inschakeling" in this lecture for the case of satellite gradiometry but of course also rather difficult. I shall therefore talk only about the required ingredients, the principle, the geometry of the gravitational field, the observability of the gradients, and the model of gravity parameter recovery. The actual definition of the required experiments and model is left to the reader. In the last chapter only a number of questions are raised referring to the feasibility of the gradiometric experiments and to the adequacy of the mathematical model.

2. PRINCIPLES OF GRADIOMETRY.

We know from inertial surveying that the accelerometers of a moving measurement unit observe the so-called specific force, which is the sum of inertial and gravitational accelerations. And the two contributions can only be separated approximately by introducing a model of the earth's gravitational field. The equivalence of gravitation and inertia is a well established phenomenon, and a cornerstone of Einstein's development of the theory of general relativity. However, this principle holds only for a single point (proof mass) or for a differential region around it, so small that the gravitational field can still be considered uniform. In a somewhat larger region it becomes possible to distinguish between gravitational and inertial effects, as the following example shows, compare (Carroll & Savet, 1959).

With good approximation the earth can be considered a uniform, spherical mass and hence its gravitational field a central force field. A point mass subject to this field and released from rest in free fall shall move along a radial straight line. Two point masses, starting from the same spherical equipotential surface at the same time shall follow to converging lines of force of the central field, compare Figure 1a. The decreasing distance of the two point masses is a measure of the existence of the central force field. Suppose the point masses are connected by a rod with sensitive springs (accelerometers) at each end, the springs would sense a horizontal compression.

Similarily, if the two proof masses are located at the same straight line of force but at some radial distance and they are again simultaneously released from rest, their distance shall slowly increase due to the different distance from the attracting central body, see Figure 1b. Again, the effect can be sensed by the rod with sensitive springs at both ends. However in the second case a tension in radial direction shall be experienced. The contraction, respectively tension would not be experienced if the proof masses would be placed in a uniform force field with parallel force vectors. Hence with this experiment the presence of the central force field can be detected and discriminated from linear acceleration.

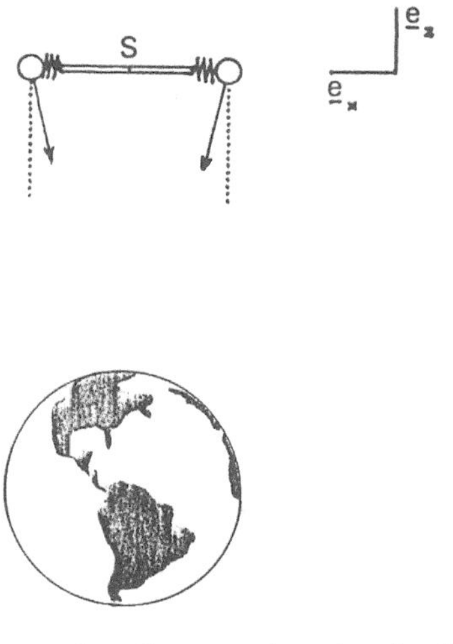

Figure 1a: Compression due to the convergence of the lines of force.

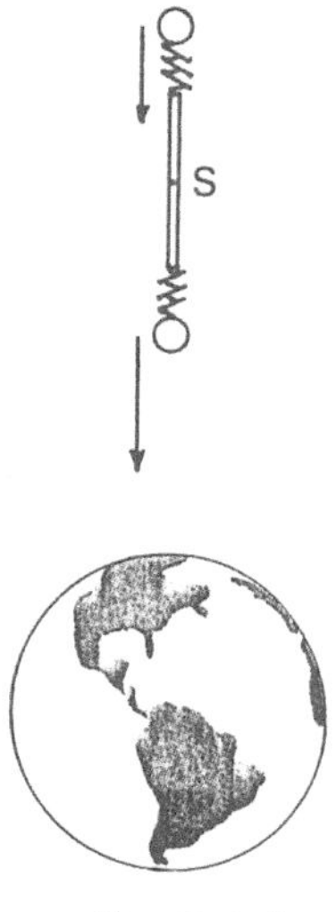

Figure 1b: Tension due to the difference in distance from the attracting body.

Expressed in formulas the situation of the above example is as follows. The acceleration vector, $\underline{\ddot{x}}$, of the center of mass, S, of the rod with two springs in a gravitational field is

$$\ddot{\underline{x}}(S) = \nabla_S V = \begin{bmatrix} V_x \\ V_y \\ V_z \end{bmatrix}_S \quad , \tag{2.1}$$

where $\nabla_S V$ is the gradient of the gravitational potential of the attracting body with components $V_x = \frac{\partial V}{\partial x}$, $V_y = \frac{\partial V}{\partial y}$, and $V_z = \frac{\partial V}{\partial z}$. In index notation eq. (2.1) becomes

$$\ddot{x}_i(S) = V_i(S) \qquad i = 1,2,3 \ , \tag{2.2}$$

where $V_i = \frac{\partial V}{\partial x_i}$ and the components refer to a local orthonormal frame with base vectors $\underline{e}_i$, i = 1,2,3 (or $\underline{e}_x$, $\underline{e}_y$, $\underline{e}_z$, whenever convenient) in the orientation shown in Figure 1 with $\underline{e}_y$ orthogonal to the paper plane pointing to the reader[1]. The gravitational acceleration at location P_1 (proof mass 1) is

$$\ddot{x}_i(P_1) = V_i(P_1),$$

or when expressed relatively to point S, expanded in a Taylor series truncated after the linear term[2]

$$\begin{aligned} \ddot{x}_i(P_1) &\approx V_i(S) + V_{ij}(S)dx_j \\ &= \ddot{x}_i(S) + V_{ij}(S)dx_j \quad , \end{aligned} \tag{2.3}$$

where $V_{ij} = \frac{\partial^2 V}{\partial x_i \partial x_j}$ and dx_j the coordinate differences between P_1 and S. Analogously we find for P_2 (proof mass 2)

$$\ddot{x}_i(P_2) = \ddot{x}_i(S) + V_{ij}(S)dx_j$$

and for the acceleration difference between the proof masses at P_1 and P_2

$$\begin{aligned} d\ddot{x}_i &= \ddot{x}_i(P_2) - \ddot{x}_i(P_1) \\ &= V_{ij}(S)(dx_j(P_2,S) - dx_j(P_1,S)) \\ &= V_{ij}(S)dx_j \quad . \end{aligned} \tag{2.4}$$

---

[1] Since all quantities are expressed relative to orthonormal triads, no distinction between co- and contra-variant differentiation is required. All tensorial quantities can be expressed using solely sub-indices.

[2] Einstein summation convention over repeated indices is applied.

Now dx_j denotes the coordinate difference between P_1 and P_2. Hence by measuring the distance components dx_j and, by means of the springs, the forces $md\ddot{x}_i$ it should be feasible to derive the components V_{ij}. V_{ij} is called gravitational tensor. Often the components V_{ij} are referred to as gravity gradients, and a device with which the components or linear combinations of them are measured is called a gradiometer.

The basic properties of the gravitational field, namely the Laplace equation $\Delta V = \nabla^2 V = -4\pi G\rho$ (G ... gravitational constant, and ρ ... density) which becomes zero outside a mass distribution, and $\nabla \times \nabla V = 0$ (conservative field) can be translated into

$$\sum_{i=1}^{3} V_{ii} = -4\pi G\rho \qquad \text{and} \qquad V_{ij} = V_{ji} \quad . \tag{2.5a-b}$$

We see V_{ij} is symmetric and has only five independent components. Because of its tensor character V_{ij} can be transformed from the $\underline{e}_i$-triad to an arbitrary $\underline{e}_{i'}$-triad

$$V_{i'j'} = A_{i'i} \; V_{ij} \; A_{jj'} \tag{2.6}$$

with the orthogonal transformation $A_{i'i} = A_{i'i}(\alpha,\beta,\gamma)$, a function of the Eulerian angles α,β,γ.

We shall give an example. Let us assume the rod is of length 2d and is oriented along the z-axis, Figure 1b, $dx = dy = 0$. The earth is approximated by a homogeneous sphere of mass M and radius R, and potential $V = \frac{GM}{R}$. In this case the components of V_{ij} are

$$V_{ij} = \frac{GM}{r^3} \begin{bmatrix} -1 & 0 & 0 \\ 0 & -1 & 0 \\ 0 & 0 & 2 \end{bmatrix} \quad , \tag{2.7}$$

where r is the radial distance of S from the center of mass of the earth. Thus eq. (2.4) yields

$$d\ddot{x}_i = \frac{GM}{r^3} \begin{bmatrix} -1 & 0 & 0 \\ 0 & -1 & 0 \\ 0 & 0 & 2 \end{bmatrix} \begin{bmatrix} 0 \\ 0 \\ d \end{bmatrix} = 2\,\frac{GM}{r^3} \begin{bmatrix} 0 \\ 0 \\ 1 \end{bmatrix} d \quad , \tag{2.8}$$

an acceleration directed away from S, causing a tension of the spring. Rotated about the y-axis by α we find with eq. (2.6)

$$V_{i'j'} = \frac{GM}{r^3} \begin{bmatrix} \cos\alpha & 0 & -\sin\alpha \\ 0 & 1 & 0 \\ \sin\alpha & 0 & \cos\alpha \end{bmatrix} \begin{bmatrix} -1 & 0 & 0 \\ 0 & -1 & 0 \\ 0 & 0 & 2 \end{bmatrix} \begin{bmatrix} \cos\alpha & 0 & \sin\alpha \\ 0 & 1 & 0 \\ -\sin\alpha & 0 & \cos\alpha \end{bmatrix}$$

$$= \frac{GM}{r^3} \begin{bmatrix} -1 + 3\sin^2\alpha & 0 & -\frac{3}{2}\sin 2\alpha \\ 0 & -1 & 0 \\ -\frac{3}{2}\sin 2\alpha & 0 & -1 + 3\cos^2\alpha \end{bmatrix} . \tag{2.9}$$

Now the components of $d\ddot{x}_i$ become

$$d\ddot{x}_i = \frac{GM}{r^3} \begin{bmatrix} -\frac{3}{2}\sin 2\alpha \\ 0 \\ -1 + 3\cos^2\alpha \end{bmatrix} d . \tag{2.10}$$

For $\alpha = 0^0$ or 180^0 the original situation of eq. (2.8) is obtained, for $\alpha = 90^0$ or 270^0, which corresponds to the horizontal orientation of Figure 1a, the components become $\frac{GM}{r^3}\, d\{0,0,-1\}$, an acceleration directed towards S, which causes the contraction of the spring. We further observe that the tension is two times the contraction and that for $\alpha = \{54^0.7,\ 125^0.3,\ 234^0.7,\ 305^0.3\}$ no force at all acts on the springs in the direction of their sensitivity axis. The acceleration vectors relative to the undeformed state are displayed in Figure 2.2.

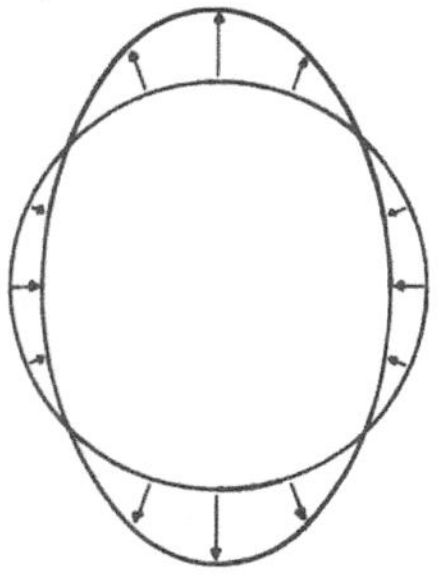

Figure 2.2: Tidal ellipse.

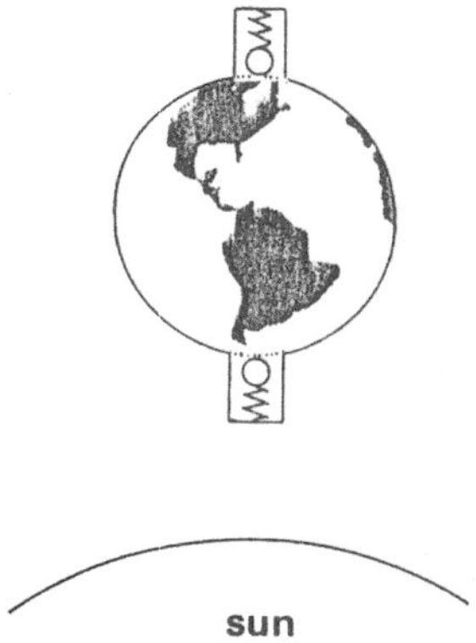

Figure 2.3: Tidal acceleration of the sun acting on two gravimeters, an analogon of the "rod + two springs" experiment.

Figure 2.2 is very well known from the case of the earth tides. Is there an analogy between our experiment with the rod with two springs and the phenomenon of earth tides? If we would place a pair of gravimeters (i.e. two measuring springs) one time along the line connecting the earth's center of mass with the sun (or the moon) on opposite sides at the earth's surface and one time under an angle of 90^0, compare Figure 2.3, our experiment would resemble that of the rod with two strings, whereby the sun (or the moon) takes the place of the gravitating body. However there are also a number of important differences between these two experiments. (1) Whereas the rod is assumed to be rigid, the earth and its hydrosphere and atmosphere are deformable. This is why one can observe the tides of the oceans and this is why there exists an observable difference between the tidal accelerations one would expect to measure and the actually observed signal. (2) The earth is rotating. This affects the observable accelerations. In addition, the gravitational tensor V_{ij} of the sun and moon influences the rotational motion of the earth, at least if the earth is oblate. (3) The effect of the potential V of sun and moon is by orders of magnitude dominated by the gravitational potential of the earth itself. One could say the self-gravitation of the earth plays the major role.

If we would like to simulate the tidal influence of sun and moon on the earth, the rod with the two measuring springs would have to be modified by (1) probably including some Maxwell type body  into the rod, (2) putting the device into rotation, (3) and adding to it some considerable mass. Despite these differences with our well-known phenomenon tides, it is customary to denote $d\ddot{x}_i$ of eq. (2.4) <u>tidal acceleration</u> and V_{ij} the tensor of the <u>tidal potential</u>.

With $g = \frac{GM}{r^2}$ eq. (2.10) can be re-written to give

$$\frac{d\ddot{x}_i}{d} = \frac{g}{r} \begin{bmatrix} -\frac{3}{2}\sin 2\alpha \\ 0 \\ -1 + 3\cos^2\alpha \end{bmatrix} . \tag{2.11}$$

This form indicates that it is likely that our experiment "rod + 2 springs" yields information about

- g: gravitational attraction,
- r: position, and
- α: the orientation of our device with respect to the gravitating body.

A word on the order of magnitude of the gravitational tensor components V_{ij} of the earth: With $GM \approx 398600 \cdot 10^9\ m^3 s^{-2}$ and $R \approx 6.371 \cdot 10^6$ m it is approximately, see eq. (2.7)

$$V_{xx} = V_{yy} = -1540 \cdot 10^{-9}\ s^{-2} = -1540\ \text{E.U.},$$
$$V_{zz} = 3080\ \text{E.U.},$$
$$V_{xy} = V_{xz} = V_{yz} = 0,$$

where 1 Eötvös Unit (E.U.) = $10^{-9}\ s^{-2}$. Hence the measurable acceleration difference over a base line of 1 m becomes $d\ddot{x} = 3080 \cdot 10^{-9} \frac{m}{s^2}$ or approximately $3 \cdot 10^{-7} \cdot g$, a quantity indeed very small and requiring extremely sensitive equipment to become observable with a sufficient number of significant digits.

3. GEOMETRY OF THE TIDAL FIELD.

It has been shown that the components of the gravitational tensor V_{ij} can be measured, in principle, with the rather simple device "rod + two springs". In order to keep the derivations simple the gravitating body was assumed to be a homogeneous sphere. Actually, however, for the purpose of gravity field determination one is interested in the deviations of the actual field from such an idealization. We shall therefore take a closer look into the properties of V_{ij}. A very suitable way to do this, is to study the geometrical structure of the gravitational field.

Gradiometry is probably the most beautiful example of the interrelation of geometrical and dynamic concepts in geodesy. No wonder that Prof. Marussi was pioneering this subject in a series of well-known publications, (Marussi, 1979, 1984, and 1985). In this chapter the following question shall be addressed: Assume the components V_{ij} are observable, what do they tell us in terms of the geometry of the earth's gravitational field?

The basic relation between the gravitational acceleration difference components $d\ddot{x}_i$ and the tensor V_{ij} was

$$d\ddot{x}_i = V_{ij} dx_j \quad . \tag{2.4}$$

The components are now expressed in the local astronomical triad with $\underline{e}_{i=1} = \underline{e}_x$ directing north, $\underline{e}_{i=2} = \underline{e}_y$ directing east, and $\underline{e}_{i=3} = \underline{e}_z$ pointing direction zenith. In view of property (2.5b) $d\ddot{x}_i$ can be considered the gradient of the potential

$$dV = \tfrac{1}{2} V_{ij} dx_i dx_j \quad , \tag{3.1}$$

the so-called <u>tidal potential</u>. Eq. (3.1) represents a second-order form. The relationship between the second derivatives V_{ij} and the geometrical parameters defining the second-order surface and its orthogonal trajectories is, following e.g. (Marussi, 1979):

$$k_1 = \frac{1}{r_1} = -\frac{V_{11}}{g} \qquad \text{north-south curvature,}$$

$$k_2 = \frac{1}{r_2} = -\frac{V_{22}}{g} \qquad \text{east-west curvature,}$$

$$t_1 = t_2 = -\frac{V_{12}}{g} \qquad \text{torsion,}$$

$$f_1 = -\frac{V_{13}}{g} \qquad \text{curvature of the plumb line (xz-plane),}$$

$$f_2 = -\frac{V_{23}}{g} \qquad \text{curvature of the plumb line (yz-plane),}$$

$$H = k_1 + k_2 = \frac{V_{33}}{g} \qquad \text{mean curvature (assumed: } \rho = 0\text{).}$$

Furtheron are r_1 and r_2 the N-S respectively E-W radius of curvature and g the scalar gravity. With these expressions V_{ij} becomes

$$V_{ij} = -g \begin{bmatrix} k_1 & t_1 & f_1 \\ t_1 & k_2 & f_2 \\ f_1 & f_2 & -H \end{bmatrix} \quad . \tag{3.2}$$

When transformed, for example, to the $\underline{e}_c$, c = 1,2,3 triad, (c for curvature) with $\underline{e}_{c=1}$ and $\underline{e}_{c=2}$ aligned with the principal directions of the equipotential surface and $\underline{e}_{c=3} = \underline{e}_{i=3}$, the torsion t_1 becomes zero, k_1 and k_2 become the principal curvatures, and f_1 and f_2 are expressed in the new (xz)- and (yz)-planes. The tidal potential expressed in the $\underline{e}_c$-system is with eqs. (3.1) and (3.2)

$$2dV = -g(k_1dx_1^2 + k_2dx_2^2 - Hdx_3^2 + 2f_1dx_1dx_3 + 2f_2dx_2dx_3). \tag{3.3}$$

For a classification of the second-order surface (3.3), it is preferable, to transform it from the $\underline{e}_c$-triad (or $\underline{e}_i$-triad) to the system of eigen vectors of V_{ij}.

The idea is to search for components μ_i such that μ_i is transformed into itself by the transformation $V_{ij}\mu_j$, which means that

$$V_{ij}\mu_j = \lambda\mu_i \quad ,$$

with λ an arbitrary scalar, or

$$(V_{ij} - \lambda\delta_{ij})\mu_j = 0 \quad . \tag{3.4}$$

In eq. (3.4) δ_{ij} is the Kronecker delta. For i = 1,2,3 eq. (3.4) is a (3 × 3) linear system of equations. A trivial solution of (3.4) is $\mu_1 = \mu_2 = \mu_3 = 0$. Condition for any non-trivial solution is zero determinant, i.e.

$$|V_{ij} - \lambda\delta_{ij}| = 0 \quad .$$

Evaluation of the determinant leads to the cubic polynomial in λ:

$$\lambda^3 - I_1\lambda^2 + I_2\lambda + I_3 = 0 \quad . \tag{3.5}$$

The coefficients I_i are the <u>invariants of V_{ij}</u>:

$$I_1 = \sum_{i=1}^{3} V_{ii} \quad , \qquad \text{trace}$$

$$I_2 = \tfrac{1}{2} \sum_{i=1}^{3} \sum_{j=1}^{3} (V_{ij}V_{ji} - V_{ii}V_{jj}) \quad , \tag{3.6a-c}$$

$$I_3 = \frac{1}{6} \sum_{i=1}^{3} \sum_{j=1}^{3} \sum_{k=1}^{3} (2V_{ij}V_{jk}V_{ki} - 3V_{ij}V_{ji}V_{kk} + V_{ii}V_{jj}V_{kk})$$

$$= |V_{ij}| \quad . \qquad \text{determinant}$$

In our case we know from eq. (2.5a) that for $\rho = 0$ (space outside any mass distribution)

$$I_1 = 0 \quad . \tag{3.7}$$

A tensor with this property is called deviator. For I_2 and I_3 we find

$$I_2 = K + C + f_1^2 + f_2^2 \tag{3.8}$$

and

$$I_3 = -KH - f_1^2 k_2 - f_2^2 k_1 \quad , \tag{3.9}$$

where $K = k_1 k_2$ is the Gaussian curvature and $C = k_1^2 + k_2^2$ the Casarati curvature.

The eigenvalues of a symmetric tensor with real elements are real. They belong to the corresponding eigen vectors $\underline{e}_p$ (p for principal axis), p = 1,2,3, which form an orthonormal triad. Transformed to the $\underline{e}_p$-triad the gravitational tensor takes the diagonal form

$$V_{ij} = \begin{bmatrix} \lambda_1 & 0 & 0 \\ 0 & \lambda_2 & 0 \\ 0 & 0 & \lambda_3 \end{bmatrix} \quad . \tag{3.10}$$

This is the form V_{ij} takes from the very beginning in the case of a central field, compare chapter 2. Again Laplace equation (2.5a) yields $\sum_{p=1}^{3} \lambda_p = 0$. The second-order form of the tidal potential (3.3), when expressed in the $\underline{e}_p$-triad becomes

$$2dV = \lambda_1 dx_1^2 + \lambda_2 dx_2^2 + \lambda_3 dx_3^2 \quad , \tag{3.11}$$

with λ_1 and $\lambda_2 < 0$ and $\lambda_3 > 0$. For $dV > 0$ the equipotential surface of the tidal potential becomes a hyperboloid of two sheets, for $dV < 0$ one of one sheet, and for $dV = 0$ a cone. Hence we could represent the tidal field in any arbitrary point in space by its equipotential surfaces, families of hyperboloids, compare Figure 3.1.

The tidal acceleration vector $d\ddot{x}_i$, eq. (2.4), which is the gradient of dV becomes, when expressed in the e_i-triad:

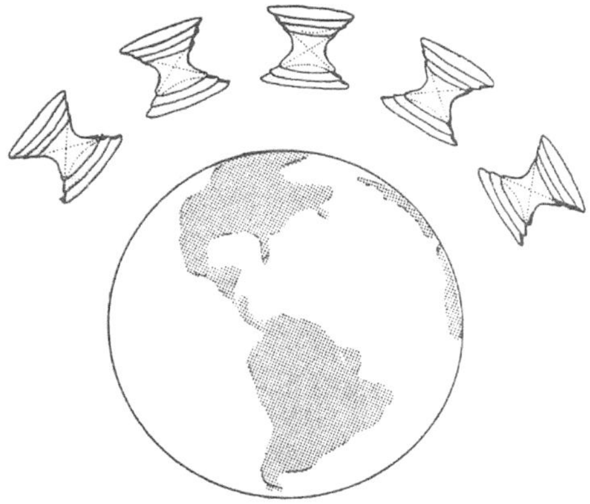

Figure 3.1: Local gravity field representation in terms of hyperboloids associated to the tidal potential.

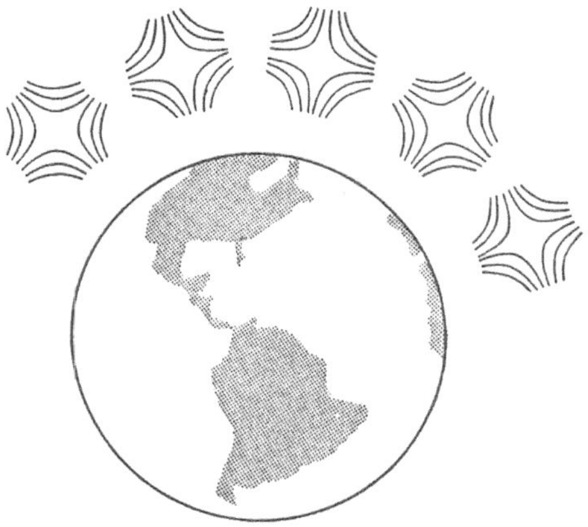

Figure 3.2: Local gravity field representation in terms of lines of force.

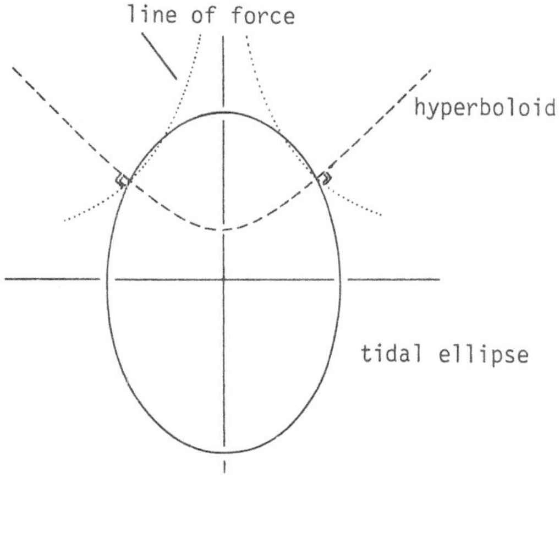

Figure 3.3: Second-order tidal surface, line of force and tidal ellipse.

$$d\ddot{x}_i = dV_i = -g \begin{bmatrix} k_1 dx_1 + t_1 dx_2 + f_1 dx_3 \\ t_1 dx_1 + k_2 dx_1 + f_2 dx_3 \\ f_1 dx_1 + f_2 dx_2 - H dx_3 \end{bmatrix} \quad . \tag{3.12}$$

The vectors $d\ddot{x}_i$ are orthogonal to the equipotential surfaces dV = const. They are the basis of a second possible geometrical representation of the local tidal field, in terms of lines of force, the orthogonal trajectories to dV = const., see Figure 3.2.

The geometrical connection of second-order tidal surface, lines of force, and the tidal ellipse is given in Figure 3.3. The tidal ellipse results from the tidal deformation of an originally undeformed spherical equipotential surface, in chapter 2 represented by the rigid rod. The spherical equipotential surface could be due to the gravitational field of a spherical proof mass placed in the earth's gravitational field.

Although these geometrical considerations are in the first place meant to give a better insight into the gradiometer problem, one could nevertheless pose the question, whether they could not also be of immediate practical use. We shall therefore consider two cases.

CASE 1: Assume all components V_{ij} are observable, but the orientation of the measurement frame is not known.

Choice A: From eqs. (3.6) the invariants I_i can be computed. I_1 has to be zero, eq. (2.5a). This imposes a condition on the diagonal components V_{ii}. I_2 and I_3 are invariant under coordinate transformation. Hence one could use these derived quantities for gravity field analysis. A problem is the high non-linearity of I_2 and I_3, compare eqs. (3.6). In order to get an idea about their order of magnitude consider the spherical gravitational field of eq. (2.7). It is $V_{xx} = V_{yy} = -\frac{GM}{r^3}$, $V_{zz} = 2\frac{GM}{r^3}$, and $V_{xy} = V_{xz} = V_{yz} = 0$. Thus, eq. (3.6a-c) yield

$$\begin{aligned} I_1 &= 0 \\ I_2 &= 3\left(\frac{GM}{r^3}\right)^2 \\ I_3 &= 2\left(\frac{GM}{r^3}\right)^3 \quad . \end{aligned} \tag{3.9}$$

Choice B: One could as well transform the components V_{ij}, observed in an arbitrary system, into the eigen vector (= principal axis) system $\underline{e}_p$, as shown above. In $\underline{e}_p$ the tensor V_{ij} takes diagonal form. The orientation of the $\underline{e}_p$-triad is not very different from the local astronomical $\underline{e}_i$-triad, for we saw already in eq. (2.7) that in spherical approximation V_{ij} takes diagonal form, which means that in this case $\underline{e}_i$, $\underline{e}_c$, and $\underline{e}_p$ coincide. The deviations of the actual mass distribution from that of a homogeneous sphere are only of the order of the earth's flattening ($\approx \frac{1}{300}$). Introducing an ellipsoidal approximation the uncertainties in orientation can be reduced to the 10"-level.

CASE 2: Bocchio (1982) discussed the case $I_3 = |V_{ij}| = 0$, the so-called singularity problem. For $I_3 = 0$ the second-order tidal surface degenerates to a cylinder. A practical consequence would be the coincidence of gravitational vectors of different points. (What would this mean in terms of astronomical positioning?)

A very extreme case of this nature is the complete disappearence of the V_{ij}. This case is discussed in a beautiful paper by Forward (1982). However so academic it may seem, it has a relevant practical background. As is well known micro-gravity research has become a very important branch of space science. It takes advantage of the almost complete absence of gravitational attraction inside a space ship for chemical, biological, medical or material science experiments. However, as we know from chapter 2, perfect zero-gravity is attained only in one single point inside the space ship, in all other points the small but not negligible tidal gravity field is present, see e.g. (Olsen & Mockovciak, 1981). In order to reduce this residual field one would have to generate a piece of "flat space". How to achieve this is discussed in (Forward, ibid).

Summarizing, the point of departure was eq. (2.4) that connects the gravitational tensor V_{ij} and displacement dx_j with acceleration $d\ddot{x}_i$. V_{ij} bears all the local geometrical information of the potential surfaces. The acceleration vector $d\ddot{x}_i$ is the gradient of the tidal potential dV. The associated second-order surfaces dV = const. are hyperboloids. They are analyzed easiest by transforming dV into the system of principal axes. Linear mappings completely analogous to eq. (2.4) exist also for the tensors of inertia, stress, and strain. Especially the strain tensor is research subject of geodesists in the field of deformation analysis, we refer to the review article by Dermanis & Livieratos (1983). Question: Of what type are the second-order surfaces associated to the inertia, stress, and strain tensors, respectively?

Marussi (1985) derives the second-order surface of the tidal potential also for the case of a rotating coordinate system, where apparent forces have to be taken into account. We did not yet discuss such a situation. When we discussed the tidal acceleration acting upon our "rod + 2 springs" under various orientations α in chapter 2, the gravitating body was so-to-say assumed to rotate about our device. The situation in a moving coordinate system shall be treated in the next chapter.

4. OBSERVING GRAVITY GRADIENTS.

In the preceeding chapter the gravitational tensor V_{ij} was assumed to be given and we looked into the type of information these components carry. Now we shall turn to the question of how to obtain the V_{ij}. However no attempt shall be made to address the problems of instrument design. For the current state of art of gradiometer design we refer to (Spaceborne Gravity Gradiometers, 1983). We shall rather try to find out from what type of observables V_{ij} can be derived. In a certain sense this was also the objective of chapter 2, but here less simplifications shall be introduced in the model. A gradiometer could be based on a variety of princples and operate in various environments, it could for example be placed on earth or orbit in free fall around the earth, it could be rotating or space fixed, measure forces or torques. Since I could not find one general line for presenting the various situations, I selected a number of cases, that seem to be representative.

One of the most important simplifications of the "rod + 2 springs" experiment in chapter 2 was, that the accelerations $d\ddot{x}_i$ were assumed to be expressed in an inertial frame. We shall now look into the situation of a rotating (moving) frame. In order to keep the notation pragmatic, the indices i, j, k shall from now on refer to the moving frame with base vectors $\underline{e}_i$, $i = 1,2,3$ and I, J, K refer to the inertial frame with base vectors $\underline{e}_I$, $I = 1,2,3$. The acceleration components $\ddot{x}_I$ in the inertial system are related to those in the moving system by

$$R_{iI}\ddot{x}_I = \ddot{x}_i + 2\Omega_{ij}\dot{x}_j + \dot{\Omega}_{ij}x_j + \Omega_{ij}\Omega_{jk}x_k + R_{iI}\ddot{b}_I \quad , \tag{4.1}$$

compare (Moritz, 1968) or the short derivation of Appendix A. In eq. (4.1) it is

R_{iI} .... the (instantaneous) orthogonal transformation from $\underline{e}_I$ to $\underline{e}_i$,

$\ddot{b}_I$ .... the inertial acceleration of the origin of the moving system, and

$\Omega_{ij} = R_{iI}\dot{R}_{Ij}$ .... the Cartan transformation with elements ω_I the angular velocity components of the moving triad $\underline{e}_i$. It is by definition

$$\Omega_{ij} = \begin{bmatrix} 0 & -\omega_3 & \omega_2 \\ \omega_3 & 0 & -\omega_1 \\ -\omega_2 & \omega_1 & 0 \end{bmatrix} ,$$

furtheron the inertial accelerations

$-2\Omega_{ij}\dot{x}_j$ .... the Coriolis acceleration,

$-\dot{\Omega}_{ij}x_j$ .... the acceleration due to a change of the angular velocity vector ω_I, and

$-\Omega_{ij}\Omega_{jk}x_k$ .... the centrifugal acceleration.

According to Newton's second law the change in linear momentum ($m\dot{x}_I$) of a point mass m is equal to the applied forces ($F_I = F_I^{(1)} + F_I^{(2)} + \ldots$):

$$m\ddot{x}_I = F_I \quad . \tag{4.2}$$

The forces could be either solely the gravitational force, $F_I^{(1)} = G_I$, or also for example atmospheric drag and solar radiation in the case of a free falling proof mass, or some spring force that confines the free motion of a proof mass. In terms of forces per unit mass, we define the gravitational acceleration to be

$$V_I = \frac{\partial V}{\partial x_I} = \frac{1}{m} G_I \quad ,$$

and the so-called specific force

$$f_I = \frac{1}{m}(F_I^{(2)} + F_I^{(3)} + \ldots) \quad .$$

Combining eqs. (4.1) and (4.2) we find

$$R_{iI}(V_I + f_I) =$$

$$V_i + f_i = \ddot{x}_i + 2\Omega_{ij}\dot{x}_j + \dot{\Omega}_{ij}x_j + \Omega_{ij}\Omega_{jk}x_k + R_{iI}\ddot{b}_I \quad . \tag{4.3}$$

Important is, that only the motion of the proof mass relative to the measurement frame is observable. This is another way to put the principle of equivalence.

The observables can be divided into two main classes: kinematical and dynamical. Under kinematical measurements we understand observations of distances, distance changes, angles a.s.o. aiming for a description of $\ddot{x}_i$. In the dynamical case certain controlled forces are applied to the proof mass in order to constrain its motion.

Example "kinematical": The absolute measurement of gravity with a free fall apparatus. No forces are applied to the proof mass ($f_i = 0$), $\ddot{x}_i$ is determined interferometrically, the contribution of the inertial accelerations remains very small. The moving system is attached to the apparatus, therefore for a system at rest the origin is connected with a point 0 at the earth's surface. Its acceleration is the centrifugal acceleration due to the angular velocity of the earth. Thus, the observable is according to (4.3) $\ddot{x}_i = V_i - R_{iI}\ddot{b}_I$, the difference between V_i, the gravitational gradient, and the centrifugal acceleration of 0. This difference is by definition gravity. (Compare problem 6.14 of (Spiegel, 1967)).

Example "dynamical": Relative gravity measurement with a spring gravimeter. Again the moving frame is attached to the housing of the gravimeter, therefore the acceleration of the origin of the system is equal to the centrifugal acceleration due to the earth's rotation of the respective surface point. In point A the gravitational acceleration acting on the proof mass is balanced by the spring force per unit mass $-\frac{1}{m} ks_0$ with, k, spring modulus, and s_0 length of the stretched spring.

At point B the applied spring force would be $-\frac{1}{m}k(s_0+ds)$ with ds the additional displacement of the spring. Actually the proof mass is either kept in its original position by means of a feed back loop or brought back manually with the measuring skrew. The applied force per unit mass is $-\frac{1}{m}kds$. Since the position of x_i remains unchanged and the inertial accelerations are very small, we conclude from eq. (4.3) that $-\frac{1}{m}kds$ is equal to the change in gravitational acceleration from A to B minus the change in centrifugal acceleration of the origin of the moving triad from A to B, which means that the gravity difference between A and B is measured:
$df_i = -\frac{1}{m}kds = -d(V_i - R_{iI}\ddot{b}_I)$.

With these preparations we can discuss the first two cases. The first one is a kinematical approach to gradiometry, the second one a dynamical approach. In both cases eq. (4.3) stays central.

CASE ONE: Relative Motion of Particles in Free Fall.

Consider two proof masses A and B in free fall, solely under the influence of the gravitational field of the earth. They move in an orbit around the earth and are close together. The acceleration vectors of the two masses are

$$\ddot{x}_I(A) = V_I(A) \quad ,$$

and

$$\ddot{x}_I(B) = V_I(B) \quad .$$

We attach to B a local orthonormal triad $\underline{e}_i$ with $\underline{e}_{i=1} = \underline{e}_x$ along-track, $\underline{e}_{i=2} = \underline{e}_y$ cross-track, and $\underline{e}_{i=3} = \underline{e}_z$ radial. We shall measure the position of A relative to B in the triad $\underline{e}_i$ (so-to-say sitting on B). For simplicity it is assumed that B moves along a perfectly circular orbit. In accordance with Keplers 3rd law the orbit period is

$$\bar{n} = \sqrt{\frac{GM}{r^3}} \quad ,$$

with, r, the radial distance of B from the earth's center of mass. Consequently the $\underline{e}_i$-triad rotates with one cycle/revolution about the $\underline{e}_y$-axis. With the components $\omega_I = \{0,\bar{n},0\}$ of the angular velocity vector it is,

$$\Omega_{ij} = \begin{bmatrix} 0 & 0 & \bar{n} \\ 0 & 0 & 0 \\ -\bar{n} & 0 & 0 \end{bmatrix} \quad .$$

Applying (4.1) it is

$$R_{iI}\,V_I(A) = \ddot{x}_i(A) + 2\Omega_{ij}\dot{x}_j(A) + \Omega_{ij}\Omega_{jk}x_k(A) + R_{iI}V_I(B) \quad . \qquad (4.4)$$

In (4.4) x_i are the components of the position vector of A relative to B and expressed in the $\underline{e}_i$-triad. We expand $V_I(A)$ in a Taylor series relative to B and truncate the series after the linear term:

$$V_I(A) = V_I(B) + V_{IJ}(B)x_J \quad . \tag{4.5}$$

Inserted into (4.4) we find in component form

$$\begin{aligned}
\ddot{x} + 2\bar{n}\dot{z} - \bar{n}^2 x &= V_{xx}x + V_{xy}y + V_{xz}z, \\
\ddot{y} \qquad\qquad &= V_{yx}x + V_{yy}y + V_{yz}z, \\
\ddot{z} - 2\bar{n}\dot{x} - \bar{n}^2 z &= V_{zx}x + V_{zy}y + V_{zz}z.
\end{aligned} \tag{4.6}$$

Thereby we used, that $R_{iI}V_{IJ}x_J = R_{iI}V_{IJ}R_{Jj}R_{jJ}x_J = V_{ij}x_j$; furtheron it is $x_j = \{x,y,z\}$. Eq. (4.6) is a system of three coupled second-order differential equations. If $\ddot{x}_i$, $\dot{x}_i$, and x_i can be derived from measurements, eq. (4.6) represents a system of three equations in the five independent gradiometric unknowns V_{ij}. Additional proof masses close to B, but in independent directions could provide the required additional information to solve for V_{ij}.

However one could also proceed one step further. Assume V_{ij} is split into the gravitational tensor U_{ij} derived from a spherical reference potential U, eq. (2.7), and into the unknown tensor T_{ij} of the disturbing potential T. Then the right-hand side of eq. (4.6) can be split into

$$V_{ij}x_j = U_{ij}x_j + T_{ij}x_j \quad .$$

With U_{ij}, eq. (2.7)

$$U_{ij} = \bar{n}^2 \begin{bmatrix} -1 & 0 & 0 \\ 0 & -1 & 0 \\ 0 & 0 & 2 \end{bmatrix} \quad ,$$

we obtain now

$$\begin{aligned}
\ddot{x} + 2\bar{n}\dot{z} \qquad &= T_{xx}x + T_{xy}y + T_{xz}z = g_x \; , \\
\ddot{y} + \bar{n}^2 y \qquad &= T_{yx}x + T_{yy}y + T_{yz}z = g_y \; , \\
\ddot{z} - 2\bar{n}\dot{x} - 3\bar{n}^2 z &= T_{zx}x + T_{zy}y + T_{zz}z = g_z \; .
\end{aligned} \tag{4.7}$$

This is again a system of three coupled second-order differential equations. The homogeneous system ($g_i = 0$) describes the relative motion of the two proof masses in the gravitational field of a homogeneous sphere. The solution of the homogeneous system is

$$x = \frac{2\dot{z}^0}{\bar{n}} \cos \bar{n}t + (\frac{4\dot{x}^0}{\bar{n}} + 6z^0)\sin \bar{n}t - (3\dot{x}^0 + 6\bar{n}z^0)t + x^0 - 2\frac{\dot{z}^0}{\bar{n}} ,$$

$$y = y^0 \cos \bar{n}t + \frac{\dot{y}^0}{\bar{n}} \sin \bar{n}t , \tag{4.8}$$

$$z = -(2\frac{\dot{x}^0}{\bar{n}} + 3z^0)\cos \bar{n}t + \frac{\dot{z}^0}{\bar{n}} \sin \bar{n}t + 2\frac{\dot{x}^0}{\bar{n}} + 4z^0 ,$$

with $\dot{x}_i^0$ and x_i^0 the initial velocity and position vector, respectively.

Eq. (4.7) would apply, for example, to the motion of a proof mass (e.g. an astronaut or AUSTRIAnaut) inside the space shuttle, whereby g_i would also contain the surface forces per unit mass acting on its shield. An example based only on the homogeneous solution (4.8) is given in Figure 4.1.

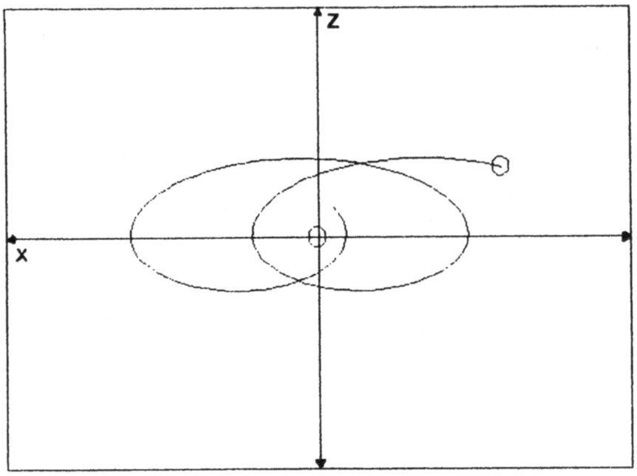

Figure 4.1: Relative motion of two test particles in free fall (scale of plot -200 m to +200 m, altitude 200 km, initial relative state components: $z^0 = 20$ m, $x^0 = -10$ m, $\dot{z}^0 = -0.06\ \frac{m}{s}$, $\dot{x}^0 = -0.0425\ \frac{m}{s}$)

It applies, with good approximation, to the relative motion of two drag-free satellites and can also be employed to the description of the actual satellite orbit relative to the approximate computed trajectory. Some examples from the literature, where (4.7) is treated are

- (Brouwer & Clemence, 1961): Hill's description of the motion of the moon;
- (Kaplan, 1976): the description of the relative motion between a so-called chase and a target satellite, and rendez-vous problems;
- (Bauer, 1982): the influence of the oblateness of the earth, the ellipticity of the orbit, and the self-gravitation on experiments inside a space laboratory;
- (Marussi & Chiaruttini, 1985): the description of the motion of a particle inside a gravitationally stabilized satellite; and
- (Colombo, 1984) or (Betti & Sansó, 1986): basic model for satellite to satellite tracking in the low-low mode.

CASE TWO: Acceleration Differences in a Moving Frame.

Next we consider a sophisticated version of the "rod + 2 springs" experiment. Four proof masses are placed at the corner points of a tetrahedron, at the origin and the end points of three orthonormal unit vectors parallel to the local triad $\underline{e}_i$, compare Figure 4.2a. The four proof masses are placed in free fall in the earth's gravitational field in an orbit around the earth. Despite the action of the non-uniform gravitational field the four proof masses are confined to their relative position, i.e. the tetrahedron configuration remains unchanged. This shall be realized in the following manner. Each proof mass is enclosed in a spherical chamber. The proof mass is kept in a levitated position by applying a magnetic field to it by means of three orthogonal pairs of magnetic coils, in principle similar to a superconducting gravimeter or to a drag-free satellite, compare Figure 4.2b. Any deviation from the zero

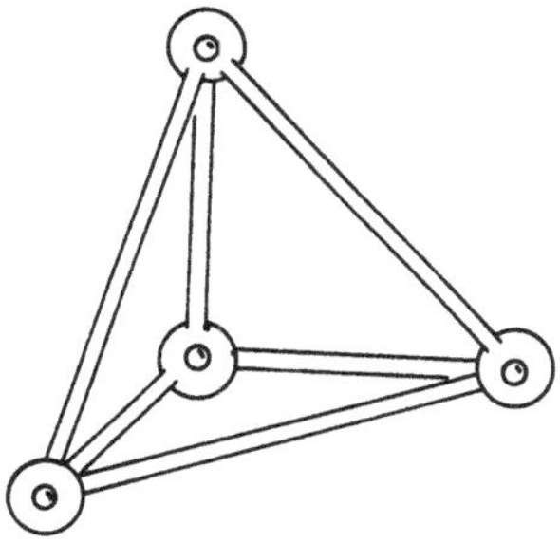

Figure 4.2a: Tetrahedron with four proof masses.

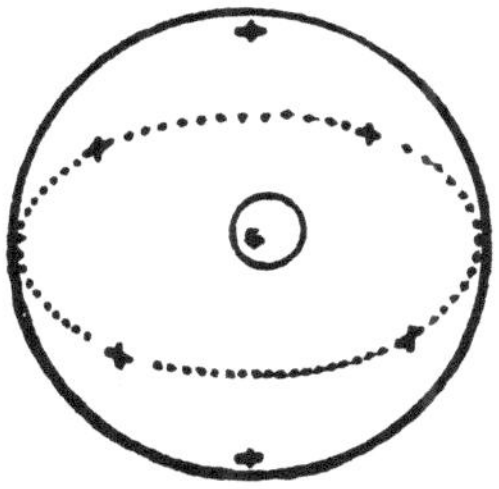

Figure 4.2b: Sphere with proof mass and three orthogonal pairs of magnetic coils.

position is corrected in a feed back loop without time delay. The complete device - a frame with four spherical chambers with levitated proof masses - resembles a satellite gradiometer. We denote the four proof masses by A, B, C, and D. In free fall the acceleration of the center of mass O is equal to the gravitational gradient, i.e. $R_{iI}\ddot{x}_I = V_i$. The measured specific force components f_i at B (the forces per unit mass, that keep B levitated) are with eq. (4.3)

$$f_i + V_i(B) = \dot{\Omega}_{ij}x_j + \Omega_{ik}\Omega_{kj}x_j + V_i(0) \quad ,$$

because $\ddot{x}_i = \dot{x}_i = 0$ in $\underline{e}_i$ due to the feed back mechanism. Expanding $V_i(B)$ relative to 0, we obtain

$$f_i + V_i(0) + V_{ij}(0)x_j = \dot{\Omega}_{ij}x_j + \Omega_{ik}\Omega_{kj}x_j + V_i(0)$$

or

$$f_i = (-V_{ij} + \dot{\Omega}_{ij} + \Omega_{ik}\Omega_{kj})x_j \quad . \tag{4.9}$$

Considering only B the following two restrictions are encountered:

(1) The x_j represent the coordinates of B relative to the center of mass 0. Usually 0 cannot be determined very well.

(2) If the gradiometer would not be in free fall, but some small surface forces would act on it an additional unknown acceleration component $R_{iI}\ddot{b}_I$ would have to be added at the right-hand side.

We can write down the same formula (4.9) for all four proof masses. Taking the difference of pairs, e.g. A and B we obtain

$$f_i(B) - f_i(A) = (-V_{ij} + \dot{\Omega}_{ij} + \Omega_{ik}\Omega_{kj})\delta x_j \quad . \tag{4.10}$$

This formula looks identical to (4.9), with δx_j the coordinate differences between A and B, but does not have the two restrictions from above, for (1) the coordinate differences between A and B can be determined very well, and (2) even if surface forces would act on the gradiometer, they would cancel out, when taking the difference (4.10).

Eq. (4.10) is fundamental in gradiometry. It shows that non-gravitational forces can be eliminated. However it also shows, that in case Ω_{ij} or $\dot{\Omega}_{ij} \neq 0$ (no inertial stabilization) the gravitational tensor is not directly observable. Instead, by taking the specific force differences between the four proof masses in various combinations, the following nine components can be derived from (4.10):

$$\frac{f_i(B,C, \text{ or } D) - f_i(A)}{\delta x_j} = \Lambda_{ij} = -V_{ij} + \dot{\Omega}_{ij} + \Omega_{ik}\Omega_{kj} \quad . \tag{4.11}$$

Example: Assuming that AB is parallel to $\underline{e}_x$, AC to $\underline{e}_y$, and AD to $\underline{e}_z$, derive the diagonal element Λ_{xx}: $\Lambda_{xx} = (f_x(B) - f_x(A))/\delta x$. This is equivalent to the sensitivity axis of the two springs being aligned with the rod in our "rod + 2 springs" experiment. Derive the off-diagonal element Λ_{yx}: $\Lambda_{yx} = (f_y(B) - f_y(A))/\delta x$, being equivalent to the sensitivity axis of the springs being orthogonal to the rod.
Is it possible to separate the gravitational tensor V_{ij} from the rotational components? In (Moritz, 1968) and (Forward, 1981) it is shown, that by differentiating (4.11) only the third derivatives $\frac{\partial^3 V}{\partial x_i \partial x_j \partial x_k} = V_{ijk}$ remain. The rotational part $\dot{\Omega}_{ij} + \Omega_{ik}\Omega_{kj}$ is uniform in space, its derivative is zero. Hence gravitation and rotation are separated. The problem with this approach is, that it requires the differencing of already very small quantities, and consequently an enormous relative precision of the sensors. Hence this approach cannot be taken in practise, at least for the time being.
What about the structure of the three tensors V_{ij}, $\dot{\Omega}_{ij}$, and $\Omega_{ik}\Omega_{kj}$? From eq. (2.5a-b)

we know that V_{ij} is symmetric with $\sum_{i=1}^{3} V_{ii} = 0$ for $\rho = 0$; $\Omega_{ik}\Omega_{kj}$ gives

$$\Omega_{ik}\Omega_{kj} = \begin{bmatrix} -(\omega_y^2+\omega_z^2) & \omega_x\omega_y & \omega_z\omega_x \\ \omega_x\omega_y & -(\omega_z^2+\omega_x^2) & \omega_y\omega_z \\ \omega_z\omega_x & \omega_y\omega_z & -(\omega_x^2+\omega_y^2) \end{bmatrix} , \tag{4.12}$$

symmetric as well, with the sum of the diagonal elements $-2(\omega_x^2+\omega_y^2+\omega_z^2)$, (minus two times the squared total angular velocity), whereas

$$\dot{\Omega}_{ij} = \begin{bmatrix} 0 & \dot{\omega}_z & -\dot{\omega}_y \\ -\dot{\omega}_z & 0 & \dot{\omega}_x \\ \dot{\omega}_y & -\dot{\omega}_x & 0 \end{bmatrix} \tag{4.13}$$

is anti-symmetric. Using this difference in structure, it is

$$\tfrac{1}{2}(\Lambda_{ij} + \Lambda_{ji}) = V_{ij} + \Omega_{ik}\Omega_{kj} \tag{4.14}$$

and

$$\tfrac{1}{2}(\Lambda_{ij} - \Lambda_{ji}) = \dot{\Omega}_{ij} \quad . \tag{4.15}$$

Thus by simple manipulations the symmetric and the anti-symmetric part can be separated. Operationally this implies, that whenever some initial angular velocity vector is given, Ω_{ij}^0, e.g. derived from a star-tracker, Ω_{ij} can be computed at any instant by

$$\Omega_{ij}(t) = \int \dot{\Omega}_{ij}\, dt + \Omega_{ij}^0 \quad ,$$

and one can separate V_{ij} from $\Omega_{ik}\Omega_{kj}$. Additional information comes from the sum of the diagonal terms of (4.14), which yields the length of the instantaneous angular velocity vector.

Conclusions:

- By taking the differences of the measured specific forces, required to keep the four proof masses in their relative position, in various combinations linear non-gravitational accelerations are eliminated, eq. (4.11).
- If the device "tetrahedron + 4 proof masses" is space stabilized these differences yield immediately the components of the gravitational tensors.
- In the case of a rotating gradiometer frame, the influence of the rotation could be eliminated by taking the differences of differences. In practise this approach is likely to fail due to the limited resolution of the sensors.

- With some moderate additional support, e.g. coming from a star tracker, it is nevertheless possible to separate the gravitational tensor from the rotational part.

Hence differential accelerometers are capable of separating gravitation from linear and angular acceleration. This is true for gradiometers in free fall, as well as for airborne or terrestrial gradiometers.

Cases One and Two were based on Newton's second law, (4.2), the change of linear momentum due to external forces F_I. One could as well look into the change of angular momentum $\frac{d}{dt} L_I$ due to applied torques N_I:

$$N_I = \frac{d}{dt} L_I, \qquad (4.16)$$

especially the torque exerted by the tidal field. This leads to a second category of cases.

CASE THREE: Pendulous Gradiometer.

An alternative to the "rod + 2 springs" experiment would be the measurement of the rotation induced by the local tidal field. Historically seen, this was the starting point of gradiometry, for the torsion balance belongs to this category, (Selényi, 1953). Again a very simple experiment is chosen to study the principle:

Compute the torque exerted by the tidal field at the point of suspension of a horizontal bar with two proof masses m_A and m_B at its ends, $m_A = m_B = m$. The situation is given in Figure 4.3a. The point of suspension is S,2ℓ is the total length

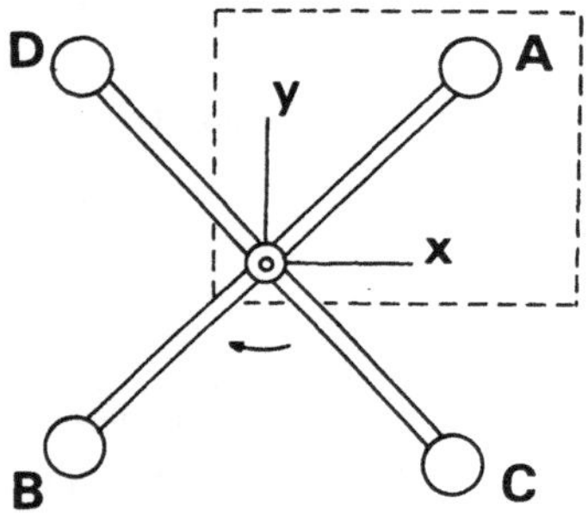

Figure 4.3a: Pendulous gradiometer

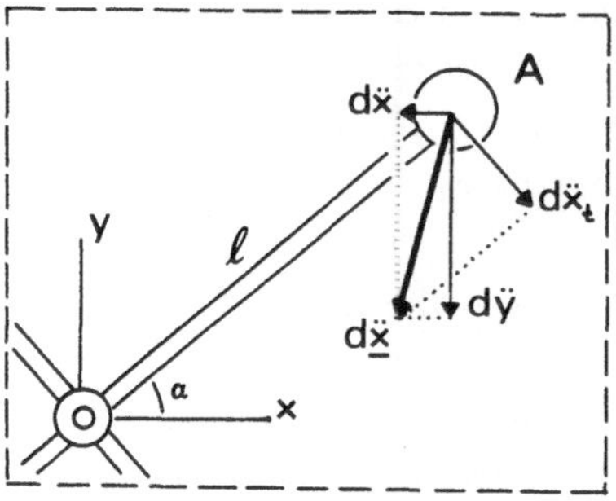

Figure 4.3b.

of the bar. The tidal acceleration relative to suspension point S is

$$d\ddot{x}_i = V_{ij} dx_j \quad . \qquad (2.4)$$

In the local triad $\underline{e}_i$ with $\underline{e}_z$ vertical, we define (compare Figure 4.3b)

$$dx = \ell \cos \alpha$$
$$dy = \ell \sin \alpha \qquad (4.17)$$
$$dz = 0 \quad .$$

The torque is exerted by the transversal force f_t. At proof mass m_A it is

$$f_t = md\ddot{x}_t$$
$$= m(d\ddot{y} \cos \alpha - d\ddot{x} \sin \alpha), \qquad (4.18)$$

(again, see Figure 4.3b) or with eqs. (2.4) and (4.17)

$$f_t = m[(V_{xy}dx + V_{yy}dy)\cos \alpha - (V_{xx}dx + V_{xy}dy)\sin \alpha)]$$
$$= m\ell[V_{xy} \cos^2 \alpha + V_{yy} \sin \alpha \cos \alpha - V_{xx} \sin \alpha \cos \alpha - V_{xy} \sin^2 \alpha] \qquad (4.19)$$
$$= m\ell[\frac{V_{yy}-V_{xx}}{2} \sin 2\alpha + V_{xy} \cos 2\alpha] \quad .$$

The torque about the z-axis, induced by f_t at A is $\ell \cdot f_t$. With the analogous equation for m_B at B, the total torque N_z becomes

$$N_z = 2m\ell^2[\frac{V_{yy}-V_{xx}}{2} \sin 2\alpha + V_{xy} \cos 2\alpha] \quad . \qquad (4.20)$$

The (gravitational) torque shall produce a damped harmonic oscillation of the bar about $\underline{e}_z$. The latter can be expressed, following (Misner et al., 1970), as

$$2m\ell^2(\ddot{\beta} + \dot{\beta}/d_0 + \omega_0^2\beta) \quad , \qquad (4.21)$$

where $2m\ell^2$ .... moment of inertia in z-direction,
β .... angular displacement due to N_z,
d_0 .... decay time of damped oscillation,
ω_0 .... angular frequency of free oscillation, and
$2m\ell^2\omega_0^2$ .... torsional spring constant.

(Angle α should more correctly be replaced by $\alpha+\beta$, but due to the small size of β one may leave (4.20) as it is.)

Equating (4.20) and (4.21) a forced second-order differential equation in β is obtained. Measuring β in a number of azimuths α the components $V_{yy}-V_{xx}$ and V_{xy} can be derived. This is the model of the classical torsion balance of the first kind, compare (Topercer, 1960). Other versions exist - torsion balances of the second kind -, where one or both proof masses are vertically displaced upwards or downwards from the xy-plane of the horizontal bar, compare for example (Jung, 1961). These types provide V_{xz} and V_{yz} too.

We return to eqs. (4.20) and (4.21). Let us assume a second horizontal bar with proof masses m_C and m_D is placed in the xy-plane at an angle of 90^0. It shall produce a torque $N_z^{CD} = -N_z^{AB}$. Thus, the differential torque $N_z^{AB} - N_z^{CD}$ becomes

$$N_z^{AB} - N_z^{CD} = 4m\ell^2[\frac{V_{yy}-V_{xx}}{2} \sin 2\alpha + V_{xy} \cos 2\alpha] \quad . \tag{4.22}$$

Finally the two crossed bars are rotated about $\underline{e}_z$ with constant angular velocity ω. With (4.21) and (4.22) this situation is described by

$$\ddot{\beta} + \dot{\beta}/d_0 + \omega_0^2\beta = (V_{yy}-V_{xx})\sin 2\omega t + 2V_{xy} \cos 2\omega t \quad . \tag{4.23}$$

Model (4.23) applies to the rotating gradiometer concepts of Forward and of Metzger, compare e.g. (Pelka & De Bra, 1979). A triplet of gradiometers rotating in three mutually orthogonal planes is required for the derivation of all tensor components.

The solution of differential equation (4.23) is

$$\beta(t) = Ce^{-\frac{1}{2d_0}t} \cos(\omega_1 t + \delta) + \bar{C} \cos(2\omega t + \bar{\delta}) \quad . \tag{4.24}$$

Amplitude C and phase angle δ belong to the free oscillation of the system ($N_z = 0$) and are determined from the initial conditions; ω_1 is the eigen frequency of the damped system, $\omega_1 = \sqrt{\omega_0^2 - (1/2d_0)^2}$. Amplitude $\bar{C}$ and phase angle $\bar{\delta}$ characterize the particular solution and are functions of the gradiometric unknowns:

$$\bar{C} = \sqrt{\frac{(V_{yy}-V_{xx})^2 + 4V_{xy}^2}{(\omega_0^2-(2\omega)^2) + (2\omega)^2(1/d_0)^2}} \quad , \tag{4.25}$$

and

$$\bar{\delta} = \frac{(V_{yy}-V_{xx})2\omega \frac{1}{d_0} + 2V_{xy}(\omega_0^2-(2\omega)^2)}{-2V_{xy}2\omega \frac{1}{d_0} + (V_{yy}-V_{xx})(\omega_0^2-(2\omega)^2)} \quad . \tag{4.26}$$

We see from (4.25) that resonance occurs for $2\omega = \omega_0$. Hence, by proper tuning the angular velocity ω of the system with its eigen frequency ω_0, the gradiometric observable can be magnified. Complete resonance would require zero damping ($d_0 \to \infty$). For small damping the phase angle $\bar{\delta}$ tends toward $\frac{2V_{xy}}{V_{yy}-V_{xx}}$.

The derivation of models (4.20) and (4.21) was carried out in a rather pragmatic way, in particular assuming the idealized construction of a weightless bar with two spherical point masses at its ends. We shall now in short repeat the derivation, trying to stay more general and keeping a closer connection with the starting point, the expression for the change in angular momentum (expressed in a local non-rotating frame $\underline{e}_i$)

$$N_i = \frac{d}{dt} L_i \quad . \tag{4.16}$$

In vector notation $\underline{N} = \underline{x} \times \underline{F}$ or in index notation

$$N_i = \epsilon_{ijk} x_j F_k$$

with permutation symbol ϵ_{ijk}. For a continuous mass distribution and with tidal force $V_{ij} dx_j dm$ acting on dm the above formula yields

$$N_i = \int_m \epsilon_{ijk}\, dx_j\, V_{k\ell}\, dx_\ell\, dm \quad . \tag{4.27}$$

Still assuming that our measuring device consists of two pendulous arms of length 2ℓ and orthogonal to each other, eq. (4.27) becomes with $dx_i = \{dx, dy, 0\}$

$$N_i = \begin{bmatrix} \int (V_{zx} dx dy + V_{zy} dy^2) dm \\ \int (V_{zx} dx^2 + V_{zy} dx dy) dm \\ \int (V_{xy} dx^2 + (V_{yy} - V_{xx}) dx dy - V_{xy} dy^2) dm \end{bmatrix} ,$$

or with the usual definition of the moments of inertia

$$N_i = \begin{bmatrix} V_{zx} I_{xy} + V_{zy} I_{yy} \\ V_{zx} I_{xx} + V_{zy} I_{xy} \\ V_{xy} I_{xx} + (V_{yy} - V_{xx}) I_{xy} - V_{xy} I_{yy} \end{bmatrix} . \tag{4.28}$$

We transform the inertia tensor I_{ij} from the space-fixed local triad $\underline{e}_i$ to a body fixed system $\underline{e}_{i'}$, rotating with the gradiometer and with the x- and y-axis aligned with the - undisturbed - gradiometer arms,

$$I_{i'j'} = R_{i'i}(\omega t) I_{ij} R_{jj'}(\omega t),$$

and find with (4.28) for the torque component N_z

$$\begin{aligned} N_z &= V_{xy}(I_{x'x'} \cos^2 \omega t + I_{y'y'} \sin^2 \omega t) + (V_{yy} - V_{xx})(I_{x'x'} - \\ &\quad I_{y'y'}) \sin \omega t \cos \omega t - V_{xy}(I_{x'x'} \sin^2 \omega t + I_{y'y'} \cos^2 \omega t) \\ &= V_{xy}(I_{x'x'} - I_{y'y'})(\cos^2 \omega t - \sin^2 \omega t) + (V_{yy} - V_{xx})(I_{x'x'} - \\ &\quad I_{y'y'}) \sin \omega t \cos \omega t \\ &= (I_{x'x'} - I_{y'y'})[\frac{V_{yy} - V_{xx}}{2} \sin 2\omega t + V_{xy} \cos 2\omega t] \quad , \end{aligned} \tag{4.29}$$

with $I_{x'x'}$, $I_{y'y'}$, $I_{z'z'}$ the principal moments of inertia. The change in angular momentum, eq. (4.21), becomes

$$I_{z'z'}(\ddot{\beta} + \dot{\beta}/d_0 + \omega_0^2\beta) \quad . \tag{4.30}$$

We see that for $I_{x'x'} = I_{z'z'} = 2m\ell^2$ and $I_{y'y'} = 0$ (one weightless horizontal bar with two spherical proof masses) equations (4.29) and (4.30) reduce to (4.20) and (4.21), respectively. The more general case (4.29) with (4.30) is treated in (Heitz, 1980).

CASE FOUR: Gravity Gradient Stabilization.

In the preceeding case the gravitationally induced torque has been discussed as a means to measure the tensor components V_{ij}. The same torque can be employed in order to attain a certain desirable orientation of a spacecraft. This is called gravity gradient stabilization. It has first been proposed around thirty years ago, compare for instance (Roberson, 1958) and has been applied for example in the case of the GEOS-3 satellite.

Now we are not aiming at a determination of the gravitational tensor components, but use them in order to achieve a favorable orientation. Again we start from

$$N_i = \int_m \epsilon_{ijk}\, dx_j\, V_{k\ell}\, dx_\ell\, dm \quad . \tag{4.27}$$

For a satellite with arbitrary mass distribution the components become

$$\begin{aligned} N_x &= V_{zx}I_{xy} + V_{zy}I_{yy} + V_{zz}I_{zy} - V_{yx}I_{xz} - V_{yy}I_{yz} - V_{yz}I_{zz} \quad , \\ N_y &= V_{xx}I_{xz} + V_{xy}I_{yz} + V_{xz}I_{zz} - V_{zx}I_{xx} - V_{zy}I_{xy} - V_{zz}I_{xz} \quad , \\ N_z &= V_{xy}I_{xx} + V_{yy}I_{xy} + V_{zy}I_{xz} - V_{xx}I_{xy} - V_{xy}I_{yy} - V_{xz}I_{zy} \quad . \end{aligned} \tag{4.31}$$

Specialized to the triad $\underline{e}_p$ of principal axes of the gravitational tensor V_{ij}, the torque components become

$$\begin{aligned} N_x &= (V_{zz}-V_{yy})I_{yz} \quad , \\ N_y &= (V_{xx}-V_{zz})I_{xz} \quad , \\ N_z &= (V_{yy}-V_{xx})I_{xy} \quad . \end{aligned} \tag{4.32}$$

In spherical approximation with $V_{xx} = V_{yy} = -\frac{GM}{r^3}$, and $V_{zz} = 2\frac{GM}{r^3}$, eq. (4.32) reduces to

$$N_i = 3\frac{GM}{r^3}\begin{bmatrix} I_{yz} \\ -I_{xz} \\ 0 \end{bmatrix} \quad . \tag{4.33}$$

Conclusion: For a satellite with its principal axes of inertia aligned with the principal axes of V_{ij} (i.e. $I_{xy} = I_{yz} = I_{xz} = 0$), no gravitational torque exists. The satellite is gravity gradient stable. Of course, we have to keep in mind that the $\underline{e}_p$-triad does not exactly coincide with the local astronomical triad, compare chapter 3. In addition, since $V_{yy}-V_{xx} \approx 0$ even without spherical approximation, any rotation about the z-axis can hardly be stabilized by means of the tidal field.

We remain in spherical approximation with $\underline{e}_p = \underline{e}_i$ and $\underline{e}_x$ along track, $\underline{e}_y$ cross track, and $\underline{e}_z$ radial, and assume that there exists a small misalignment with respect to the body fixed principal moment of inertia triad $\underline{e}_{i'}$. The misalignment can be expressed by three small angles ψ_x (rotation about $\underline{e}_x$: roll), ψ_y (rotation about $\underline{e}_y$: pitch), and ψ_z (rotation about $\underline{e}_z$: yaw). Hence, it is

$$\underline{e}_i = \underline{e}_{i'} - \Omega_{ii'}\underline{e}_{i'} = (E-\Omega)_{ii'}\underline{e}_{i'} \quad , \tag{4.34}$$

with E the identity transformation, where

$$\Omega_{ii'} = \begin{bmatrix} 0 & \psi_z & -\psi_y \\ -\psi_z & 0 & \psi_x \\ \psi_y & -\psi_x & 0 \end{bmatrix} \quad .$$

With (2.6) we find

$$V_{i'j'} = (E-\Omega)_{i'i}\, V_{ij}(E-\Omega)_{jj'} \quad , \tag{4.34}$$

or in components, neglecting second-order terms

$$V_{i'j'} = \begin{bmatrix} V_{xx} & (V_{xx}-V_{yy})\psi_z & (V_{zz}-V_{xx})\psi_y \\ (V_{xx}-V_{yy})\psi_z & V_{yy} & (V_{yy}-V_{zz})\psi_x \\ (V_{zz}-V_{xx})\psi_y & (V_{yy}-V_{zz})\psi_x & V_{zz} \end{bmatrix} \quad .$$

Thus, with eq. (4.31) expressed in the body-fixed $\underline{e}_{i'}$-triad, it is

$$N_{i'} = \begin{bmatrix} V_{z'y'}(I_{y'y'}-I_{z'z'}) \\ V_{x'z'}(I_{z'z'}-I_{x'x'}) \\ V_{x'y'}(I_{x'x'}-I_{y'y'}) \end{bmatrix} = \begin{bmatrix} (V_{yy}-V_{zz})\psi_x(I_{y'y'}-I_{z'z'}) \\ (V_{zz}-V_{xx})\psi_y(I_{z'z'}-I_{x'x'}) \\ (V_{xx}-V_{yy})\psi_z(I_{x'x'}-I_{y'y'}) \end{bmatrix} \quad . \tag{4.35}$$

Inserting V_{xx}, V_{yy}, and V_{zz} in spherical approximation and remembering that $\frac{GM}{r^3} = \bar{n}^2$, the torque components N_i become

$$N_{i'} = 3\bar{n}^2 \begin{pmatrix} -\psi_x(I_{y'y'}-I_{z'z'}) \\ \psi_y(I_{z'z'}-I_{x'x'}) \\ 0 \end{pmatrix} . \tag{4.36}$$

When expressed in the rotating body fixed (principal moment of inertia) triad $\underline{e}_{i'}$, eq. (4.16) takes the form

$$\begin{aligned} N_{x'} &= I_{x'x'}\dot{\omega}_{x'} + \omega_{y'}\omega_{z'}(I_{z'z'}-I_{y'y'}) , \\ N_{y'} &= I_{y'y'}\dot{\omega}_{y'} + \omega_{x'}\omega_{z'}(I_{x'x'}-I_{z'z'}) , \\ N_{z'} &= I_{z'z'}\dot{\omega}_{z'} + \omega_{x'}\omega_{y'}(I_{y'y'}-I_{x'x'}) , \end{aligned} \tag{4.37}$$

the famous Euler equations, compare e.g. (Moritz, 1982; p. 179). The components of the angular velocity vector $\omega_{i'}$ are composed of the orbit angular velocity $\dot{\theta}$ and the changes in ψ_x, ψ_y, and ψ_z. For an orbit with small eccentricity e, the angular velocity $\dot{\theta}$ can be written as

$$\dot{\theta} \approx \bar{n}(1 + 2e \cos \bar{n}t)$$

and

$$\ddot{\theta} \approx -2\bar{n}\ e \sin \bar{n}t \quad .$$

Furtheron with $\omega_{i'} = \omega^0_{i'} + \dot{\psi}_{i'}$ the components of $\omega_{i'}$ become $\omega_{i'} = \{\dot{\psi}_x + \bar{n}\psi_z, \dot{\psi}_y + \dot{\theta}, \dot{\psi}_z - \bar{n}\psi_x\}$, where we used $\omega^0_i = \{0,\dot{\theta},0\}$, and $\omega^0_{i'} = \{\bar{n}\psi_z,\dot{\theta},-\bar{n}\psi_x\}$. Inserted into eq. (4.37) leads together with eq. (4.36) to the final system

$$\begin{aligned} & I_{x'x'}(\ddot{\psi}_x+\bar{n}\dot{\psi}_z) + (\bar{n}\dot{\psi}_z-\bar{n}^2\psi_x)(I_{z'z'}-I_{y'y'}) = 3\bar{n}^2\psi_x(I_{z'z'}-I_{y'y'}) , \\ & I_{y'y'}(\ddot{\psi}_y-2\bar{n}^2 e \sin \bar{n}t) = -3\bar{n}^2\psi_y(I_{x'x'}-I_{z'z'}) , \\ & I_{z'z'}(\ddot{\psi}_z-\bar{n}\dot{\psi}_x) + (\bar{n}\dot{\psi}_x+\bar{n}^2\psi_z)(I_{y'y'}-I_{x'x'}) = 0 . \end{aligned} \tag{4.38a-c}$$

The result is a set of three second-order differential equations in the alignment angles ψ_x, ψ_y, and ψ_z. The pitch motion, eq. (4.38b), represents the deviation from the vertical axis. It is de-coupled from the two other equations and can be written as

$$\ddot{\psi}_y + 3\bar{n}^2 \frac{I_{x'x'}-I_{z'z'}}{I_{y'y'}} \psi_y = 2\bar{n}^2 e \sin \bar{n}t \quad . \tag{4.39}$$

With $\sigma_y = \frac{I_{x'x'}-I_{z'z'}}{I_{y'y'}}$ the solution of (4.39) becomes

$$\psi_y = C\cos(\sqrt{3\sigma_y}\,\bar{n}t + \delta) + \frac{2e}{3\sigma_y - 1}\sin\bar{n}t \quad , \tag{4.40}$$

where C and δ are the amplitude and phase angle of the homogeneous solution. Critical is the resonance case $\sigma_y = \frac{1}{3}$. For GEOS-3 it was e = 0.0054 and σ_y = 0.984, compare (Wertz, 1978), which results in a sinusoidal vertical alignment error of 0.3^o. In order to achieve gravity gradient stabilization extendable booms with an end mass are used, in order to approximate a dumb-bell configuration ("bar with two spherical masses"), compare Figure 4.4. For the ideal dumb-bell configuration of case Three in free fall, we would have $I_{x'x'} = I_{y'y'} = 2m\ell^2$, and $I_{z'z'} = 0$. It follows $\sigma_y = 1$ and a sinusoidal amplitude of e, the orbit eccentricity.

The derivation of case Four follows (Kaplan, 1976; ch. 5.5).

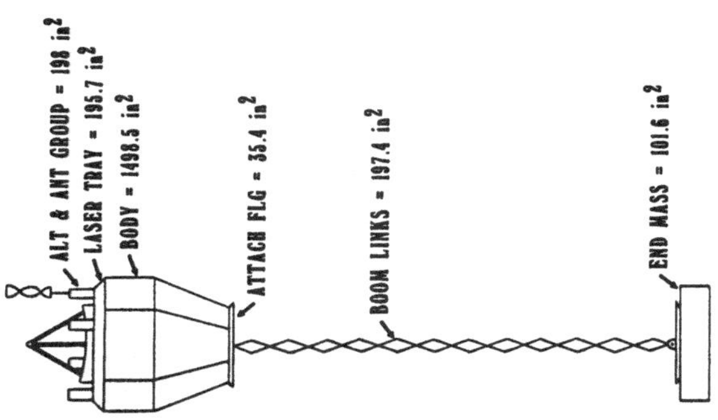

Figure 4.4: GEOS-3 with boom and end mass for the purpose of gradient stabilization.

Outlook: In four cases the observability (or use) of the gravitational tensor components was discussed. Two cases treated the change in linear momentum caused by the tidal force, two cases the change in angular momentum.
Possibly a more general line, common to all four cases, could have been attained in considering the exchange of various forms of energies in a closed system, consisting of earth, sun, moon, and a number of test particles. This approach has for example been taken in (Ilk, 1983a and b).

5. SATELLITE GRADIOMETRY.

By now we can be rather confident, that the components of the gravitational tensor are observable, not affected for example by the linear acceleration of the gradiometer or the apparent forces due to the rotation of the instrument in space. We therefore proceed to the use of a gradiometer in a satellite. Since the model will be derived in a way analogous to what we know from the geodetic boundary value problem (b.v.p.), a short excursion to the latter shall be taken. The fundamentals of the b.v.p., for example the derivation of the Stokes' formula are very well known. Hence all derivations can be kept very short.

Assume the tensor components W_{ij} of the gravity potential W at a surface point P are given in the local astronomical triad $\underline{e}_i$. The orthonormal base vectors are $\underline{e}_{i=1} = \underline{e}_x$ directing north, $\underline{e}_{i=2} = \underline{e}_y$ directing east, and $\underline{e}_{i=3} = \underline{e}_z$ pointing to the zenith. We introduce a known approximate or normal field U with tensor components U_{ij}. Then it is

$$W_{ij}(P) = U_{ij}(P) + T_{ij}(P) \tag{5.1}$$

with T the disturbing potential. Secondly, since the coordinates of the measurement point P are usually only known approximately (P'), and therefore the components U_{ij} at P cannot be computed, $U_{ij}(P)$ is expressed relative to $U_{ij}(P')$ using a Taylor expansion truncated after the linear term

$$U_{ij}(P) \approx U_{ij}(P') + U_{ijk}(P')\Delta x_k \quad , \tag{5.2}$$

with $U_{ijk} = \frac{\partial U_{ij}}{\partial x_k}$ and Δx_k the coordinate differences between P and P'. Inserted into (5.1) and remembering that $T_{ij}(P') \approx T_{ij}(P)$ the linear gradiometer model becomes:

$$\Delta\Gamma_{ij} = W_{ij}(P) - U_{ij}(P') = U_{ijk}(P')\Delta x_k + T_{ij}(P') \quad . \tag{5.3}$$

with the gradiometric anomalies (observable minus computed) $\Delta\Gamma_{ij}$. In spherical approximation the coefficients U_{ijk} of Δx_k become

$$\begin{pmatrix} \Delta\Gamma_{xx} \\ \Delta\Gamma_{xy} \\ \Delta\Gamma_{xz} \\ \Delta\Gamma_{yy} \\ \Delta\Gamma_{yz} \\ \Delta\Gamma_{zz} \end{pmatrix} = \frac{3GM}{r^3} \begin{pmatrix} 0 & 0 & \frac{1}{r} \\ 0 & 0 & 0 \\ \frac{1}{r} & 0 & 0 \\ 0 & 0 & \frac{1}{r} \\ 0 & \frac{1}{r} & 0 \\ 0 & 0 & -\frac{2}{r} \end{pmatrix} \begin{pmatrix} \Delta x \\ \Delta y \\ \Delta z \end{pmatrix} + \begin{pmatrix} T_{xx} \\ T_{xy} \\ T_{xz} \\ T_{yy} \\ T_{yz} \\ T_{zz} \end{pmatrix} \quad . \tag{5.4a-f}$$

Naturally the sum of the main diagonals $\sum_i \Delta\Gamma_{ii}$ has to be $2\omega^2$. We observe, for example, that in the coordinate corrections Δx_i the diagonal terms are de-linked from the off-diagonals.

In deriving the solution of the geodetic b.v.p. in spherical approximation the same type of linearization leads for the four classical observables Φ, astronomical latitude, Λ, astronomical longitude, C, potential difference, and g, scalar gravity to

$$\begin{bmatrix} \Delta\Phi \\ \Delta\Lambda \\ \Delta C \\ \Delta g \end{bmatrix} = \begin{bmatrix} 0 \\ 0 \\ -\Delta W_0 \\ 0 \end{bmatrix} + \begin{bmatrix} \frac{1}{r} & 0 & 0 \\ 0 & \frac{1}{r\cos\varphi} & 0 \\ 0 & 0 & -\gamma \\ 0 & 0 & -\frac{2\gamma}{r} \end{bmatrix} \begin{bmatrix} \Delta x \\ \Delta y \\ \Delta z \end{bmatrix} + \begin{bmatrix} -\frac{1}{\gamma} T_x \\ -\frac{1}{\gamma\cos\varphi} T_y \\ T \\ -T_z \end{bmatrix} . \quad (5.5a\text{-}d)$$

In this context the datum unknown ΔW_0 may be neglected. For details it is referred e.g. to (Moritz, 1980; ch. 27) or (Rummel, 1984). In Δx_i the anomalies ΔC and Δg are de-linked from $\Delta\Phi$ and $\Delta\Lambda$. Hence elimination of Δz from (5.5c) leads via the Bruns equation

$$\Delta z = N = \frac{T-\Delta C}{\gamma} \quad (5.6)$$

to the well-known fundamental equation ($T_z = T_r$)

$$(\Delta g - \frac{2}{r}\Delta C) = -(\frac{\partial}{\partial r} + \frac{2}{r})T \quad . \quad (5.7)$$

Most of the time ΔC is assumed to be zero. This is equivalent with a certain - very convenient - choice of the coordinates of the approximate point P'. However one could as well use any other approximate point P' that leaves $\Delta C \neq 0$ or for example instead of $\Delta C = 0$ use $\Delta g = 0$, compare (Krarup, 1973), (Grafarend, 1979), or (Rummel & Teunissen, 1982). In any case this question has nothing to do with the elimination of Δz from eq. (5.5). In eq. (5.7) the left hand side is assumed to be known e.g. globally or in a number of points locally and T is the unknown disturbing potential function. T can be solved for by introducing some smart representation of it for example

(1) with a reproducing kernel and apply collocation or

(2) by an expansion in terms of spherical harmonics

$$T(P') = \frac{\delta(GM)}{r_{P'}} + \frac{GM_0}{R} \sum_{n=2}^{\infty} \sum_{m=0}^{n} (\frac{R}{r_{P'}})^{n+1} [\Delta\bar{C}_{nm} \cos m\lambda_{P'} + \Delta\bar{S}_{nm} \sin m\lambda_{P'}] \bar{P}_{nm}(\sin\varphi_{P'}), \quad (5.8)$$

with the unknowns $\delta(GM)$, correction to GM, and $\Delta\bar{C}_{nm}$, $\Delta\bar{S}_{nm}$, unknown potential coefficients of T. This representation together with (5.7) could directly be used as an adjustment model with the summation truncated at some finite degree $n = \bar{N}$ and the anomalies given globally either as mean block averages or grid values, compare (Colombo, 1981). Alternatively, one could derive from eqs. (5.7) and (5.8)

(3) an analytical solution. In this latter case $(\Delta g - \frac{2}{r}\Delta C)$ is assumed to be given as a continuous function all over the earth.

The analytical approach leads to the well-known Stokes formula, compare (Heiskanen and Moritz, 1967):

$$T(P) = \frac{\delta(GM)}{r_P} + \frac{R}{4\pi} \int_\sigma St(\psi_{PQ})(\Delta g(Q) - \frac{2}{r}\Delta C(Q))d\sigma_Q \tag{5.9}$$

or with (5.6)

$$\Delta z(P) = N(P) = \frac{\delta(GM)}{R\gamma} - \frac{\Delta C}{\gamma} + \frac{R}{4\pi\gamma} \int_\sigma St(\psi_{PQ})(\Delta g(Q) - \frac{2}{r}\Delta C(Q))d\sigma_Q \quad . \tag{5.10}$$

In eqs. (5.9) and (5.10) $St(\psi)$ is the well-known integral kernel with its spectral form

$$St(\psi) = \sum_{n=2}^{\infty} \frac{2n+1}{n-1} P_n(\cos\psi) \tag{5.11}$$

or the closed analytical expression

$$St(\psi) = \frac{1}{\sin\frac{\psi}{2}} + 1 - 5\cos\psi - 6\sin\frac{\psi}{2} - 3\cos\psi \ln(\sin\frac{\psi}{2} + \sin^2\frac{\psi}{2}) \quad . \tag{5.12}$$

When merging the linear model of gradiometry (5.4) with the classical one, eq. (5.5) the coordinate unknowns Δx_i can be eliminated in various ways, and one could obtain a variety of linear combinations of anomalies expressed as linear functionals of T. Two rather logical combinations, obtained again from the elimination of Δz are

$$(\Delta\Gamma_{zz} - \frac{6}{r^2}\Delta C) = (\frac{\partial^2}{\partial r^2} - \frac{6}{r^2})T \quad , \tag{5.12}$$

(with $T_{zz} = T_{rr}$) and

$$(\Delta\Gamma_{zz} - \frac{3}{r}\Delta g) = (\frac{\partial^2}{\partial r^2} + \frac{3}{r}\frac{\partial}{\partial r})T \quad . \tag{5.13}$$

These combinations lead to new "fundamental" equations. Their analytical solution is discussed in (Heck, 1982). In the case of eq. (5.12) the spectral form is

$$H(\psi) = \sum_{n=2}^{\infty} \frac{2n+1}{(n+4)(n-1)} P_n(\cos\psi) \quad . \tag{5.14}$$

The analytical expression can be found by the method described in (Tscherning, 1972) or (Moritz, 1980; ch. 23). It yields

$$H(\psi) = \frac{1}{5}\{\frac{71}{12} - \frac{22}{5}\cos\psi - \frac{35}{2}\cos^2\psi + (35\cos^2\psi + \frac{35}{3}\cos\psi - \frac{32}{3})\sin\frac{\psi}{2} + \frac{7}{2}\cos\psi(5\cos^2\psi - 3)\ln(1 + \frac{1}{\sin\psi}) - 3\cos\psi \cdot \ln(\sin\frac{\psi}{2} + \sin^2\frac{\psi}{2})\} \quad . \tag{5.15}$$

Analogously one obtains for combination (5.13)

$$G(\psi) = \sum_{n=2}^{\infty} \frac{2n+1}{(n+1)(n-1)} P_n(\cos\psi)$$

$$= 1 - 3\sin\frac{\psi}{2} - \frac{7}{4}\cos\psi - \frac{3}{2}\cos\psi \ln(\sin\frac{\psi}{2} + \sin^2\frac{\psi}{2})$$

$$+ \tfrac{1}{2}\ln(1 + \frac{1}{\sin\frac{\psi}{2}}) \quad . \tag{5.16}$$

Hence the vertical displacement (= height anomaly) becomes from (5.12)

$$\Delta z(P) = N(P) = \frac{\delta(GM)}{R\gamma} - \frac{\Delta C}{\gamma} + \frac{R}{4\pi\gamma} \int_\sigma H(\psi_{PQ})(\Delta\Gamma_{zz}(Q) - \frac{6}{r^2}\Delta C(Q))d\sigma_Q \tag{5.17}$$

and from (5.13)

$$\Delta z(P) = N(P) = \frac{\delta(GM)}{R\gamma} - \frac{\Delta C}{\gamma} + \frac{R}{4\pi\gamma} \int_\sigma G(\psi_{PQ})(\Delta\Gamma_{zz}(Q) - \frac{3}{r}\Delta g(Q))d\sigma_Q \quad . \tag{5.18}$$

Of course precondition for the applicability of these formulas would be the availability of a global set of gradiometric anomalies, besides of gravity or potential anomalies. *)

After this excursion we return to the case of satellite gradiometry. No gradiometric satellite mission has been carried out until now. The purpose of such a mission would be to determine with rather high precision the earth's gravitational field with a resolution of about 200 or 300 in terms of the maximum degree $\bar{N}$ of a spherical harmonic expansion, (5.8). This means that details of the gravity field with a lateral extension of 50 to 100 km would still be discernible. Gradiometry is selected for this purpose as a candidate technique, because the high sensitivity of a gradiometer for local details to a certain extent counterbalances the attenuation of the gravity field with increasing satellite altitude.

Let us return to the linear model, eq. (5.4). In the case of a satellite gradiometer the observation point P is located at the satellite. All observation points P make up the actual orbit trajectory. The approximate points P' are located at the approximate orbit, which is computed from the chosen normal gravity field and an initial state vector. The coordinate displacements Δx_i represent now the separation of the approximate from the actual orbit. And all quantities are now expressed in the local triad with $\underline{e}_{i=1} = \underline{e}_x$ along track, $\underline{e}_{i=2} = \underline{e}_y$ cross-track, and $\underline{e}_{i=3} = \underline{e}_z$ radial from the earth's center of mass instead of in the astronomical triad. (The question of whether and how it is possible to transform the components measured in the instrument frame to the $\underline{e}_i$-triad shall not be discussed here.)

We observe that $\Delta\Gamma_{xx}$, $\Delta\Gamma_{yy}$, and $\Delta\Gamma_{zz}$ depend on the radial orbit uncertainty Δz. Analoguous to the geodetic b.v.p. Δz is eliminated from the equations by one of the three combinations

*) Compare APPENDIX C

$$\begin{aligned} \Delta\Gamma_{xx} - \Delta\Gamma_{yy} &= \\ 2\Delta\Gamma_{xx} + \Delta\Gamma_{zz} &= \\ -2\Delta\Gamma_{yy} - \Delta\Gamma_{zz} &= 2T_{xx} + T_{zz} \quad . \end{aligned} \tag{5.19}$$

One could say, we have eliminated the orbit. This is a genuine feature of satellite gradiometry, that follows from the simultaneous measurement of independent gravity field components.

Hence it is possible to derive from either $\Delta\Gamma_{xy}$ or from one of the linear combinations (5.19) an improved gravity field representation. The improved gravity field model is applied to improve the orbits (= approximate points P'). The anomalies $\Delta\Gamma_{ij}$ can be re-computed using the improved (normal-)gravity field and the improved orbit and the procedure starts finally again with a second iteration. We shall now explain the main characteristics of this procedure.

If we choose as gravity field representation the expansion of T into spherical harmonics, expression (5.19) selecting $2\Delta\Gamma_{xx} + \Delta\Gamma_{zz}$ becomes

$$\begin{aligned} 2\Delta\Gamma_{xx} + \Delta\Gamma_{zz} = \frac{GM_0}{R} \sum_{n=2}^{\bar{N}} \sum_{m=0}^{n} & \Delta\bar{C}_{nm}\left[2\frac{\partial^2}{\partial x^2} + \frac{\partial^2}{\partial z^2}\right]\left[\left(\frac{R}{r_P}\right)^{n+1} \cos m\lambda\, \bar{P}_{nm}(\sin\varphi)\right] \\ & + \Delta\bar{S}_{nm}\left[2\frac{\partial^2}{\partial x^2} + \frac{\partial^2}{\partial z^2}\right]\left[\left(\frac{R}{r_P}\right)^{n+1} \sin m\lambda\, \bar{P}_{nm}(\sin\varphi)\right] \quad . \end{aligned} \tag{5.20}$$

For the computation the derivatives in the $\underline{e}_i$ system are transformed into the curvilinear $\{\varphi,\lambda,r\}$-system, in which the spherical harmonics are expressed. The transformation equations are given in Appendix B. In (5.20) $2\Delta\Gamma_{xx} + \Delta\Gamma_{zz}$ are the observables in each sampling point, the $\bar{N}^2$ coefficients $\Delta\bar{C}_{nm}$ and $\Delta\bar{S}_{nm}$ are the unknowns. The computation of the unknowns represents a straightforward adjustment problem, in principle. However for $\bar{N}$ = 200 to 300 the number of unknowns becomes so large, that the resulting system of linear equations can only be solved if some favorable symmetries exist. The symmetry is lacking, whenever the several million of observations taken during a six months mission are taken as they are and where they are.

Colombo (1981) showed that for regular $\{\Delta\varphi,\Delta\lambda\}$-data grids or regular coverage with block averages on a sphere the observation equations of the form (5.20) result in favorable matrix structures and can be solved in a very efficient manner. We call his approach Fast Harmonic Analysis on a Sphere (FHAS) in analogy to FFT. In order to achieve a regular structure with the gradiometer data, the scattered gradiometer anomalies, $\Delta\Gamma_{ij}$, are transformed into block averages independently for each component in the following manner: Let us assume two geo-centric concentric spheres, one inside and one outside the orbits, in such a way that all observations are contained between these two spheres. The area between the two spheres is divided into equi-angular cells of surface area $r^2 \cos\varphi\Delta\varphi\Delta\lambda$ and thickness Δr, see Figure 5.1a and b, with Δr typically less than 10 km. Inside each of the blocks the averages are taken.

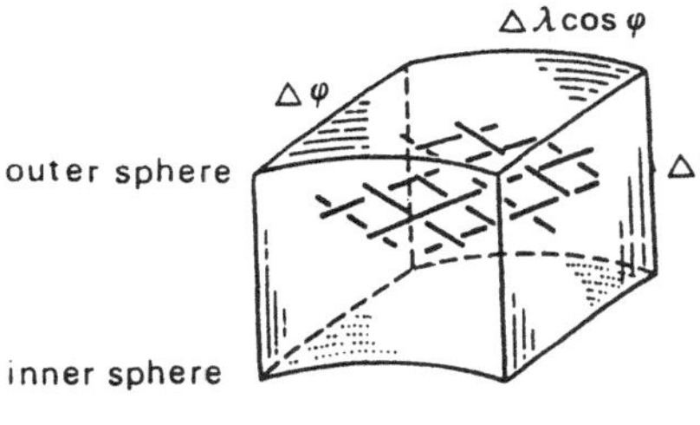

Figure 5.1a: Cell $r^2\cos\phi\Delta\phi\lambda\Delta r$ containing pieces of several satellite arcs.

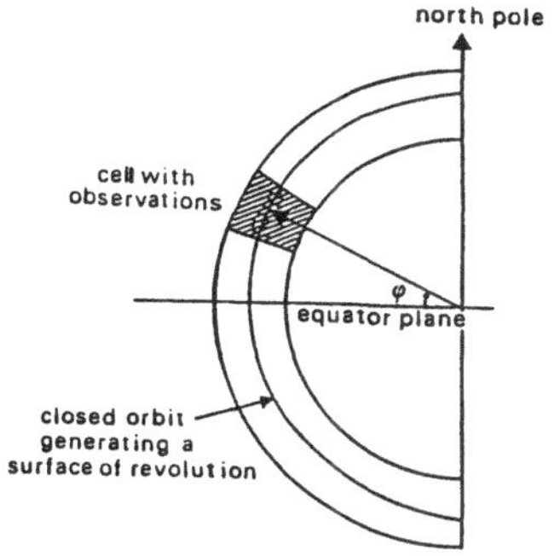

Figure 5.1b: Inner and outer sphere, cell with observations, and surface of revolution used for the linearization.

The linear model, eq. (5.20), holds now for the block averages (with the spherical harmonics averaged over the block). In order to attain the symmetry in the normal equations, the series expansion (5.8) is written in a different sequence with the summation over m interchanged with that over n:

$$\begin{aligned} T(P') &= \frac{\delta(GM)}{r_{P'}} + \frac{GM_0}{R} \sum_{m=0}^{\bar{N}} \Big(\sum_{n=m}^{\bar{N}} \Big(\frac{R}{r_{P'}}\Big)^{n+1} \Delta\bar{C}_{nm}\, \bar{P}_{nm}(\sin\varphi_{P'})\Big)\cos m\lambda_{P'} + \\ &\quad \Big(\sum_{n=m}^{\bar{N}} \Big(\frac{R}{r_{P'}}\Big)^{n+1} \Delta\bar{S}_{nm}\, \bar{P}_{nm}(\sin\varphi_{P'})\Big)\sin m\lambda_{P'}\Big) \\ &= \frac{\delta(GM)}{r_{P'}} + \frac{GM_0}{R} \sum_{m=0}^{\bar{N}} A_m \cos m\lambda_{P'} + B_m \sin m\lambda_{P'} \quad . \end{aligned} \tag{5.21}$$

We see T can now be expressed as a Fourier series in λ with coefficients

$$\begin{aligned} A_m &= \sum_{n=m}^{\bar{N}} \Big(\frac{R}{r_{P'}}\Big)^{n+1} \Delta\bar{C}_{nm}\, \bar{P}_{nm}(\sin\varphi_{P'}), \\ B_m &= \sum_{n=m}^{\bar{N}} \Big(\frac{R}{r_{P'}}\Big)^{n+1} \Delta\bar{S}_{nm}\, \bar{P}_{nm}(\sin\varphi_{P'}) \quad . \end{aligned} \tag{5.22a-b}$$

Observing in addition the symmetry respectively anti-symmetry of even, respectively odd degree associated Legendre function $\bar{P}_{nm}(\sin\varphi)$, arrangement of the right-hand side of (5.20) according to (5.21) produces the block-diagonal structure of the normal matrix displayed in Figure 5.2. The size of the largest sub-block is now only $\bar{N} \times \bar{N}$.

This is how the improved gravity field parameters $\Delta\bar{C}^1_{nm}$, and $\Delta\bar{S}^1_{nm}$ are derived in a first iteration ("1"). They give $T^{(1)}_{ij}$. Solving eq. (5.4a-f) for Δx_i gives for example

$$\Delta x^{(1)} = (\frac{3GM}{r^4})^{-1}(\Delta\Gamma_{xz}-T_{xz}^{(1)})$$

$$\Delta y^{(1)} = (\frac{3GM}{r^4})^{-1}(\Delta\Gamma_{yz}-T_{yz}^{(1)}) \qquad (5.23a\text{-}c)$$

$$\Delta z^{(1)} = -(\frac{6GM}{r^4})^{-1}(\Delta\Gamma_{zz}-T_{zz}^{(1)}) \quad .$$

These are preliminary orbit corrections. Inserted into the left hand side of the Hill's equations (4.8) improved initial state components $\{x_0^{(1)}, y_0^{(1)}, z_0^{(1)}, \dot{x}_0^{(1)}, \dot{y}_0^{(1)}, \dot{z}_0^{(1)}\}$ are computed. The up-dated initial state together with the improved gravity field approximation yields an improved orbit. Finally, from the new orbit and the improved gravity field parameters new anomalies $\Delta\Gamma_{ij}$ are obtained, with which a new iteration loop can be started. The steps of the procedure are summarized in Table 5.1.

The complete procedure and its numerical verification with a zonal gravity field up to $\bar{N}$ = 300 are described in (Rummel and Colombo, 1985).

1	Defining the equal angular cells and forming the cell averages.
2	Pre-adjustment of the diagonal anomalies $\Delta\Gamma_{ii}$ using Laplace' equation.
3	Forming one of the linear combinations of eq. (5.19).
4	Gravity parameter adjustment based on the chosen linear combination and on $\Delta\Gamma_{xy}$ (the coefficients of the $\Delta\bar{C}_{nm\alpha}$ are evaluated at the surface of revolution generated by a closed orbit).
5	Up-date of the nominal orbit from the estimated $\Delta C_{nm\alpha}$ and correction of the gradient anomalies for the new nominal orbit.
6	Computation of the $\Delta z^{(1)}$, eq. (5.23c).
7	Improved initial state vector elements $\Delta z(t_0)$, $\Delta\dot{z}(t_0)$, and $\Delta\dot{x}(t_0)$, eq. (4.8).
8	Up-date of the nominal orbit and up-date of the anomalies by means of the improved orbit and the estimated $\Delta\bar{C}_{nm\alpha}$.
9	Repeat all steps from step 3 on until the iteration converges.
10	Determination of the $\Delta\hat{x}$ and $\Delta\hat{y}$ (along and cross track) from $\Delta\tilde{\Gamma}_{xz}$ and $\Delta\tilde{\Gamma}_{yz}$, respectively, using the improved gravity field and orbit parameters.
11	Final orbit adjustment based on the $\Delta\hat{x}$, $\Delta\hat{y}$, and $\Delta\hat{z}$ using equations (4.8).
12	Final gravity parameter adjustment based on all five independent satellite gradiometer components.

Table 5.1: The steps of the processing procedure (Rummel & Colombo, 1985).

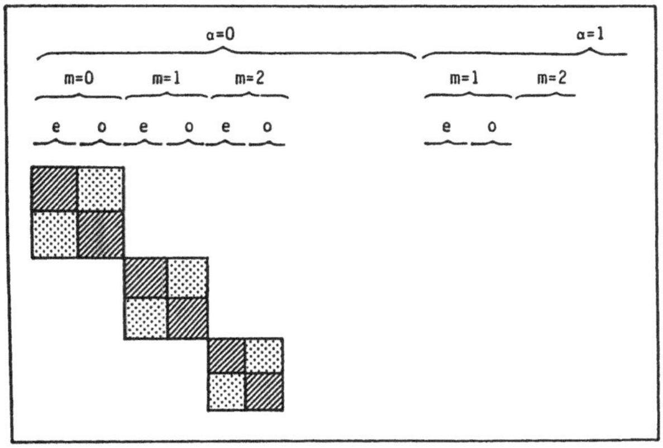

Figure 5.2: Structure of the normal matrix. Light dotted areas close to zero (α = 0: sin-terms, α = 1 cos-terms, e: even, o: odd).

6. PHYSICAL MODEL OF GRADIOMETRY.

At this point we could be rather satisfied, everything seems to be settled: the gravity gradients are observable, the tensor components V_{ij} have a nice geometrical interpretation, and obviously the gravity field can be determined from the observed gradients. However, one could claim somewhat provocative, that all of this has very little to do with geodesy, for on the one hand we may leave it to the instrument designer to worry about the observability of gravity gradients and on the other hand any interested mathematician could work out the mathematical models. It is the task of the engineer (the geodesist) to merge observations and model to one unit, and to derive from this combination a product, that has to meet certain beforehand defined standards, and the quality of which can be controlled by means of independent experiments. Although this could be adopted as a general principle to geodetic work, it can be very difficult to follow this line very consequent. This is the reason, why instead of placing these lectures into the general frame of experiment and model selection and connection, only a number of related problems are raised.

What is the objective of satellite gradiometry? A typical answer, one comes across in various documents, is "the determination of the detailed structure of the earth's gravity field with a precision of approximately ±10 cm, when expressed in terms of geoid heights, or ±5 mgal in terms of gravity anomalies and with a resolution of degree and order $\bar{N}$ = 200 to 300, in terms of a spherical harmonic expansion of the gravity field". This would open completely new terrains inside geodesy, but also contribute significantly to solid earth physics and especially to physical oceanography. Now, that we have defined the purpose of satellite gradiometry, we have to select an experiment (gradiometer measurements with the required precision), define the sampling rate of the measurements, and choose an adequate mathematical model, that connects the observables with the parameters to be estimated. Hence measurement precision and all parts of the models have to be such that the objectives of such a mission will be met. For the role of the stochastic model in this context I refer to (Baarda, 1967) or to the introduction in (Rummel, 1984). In the case of satellite gradiometry, employing some simple spectral analysis, it was shown, in a number of studies, that the above stated objectives could be met by (1) using a gradiometer with a precision of somewhere around 0.01 E.U., (2) selecting a low altitude (≈ 200 km) satellite orbit, and (3) observing at a high sampling rate (≈ 1 s). All of these analyses were based upon some simplified idea about the spectrum of the observable gravity gradient signal and of the measurement noise, compare for an example Figure 6.1.

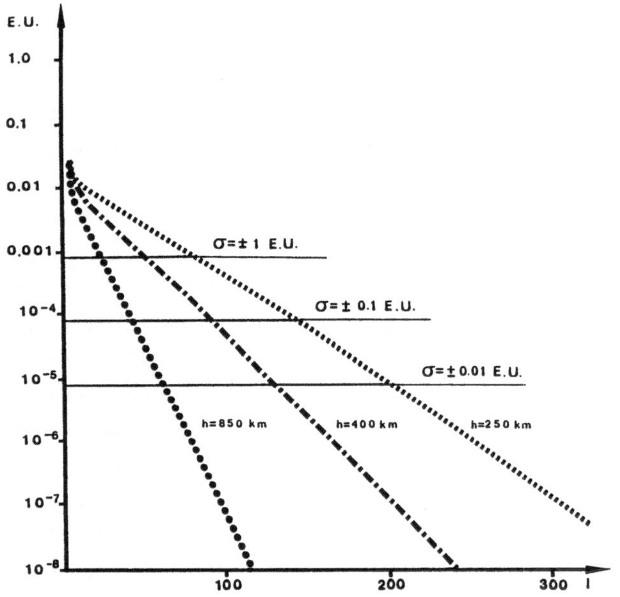

Figure 6.1: Spectrum (degree-order standard deviation) of vertical gradient T_{zz} at various altitudes versus noise spectra.

These studies have the character of a priori error propagations. Based upon their outcome one takes - in a second phase - usually a closer look into the required instrument design and analysis model. The two aspects, required measurement precision and model choice, shall be illustrated with a simple, not in every respect very serious

EXAMPLE: Determine the shape of the earth in radial direction with a precision of 5 cm from one combined absolute gravity and vertical gradient measurement.

The required precision of dr = 5 cm corresponds to a relative precision of

$\frac{dr}{r} \approx 10^{-8}$ relative to the earth's radius, $r \approx 6370$ km. We choose the gravity model of a spherical, homogeneous sphere with potential $V = \frac{GM}{r}$. With $\underline{e}_z$ vertical (or radial outwards), it is

$$V_z = -\frac{GM}{r^2} = -g$$

with, g, scalar gravity, and

$$V_{zz} = \frac{2GM}{r^3} = \frac{2g}{r} \quad .$$

Thus our model for the determination of the radial distance r (= shape of the earth) from vertical gradient, V_{zz}, and scalar gravity, g, becomes

$$r = \frac{2g}{V_{zz}} \quad .$$

<u>Required measurement precision</u>: Differentiation yields

$$\left|\frac{dr}{r}\right| = \left|\frac{dV_{zz}}{V_{zz}}\right| + \left|\frac{dg}{g}\right| \quad ,$$

which means that the vertical gradient as well as the scalar gravity have to be determined to better than 10^{-8}. For the gravity measurement, we employ a modern free fall apparatus. From $g = \frac{2s}{t^2}$ and s, distance of free fall, and t, time interval of free fall, we find

$$\frac{ds}{s} < 10^{-8} \qquad \text{and} \qquad 2\,\frac{dt}{t} < 10^{-8} \quad .$$

These relative precisions in time and distance are achieved with the modern equipment. We derive the vertical gradient V_{zz} from two simultaneous free fall measurements carried out at points A and B with vertical distance z = 1 m. Hence our measurement model is

$$V_{zz} = \frac{g(B)-g(A)}{z} = \frac{\Delta g}{z} \quad .$$

For the relative precision of V_{zz} we find

$$\left|\frac{dV_{zz}}{V_{zz}}\right| = \left|\frac{dg(B)}{\Delta g}\right| + \left|\frac{dg(A)}{\Delta g}\right| + \left|\frac{dz}{z}\right| \quad .$$

With $V_{zz} \approx 3000$ E.U. the required 10^{-8} relative precision would imply a measurement precision of $0.3 \cdot 10^{-4}$ E.U. Whereas it seems possible to determine $\frac{dz}{z}$ to 10^{-8}, the required 10^{-8} for $\frac{dg}{\Delta g}$ seems very difficult to obtain. It would imply with $\Delta g = 0.3\,\frac{\text{mgal}}{\text{m}} \cdot 1$ m that dg has to be measured precise to $\pm\, 0.3 \cdot 10^{-8}$ mgal or

$0.3 \cdot 10^{-13} \frac{m}{s^2}$. However let us assume here, it is feasible to determine $\frac{dV_{zz}}{V_{zz}}$ with 10^{-8}.

Model requirements: With the above assumption, r can be derived in our model with a relative precision of 10^{-8}. The problem is solved. However, when comparing the derived radial distance with independent measurements, e.g. from geometric satellite geodesy, we shall see, that the discrepancies can be easily of the order of 20 km. This is the result, as we know, of the poorly chosen model. The simple spherical model is only valid to an accuracy of the order of the earth's flattening, which means to $\frac{r}{300}$. And certainly an adequate model would look much more complicated, and certainly hundreds of thousands measurements would be required instead of only two.

The purpose of this example was to show that both the experiment and the model of analysis need careful selection and that the considerations in both cases are determined by the a priori defined requirements. Hence "adequate" means to find the proper balance between on the one hand not forgetting essential elements in the experiment and model, and on the other hand - in view of economy and robustness - not including superfluous parts.

Now we turn to a few selected aspects related to the adequate choice of experiment and analysis model for the case of satellite gradiometry. The requirements in terms of precision and resolution were defined above. From spectral analysis follows that in order to attain a resolution of $\bar{N} > 200$ the gradiometer precisions at a satellite altitude of below 250 km has to be below 10^{-2} E.U. This agrees well with the envisaged measurement precisions for the satellite gradiometers under development, that range from 10^{-2} E.U. to 10^{-4} E.U., depending on whether cryogenic cooling is applied or not. However the spectral studies usually consider only the dominant component, V_{zz}; in addition, very little is known about the actual gradiometric spectrum, especially its higher frequencies. Considering the tensor components V_{ij} in spherical approximation, chapter 2, and in ellipsoidal approximation, we would expect the observable gradiometer components at 200 km altitude to be: $V_{zz} \approx$ 2800 E.U., $V_{xx} \approx V_{yy} \approx$ 1400 E.U., $V_{xz} \approx$ 10 E.U. (of the order of the flattening) and V_{xy} and $V_{yz} \approx 10^{-5}$ to $10^{-6} \cdot V_{zz} \approx 2.8 \cdot 10^{-2}$ E.U. Since actual terrestrial measurements yield values typically between 1 E.U. and 100 E.U. for V_{xy} and V_{yz}, probably due to local effects, we assume the actual values at satellite altitude to be of the order of 0.1 E.U. for these components. This results in the following approximate relative precisions $d \ln V_{ij} = \frac{dV_{ij}}{V_{ij}}$:

σ \ $d \ln V_{ij}$	V_{zz}	$V_{xx} \approx V_{yy}$	V_{xz}	$V_{xy} \approx V_{yz} \approx V_{xx} - V_{yy}$
10^{-2} E.U.	$4 \cdot 10^{-6}$	$7 \cdot 10^{-6}$	10^{-3}	10^{-1}
10^{-4} E.U.	$4 \cdot 10^{-8}$	$7 \cdot 10^{-8}$	10^{-5}	10^{-3}

Hence we see, the relative precision of the various components differ considerably. Are the envisaged measurement precisions adequate? We do not know. First of all, the planned measurement precisions of 10^{-2} to 10^{-4} E.U. are difficult to achieve. In (Reinhardt et al., 1982) it is described what a 10^{-3} to 10^{-4} E.U. precision, measured by differential accelerometers, would imply in terms of displacement requirement of the sensor mass, dynamic range, and calibration.

One of their conclusions is that a calibration at this precision level shall not be possible. This means also that the idea of orbit determination from gradiometry has to be reconsidered. An improvement of GM_0, known already to 10^{-8}, could anyway not be expected.

The gradiometer components are derived by linearization, compare eq. (2.3). Thereby V_{ij} is considered constant over the small range of the gradiometer baseline. The neglected second-order term is $\frac{1}{2}V_{ijk}dx_j dx_k$. Considering the maximum effect, which is obtained for the term V_{zz}, we find for the second order contribution in spherical approximation $-\frac{1}{2}\ 3\ V_{zz}\ \frac{\Delta z}{r}$. For $\Delta z = 1$ m this is $-V_{zz} \cdot 2.3 \cdot 10^{-7}$. Hence we see that for gradiometers with a relative precision of better than 10^{-6} this effect has to be taken into account.

Is the model adequate?

In chapter 5 a linear model has been derived. The assumption underlying the linearization is, that the approximate gravitational field describes the major part of the field and that the unknown gravity parameters can be determined from the anomalous quantities (observable minus computed) in a few iterations. At least in the case of terrestrial gradiometry an adequate reference model for the linearization of the off-diagonal components and for the difference quantities ($V_{xx} - V_{yy}$, $2V_{xx} + V_{zz}$, or $2V_{yy} + V_{zz}$) does not exist. Proper linearization would require the introduction of a geophysical model for the mass distribution in the vicinity of the observation points. Due to the attenuation effect with altitude one could hope that in the case of satellite gradiometry a valid linearization can be achieved with a low degree and order reference field. The problem there is that the anomalies $\Delta\Gamma_{ij}$, at least for some components, become very small (remember for example, that in chapter 5 the anomaly combination $2\Delta\Gamma_{xx} + \Delta\Gamma_{zz}$ played a central role). Even for the dominant $\Delta\Gamma_{zz}$ component the Tscherning-Rapp model yields a signal r.m.s. value of only 0.3 E.U. at 250 km altitude (referred to an ellipsoidal reference field). This aspect deserves certainly additional considerations, because for some components the signal anomalies might even remain below the level of the measurement errors.

If the gradiometers can indeed not be calibrated with the required precision, the entire model will have to be written e.g. in terms of ratios relative to a chosen initial point. Thereby it is still assumed that the relative scale between the components is known. Much more critical would be a situation, in which the scale cannot be kept constant, as some instrument designers expect.

All what has been said about model adequacy until now referred to presumably weak spots of the basic chosen model. The model itself has not been questioned.

The model was derived starting from Newtons law of the change in linear momentum and angular momentum. However, can the model be based at all on Newtonian mechanics?

In (Moritz, 1968) the fundamental gradiometric equations are derived in the frame of general relativity, consistently limited to the linear approximation. In the language of space-time geometry, the space components of the equation of the geodesic deviation (relative motion) between two test particles, expressed in a local (Lorentz) frame, are

$$\frac{d^2\xi^i}{d\tau^2} + c^2 R^i_{ojo}\, \xi^j = 0 \qquad i = 1,2,3 \quad , \tag{6.1}$$

as derived in (Misner et al., 1973, p. 37). The ξ^i are the coordinate differences between the two particles, c is the velocity of light, and the R^i_{ojo} express nine elements of the Riemann curvature tensor, R^i_{kjl}. (Since we deal now with curvi-linear coordinates, distinction between contra- and co-variant components is necessary.) In Newtonian approximation we obtain $-c^2 R^i_{ojo} = V_{ij}$ and $d^2\xi^i/d\tau^2 = d\ddot{x}_i$, and eq. (6.1) becomes eq. (2.4).

Although somewhat premature, three problems concerning the adequacy of the Newtonian model are mentioned.

- We express the gravitational potential in spherical harmonics:

$$V = \frac{GM}{r}\,[1 + \sum_{n=2}^{\infty} \sum_{m=0}^{n} (\frac{R}{r})^n (\bar{C}_{nm} \cos m\lambda + \bar{S}_{nm} \sin m\lambda)\bar{P}_{nm}(\sin \varphi)] \quad . \tag{6.2}$$

Taking Kaula's rule of thumb $\sigma\{\bar{C}_{nm},\bar{S}_{nm}\} \approx \pm \frac{10^{-5}}{n^2}$, it follows, that for $n \approx 300$ the size of the coefficients is approximately 10^{-10}. They have to be determined from gradiometry. The neglected quadratic terms in deriving the linear approximation in (Moritz, ibid) are of the order of magnitude of $c^{-2}V \approx 10^{-9}$. Does this term cause a distortion in our estimated coefficients?

In a recent dissertation by Theiss (1984), equation (6.1) is studied under various assumptions, concerning the tide generating body. Two conclusions are of immediate consequence for us. According to Theiss (ibid):

- the angular momentum of the earth produces a secular contribution to the tidal acceleration of the order of 10^{-7} (after a period of 20 days);
- the oblate earth produces a very significant contribution to the tidal acceleration of two test particles in an inclined orbit.

The purpose of this last chapter was to leave us not with too much confidence in the sense that everything is sorted out in satellite gradiometry, and to draw some attention to the difficulties the proper formulation of a complete physical model might pose. Especially the last issue, the adequacy of the Newtonian model in gradiometry, should be seen as a challenge to look into refined models of curved space-time, motivated by the immediate practical need for satellite gradiometry missions in the 1990ies.

APPENDIX A:

It is in P

$$\underline{x}(P) = x_I \underline{e}_I = x_i \underline{e}_i$$

and with the orthogonal transformation R_{iI}

$$\underline{e}_i = R_{iI} \underline{e}_I \quad .$$

It follows

$$x_I = R_{Ii} x_i \quad ,$$

$$\dot{x}_I = R_{Ii} \dot{x}_i + \dot{R}_{Ii} x_i \quad , \qquad \text{(A.1a-c)}$$

$$\ddot{x}_I = R_{Ii} \ddot{x}_i + 2\dot{R}_{Ii} \dot{x}_i + \ddot{R}_{Ii} x_i \quad .$$

From

$$R_{iI} R_{Ij} = E_{ij}$$

with E_{ij} the unit transformation, follows

$$R_{iI} \dot{R}_{Ij} + \dot{R}_{iI} R_{Ij} = 0$$

or defining

$$R_{iI} \dot{R}_{Ij} = \Omega_{ij} \quad : \qquad \text{(A.2)}$$

$$\Omega_{ij} = -\Omega_{ji} \quad ,$$

(skew-symmetric). From the definition of Ω_{ij} we see, that

$$\begin{aligned} \dot{\Omega}_{ij} &= R_{iI} \ddot{R}_{Ij} + \dot{R}_{iI} \dot{R}_{Ij} \\ &= R_{iI} \ddot{R}_{Ij} + \dot{R}_{iI} R_{Ik} R_{kJ} \dot{R}_{Jj} \qquad \text{(A.3)} \\ &= R_{iI} \ddot{R}_{Ij} - \Omega_{ik} \Omega_{kj} \quad . \end{aligned}$$

(A.2) and (A.3) inserted in (A.1) yield

$$R_{iI}x_I = x_i \quad ,$$

$$R_{iI}\dot{x}_I = \dot{x}_i + \Omega_{ij}x_j \quad , \qquad \text{(A.4a-c)}$$

$$R_{iI}\ddot{x}_I = \ddot{x}_i + 2\Omega_{ij}\dot{x}_j + \dot{\Omega}_{ij}x_j + \Omega_{ij}\Omega_{jk}x_k \quad .$$

APPENDIX B:

It is in the local (spherical) $\underline{e}_i$-triad ($\underline{e}_z$ radial):

$$U_{xx} = \frac{1}{r^2} U_{\varphi\varphi} + \frac{1}{r} U_r$$

$$U_{xy} = \frac{1}{r^2 \cos\varphi} U_{\varphi\lambda} + \frac{\sin\varphi}{r^2 \cos^2\varphi} U_\lambda$$

$$U_{xz} = \frac{1}{r} U_{\varphi r} - \frac{1}{r^2} U_\varphi$$

$$U_{yy} = \frac{1}{r^2 \cos^2\varphi} U_{\lambda\lambda} - \frac{\tan\varphi}{r^2} U_\varphi + \frac{1}{r} U_r$$

$$U_{yz} = \frac{1}{r\cos\varphi} U_{\lambda r} - \frac{1}{r^2 \cos\varphi} U_\lambda$$

$$U_{zz} = U_{rr}$$

cf. (Reed, 1973; ch. 3) or (Tscherning, 1976).

APPENDIX C

by R. Rummel and P.J.G. Teunissen:

Naturally also the overdetermined b.v.p. with C, g, and Γ can be considered. In this case the linear models, eqs. (5.4) and (5.5), are merged. Applying basically a least-squares adjustment approach to this linear system one obtains

$$d\hat{\bar{C}}_{nm} = \frac{1}{(n-1)E} \{p_W p_g (dg_{nm\alpha} - 2dW_{nm\alpha}) + p_g p_\Gamma (n+1) \cdot \cdot (2d\Gamma_{nm\alpha} - 3dg_{nm\alpha}) + p_\Gamma p_W \tfrac{1}{2}(n+4)(-3dW_{nm\alpha} - d\Gamma_{nm\alpha})\} \qquad (C.1)$$

with

$$E = \{p_W p_g + p_g p_\Gamma (1+n)^2 + p_\Gamma p_W (-\tfrac{1}{2}(n+4))^2\} \quad .$$

In eq. (C.1) it is

p_W, p_g, p_Γ ..... a priori weights of potential, gravity and vertical gradient,

$d\bar{C}_{nm\alpha}$, $dW_{nm\alpha}$, $dg_{nm\alpha}$, and $d\Gamma_{nm\alpha}$ dimensionless coefficients derived from the expansion of the corresponding quantities T/U_0, $\Delta W/U_0$, $\Delta g/\gamma_0$, and $\Delta\Gamma/\Gamma_0$.

From (C.1) the cases {C,g}, {C,Γ}, and {g,Γ} follow by specializing the weights.

A derivation of this result is given in a paper by the authors, to be presented at the intern. symposium FIGURE AND DYNAMICS OF THE EARTH, MOON, AND PLANETS in Prague, 1986.

ACKNOWLEDGEMENT.

I greatfully acknowledge the careful and fast typing by Wil Coops, that compensated by delays in delivering the draft.

LITERATURE.

Baarda, W.: Statistical Concepts in Geodesy, Netherlands Geodetic Commission, New Series, 2, 4, Delft, 1967.
Bauer, H.F. Environmental Effects on Micro-Gravity Experiments, Z. Flugwiss. Weltraumforsch., 6, 3, 184-194, 1982.
Betti, B., F. Sansõ: A Possible Use of the Results of Hipparcos Project in Satellite to Satellite Tracking, manuscripta geodaetica, 1986 (in print).
Bocchio, F.: Geodetic Singularities, Rev. Geoph. Space Physics, 20, 3, 399-409, 1982.
Brouwer, D., G.M. Clemence, Celestical Mechanics, Academic Press, New York, 1961.
Carroll, J.J., P.H. Savet: Gravity Difference Detection, Aero/Space Engineering, 44-47, 1959.
Colombo, O.L.: Numerical Methods for Harmonic Analysis on the Sphere, Dept. Geodetic Science, 310, The Ohio State University, Columbus, 1981.
Colombo, O.L.: The Global Mapping of Gravity with Two Satellites, Netherlands Geodetic Commission, New Series, 7, 3, 1984.
Dermanis, A., E.Livieratos: Applications of Deformation Analysis in Geodesy and Geodynamics, Rev. Geophys. Space. Physics, 21, 1, 41-50, 1983.
Forward, R.L.: Gravity Sensors and the Principle of Equivalence, IEEE Transactions on Aerospace and Electronic Systems, AES-17, 4, 511-519, 1981.
Forward, R.L.: Flattening Spacetime Near the Earth, physical review D, 26, 4, 735-744, 1982.
Grafarend, E.: The Bruns Transformation and a Dual Set-up of Geodetic Observational Equations, U.S. Dept. Commerce, NOAA-NGS, Rockville, Md., 1979.
Heck, B.: On Various Formulations of the Geodetic Boundary Value Problem Using the Vertical Gradient of Gravity, in: proc. Intern. Symp. "Figure of the Earth, the Moon and Other Planets", Prague, 1982.
Heiskanen, W., H. Moritz: Physical Geodesy, Freeman & Comp., San Francisco, 1967.
Heitz, S.: Mechanik fester Körper, Band 1, Dümmler, Bonn, 1980.
Ilk, K.H.: On the Dynamics of a System of Rigid Bodies, manuscripta geodaetica, 8, 2, 139-198, 1983a.
Ilk, K.H.: Ein Beitrag zur Dynamik ausgedehnter Körper, Gravitationswechselwirkung, Deutsche Geodätische Kommission, C-288, München, 1983b.
Jung, K.: Schwerkraftverfahren in der Angewandten Geophysik, Akademische Verlagsgesellschaft, Leipzig, 1961.
Kaplan, M.: Modern Spacecraft Dynamics & Control, John Wiley & Sons, New York, 1976.
Krarup, T.: Letters on Molodensky's Problem, I-IV, Communication to the members of IAG-special study group 4.31, 1973.
Marussi, A.: The Tidal Field of a Planet and the Related Intrinsic Reference Systems, Geophys. J.R. astr. Soc., 56, 409-417, 1979.
Marussi, A.: Microgravitation in Space, Geophys. J.R. astr. Soc., 76, 691-695, 1984.
Marussi, A., Cl. Chiaruttini: The Motion of a Free Particle and of a Spherical Pendulum in the Microgravitational Field of a Gravitationally Stabilized Satellite in a Circular Orbit in a Central Field, in: Marussi, A.: Intrinsic Geodesy, 179-189, Springer, Berlin, 1985.
Misner, Ch.W., K.S. Thorne, J.A. Wheeler: Gravitation, Freeman and Comp., San Francisco, 1970.
Moritz, H.: Kinematical Geodesy, Deutsche Geodätische Kommission A-59, München, 1968.
Moritz, H.: Advanced Physical Geodesy, Wichmann, Karlsruhe, 1980.
Moritz, H.: Variational Methods in Earth Rotation, in: Geodesy and Geodynamics, eds.: Moritz, H. & H. Sünkel, 167-226, Graz, 1982.
Moritz, H.: Inertia and Gravitation in Geodesy, in: proc. of the 3rd Intern. Symposium on Inertial Technology for Surveying and Geodesy, vol. I, Banff, 1986.
Olsen, R.E., J. Mockovciak, Jr.: Operational Factors Affecting Microgravity Levels in Orbit, Journ. Spacecraft, 18, 2, 141-144, 1981.
Pelka, E.J., D.B. De Bra: The Effects of Relative Instrument Orientation upon Gravity Gradiometer System Performance, J. Guidance and Control, 2, 1, 18-24, 1979.
Reed, G.B.: Application of Kinematical Geodesy for Determining the Short Wave Length Components of the Gravity Field by Satellite Gradiometry, Dept. Geodetic Science, 201, The Ohio State University, 1973.
Reinhardt, V.S., F.O. Vonbun, J.P. Turneaure: A Supersensitive Accelerometer for Spacecraft Gradiometry, proc.: IEEE Position Location and Navigation Symposium, Atlantic City, 1982.

Roberson, R.E.: Gravitational Torque on a Satellite Vehicle, Journ. Franklin Inst., 265, 1, 13-22, 1958.
Rummel, R., P.J.G. Teunissen: A Connection Between Geometric and Gravimetric Geodesy, Some Remarks on the Role of the Gravity Field, in: Baarda Festbundel, vol. 2, 603-621, 1982.
Rummel, R., O.L. Colombo: Gravity Field Determination from Satellite Gradiometry, bulletin géodésique, 59, 233-246, 1985.
Rummel, R.: From the Observational Model to Gravity Parameter Estimation, in: proc. Local Gravity Field Approximation, 67-106, Beijing, 1984.
Selényi, P. (ed.): Roland Eötvös Gesammelte Arbeiten, Akademiai Kiado, Budapest, 1953.
Spaceborne Gravity Gradiometers, proc. workshop held at NASA Goddard Space Flight Center, Greenbelt, Md., 1983.
Spiegel, M.R.: Theoretical Mechanics, Schaum's Outline Series, McGraw-Hill, New York, 1967.
Theiss, D.S.: Neue Gravitative Effekte rotierender Massen Möglichkeiten für weitere Tests der Allgemeinen Relativitätstheorie, Dissertation, Köln, 1984.
Toperczer, M.: Lehrbuch der allgemeinen Geophysik, Springer, Wien, 1960.
Tscherning, C.C.: Representation of Covariance Functions Related to the Anomalous Potential of the Earth Using Reproducing Kernels, Internal Report 3, Dan. Geod. Inst., Copenhagen, 1972.
Tscherning, C.C.: Comparison of the Second-Order Derivatives of the Normal Potential Based on the Representation by a Legendre Series, manuscripta geodaetica, 1, 2, 71-92, 1976.
Wertz, J.R. (ed.): Spacecraft Attitude Determination and Control, D. Reidel, Dordrecht, 1978.

GLOBAL GEOPOTENTIAL SOLUTIONS

by

R.H. Rapp

Department of Geodetic Science and Surveying
The Ohio State University
1958 Neil Avenue
Columbus, Ohio 43210, U.S.A.

Lecture Notes in Earth Sciences, Vol. 7
Mathematical and Numerical Techniques in Physical Geodesy
Edited by H. Sünkel

1.0 INTRODUCTION

Since 1978 a number of high degree (ℓ_{max} = 180) spherical harmonic solutions of the earth's gravity field have been computed. These models have shown their value in a number of applications (Tscherning, 1983). Lower degree fields have also been developed through the analysis of satellite data and at times with the combination of terrestrial gravity data. I show in Table 1 a list of some solutions that are used in various investigations at this time.

Table 1

Global Geopotential Models			
Name	Author	Date	N max
GEM9	Lerch et al	1977	20(+)
Rapp78	Rapp	1978	180
GEM10B	Lerch et al	1981	36
GEM10C	Lerch et al	1981	180
Rapp81	Rapp	1981	180
GEML2	Lerch et al	1982	20(+)
Hajela84	Hajela	1984	250
GRIM3-L1	Reigber	1985	36
GPM2	Wenzel	1985	200

As high degree potential coefficient fields become more widely used it is increasingly important to know the assumptions related to current models and how such models are computed. The primary purpose of this paper is to describe the combination of satellite and terrestrial data that leads to the high degree potential coefficient models.

2.0 THE ANALYSIS OF A GLOBAL TERRESTRIAL FIELD

The standard representation of the gravitational potential is taken as follows:

$$V(r, \theta, \lambda) = \frac{kM}{r}\left[1 + \sum_{\ell=2}^{\infty} \left(\frac{a}{r}\right)^{\ell} \sum_{m=0}^{\ell} (C_{\ell m}\cos m\lambda + S_{\ell m}\sin m\lambda)P_{\ell m}(\cos\theta)\right] \quad (1)$$

where:

- r, θ, λ are the polar coordinates of the point at which V is to be determined;
- kM geocentric gravitational constant;
- a scaling parameter associated with the potential coefficients;
- C, S fully normalized potential coefficients;
- $P_{\ell m}$ fully normalized associated Legendre functions.

A more compact form for (1) may be obtained with the following substitutions:

$$C^{\alpha}_{\ell m} = \begin{Bmatrix} C_{\ell m}, & \alpha=0 \\ S_{\ell m}, & \alpha=1 \end{Bmatrix} \quad (2)$$

$$Y^{\alpha}_{\ell m}(\theta, \lambda) = \begin{Bmatrix} P_{\ell m}(\cos\theta)\cos m\lambda, & \alpha=0 \\ P_{\ell m}(\cos\theta)\sin m\lambda, & \alpha=1 \end{Bmatrix} \quad (3)$$

In this case (1) becomes:

$$V(r, \theta, \lambda) = \frac{kM}{r}\left[1 + \sum_{\ell=2}^{\infty} \left(\frac{a}{r}\right)^{\ell} \sum_{m=0}^{\ell} \sum_{\alpha=0}^{1} C^{\alpha}_{\ell m} Y^{\alpha}_{\ell m}(\theta, \lambda)\right] \quad (4)$$

We define the disturbing potential T at the point r, θ, λ:

$$T = V - U \quad (5)$$

where U is a reference potential, usually that implied by a rotational, symmetric, equipotential ellipsoid. Assuming the mass of the reference ellipsoid and the earth are the same we have:

$$T(r, \theta, \lambda) = \frac{kM}{r} \sum_{\ell=2}^{\infty} \left(\frac{a}{r}\right)^{\ell} \sum_{m=0}^{\ell} \sum_{\alpha=0}^{1} C^{\alpha}_{\ell m} Y^{\alpha}_{\ell m}(\theta, \lambda) \quad (6)$$

where we will understand for future discussions that the C^{0}_{2i} coefficients have the reference field coefficients removed. This is done usually to i=3.

In a spherical approximation, a gravity anomaly, can be expressed as follows (Heiskanen and Moritz, 1967, (2-154), p. 89):

$$\Delta g(r, \theta, \lambda) = -\frac{\partial T}{\partial r} - \frac{2}{r} T(r, \theta, \lambda) \quad (7)$$

Using (6), (7) becomes

$$\Delta g(r, \theta, \lambda) = \frac{kM}{r^2} \sum_{\ell=2}^{\infty} (\ell-1)\left(\frac{a}{r}\right)^{\ell} \sum_{m=0}^{\ell} \sum_{\alpha=0}^{1} C^{\alpha}_{\ell m} Y^{\alpha}_{\ell m}(\theta, \lambda) \quad (8)$$

Formally we consider (7) to be the radial component of the total gravity anomaly. A more precise formulation will be considered in Section 3.1.

We now evaluate (8) on the spherical surface of radius a:

$$\Delta g(a, \theta, \lambda) = \gamma \sum_{\ell=2}^{\infty} (\ell-1) \sum_{m=0}^{\ell} \sum_{\alpha=0}^{1} C^{\alpha}_{\ell m} Y^{\alpha}_{\ell m}(\theta, \lambda) \quad (9)$$

where

$$\gamma = \frac{kM}{a^2} \quad (10)$$

Equation (9) allows us to calculate anomalies on a spherical surface of radius a.

2.1 THE USE OF ORTHOGONALITY RELATIONSHIPS FOR COEFFICIENT DETERMINATIONS

Now assume that we are given the Δg values on the sphere and that we are to find the potential coefficients. We use the orthogonality relationships described in Heiskanen and Moritz (1967, section 1-13) to find:

$$C^{\alpha}_{\ell m} = \frac{1}{4\pi\gamma(\ell-1)} \iint_{\sigma} \Delta g(a, \theta, \lambda)\, Y^{\alpha}_{\ell m}(\theta, \lambda)\, d\sigma \tag{11}$$

If the Δg values are given as a continuous function on our sphere we can calculate the potential coefficients. Recall that (11) is an approximation because we assumed a spherical boundary condition and we assumed that we knew Δg on the sphere of radius a. Note also that we have not expressed concern about the convergence of (1) at the surface of the earth. Such discussions may be found in Jekeli (1981), Sjoberg (1980) and others.

The actual evaluation of (11) is carried out by replacing the integration by a summation over point anomalies on a grid or a set of mean anomalies. A mean anomaly can be computed from (9) as follows (Colombo, 1981, p. 3):

$$\overline{\Delta g} = \frac{\gamma}{\sigma_{ij}} \sum_{\ell=2}^{\infty} (\ell-1) \sum_{m=0}^{\ell} \sum_{\alpha=0}^{1} C^{\alpha}_{\ell m} \iint_{\sigma_{ij}} Y_{\ell m}(\theta,\lambda) d\sigma \tag{12}$$

where σ_{ij} is the area of the block for which the mean anomaly is being computed and (with Δ = polar distance increment);

$$\sigma_{ij} = \Delta\lambda(\cos\theta_i - \cos(\theta_i+\Delta)) \tag{13}$$

Consider the evaluation of (11) with point gravity values given on a grid defined by θ_i, λ_j. Let the grid interval be the same in θ and λ and set the number of parallel rows to be N and the number of longitude blocks per row to be 2N. Then (11) can be written (Colombo, 1981, p. 4):

$$C^{\alpha}_{\ell m} = \frac{1}{4\pi\gamma(\ell-1)} \sum_{i=0}^{N-1} \sum_{j=0}^{2N-1} Y^{\alpha}_{\ell m}(\theta_i, \lambda_j)\Delta g(\theta_i, \lambda_j)\sigma_{ij} \tag{14}$$

The highest degree ℓ coefficients that can be determined from (10) can be determined from the Nyquist frequency analogy in Fourier analysis. If Δ is the grid spacing we have:

$$\ell_{max} = \frac{180^\circ}{\Delta^\circ} \tag{15}$$

Now assume that mean anomalies are given such that there are N values in latitude and 2N values in longitude. These values are considered to be equiangular. Then equation (11) can be approximated by:

$$C^{\alpha}_{\ell m} = \frac{1}{4\pi\gamma(\ell-1)q_\ell} \sum_{i=0}^{N-1} \sum_{j=0}^{2N-1} \overline{\Delta g}_{ij} \iint_{\sigma_{ij}} Y^{\alpha}_{\ell m}(\theta, \lambda) d\sigma \tag{16}$$

where q_ℓ is a quantity introduced to reduce the approximation in (16). This approximation arises from two sources:

A. Sampling Error - The error caused by the finite size of the block in which the anomaly is given.

B. Smoothing Problem - The problem arises from the averaging process used in determining the mean anomalies.

The averaging inherently dampens the higher frequencies making the recovery of such frequencies more difficult than the lower frequencies.

A number of different procedures can be used to estimate q_ℓ. One procedure utilizes the following relationship given by Meissl (1971, p. 22)

$$\frac{1}{\sigma_{ij}} \iint_{\sigma_{ij}} Y^{\alpha}_{\ell m}(\theta, \lambda) d\sigma = \beta_\ell(\psi_0)\, Y^{\alpha}_{\ell m}(\theta_i, \lambda_i) \tag{17}$$

where θ_i, λ_i is the location of the center of a spherical cap of radius ψ_0. The area of the cap is σ_{ij} and β_ℓ is the Pellinen smoothing operator. β_ℓ can be computed using the following formula:

$$\beta_\ell = \frac{1}{1-\cos\psi_0} \frac{1}{\sqrt{2\ell+1}} \left[P_{\ell-1}(\cos\psi_0) - P_{\ell+1}(\cos\psi_0)\right] \tag{18}$$

or using a recurrence procedure given by Sjoberg (1980). Although (17) holds only for a circular cap it is also a good approximation for a rectangular block when the area of the block and of the cap are made to be equal (Katsambalos, 1979). If θ is the block side of the equator, the corresponding ψ_0 values are computed from:

$$\sin\left(\frac{\psi_0}{2}\right) = \left(\frac{\theta\sin\theta}{4\pi}\right)^{1/2} \tag{19}$$

For example, if $\theta=1^\circ$, $\psi_0=0^\circ.564$. The above relationship and consequently (17) is only approximate since the area of the equiangular blocks is a function of latitude.

If we now solve (17) for $P_{\ell m}(\cos\theta)(\cos m\lambda, \sin m\lambda)$, and substitute this into (14) we have (taking Δg to be the mean anomally ($\overline{\Delta g}$) of the block):

$$C^{\alpha}_{\ell m} = \frac{1}{4\pi\gamma\beta_\ell(\ell-1)} \sum_{i=0}^{N-1} \sum_{j=0}^{2N-1} \overline{\Delta g}(\theta_i, \lambda_j) \iint_{\sigma_{ij}} Y^{\alpha}_{\ell m}(\theta, \lambda) d\sigma \tag{20}$$

Comparing (16) and (20) we see that q_ℓ is at least approximately equal to β_ℓ. We then view β_ℓ as a de-smoothing operator that tries to take into account that frequencies are damped out in taking the average to obtain the mean anomaly.

Colombo (1981, p. 76) investigated optimum quadrature weights (as we might call the q_ℓ values) and suggested the following:

$$\begin{aligned} q_\ell &= \beta_\ell^2; \quad 0 \leq \ell \leq N/3 \\ q_\ell &= \beta_\ell; \quad N/3 < n < N \\ q_\ell &= 1; \quad \ell > N \end{aligned} \tag{21}$$

where $N = 180^\circ/\theta^\circ$. These suggestions for q_ℓ are not firmly defined but were suggested by numerical tests carried out by Colombo with zero data noise. Because of this we will describe in a future section some tests used to consider the use of various q_ℓ values as well as the point integration procedure (equation 14) used with mean anomaly computations.

2.2 THE USE OF FOURIER ANALYSIS IN POTENTIAL COEFFICIENT DETERMINATIONS

We now seek to reformulate (20) so that computations can be made substantially more efficient. The following discussion is based on that of Gleason (1985). We start by integrating terms that are part of $Y_{\ell m}^{\alpha}$. We have:

$$\int_{\lambda_j}^{\lambda_j+\Delta\lambda} \cos m\lambda d\lambda = A(m)\cos(mj\Delta\lambda) + B(m)\sin(mj\Delta\lambda)$$
$$\int_{\lambda_j}^{\lambda_j+\Delta\lambda} \sin m\lambda d\lambda = -B(m)\cos(mj\Delta\lambda) + A(m)\sin(mj\Delta\lambda) \quad (22)$$

where

$$A(m) = \frac{\sin(m\Delta\lambda)}{m} \quad \text{if } m \neq 0$$
$$= \Delta\lambda \quad \text{if } m = 0 \quad (23)$$

$$B(m) = \frac{\cos(m\Delta\lambda)-1}{m} \quad \text{if } m \neq 0$$
$$= 0 \quad \text{if } m = 0 \quad (24)$$

where $j = 0, 1, 2, \ldots, 2N-1$. We next define the integral of the associated Legendre function as:

$$I_{\ell m}^{i}(\theta) = \int_{\theta_i}^{\theta_{i+1}} P_{\ell m}(\cos\theta)\sin\theta d\theta \quad (25)$$

Equation (16) can now be written

$$\begin{Bmatrix} C_{\ell m} \\ S_{\ell m} \end{Bmatrix} = \frac{1}{4\pi\gamma(\ell-1)q_\ell} \sum_{i=0}^{N-1} \left[I_{\ell m}^{i}(\theta) \left[\begin{Bmatrix} A(m) \\ -B(m) \end{Bmatrix} \sum_{j=0}^{2N-1} \overline{\Delta g}_{ij}\cos(mj\Delta\lambda) + \right.\right.$$
$$\left.\left. + \begin{Bmatrix} B(m) \\ A(m) \end{Bmatrix} \sum_{j=0}^{2N-1} \overline{\Delta g}_{ij}\sin(mj\Delta\lambda) \right]\right] \quad (26)$$

The summations on j in (26) can be compared to a discrete complex Fourier transform sequence X(k) where k is the wavenumber. We write for P complex numbers:

$$X(k) = \sum_{\ell=0}^{P-1} y(\ell)\left[\cos\left(\frac{2\pi}{P}k\ell\right) + i\sin\left(\frac{2\pi}{P}k\ell\right)\right] \quad (27)$$

where $k = 0, 1, 2, \ldots, P-1$. Specifically $y(\ell)$ is real only and equal to $\overline{\Delta g}_{ij}$. We also associate P with 2N; k with m; ℓ with j and $2\pi/P$ with $\Delta\lambda$.

Now let $X_g^i(m)$ be the complex Fourier transform sequence of the 2N mean anomalies along the ith colatitude band. We would have:

$$\mathrm{REAL}(X_g^i(m)) = \sum_{j=0}^{2N-1} \overline{\Delta g}_{ij}\cos(mj\Delta\lambda) \quad (28)$$

$$\mathrm{IMAG}(X_g^i(m)) = \sum_{j=0}^{2N-1} \overline{\Delta g}_{ij}\sin(mj\Delta\lambda) \quad (29)$$

comparing this with (26) we can write:

$$\left\{ \begin{matrix} C_{\ell m} \\ S_{\ell m} \end{matrix} \right\} = \frac{1}{4\pi\gamma(\ell-1)q_\ell} \sum_{i=0}^{N-1} I_{\ell m}^{i}(\theta) \left[\left\{ \begin{matrix} A(m) \\ -B(m) \end{matrix} \right\} \text{REAL}(X_g^i(m)) + \right.$$

$$\left. + \left\{ \begin{matrix} B(m) \\ A(m) \end{matrix} \right\} \text{IMAG}(X_g^i(m)) \right] \qquad (30)$$

The critical step in the evaluation of equation (30) is the computation of the discrete Fourier transform of an input set of mean gravity anomalies in a latitude belt, which are regarded as complex numbers, where the imaginary point is zero. This procedure enables an extremely efficient computation of high degree spherical harmonic expansions of a global anomaly field. Additional efficiencies are gained recognizing:

$$I_{\ell m}^{i}(\theta) = I_{\ell m}^{i}(-\theta) \quad \text{if } \ell + m \text{ is even}$$

and

$$I_{\ell m}^{i}(\theta) = - I_{\ell, m}^{i}(\theta) \quad \text{if } \ell + m \text{ is odd.}$$

Consequently it is necessary to compute the integrated Legendre functions only for the northern hemisphere.

We should note here that the Fourier transform also plays an important role in the calculation of gravity anomalies, on a global grid, from a set of potential coefficients. This procedure enables a very fast calculation of a global set of values from high degree expansion. Specific details may be found in Colombo (1981) and Gleason (1985).

2.3 THE USE OF LEAST SQUARES FITTING FOR POTENTIAL COEFFICIENT DETERMINATION

We now assume that mean gravity anomalies are to be represented by a truncated spherical harmonic expansion. In this case the relationship between the anomalies and coefficients could be represented by (8) for point values or by (12) for mean values in a spherical approximation, replacing the summation to ∞ by a summation to N_{max}. N_{max} may be set by a $180°/\theta°$ rule where θ is the grid spacing of the data.

The basic method is to regard (8) or (12) as an observation equation with a residual attached to the "observed" anomaly. A least squares solution is then made to determine the potential coefficients. In doing this the accuracy estimate of the individual anomalies may be taken into account for weighting purposes, or all weights may be set to one. Note here that the coefficients can be estimated by fitting to only the known mean anomalies. This is in contrast to the orthogonality relationship where a global coverage is required and the accuracy estimate of the data are not considered.

Colombo (1981, p. 11) has discussed the maximum number of coefficients that can be determined when the total number of data points in a equiangular grid on the sphere is $2N^2$. He argues that this maximum number is N^2 which implies a maximum expansion to (N-1). Higher degreee coefficients will be subject to aliasing effects.

This fitting procedure may have some disadvantages because we are fitting a finite number of coefficients to data that potentially has almost all frequencies represented in it. This causes, in some cases, the higher frequency information to be absorbed by the coefficients being solved for. Descrochers (1971) carried out tests to demonstrate this error. This will not happen if the normal equations of the adjustment are diagonal as discussed by Rapp (1969) and Wenzel (1985).

2.4 POTENTIAL COEFFICIENT DETERMINATION BY OPTIMAL ESTIMATION

The discussion in Section 2.2 did not take into account the sampling error nor the noise in the data. Specifically Colombo (1981) sought an optimum set of quadrature weights that would minimize the sum of the square of the error due to the sampling error (finite block size) and the propagated errors of the a priori mean anomalies. Additional discussion on the optimal estimation procedure may be found in Hajela (1984) and in unpublished reports by Gleason (1985, private communication). Only an outline of the procedures is given here.

Let the global mean gravity anomaly vector, $\overline{\underline{\Delta g}}$ be expressed as the sum of a signal vector $\underline{z}$ and a noise vector $\underline{n}$.

$$\overline{\underline{\Delta g}} = \underline{z} + \underline{n} \tag{31}$$

Let the potential coefficient vector $[C_{\ell m}^{\alpha}]$ be defined as $\underline{c}$. Let F be a linear operator that will determine $\underline{c}$ from $\overline{\underline{\Delta g}}$. We have:

$$\underline{\hat{c}} = \underline{F}(\underline{z} + \underline{n}) \tag{32}$$

where $\underline{\hat{c}}$ is the estimated values of $\underline{c}$. The error in the estimation is $\underline{e}$.

$$\underline{e} = \underline{c} - \underline{\hat{c}} = \underline{c} - \underline{F}(\underline{z} + \underline{n}) = (\underline{c} - \underline{Fz}) - (\underline{Fn}) \tag{33}$$

We define the sampling error to be $\underline{e}_s$ and the propagated noise error to be $\underline{e}_n$. From (33) we have:

$$\begin{aligned} \underline{e}_s &= \underline{c} - \underline{Fz} \\ \underline{e}_n &= \underline{Fn} \end{aligned} \tag{34}$$

Note from (33) that even if the noise were zero, there would still be an error in the estimate of the coefficients.

The error covariance matrix of the estimated $\underline{e}$ is:

$$E_T = E_s + E_n = M\{\underline{e}_s\, \underline{e}_s^T\} + M\{\underline{e}_n\, \underline{e}_n^T\} \tag{35}$$

where M is an averaging operator (Moritz, 1980). In writing (35) we assume the sampling error and the noise error are independent. Substituting (34) into (35) we have:

$$E_T = C - 2C_{cz}F^T + F(C_{zz} + D)F^T \tag{36}$$

where

$$C = M\{\underline{c}\ \underline{c}^T\},\ C_{cz} = M\{\underline{c}\ \underline{z}^T\}\quad C_{zz} = M\{\underline{z}\ \underline{z}^T\} \tag{37}$$
$$D = M\{\underline{n}\ n^T\}$$

C represents the covariance matrix of the potential coefficients; C_{cz} is the cross covariance between the potential coefficients and the given mean anomalies, C_{zz} is the signal covariance matrix of the given mean anomalies, and D is the error covariance of these anomalies.

We seek the optimal estimator F. The optimum estimate is defined (Colombo, 1981, p. 36) such that the sum of the squares of the individual coefficient errors is a minimum. The sum is the trace of E_T given in (36). To find the minimum, this trace is differentiated with respect to F. We have:

$$\frac{1}{2}\frac{\partial[E_T]_{TR}}{\partial F} = -C_{cz}^T + (C_{zz} + D)F^T \tag{38}$$

Setting this to zero we have the "best possible linear estimator".

$$\underline{F} = C_{cz}(C_{zz} + D)^{-1} \tag{39}$$

The coefficients would then be (from (23))

$$\hat{\underline{c}} = C_{cz}(C_{zz} + D)^{-1}\overline{\underline{\Delta g}} \tag{40}$$

Substituting $\underline{F}$ into (27) we have:

$$E_T = C - C_{cz}(C_{zz} + D)^{-1}C_{cz}^T \tag{41}$$

Assume that we are given a complete set of 64,800 1°x1° mean anomalies. This then implies that the matrix to be inverted (or system of equations to be solved) is 64800x64800. Even though this is a symmetric matrix inversion (or solution) it is <u>almost</u> impossible to implement such a rigorous solution.

Colombo (ibid, section 2.11) has shown that, with some assumptions the large inversion can be very much simplified. These assumptions are:

1. The mean anomaly covariance function is isotropic (i.e. dependent only on the separation between the blocks).
2. The longitude increment of the grid is a constant.
3. The noise is uncorrelated and the variances of the anomalies are constant along parallels.

Assumption 1 is normally made in collocation solutions. Assumption 2 is fulfilled by equiangular blocks. Assumption 3 is an approximation. These assumptions lead to a special structure for $(C_{zz} + D)$ which allows for a simplified inversion procedure.

Let $N\theta$ be the number of parallels and $N\lambda$ be the number of meridians in the equal angular grid. Let N_{max} be the maximum degree being sought. Then the inversion of $(C_{zz} + D)$ can be carried out by the inversion of a set of $N\theta \mathrm{x} N\theta$ matrices, called R(m), for m = 0 to N_{max}. The elements of R(m) are related to the discrete Fourier transform of covariances of anomaly blocks in a pair of latitude bands. The computation of the elements of R(m) requires the evaluation of the following integrals (Colombo, ibid, eq. 2.63)

$$I_{\ell t}^{i} = \sqrt{\frac{c_\ell}{2\ell+1}}\,\frac{1}{\sigma_{ij}}\int_{\theta_i}^{\theta_i+\Delta\theta} P_{\ell t}(\cos\theta)\sin\theta d\theta \tag{42}$$

where c_ℓ is the a priori anomaly degree variance and $\Delta\theta$ is the co-latitude increment.

The elements $(c^{i}_{\ell m\alpha})$ of the cross-covariance matrix C_{cz} are computed from (Colombo, ibid, eq. 2.59):

$$\underline{c}_{\ell m\alpha},\ \underline{z} = \frac{c_\ell}{2\ell+1}\Big[\ldots \frac{1}{\sigma_{ij}}\int_{\theta_i}^{\theta_i+\Delta\theta} P_{\ell m}(\cos\theta)\sin\theta d\theta \int_{\lambda_i}^{\lambda_j+\Delta\lambda} \underline{c}^{\alpha T}_{m} d\lambda \ \ldots\Big]^T \qquad (43)$$

$i=0$ to $(N_\theta-1)$

where

$$\underline{c}^{\alpha}_{m} = \Big[\ldots \left\{ \begin{matrix} \cos mj\Delta\lambda \\ \sin mj\Delta\lambda \end{matrix} \right\} \ldots\Big]^T; \quad j = 0 \text{ to } (N_\lambda-1) \qquad (44)$$

Let (43) be written in the following form which defines $k^{i}_{\ell m}$:

$$\underline{c}_{\ell m\alpha,z} = \Big[\ldots k_{\ell m}{}^{i} \int_{\lambda_j}^{\lambda_j+\Delta\lambda} \underline{c}^{\alpha T}_{m} d\lambda \ldots\Big]^T \qquad (45)$$

We also define a set of $X_{\ell m}{}^{i}$ values as follows:

$$\underline{X}_{\ell m}{}^{i} = (R(m) + W)^{-1}\underline{k}_{\ell m}{}^{i} \qquad (46)$$

where W is the diagonal matrix representing the average anomaly variance in each of the N latitude bands.

With the above information Colombo (ibid, equation (2.61) shows that the estimate of the potential coefficients $\hat{c}$, from (40) can be represented in the following quadrature form:

$$C_{\ell m}^{\alpha} = \frac{1}{\gamma(\ell-1)}\sum_{i=0}^{N-1} X_{\ell m}{}^{i} \sum_{j=0}^{2N-1} \Big[\left\{ \begin{matrix} A(m) \\ -B(m) \end{matrix} \right\} \cos mj\Delta\lambda + \left\{ \begin{matrix} B(m) \\ A(m) \end{matrix} \right\} \sin mj\Delta\lambda\Big] \overline{\Delta g}_{ij} \Big|_{\alpha=1}^{\alpha=0} \qquad (47)$$

where A(m) and B(m) are defined in equation (23) and (24). Note the similarity in form of (47) and (26). Hajela (1984, p. 20) gives a form of (47) that takes into account the symmetries of $X^{i}_{\ell m}$.

The $X_{\ell m}$ values might be viewed as optimum quadrature weights that take into account the sample size and the data noise.

The total error variance of the potential coefficients is found by subsitituting into (31). Because of the assumptions made the variance of $\overline{C}^{1}_{\ell m}$ is equal to $C^{2}_{\ell m}$. Hajela (1984, eq. (3.8)) gives:

$$\sigma^2_{c_{\ell m}} = \sigma^2_{s_{\ell m}} = \frac{1}{\gamma^2(\ell-1)^2}\left[\frac{c_\ell}{2\ell+1} - \left\{ \begin{matrix} \Delta\lambda^2 & \text{if } m=0 \\ \frac{1-\cos m\Delta\lambda}{m^2} & \text{if } m\neq 0 \end{matrix} \right\} 2N\sum_{i=0}^{N-1} k_{\ell m}{}^{i} X_{\ell m}{}^{i}\right] \qquad (48)$$

This equation can also be broken up into the finite size (sampling) error and the propagated noise components.

The computation of the R(m) matrices is done through a Fourier transform of elements that depend on $I_{\ell t}{}^{i}$ defined in (33). A summation is involved in these evaluations that has to be determined empirically (Colombo, ibid, section 4.2).

Colombo (ibid) originally developed a number of programs to implement the optimal estimation procedure. They were tested for area means of 30°x30°; 10°x10°; and 5°x5°. Hajela (ibid) extended the programs and equations of Colombo so that they could be applied to 1°x1° anomalies. This was a very large task considering the large number of computations and programs that had to be implemented.

2.4 APPLICABLE COMPUTER PROGRAMS

In this section I would like to briefly review some of the software that exists for the evaluation of the equations described in previous sections.

One of the fundamental quantities in all these discussions is the generation of the "point" fully normalized associated Legendre functions and the integrals of these functions. An analysis of "point" generation techniques has been given by Singh (1982, 1984). A Fortran program for $P_{\ell m}$ and it's derivative is given in Colombo (ibid). Paul (1978) has given a valuable recurrence procedure for computing the needed integrals. Gleason (1985a) has presented some alternate algorithms based on the Clenshaw summation and has made suggestions on the improvement of the Paul procedure for certain latitudes.

The evaluation of the gravity anomaly (in (8)) is relatively straight forward for a few points but can become quite expensive when many points are to be evaluated when the maximum degree of the exansion is high (for example, 180). A program to carry out efficient computations of this nature is due to Rizos (1979). The Rizos program is designed to compute anomalies (or geoid undulations) with a specified grid spacing in a specified region that may be global. The program computes the associated Legendre functions for the latitude belts needed for the specific computation. Some timing tests for this program are described in Tscherning et al. (1983). For example, the calculation of a 64800 1°x1° point grid using a field to degree 180 took 344 seconds on an Amdahl 470 V/8 computer. Programs for the point calculation of gravimetric quantities can be found in Tscherning et al (ibid).

Rizos also prepared an efficient program for the evaluation of equation (14) that calculates potential coefficients from gridded anomalies given on a global basis. This program takes into account latitude, longitude, and $P_{\ell m}$ symmetries. This program is run at Ohio State with a number of small modifications of the original version. Given a global 1°x1° (mean) anomaly data set, it takes 66 seconds, on an IBM 3081D to generate a potential coefficient set complete to degree 180. This includes time for the generation of the associated Legendre functions.

The calculation of point and mean anomalies on a grid using precomputed associated Legendre funcitons is described in Colombo (1981). The basic subroutine is SSYNTH. This program implements a fast Fourier transform procedure and takes into account grid symmetry with respect to the equator and the corresponding even-odd symmetry of the Legendre functions or their integrals.

The evaluation of (30) can be efficiently done by subroutine HARMIN described by Colombo (ibid). Pre-computed associated Legendre function integrals are used. As described in the Colombo report the de-smoothing parameter q_ℓ, as defined by equation (21), is used. A potential coefficient set to degree 180 using 1°x1° mean anomalies can be computed in 15 seconds on an IBM 3081D.

The evaluation of the optimal estimator procedure requires a sequence of approximately six programs. Two of the programs require substantial computational effort. A complete discussion of the various programs as used by Hajela (1984) is given by Priovolos (1985). The evaluation of the elements of the R(m) matrices for all latitude bands takes 43 minutes on the IBM 3081D for 1° data and solutions to degree 250. The calculation and ordering of the $k_{\ell m}{}^{i}$ values (see (45)) takes about 8 minutes. The calculation of the optimal quadrature weights (including the inversion of N_{max}, NxN matrices where N is $180°/\theta°$) and the standard deviations of the potential coefficients takes 50 minutes.

It is very clear that the optimum estimator requires substantially more time than the HARMIN procedure. However, it does provide us with the accuracy estimates of the potential coefficients; values not provided by other methods.

2.5 SOME NUMERICAL CONSISTENCY TESTS

In this section I describe some numerical tests that compare the consistency of the computation of potential coefficients from mean anomalies which have (Case 1) been computed from potential coefficients and which have not (Case 2) been computed from known coefficients.

First we define terms of comparison for two potential coefficient sets given to the same degree. We have:

(1) The root mean square undulation difference by degree and for the whole coefficient set.

A. By degree:

$$\delta N_{\ell} = \left[R^2 \sum_{m=0}^{\ell} \sum_{\alpha=0}^{1} \Delta C_{\ell m}^{2,\alpha}\right]^{1/2} \tag{49}$$

where R is a mean earth radius.

B. Cumulatively:

$$\delta N = \left[\sum_{\ell=2}^{N_{max}} \delta N_{\ell}^{2}\right]^{1/2} \tag{50}$$

(2) The root mean square anomaly difference by degree and for the whole coefficient set:

A. By degree

$$\delta g_{\ell} = \left[\gamma^2 (\ell-1)^2 \sum_{m=0}^{\ell} \sum_{\alpha=0}^{1} \Delta C_{\ell m}^{2,\alpha}\right]^{1/2} \tag{51}$$

B. Cumulatively:

$$\delta g = \left[\sum_{\ell=2}^{N_{max}} \delta g_{\ell}^{2}\right]^{1/2} \tag{52}$$

(3) The percentage difference by degree and cumulatively:

A. By degree:

$$P_\ell = \left[\frac{\sum_{m=0}^{\ell}\sum_{\alpha=0}^{1}\Delta C_{\ell m}^{2,\alpha}}{\sum_{m=0}^{\ell}\sum_{\alpha=0}^{1} C_{\ell m}^{2,\alpha}}\right]^{1/2} 100 \tag{53}$$

B. Average Percentage Difference

$$P = \frac{\sum_{\ell=2}^{N_{max}} P_\ell}{N_{max}-1} \tag{54}$$

The values of δN and of δg represent the global mean difference of the two potential coefficient functions over the sphere. The values of δN_ℓ and δg_ℓ represent the same information by degree. The percentage difference provides information that takes into account the relative magnitude of the coefficient differences to the coefficient magnitudes.

Another quantity that can be used for the comparisons of two coefficient sets is the degree correlation coefficients (ρ_ℓ) and the overall correlation coefficient ρ. The degree correlation between the coefficient sets i and j would be:

$$\rho_\ell = \frac{\sum_{m=0}^{\ell}(C_{\ell m_i}C_{\ell m_j} + S_{\ell m_i}S_{\ell m_j})}{\left[\sum_{m=0}^{\ell}\sum_{\alpha=0}^{1} C_{\ell m_i}^{2,\alpha}\right]^{1/2}\left[\sum_{m=0}^{\ell}\sum_{\alpha=0}^{1} C_{\ell m_j}^{2,\alpha}\right]^{1/2}} \tag{55}$$

The overall correlation would be:

$$\rho = \frac{\sum_{\ell=2}^{N_{max}}\sum_{m=0}^{\ell}(C_{\ell m_i}C_{\ell m_j} + S_{\ell m_i}S_{\ell m_j})}{\left[\sum_{\ell=2}^{N_{max}}\sum_{m=0}^{\ell}\sum_{\alpha=0}^{1} C_{\ell m_i}^{2,\alpha}\right]^{1/2}\left[\sum_{\ell=2}^{N_{max}}\sum_{m=0}^{\ell}\sum_{\alpha=0}^{1} C_{\ell m_j}^{2,\alpha}\right]^{1/2}} \tag{56}$$

In practice it is not clear how one should interpret small differences in ρ_ℓ and ρ between various coefficient sets. We therefore will not use these values for this paper.

For the first test we took the Rapp (1981) potential coefficients to degree 180 and computed 1°x1° mean anomalies using the SSYNTH program of Colombo. In this run all (a/r) values were set to one for consistency with the spherical approximation of formulas to be used later. The 64800 anomalies were then used to recover the implied potential coefficients in ways defined as follows:

<u>Method 1</u> - The 1°x1° anomalies were used in HARMIN (see equation (16)) using the q_ℓ values given in (21).

<u>Method 2</u> - The 1°x1° anomalies were used in the Rizos program as center point values (see equation (14)).

<u>Method 3</u> - The 1°x1° anomalies were used in the Rizos program as center point values but β_ℓ was introduced in the denominator to the left of the first summation in (14). This was done in some analogy to (16) but with no specific a priori justification.

The computed potential coefficients were then compared to the potential coefficients used in the anomaly generations. The results of the comparisons are given in Table 2.

Table 2

Comparison of Input and Output Potential Coefficient Sets

	Method 1			Method 2			Method 3		
ℓ	δN†	δg†	P†	δN†	δg†	P†	δN†	δg†	P†
2	0.1	.00	.01	0.1	.00	.01	0.0	.00	.00
10	0.1	.00	.06	0.3	.00	.12	0.1	.00	.03
30	0.2	.01	.47	0.4	.02	1.03	0.1	.00	.24
50	0.3	.02	1.21	0.7	.05	2.84	0.1	.01	.59
100	1.1	.17	10.6	1.2	.18	10.9	0.3	.05	2.84
150	1.3	.30	22.1	1.3	.30	22.6	0.5	.11	8.54
180	1.3	.35	27.3	1.4	.39	30.1	1.1	.30	23.0
to 30	0.8	.02	0.2	1.4	.04	0.38	0.4	.01	0.10
to 180	13.1	2.69	10.64	14.0	2.79	11.26	4.6	1.02	3.81

† Units of δN are cm. Units of δg are mgals. The units of P is percent.

The best method for the recovery of the coefficients from these tests would be Method 3 based on having the smallest differences between the recovered and the input potential coefficients. We note that the disagreement in all methods increases with ℓ. If one restricts the comparisons to the coefficients up to degree 30, the agreement of all methods is quite good.

A second test was carried out where starting quantities were a set of 64800 1°x1° anomalies that were the result of a combination of terrestrial and satellite data. The implied potential coefficients were computed in four ways:

Method A -The application of the optimal estimation procedure using the c_ℓ model described in Hajela (1984, p. 23) with the average latitude band standard deviations computed from the values used in the combination solution.

Method B -The use of HARMIN with the q_ℓ factors as defined in (21).

Method C -The use of the Rizos program where the mean anomalies were regarded as center point values.

Method D -The use of the Rizos program with the introduction of β_ℓ as done in Method 3 described previously.

Solutions B, C, and D were compared to the Method A solution which one might regard as the "best" solution. The comparisons are shown in Table 3.

Table 3

Comparison of the Optimally Estimated Potential Coefficient Set With Three Other Estimates

ℓ	Method B $\delta N^{\dagger}$	Method B $\delta g^{\dagger}$	Method B $P^{\dagger}$	Method C $\delta N^{\dagger}$	Method C $\delta g^{\dagger}$	Method C $P^{\dagger}$	Method D $\delta N^{\dagger}$	Method D $\delta g^{\dagger}$	Method D $P^{\dagger}$
2	3.1	.00	.18	3.1	.00	.17	3.2	.00	.18
10	1.1	.02	.51	0.9	.01	.38	1.1	.02	.50
30	1.6	.07	4.50	1.1	.05	3.23	1.5	.07	4.37
50	1.4	.11	6.63	0.7	.05	3.21	1.4	.10	6.35
100	1.0	.16	11.37	0.9	.14	10.11	2.2	.34	24.53
150	1.1	.25	24.92	1.0	.22	21.72	2.7	.62	61.05
180	1.6	.43	52.73	1.4	.40	48.05	3.7	1.01	122.45
to 30	8.0	.22	2.09	6.5	.17	1.64	7.8	0.21	2.05
to 180	16.7	2.51	13.84	13.2	2.17	11.72	29.5	5.80	30.55

† Units of δN are cm. Units of δg are mgals. The unit of P is percent.

From the results in Table 3 it appears that the original center point formulation yields the best results. If we consider the solutions up to degree 30 only, the Method D is slightly better then Method B (HARMIN). It is quite surprising to see the large percentage differences at the high degree for Method D since such magnitudes were not reached in the first sequence of tests reported in Table 2. There is a clear difference in conclusion on the agreement of the method when Table 2 and Table 3 are considered. This is probably caused by the incorporation of noise in the method A solution which tends to smooth the estimated coefficients.

The results of this section have been given to demonstrate the consistency of the various methods for potential coefficient determinations. Unfortunately no result was obtained for the weighted least squares fit of a truncated series to the anomalies. This awaits additional work which will be (almost) rigorously impossible for a high (180) degree solution due to the number of observations and unknowns. The results that we do have demonstrate the approximation in all procedures. No procedures exactly "close the loop". This is because we are dealing with data given in finite size blocks. As the size of the blocks decrease we should expect better agreement between the methods.

3.0 DOWNWARD CONTINUATION, ELLIPSOIDAL AND ATMOSPHERIC CORRECTIONS

In carrying out the computations in the previous section we made spherical approximations, assumed data was given on a sphere, and we ignored the net attraction of the atmosphere. We discuss the problems in the following sections.

3.1 DOWNWARD AND UPWARD CONTINUATION OF THE GRAVITY ANOMALIES

Free-air gravity anomaly data is normally computed using the following:

$$\Delta g = g_{OBS} - \frac{\partial \gamma}{\partial h} H - \gamma \tag{57}$$

where H is the orthometric height and γ is normal gravity on the ellipsoid. The Molodenski surface free air anomalies are normally defined as

$$\Delta g_s = g_{OBS} - \left(\frac{\partial \gamma}{\partial h} H^* + \gamma\right) \tag{58}$$

where H^* is the normal height. Since H and H^* differ by only a meter in mountainous areas (Heiskanen and Moritz, 1967, page 328) we can conclude that the gravity anomaly data we generally have to work with can be interpreted as surface free air anomalies.

For the use of the orthogonality relationship, we should reduce the given anomalies to the common surface over which the integration is taking place. Several proposals have been made for doing this. One suggestion (Rapp, 1978, 1984) is to upward continue the surface gravity anomalies to a sphere that encloses the topographic surface. This sphere is known as the Brillouin sphere (Moritz, 1980, p. 430). Cruz (1985) has suggested that the surface anomalies be downward continued to the ellipsoid. To do this we utilize a Taylor series expansion:

$$\Delta g_E = \Delta g_s - \frac{\partial \Delta g}{\partial r} h - \frac{1}{2!} \frac{\partial^2 \Delta g}{\partial r^2} h^2 \cdots \tag{59}$$

where h is the height of the point (or mean elevation of a block) above the ellipsoid. In practice h could be taken as an orthometric height. In writing (59) we recognize the concern with analytical continuation and the problem of computing the gradient expressions. Although local gradients can be computed from detailed data, we need gradients for the reduction of mean anomalies on a global basis. The most efficient way to do this is to differentiate the anomaly expression (8) assuming potential coefficients are known. We have:

$$\frac{\partial \Delta g}{\partial r} = \frac{-kM}{ar^2} \left[\sum_{\ell=2}^{\infty} (\ell-1)(\ell+2) \left(\frac{a}{r}\right)^{\ell+1} \sum_{m=0}^{\ell} \sum_{\alpha=0}^{1} C_{\ell m}^{\alpha} Y_{\ell m}^{\alpha}(\theta,\lambda)\right] \tag{60}$$

$$\frac{\partial^2 \Delta g}{\partial r^2} = \frac{kM}{a^2 r^2} \left[\sum_{\ell=2}^{\infty} (\ell-1)(\ell+2)(\ell+3) \left(\frac{a}{r}\right)^{\ell+2} \sum_{m=0}^{\ell} \sum_{\alpha=0}^{1} C_{\ell m}^{\alpha} Y_{\ell m}^{\alpha}(\theta,\lambda)\right] \tag{61}$$

The evaluation of these gradient terms can be done by simple modification of the Rizos type program or the Colombo SSYNTH program.

The root mean square value of the first derivative reduction is $\pm$ 0.4 mgal (max = 19 mgal); for the second derivative reduction, it is $\pm$ 0.01 mgal (max = 1.0 mgal) (Rapp, 1984). These computations were carried out with the Rapp (1981) potential coefficient field to degree 180. Recent calculations by Cruz (1986, private communication) have used 1°x1° mean gradient computation using an unpublished expansion to degree 300. For this case the RMS value of the first derivative is $\pm$ 0.8 mgal (max = 30 mgal); for the second derivative is $\pm$ 0.04 mgals (max = 2.1 mgals). In practice we compute the gradient correction terms and store them for application to various sets of the surface free-air anomalies.

3.2 THE SPHERICAL APPROXIMATION PROBLEM

Our fundamental anomaly equation (8) is a spherical approximation to an equation valid for the radial component of the gravity anomaly. Equation (9) led to the integration over a sphere which yielded potential coefficients. In reality we now have the gravity anomalies given on the surface of the ellipsoid. There clearly is an error if we use these anomalies in the spherical formulas.

A number of derivations have been carried out to calculate potential coefficient correction terms to take into account the spherical approximation. Recent papers include those by Pellinen (1981), Rapp (1984) and Cruz (1985). Rapp (1984) extended the derivation of Pellinen (1981) to obtain equations using geocentric latitude. Cruz (1985, private communication) gave an alternate derivation to the Pellinen/Rapp solution and extended the derivation. Neglecting terms on the order of e^4 (i.e. $O(e^4)$) we have (Heiskanen and Moritz, 1967, eq. (2-147)):

$$\Delta g = -\frac{\partial T}{\partial h} + \frac{1}{\gamma}\frac{\partial \gamma}{\partial h} T \tag{62}$$

Evaluation of the individual quantities yields

$$-\frac{\partial T}{\partial h} = -\frac{\partial T}{\partial r} - e^2 \sin\bar{\phi}\cos\bar{\phi}\,\frac{1}{r}\frac{\partial T}{\partial \bar{\phi}} \tag{63}$$

$$\frac{1}{\gamma}\frac{\partial \gamma}{\partial h} T = -\frac{2}{r} T - e^2(2 - 3\sin^2\bar{\phi})\,\frac{T}{r} \tag{64}$$

where $\bar{\phi}$ is the <u>geocentric</u> latitude.

Comparing (62), (63), and (64) to (7) we can write:

$$\Delta g = \Delta g^{*} - \varepsilon_1 - \varepsilon_2 \tag{65}$$

where Δg^{*} is given by (7), and

$$\varepsilon_1 = e^2 \sin\bar{\phi}\cos\bar{\phi}\,\frac{1}{r}\frac{\partial T}{\partial \bar{\phi}} \tag{66}$$

$$\varepsilon_2 = e^2(2 - 3\sin^2\bar{\phi})\,\frac{T}{r} \tag{67}$$

For the analysis of Δg through the spherical formulas, we introduce the anomalies (Δg_a) on the sphere of radius a. We compute Δg_a through a Taylor series expansion about a point on the sphere such that:

$$\Delta g_E = \Delta g_a + \left.\frac{\partial \Delta g}{\partial r}\right|_a (r_E - a) + \frac{1}{2!}\left.\frac{\partial^2 \Delta g}{\partial r^2}\right|_a (r_E - a)^2 + \text{-----} \tag{68}$$

where r_E is the geocentric radius to the point on the ellipsoid. We write (68) in the following form:

$$\Delta g_E = \Delta g_a - \varepsilon_3 - \varepsilon_4 \tag{69}$$

where ε_3 and ε_4 are clearly defined from (68). Evaluate (65) on the sphere of radius a and combine with (69) to find:

$$\Delta g_a^{*} = \Delta g_E + \varepsilon_1 + \varepsilon_2 + \varepsilon_3 + \varepsilon_4 \tag{70}$$

where Δg_a^{*} is the spherical anomaly on the sphere of radius a. If we now substitute

(70) into (11) we have:

$$C_{\ell m}^{\alpha} = \frac{1}{4\pi\gamma(\ell-1)} \iint_{\sigma} (\Delta g_E + \varepsilon_1 + \varepsilon_2 + \varepsilon_3 + \varepsilon_4) Y_{\ell m}^{\alpha}(\theta,\lambda) d\sigma \tag{71}$$

where γ is given by (10). We can write:

$$C_{\ell m}^{\alpha} = \frac{1}{4\pi\gamma(\ell-1)} \iint_{\sigma} \Delta g_E Y_{\ell m}^{\alpha}(\theta,\lambda) d\sigma + \frac{1}{4\pi\gamma(\ell-1)} \iint_{\sigma} (\varepsilon_1 + \varepsilon_2 + \varepsilon_3 + \varepsilon_4) Y_{\ell m}^{\alpha}(\theta,\lambda) d\sigma \tag{72}$$

We now write:

$$C_{\ell m}^{\alpha} = C_{\ell m}^{\alpha,0} + C_{\ell m}^{\alpha,1} + C_{\ell m}^{\alpha,2} + C_{\ell m}^{\alpha,3} + C_{\ell m}^{\alpha,4} \tag{73}$$

where $C_{\ell m}^{\alpha,0}$ are the potential coefficients implied by the ellipsoidal anomalies and the other terms in (73) are ellipsoidal correction terms. The derivation for the sum of $C_{\ell m}^{\alpha,1}$, $C_{\ell m}^{\alpha,2}$ and $C_{\ell m}^{\alpha,3}$ has been carried out by Rapp (1984) and by Cruz (1985) using different procedures. The derivation of $C_{\ell m}^{\alpha,4}$ has been carried out by Cruz (1985, private communication). Only the final results are given here where we let:

$$C_{\ell m} = C_{\ell m}^{0} + \delta C_{\ell m}^{0} + C_{\ell m}^{4} \tag{74}$$

where we have dropped the α superscript for convenience. We then have:

$$\delta C_{\ell m}^{0} = e^2 (p_{\ell m} u_{\ell m} C_{\ell-2,m} + q_{\ell m} C_{\ell m} + r_{\ell m} v_{\ell m} C_{\ell+2,m}) \tag{75}$$

where

$$p_{\ell m} = -\frac{(\ell^2-\ell+2)(\ell-m-1)(\ell-m)}{2(\ell-1)(2\ell-3)(2\ell-1)}$$

$$q_{\ell m} = \frac{-2\ell^4+2\ell^2m^2-4\ell^3+2\ell m^2+9\ell^2+2m^2+11\ell-8}{2(\ell-1)(2\ell+3)(2\ell-1)} \tag{76}$$

$$r_{\ell m} = -\frac{(\ell^2+3\ell+4)(\ell+m+2)(\ell+m+1)}{2(\ell-1)(2\ell+5)(2\ell+3)}$$

$$u_{\ell m} = \sqrt{\frac{(2\ell-3)(\ell+m-1)(\ell+m)}{(2\ell+1)(\ell-m-1)(\ell-m)}}$$

$$v_{\ell m} = \sqrt{\frac{(2\ell+5)(\ell-m+1)(\ell-m+2)}{(2\ell+1)(\ell+m+1)(n+m+2)}}$$

As shown by Cruz:

$$C_{\ell m}^{4} = -\frac{e^4}{8(\ell-1)} X_{\ell m} \tag{77}$$

where

$$\begin{aligned} X_{\ell m} = {} & (\ell-5)(\ell-2)(\ell-1)K_{\ell-4,m}x_{\ell m}C_{\ell-4,m} + (\ell-3)(\ell)(\ell+1)L_{\ell-2,m}u_{\ell m}C_{\ell-2,m} \\ & + (\ell-1)(\ell+2)(\ell+3)M_{\ell m}C_{\ell m} + (\ell+1)(\ell+4)(\ell+5)N_{\ell+2,m}v_{\ell m}C_{\ell+2,m} \\ & + (\ell+3)(\ell+6)(\ell+7)P_{\ell+4,m}y_{\ell m}C_{\ell+4,m} \end{aligned} \tag{78}$$

$$K_{\ell m} = \alpha_{\ell m}\alpha_{\ell+2,m}$$

$$L_{\ell m} = \alpha_{\ell m}(\beta_{\ell m} + \beta_{\ell+2,m})$$

$$M_{\ell m} = \alpha_{\ell m}\gamma_{\ell+2,m} + \beta_{\ell m}^2 + \alpha_{\ell-2,m}\gamma_{\ell m}$$

$$N_{\ell m} = \gamma_{\ell m}(\beta_{\ell-2,m} + \beta_{\ell m})$$

$$P_{\ell m} = \gamma_{\ell-2,m}\gamma_{\ell m}$$

$$
\begin{aligned}
\alpha_{\ell m} &= \frac{(\ell-m+1)(\ell-m+2)}{(2\ell+1)(2\ell+3)} \\
\beta_{\ell m} &= \frac{(2\ell^2-2m^2+2\ell-1)}{(2\ell+3)(2\ell-1)} \\
\gamma_{\ell m} &= \frac{(\ell+m)(\ell+m-1)}{(2\ell+1)(2\ell-1)} \\
x_{\ell m} &= \sqrt{\frac{(2\ell-7)(\ell+m-3)(\ell+m-2)(\ell+m-1)(\ell+m)}{(2\ell+1)(\ell-m-3)(\ell-m-2)(\ell-m-1)(\ell-m)}} \\
y_{\ell m} &= \sqrt{\frac{(2\ell+9)(\ell-m+1)(\ell-m+2)(\ell-m+3)(\ell-m+4)}{(2\ell+1)(\ell+m+1)(\ell+m+2)(\ell+m+3)(\ell+m+4)}}
\end{aligned}
\tag{79}
$$

The effect of those correction terms is more easily judged in the undulation domain. Thus evaluate:

$$
\delta N = R \sum_{\ell=2}^{N_{max}} \sum_{m=0}^{\ell} \sum_{\alpha=0}^{1} (\delta C_{\ell m}^{\alpha,0} + C_{\ell}^{\alpha,4}) \tag{80}
$$

Considering correction terms to degree 20, the root mean square δN is 33 cm; the minimum value is -96 cm and the maximum value is 112 cm. The corrections are the largest above $\phi = 60°$ and below $\phi = -60°$.

The correction formulas derived in this section are needed when we wish to compare satellite implied potential coefficients with the coefficients implied by the terrestrial gravity data reduced to the ellipsoid (i.e. Δg_E). For example use the Δg_E anomalies in (16) to get $C_{\ell m}^{o}$ and then add the correction terms given in (74) using the $C_{\ell m}^{o}$ as the a priori set of coefficients. The ellipsoidally corrected coefficients should then be the same as the satellite derived potential coefficients. Alternately we could correct the satellite implied coefficients to get the $C_{\ell m}^{o}$ values that should be consistent with that found from the Δg_E values in (16).

3.3 ATMOSPHERIC CORRECTIONS

To take into account that the earth is surrounded by an atmosphere, and that the disturbing potential, T, should be harmonic outside the earth, it is necessary to correct the surface anomalies for atmospheric effects. The details are given in Moritz (1980, p. 422). Basically the proper anomaly to start with in our calculations is the original anomaly (Δg_s) after the atmospheric correction δg_A has been applied. The appropriate anomaly is Δg_s^o defined as follows:

$$
\Delta g_s^o = \Delta g_s + \delta g_A \tag{81}
$$

It is Δg_s^o that should be used for Δg_s in (59) and in any formula that is used to calculate potential coefficients from terrestrial gravity data.

δg_A is 0.87 mgal for blocks of zero elevation. At an elevation of 2500 m, δg_A is 0.64 mgal. To separately see the effect of δg_A on potential coefficient determinations we can substitute δg_A in (11):

$$\delta C^{\alpha}_{\ell mA} = \frac{1}{4\pi\gamma(\ell-1)} \iint_{\sigma} \delta g_A Y^{\alpha}_{\ell m}(\theta,\lambda) d\sigma \tag{82}$$

The dominant effect will be on the zero degree term. The remaining corrections can be displayed in terms of a geoid undulation correction map that would be computed from:

$$\delta N_A = R \sum_{\ell=2}^{N_{max}} \sum_{m=0}^{\ell} \sum_{\alpha=0}^{1} \delta C_{\ell mA} Y_{\ell m}(\theta,\lambda) \tag{83}$$

This is a spherical approximation to the correction formula. Rummel and Rapp (1976) have evaluated (82) and (83) with N_{max} equal to 36. Figure 1 of Rummel and Rapp (ibid) show a maximum correction of -50 cm in the Himalayan region with much smaller (~ 5 cm) corrections in the United States, Western Europe, and Australia. These correction terms would be added to the result found from potential coefficients computed from free-air anomalies without atmospheric corrections. Although the global effect is small it is important to consider in the more accurate solutions under development.

The condensation of the atmosphere into the earth (as is done for the atmospheric correction problem) causes an indirect effect, δW_A on the potential. In terms of the indirect effect of the geoid, δW_A corresponds to about 7 mm, and is thus usually neglected. It would be possible to obtain a spherical harmonic representation of δW_A.

4.0 TWO METHODS FOR THE COMBINATION OF SATELLITE AND TERRESTRIAL DATA

The use of terrestrial gravity data as the sole source of potential coefficient information has long been recognized to yield weak determinations for the long wavelengths (i.e. low degrees) because of the lack of global gravity coverage. Such information can be much more accurately determined from the analysis of the orbits of various satellites. The optimum solution attempts to find a best way to combine the terrestrial and satellite data. Many solutions for this have been discussed in the literature. For this paper we will discuss two methods that have led to actual high degree solutions. These methods were analyzed together by Rapp (1969) under various assumptions that, in some cases, are no longer needed.

We first introduce our general adjustment procedure. Define the following terms:

- F a set of functions relating observations and parameters;
- L_ℓ a set of observations;
- L_x^0 a given set (approximate or observed) of parameters;
- V_ℓ a set of residuals to be added to L_ℓ to obtain the adjusted observations, L_ℓ^a;
- V_x a set of residuals to be added to L_x^0 to obtain the adjusted parameters, L_x^a;

W the misclosure vector, $W = F(L_\ell, L_x{}^o)$.

The mathematical model is written as:

$$F = F(L_\ell{}^a, L_x{}^a) = 0 \qquad (84)$$

which is linearized to yield the observation equation:

$$B_\ell V_\ell + B_x V_x + W = 0 \qquad (85)$$

where:

$$B_\ell = \frac{\partial F}{\partial L_\ell}\;; \qquad B_x = \frac{\partial F}{\partial L_x} \qquad (86)$$

If we designate P_ℓ and P_x as the weight matrices for the observations and parameters respectively, the weighted least squares condition for solution is:

$$V_\ell{}^T P_\ell V_\ell + V_x^T P_x V_x = \text{a minimum} \qquad (87)$$

The solution for V_x is:

$$V_x = -\,(B_x^T M^{-1} B_x + P_x)^{-1} B_x^T M^{-1} W \qquad (88)$$

where

$$M = B_\ell P_\ell^{-1} B_\ell{}^T \qquad (89)$$

The observation residuals are:

$$V_\ell = -\,P_\ell^{-1} B_\ell{}^T M^{-1} (B_x V_x + W) \qquad (90)$$

The relationship between weight matrices and error variance-covariance matrices is:

$$\Sigma = m_0^2 P^{-1} \qquad (91)$$

where m_0^2 is the variance of unit weight. For example, the error variance-covariance matrix for the solution vector would be:

$$\Sigma_x = m_0^2((B_x^T M^{-1} B_x + P_x)^{-1} \qquad (92)$$

4.1 COMBINATION PROCEDURE - METHOD A

The mathematical structure of this method is established by forming the difference between a given set of potential coefficients (suitably corrected), $L_x{}^o$, and an estimate, $L_{x_c}^o$, computed through (11), in principle, using suitably corrected gravity anomalies. We write for the adjusted case:

$$F = L_x{}^a - L_{x_c}^a \qquad (93)$$

where, for a single coefficient, we have:

$$[L_x^c]^\alpha = \frac{1}{4\pi\gamma(\ell-1)} \iint_\sigma \Delta g_E Y_{\ell m}^\alpha(\theta,\lambda) d\sigma \qquad (94)$$

where $\alpha = 0, 1$, and Δg_E is

$$\Delta g_E = \Delta g_s + \delta g_A - \frac{\partial \Delta g}{\partial r} h - \frac{1}{2!}\frac{\partial^2 \Delta g}{\partial r^2} h^2 + \ldots \qquad (95)$$

Applying (86) to (93) we have:

$$B_x = \frac{\partial F}{\partial L_x{}^o} = I; \quad B_\ell = -\frac{\partial L_x{}^c}{\partial L_\ell} \qquad (96)$$

Then from (14):

$$[B_\ell]^\alpha = \frac{-1}{4\pi\gamma(\ell-1)} Y_{\ell m}^{\alpha}(\theta,\lambda)\sigma_{ij} \tag{97}$$

Considering the results described in Section 2.5 for Method 3, it would be appropriate to introduce β_ℓ in the denominator of (97). Note that L_{x^o} are the observed coefficients after the ellipticity corrections have been made. Specifically we have from (74):

$$L_{x^o} = C_{\ell m} - \delta C_{\ell m}^{o} - C_{\ell m}^{*} \tag{98}$$

The solution vector from (88) is:

$$V_x = -((B_\ell P_\ell^{-1} B_\ell^T)^{-1} + P_x)^{-1}(B_\ell P_\ell^{-1} B_\ell^T)W \tag{99}$$

where

$$W = L_{x^o} - L_{x^c} \tag{100}$$

The anomaly residuals may be computed from (90) or the equivalent expression:

$$V_\ell = P_\ell^{-1} B_\ell^T P_x V_x \tag{101}$$

The variance-covariance matrix for the solution vector is:

$$\Sigma_x = m_0^2((B_\ell P_\ell^{-1} B_\ell^T)^{-1} + P_x)^{-1}) \tag{102}$$

The adjusted potential coefficients (before ellipsoidal corrections are made) would be:

$$L_{x^a} = L_{x^o} + V_x \tag{103}$$

The corresponding adjusted potential coefficients, considering the ellipsoidal corrections (see 98) would be:

$$C_{\ell m,a} = L_{x^a} + \delta C_{\ell m}^{o} + C_{\ell m}^{*} \tag{104}$$

The adjusted anomalies would be found from:

$$L_{\ell^a} = L_{\ell^o} + V_\ell \tag{105}$$

The corresponding surface anomalies would be found from (95):

$$\Delta g_s = L_{\ell^a} - \delta g_A + \frac{\partial \Delta g}{\partial r} h + \frac{1}{2!}\frac{\partial^2 \Delta g}{\partial r^2} h^2 \tag{106}$$

A special case of this adjustment procedure occurs when all anomaly standard deviations are equal and the weights to the anomalies are given as

$$[P_\ell] = \frac{\cos\phi}{m^2} \tag{107}$$

where ϕ is the mean latitude of the block. Under these circumstances $B_\ell P_\ell^{-1} B_\ell^T$ is a diagonal matrix where a diagonal element for a coefficient of degree ℓ is:

$$[B_\ell P_\ell^{-1} B_\ell^T]_D = \frac{m^2 \Delta\phi\Delta\lambda}{4\pi(\gamma(\ell-1))^2} \equiv A \tag{108}$$

where $\Delta\phi$ and $\Delta\lambda$ is the side length, in radians, of the block in which the anomalies are given. Under these circumstances:

$$[V_x] = \frac{-[W]}{1+A[P_x]} \tag{109}$$

$$[\Sigma_x] = \frac{A}{1+[P_x]A} \tag{110}$$

where m_0 has been set to one.

The maximum degree of the above solution would correspond to the maximum degree of the a priori potential coefficients. Thus if they were given to degree 30, the number of unknowns would be $(30+1)^2 = 961$. To obtain a high degree solution we then use the adjusted anomalies in one of the procedures discussed in section 2. The coefficients so found (say to degree 180) will require ellipsoidal corrections.

The most time consuming aspect of the above procedure is the formation of the normal equation matrix. For a problem involving 64800 1°x1° anomalies and 582 unknowns, the estimated computer time for the combination solution is 4.5 hours on an IBM 4031D. Using vectorized code on a CRAY machine, the total solution time is 6.5 minutes.

The above solution has been described in standard least squares adjustment formulation. Colombo (ibid, p. 65) has described a variation of the method that involves an optimal estimation process that takes into account the special form of the normal matrices to achieve a solution. No test has been made of this technique that I am aware of.

4.2 COMBINATION PROCEDURES - METHOD B

This second method carries out a fit of a truncated spherical harmonic series to data such as mean gravity anomalies and geoid undulations as determined from satellite altimeter data. From equation (9) we repeat the expression for the radial component of the gravity anomaly:

$$\Delta g(r,\theta,\lambda) = \frac{kM}{r^2} \sum_{\ell=2}^{N_{max}} (\ell-1)\left(\frac{a}{r}\right)^{\ell} \sum_{m=0}^{\ell} \sum_{\alpha=0}^{1} C_{\ell m}^{\alpha} Y_{\ell m}^{\alpha}(\theta,\lambda) \tag{111}$$

We can also express the geoid undulation through the Bruns equation (T/γ) as:

$$N(r,\theta,\lambda) = \frac{kM}{r\gamma} \sum_{\ell=2}^{N_{max}} \left(\frac{a}{r}\right)^{\ell} \sum_{m=0}^{\ell} \sum_{\alpha=0}^{1} C_{\ell m}^{\alpha} Y_{\ell m}^{\alpha}(\theta,\lambda) \tag{112}$$

In (111) we can interpret r to be the geocentric distance to the earth's surface at which the anomalies are given. In (112) we interpret r to be the geocentric distance to the geoid with γ the theoretical value of gravity at the geoid. For completeness the ε_1 and ε_2 terms defined in (66) and (67) should be removed from (111).

We now form the general adjustment model:

$$F = L_{\ell} - f(L_x) \tag{113}$$

where L_{ℓ} is the observation of a gravity anomaly or geoid undulation, and L_x are the potential coefficients, with a priori estimates, to be determined. The elements of the B_x matrix are as follows:

A. For anomalies:

$$[B_x] = \frac{\partial F}{\partial L_x} = -\frac{kM}{r^2}(\ell-1)\left(\frac{a}{r}\right)^{\ell} Y_{\ell m}^{\alpha}(\theta,\lambda) \;;\quad \alpha = 1 \text{ or } 2; \tag{114}$$

B. For undulations:

$$[B_x] = \frac{\partial f}{\partial L_x} = \frac{kM}{r\gamma}\left(\frac{a}{r}\right)^{\ell} Y_{\ell m}^{\alpha}(\theta,\lambda) \;; \quad \alpha = 1 \text{ or } 2 \tag{115}$$

We also have:

$$B_\ell = \frac{\partial F}{\partial L_\ell} = I \tag{116}$$

so that the general adjustment model is:

$$V_x = -\,(B_x{}^T P_\ell B_x + P_x)^{-1} B_x P_\ell W \tag{117}$$

where

$$W = \left\{ \begin{array}{c} \Delta g - \Delta g_c \\ \text{or} \\ N - N_c \end{array} \right\} \tag{118}$$

where Δg is the observed anomaly; Δg_c is the anomaly computed from the a priori coefficients; N is the "observed undulation" and N_c is the geoid undulation computed from the a priori coefficients.

The adjusted potential coefficients would be:

$$L_{xa} = L_{xo} + V_x \tag{119}$$

The anomaly residuals could be computed as:

$$V = f(L_{xa}) - L_\ell \tag{120}$$

Rapp (1969) has considered a special case of this method where a global anomaly field of equal accuracy is given using a spherical approximation of (115). Then the resultant normal equations are diagonal and the results for the adjusted potential coefficients will be the same in Method A or Method B. However in the general case of unequal weights the results from the two methods will be different with the same data.

Wenzel (1985) has examined the structure of the normal equations in this method. Since the size of the normal matrix is equal to the number of potential coefficients being sought, it will be a large matrix. Solution strategies relate to using an iterative solution with a restricted band width caused by reordering of the unknowns as is done in triangulation networks.

4.3 PROS AND CONS OF METHODS A AND B

The two methods discussed in this section have been used by Rapp (1978, 1981) with Method A and by Wenzel with Method B. Specific solutions will be discussed in the next section. It will be appropriate here to briefly mention advantages and disadvantages of each method.

Method A - Advantages

1. Size of normal matrix equal to the number of a priori coefficients.
2. Anomaly residuals readily available as part of adjustment process.

Method B - Advantages

1. No reductions needed for surface anomalies.
2. If a rigorous definition of a gravity anomaly is used, no ellipsoidal corrections are needed. If the usual relationship (eq. 111) is used, the ε_1 and ε_2 terms (see (66) and (67)) are needed.
3. Different and repeat data can be used in the solution.
4. Global data is not required in principle.

Method A - Disadvantages

1. A global anomaly field is needed.
2. Downward continuation of surface anomalies to the ellipsoid is needed.
3. Ellipsoidal corrections are needed.

Method B - Disadvantages

1. A very large normal matrix needs to be inverted or solved.
2. The data is fit to a truncated series forcing neglected higher degree information into the solved for coefficients.
3. Data gaps can lead to spurious values computed from the adjusted coefficients.
4. The anomaly residuals are influenced by the neglected higher degree coefficients. The specific amount depends on the degree of truncation; the higher the better, but always staying within the Nyquist frequency relationship.

4.4 OTHER COMBINATION TECHNIQUES

The primary emphasis in this paper has been on two technqiues that have been used in practice in the development of high degree fields. Other methods have been proposed but have not been implemented. For example, various alternative techniques have been discussed Sjöberg (1981). The methods described by Sjöberg are directed towards finding an optimum set of potential coefficients given two series representations of the field and related error degree variances and covariances. This information can then be combined with or without a priori knowledge of the behavior of the potential spectrum.

Let the series expression for the anomaly expansion be represented in the form:

$$\nu = \sum_{\ell=0}^{\infty} \nu_\ell \tag{121}$$

Assume there exists two unbiased estimates of ν:

$$\nu_k = \sum_{\ell=0}^{\infty} \nu_\ell^{(k)}; \quad k = 0,\ 1 \tag{122}$$

The general estimator for the harmonic coefficients would then be:

$$\tilde{\nu} = \sum_{\ell=0}^{\infty} (p_\ell \nu_\ell^{(0)} + q_\ell \nu_\ell^{(1)}) \tag{123}$$

where p_ℓ and q_ℓ are degree weights. These weights are then derived to obtain a solution with minimum mean error variance. The specific formulas for p_ℓ and q_ℓ

depend on the information assumed available and assumptions on error correlations.

The output from this solution would be a set of harmonic coefficients which would represent an optimum combination of common coefficients in the two expansions. Adjusted anomalies would be computed from the adjusted coefficients.

Note the accuracy of individual anomalies is not considered in this method except in the computation of the error degree variances that would be associated with a set of coefficients computed from a set of anomalies determined from terrestrial data. A discussion of this method does not take into account the various ways in which the $\nu_\ell^{(k)}$ solutions may be carried out especially the one that uses terrestrial data.

5.0 DEVELOPMENT OF OSU 1986 SOLUTIONS

Rapp and Cruz (1986) have discussed the development of new high degree combination solutions. Additional and final solutions will be described in a future paper with the general ideas here.

The basic data for our latest solution was the following:

A. A set of potential coefficients called GEML2′. This is a modification of GEML2 (Lerch et al, 1982) such that the (2,1) coefficients were forced to be zero. The accuracy estimates were scaled by 0.7 following the results of Lerch (1985). In addition selected resonant coefficients of order 15 and 30 were used.

B. An improved set of 48955 1°x1° terrestrial gravity anomalies that are shown in Figure 1. These values include 5689 geophysically predicted anomalies shown in Figure 2.

C. An oceanwide set of about 35,000 1°x1° anomalies derived from Geos-3/Seasat altimeter data (Rapp, 1986).

The combination procedure used was that described in Section 4.1 (Method A). Gradient computations (see equations (60) and (61)) were based on the December 1981 OSU field with a maximum degree of 300 being used. Ellipsoidal corrections were based on the coefficients computed from the spherical approximation formula. The total number of unknowns was 582 with a 64800 anomalies adjusted. The maximum complete degree in the solution was 20 with additional terms to degree 30. All weight matrices were assumed to be diagonal.

The a priori 1°x1° field was obtained by merging the terrestrial and altimeter derived anomaly data. In the merger process terrestrial data in the oceans was preferred over the altimeter data if the terrestrial data had an accuracy $\leq \pm$ 5 mgal. One merged data set was created that excluded all but 33 geophysically predicted anomalies. A second merged data set was created that included all geophysical anomalies. The first set contained 50,562 anomalies while the second set contained 56,109 values. To form a global field the unknown blocks were filled in with anomalies computed from the a priori potential coefficients. Our intent with these

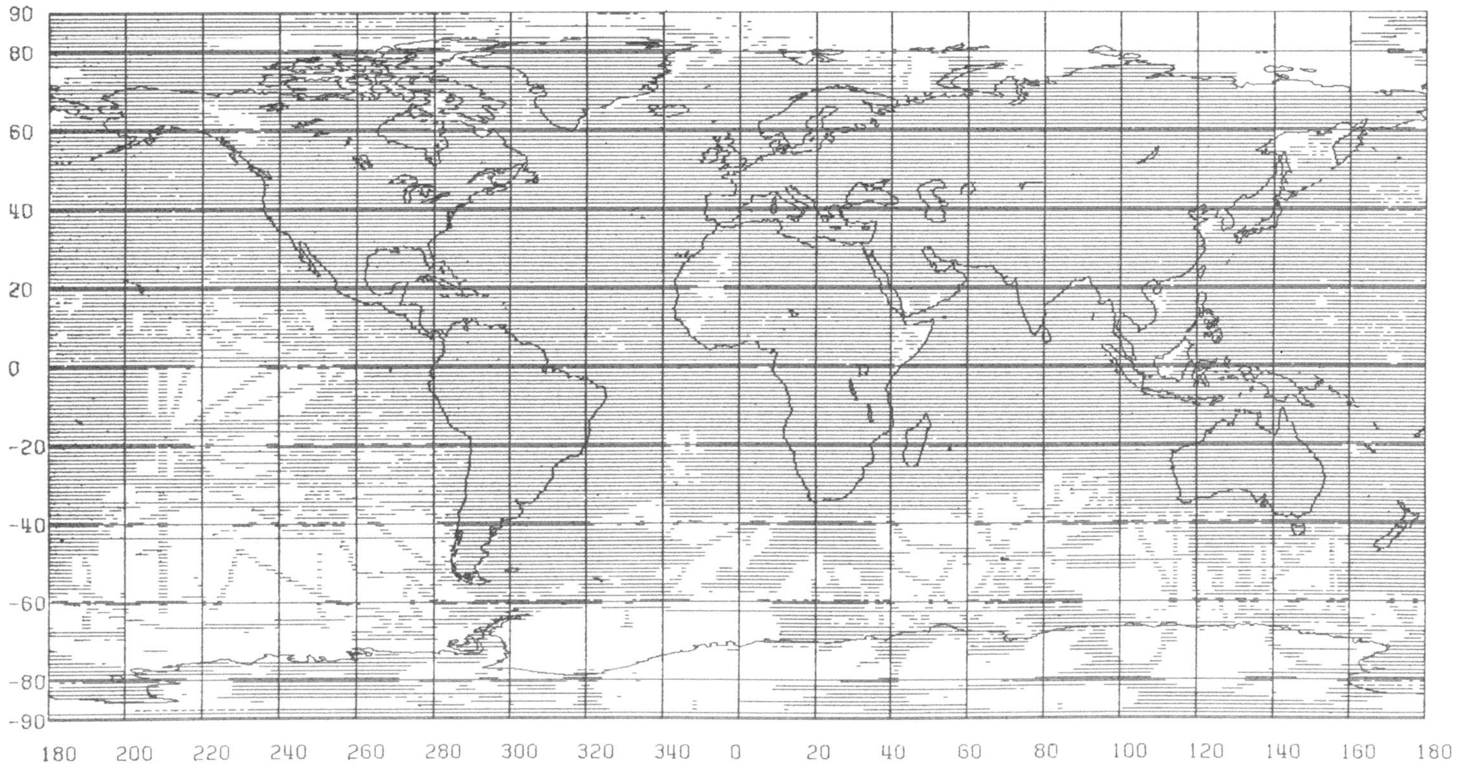

Figure 1
Location of 48955 1°x1° Terrestrial Anomalies

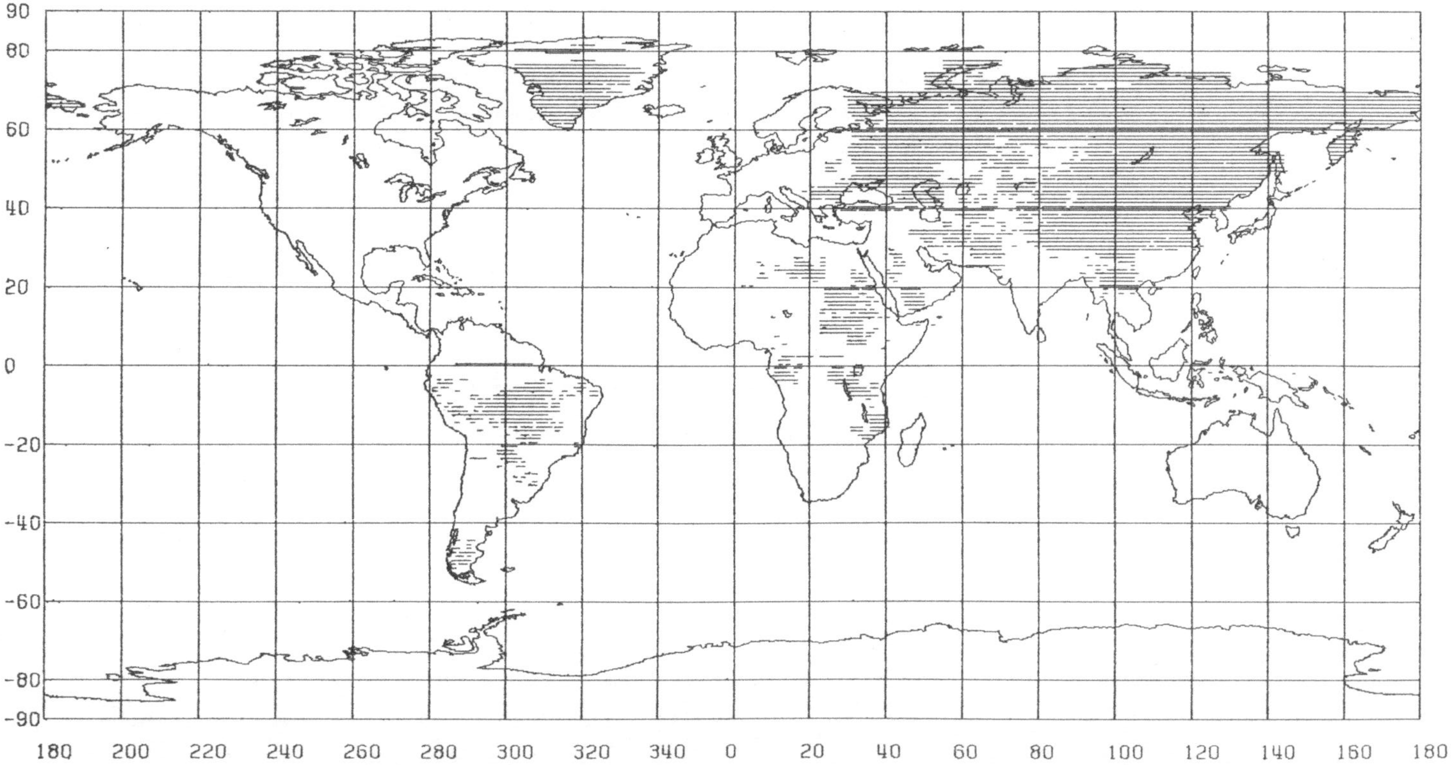

Figure 2
Location of 5689 1°x1° Geophysically Predicted Anomalies

two fields was to make two combination solutions: one with and one without, geophysical anomalies.

In carrying out our solutions a number of variations were done. One area of investigation related to the effect of weights of the terrestrial anomalies. Previous tests had indicated that a wide range of weights for the anomalies could lead to unreasonable residual distributions. For example, the accuracy of our global data set ranged from ± 1 mgal to ± 62 mgal giving a weight range from 1/1 to 1/3844. The practical consequence was the adjustment put most of the residual information in the blocks with lowest weight. After several tests we decided that one solution should be performed where the standard deviations of the anomalies should be forced to lie within the range: $8 \leq m \leq 15$ mgals. Such a range gave a reasonable weight ratio of 3.5 to 1 and maintained the average anomaly accuracy of ± 10 mgals for the global data set. For reasons to be discussed below we also carried out a solution where the accuracy range of the anomalies was specified to be $20 \leq m \leq 38$ mgals. Several solutions were made with these accuracy ranges and the results compared in several ways. The adjusted coefficients, to degree 180 were used to compute geoid undulations at a globally distributed set of Doppler stations. The comparison gave comparable undulation difference standard deviations of ± 1.68 m for the 20 to 38 mgal range and ± 1.64 m for the 8 to 15 mgal range. However, in orbit fit tests to be described in a later section, the coefficients found when using the 8 to 15 mgal range were substantially poorer than when the 20 to 38 mgal range was used. This seems natural as the higher anomaly standard deviations allow the a priori potential coefficients to dominate the solution.

It is of interest to compare these two solutions in terms of undulation and anomaly differences and percentage differences. Using the equations of section 2.5 the difference between the two solutions using different anomaly accuracy ranges are given in Table 4.

Table 4

Comparison of Potential Coefficient Sets with Anomaly Standard Deviation Range $8 \leq m \leq 15$ mgals to Alternate Solution With $20 \leq m \leq 38$ mgals

ℓ	δN†	δg†	P†
2	0.0	0.0	0.0
6	8.4	0.1	1.5
10	20.2	0.3	9.1
30	3.1	0.1	8.7
50	0.5	0.0	2.4
100	0.2	0.0	2.0
150	0.1	0.0	1.5
180	0.1	0.0	2.2
to 30	62.2	1.3	11.5
to 180	62.4	1.3	3.6

† Units of δN are cm. Units of δg are mgals. The unit of P is percent.

As would be expected the greatest change takes place at the lower degrees because it is here that the dual estimate of the potential coefficients are obtained from the satellite data and from the terrestrial data.

In carrying out the solutions with the 8-15 mgal range we noted that the accuracy of the adjusted potential coefficients was much higher than the a priori field. This implied that, if our assumptions were appropriate, the 1°x1° data was implying a much more accurate determination of the gravity field at degree values from 10 to 20 than the satellite data. To examine this we computed a set of potential coefficients from only the 1°x1° anomalies. These values were then differenced from the GEML2 coefficients. The difference anomaly degree variances were then computed and compared to the differences to be expected from the data accuracy. We found that the differences found could not be explained with our assumption of uncorrelated 1°x1° anomalies having an average standard deviation of ± 10 mgal. It could be explained if we assumed uncorrelated data having an accuracy 2.5 times poorer than the given accuracies. This could be accomplished in our combination solution if we multiplied our given accuracies by 2.5 and then restricting the range to $20 \leq m \leq 38$ mgal.

In Table 5 we show anomaly degree variance accuracy estimates as implied by the scaled GEML2 accuracies and two global 1°x1° models one with an accuracy of ± 10 mgals and the other with ± 25 mgal.

Table 5
Selected Error Anomaly Degree Variances
(Units are mgal²)

Solution	Degree 6	Degree 16
GEML2	± 0.004	± 0.514
1°x1° (±10 mgal)	± 0.032	± 0.080
1°x1° (±25 mgal)	± 0.197	± 0.50

The unrealistically low value implied by the ± 10 mgal accuracy is related to the assumption of uncorrelated noise. In fact the anomaly data is correlated. Such correlation and its role in the computation of anomaly degree variance accuracies is discussed by Weber and Wenzel (1982). At this point we have no way to treat anomaly error correlation in our solutions. Instead we have increased our anomaly accuracy estimates. Such a procedure is not desirable and in fact gives unreasonably high errors at degrees above 30. To compensate for this our accuracy estimate for unadjusted coefficients will be based on the ± 10 mgal uncorrelated noise assumption.

The above discussion should make clear that the current treatment of the anomaly errors is far from ideal. Additional work is needed to develop procedures to take into account the error correlations.

We next discuss the manner in which the high degree expansions were computed from the adjusted anomalies of the combination solutions. Our discussions in Section 2 indicated that there are a number of different procedures to recover the potential coefficients. The main comparisons to be discussed here relate to a HARMIN type solution (see equations 16 and 21) and the optimal estimation solution (see equation 47). The first type of comparison was made with the anomaly degree variances, c_ℓ. We have computed these values from the potential coefficients found from the various solutions and after the ellipsoidal corrections were made using:

$$c_\ell = \gamma^2(\ell-1)^2 \sum_{m=0}^{\infty} (\bar{C}_{\ell m}^2 + \bar{S}_{\ell m}^2) \tag{124}$$

where the c_ℓ values are interpreted to be on the surface of a sphere of radius a and γ is given by (10). The solutions to be compared have been computed from the adjusted anomalies found in the adjustment where the standard deviation range of the anomalies was 20 to 38 mgals. The following test solutions were considered.

Test 1: Application of the HARMIN subroutine.

Test 2: Application of the optimal estimation procedure with the anomaly standard deviations based on the 20 to 38 mgal range.

Test 3: Application of the optimal estimation procedure with the anomaly standard deviation based on using the standard deviation range of 8 to 15 mgal.

Test 4: Application of the optimal estimation procedure with the anomaly standard deviation based on all anomalies having a standard deviation of 1 mgal. The intent of this solution is to avoid too much smoothing caused by large anomaly errors and to simply minimize the sampling error.

Table 6

Anomaly Degree Variance Implied by Various Test Solutions

(Units are mgal²)

	Test Solution			
ℓ	1	2	3	4
2	7.59	7.59	7.59	7.59
10	9.71	9.85	9.71	9.71
30	2.60	1.78	2.43	2.59
50	2.70	1.36	2.38	2.67
100	1.90	0.39	1.57	2.28
150	1.12	0.09	0.61	1.36
180	0.97	0.08	0.33	0.95

From the results of Table 6 we see that the HARMIN solution and the optimal estimation solution with 1 mgal, uniform, standard deviation gave very similar degree variances. The optimal estimation solution (test 2) gave degree variances that were unreasonably smooth while the optimal estimation solution (test 3) gave degree variances at the higher degrees that were smaller than the other two solutions. Figure 3 shows the anomaly degree variances implied by test solutions 3 and 4 as

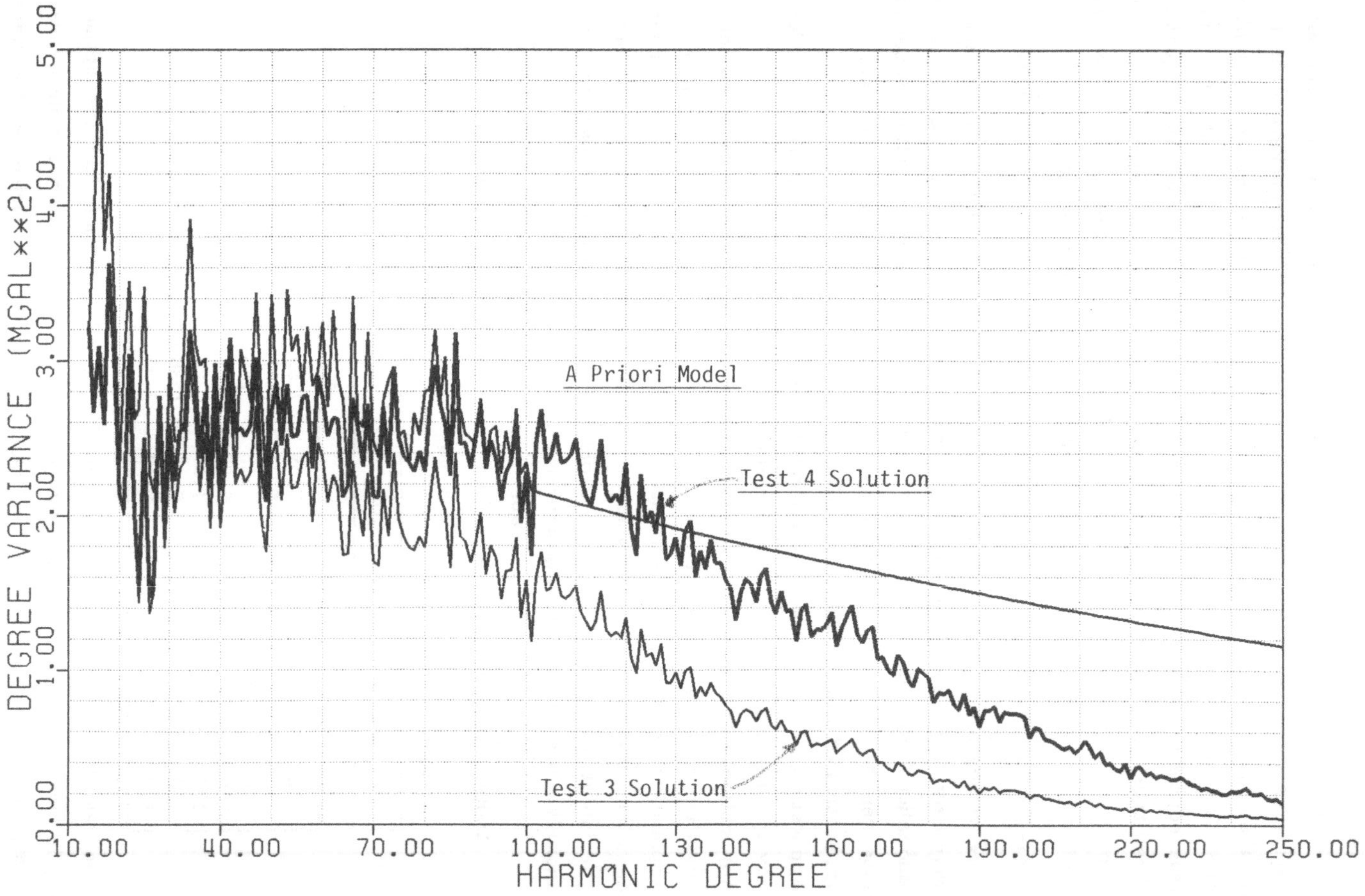

Figure 3
Anomaly Degree Variances From Two Test Solutions

well as the a priori model used in the optimal estimation solution. The smoothing effect of the anomaly noise is evident from this plot.

In order to evaluate which test solution gave more reasonable results we made comparisons with the anomaly degree variances at degree 150 and 180 implied by the altimeter spectrums discussed by Rapp (1986). Scaling the ocean spectrum to take into account the different spectra to be associated with land data, we estimated the anomaly degree variances at degree 150 and 180 to be 1.5 and 1.1 $mgal^2$ respectively. Such values agreed well with the results implied by the optimal estimation solution with the noise set at 1 mgal.

Since the optimal estimation solution is a complicated process it is of interest to compare the results of Test 1 (HARMIN) with Test 4 (optimal estimation, all anomaly standard deviations equal to 1 mgal). These comparisons are shown in Table 7.

Table 7

Comparison of Potential Coefficient Solutions from HARMIN and from Optimal Estimation (1 mgal standard error)

ℓ	δN†	δg†	P×
2	0.0	0.00	0.0
10	0.0	0.00	0.0
20	0.0	0.00	0.0
30	0.2	0.01	0.5
50	0.2	0.02	1.1
75	0.8	0.09	6.2
100	1.0	0.15	10.6
120	1.0	0.18	13.2
150	0.6	0.14	13.0
180	0.8	0.22	22.3
to 180	8.9	1.53	7.6

† Units of δN are cm. Units of δg are mgals. P is in percent.

The conclusion from Table 7 is that HARMIN gives coefficients that agree well with the coefficients found from the optimal estimation solution. The disadvantage of the HARMIN solution is that no error estimates are provided. Such values are found when the optimal estimation is carried out.

Considering the above results and a number of additional tests two final solutions for a high degree field were made. These solutions are designated OSU86C and OSU86D. Solution C is basically without geophysical anomalies (it does include 33) while solution D includes 5580 geophysical anomalies. The C and D solutions are slightly improved versions of the A and B solutions described by Rapp and Cruz (1986).

The new solutions have been computed up to degree 250 using the adjustment procedure of Method A with the anomaly range of 20 to 38 mgals. The coefficients from the optimal estimation were then merged with the adjusted coefficients from the actual adjustment. The accuracy estimates of the adjusted coefficients were taken as given from the adjustment. The accuary estimate of the unadjusted coefficients were computed from the combination of the sampling error found from the optimal estimation solution with the propagated error implied by the assumption of ± 10 mgals 1°x1° anomaly data noise. Specifically we computed

$$m(\bar{C},\bar{S}) = \sqrt{m^2(SE) + m^2(PE)} \qquad (125)$$

where m(SE) is the sampling error and m(PE) is the propagated error. From Rapp (1981) we have:

$$m(PE) = \frac{m(\Delta g)\theta}{2\gamma(\ell-1)\sqrt{\pi}} \qquad (126)$$

where $m(\Delta g)$ is the anomaly accuracy in a block of size θ (radians) and ℓ is the degree. A similar, but not as rigorous procedure was used in the development of the OSU 1981 field.

Figure 4 shows the anomaly degree variances implied by the C and D solutions and the error anomaly degree variances. From Figure 4 we see that OSU86D has slightly more power than OSU86C. The sum (to ℓ = 180) of the anomaly degree variances for the C solution is 522 mgal² while it is 547 mgal² for the D solution. The corresponding numbers when the solution is summed to 250 is 554 mgal² (C) and 580 mgal² (D). The signal to noise ratio of either the C or D solution becomes one near degree 170 indicating the unreliability of the coefficients above this degree.

Figure 5 shows the cumulative accuracy of the GEML2′ solution and the OSU86C solution in addition to the error by degree for the C solution. Note the substantial reduction in the cumulative error at degree 20 for the GEML2′ solution (± 112 cm) to the OSU86C solution (± 67 cm). To degree 180 the global accuracy of the OSU86C solution is ± 1.0 meters. This accuracy will be position dependent in practice and errors up to 10 meters may exist in areas where the gravity data does not exist or is unreliable.

The purpose of this discussion has been to describe how two new global models have been computed. A number of approximations were necessary leading to the new solutions. In the next section, comparisons of the new solutions with several other solutions will be described.

6.0 COMPARISON OF GLOBAL GEOPOTENTIAL MODELS

Table 1 of this report describes a number of high degree spherical harmonic expansions of the earth's gravity field. We now add to this list the two new OSU solutions 86C and 86D. In this section we will compare a number of these solutions in several ways.

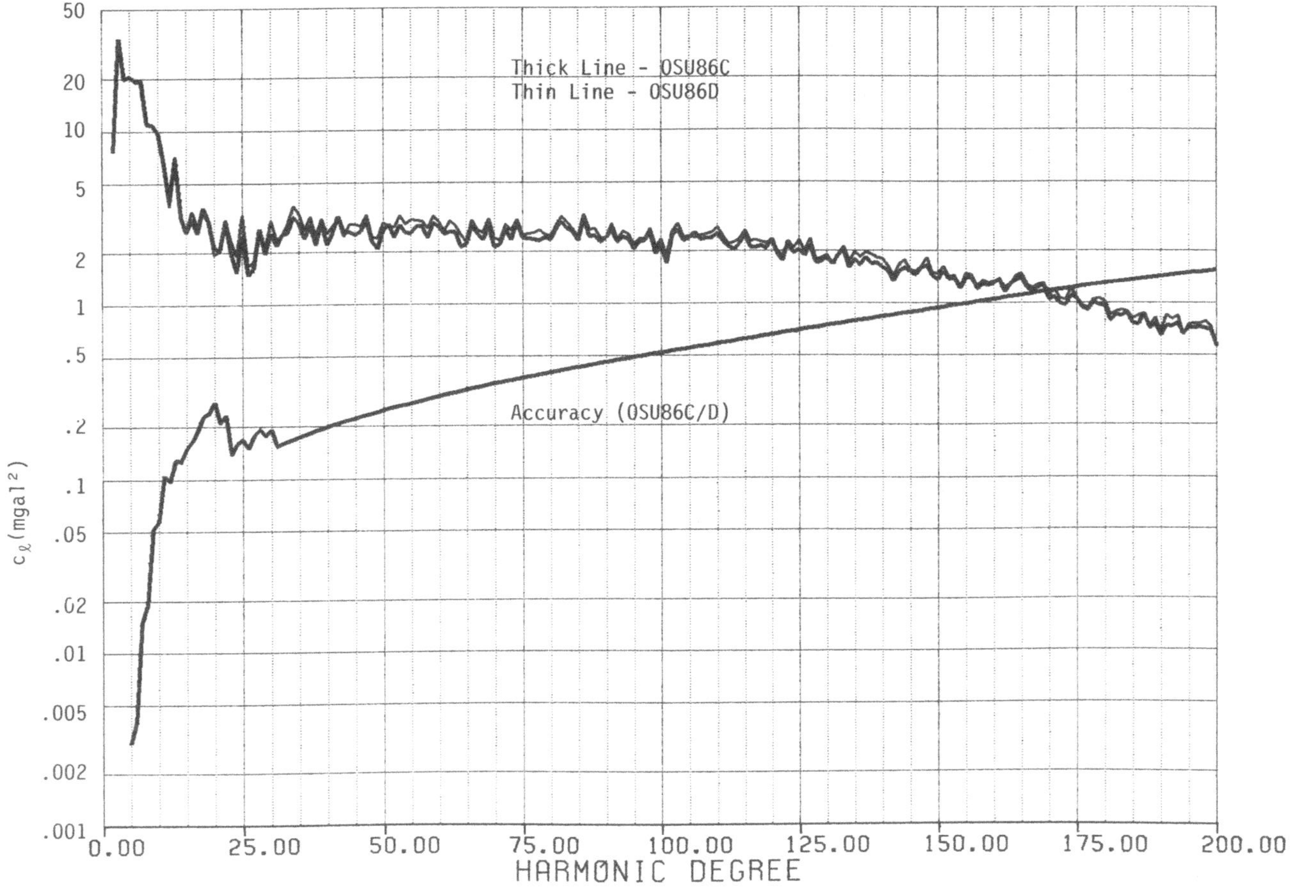

Figure 4
Anomaly Degree Variances and Their Accuracy for OSU86C and OSU86D

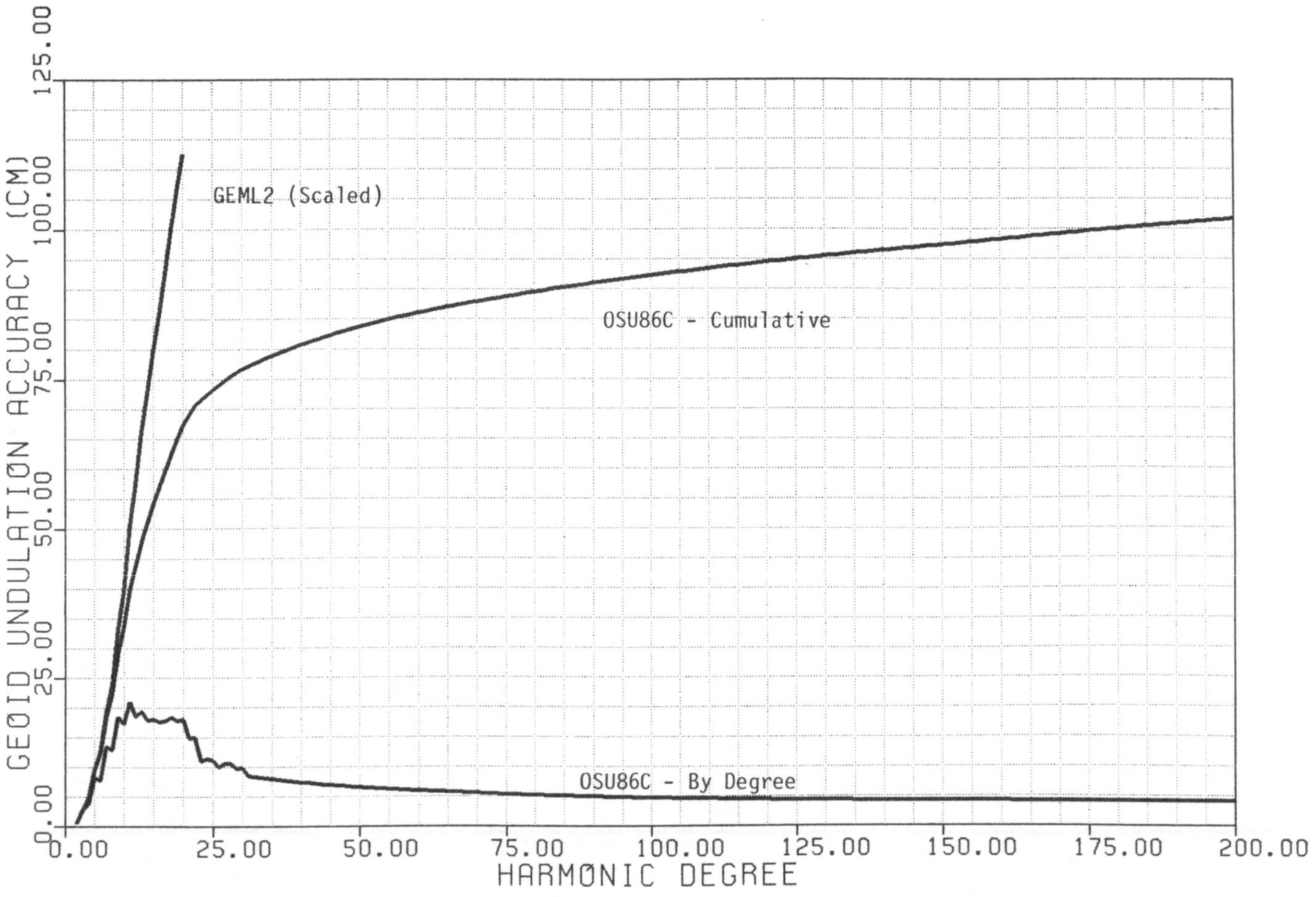

Figure 5
Geoid Undulation Accuracy for GEML2 and OSU86C Solutions

6.1 ANOMALY AND UNDULATION COMPARISONS

The first set of comparison is to indicate the global undulation (Table 8) and anomaly (Table 9) differences between several solutions.

Table 8

Potential Coefficient Differences in Terms of Geoid Undulations

(meters)

	OSU86D	OSU86C	GPM2
OSU Dec81			
to ℓ = 30	± 1.04	± 1.32	± 1.12
to ℓ = 180	1.20	1.51	1.31
GPM2			
to ℓ = 30	± 1.08	1.21	
to ℓ = 180	1.25	1.43	
OSU86C			
to ℓ = 30	± 0.56		
to ℓ = 180	0.73		

Table 9

Potential Coefficient Differences in Terms of Gravity Anomalies

(mgals)

	OSU86D	OSU86C	GPM2
OSU Dec81			
to ℓ = 30	± 2.5	± 3.0	± 2.6
to ℓ = 180	7.3	8.9	8.4
GPM2			
to ℓ = 30	± 2.6	3.0	
to ℓ = 180	8.5	9.9	
OSU86C			
to ℓ = 30	± 1.5		
to ℓ = 180	4.9		

Although the difference between the 86D and GPM2 solution is ±1 m there are large differences that reach 16 m. Figure 6 shows an undulation difference map with large discrepancies noted. A similar map is shown in Figure 7 for the gravity anomaly differences between these two solutions. The maximum anomaly difference is 185 mgal with a number of differences greater than 100 mgal as noted in the figure. These maps were made by contouring data gridded at a 2°x2° interval. Plots using 1°x1° gridded data would show more high frequency information. The points that are labeled are based on 1°x1° data points.

The differences between the OSU86 and GPM2 depend on the data used and the way data in gap areas was treated. For example, the large anomaly difference at $\phi = 5°$, $\lambda = 46°$ is primarily caused by a change in the anomaly estimate in this area.

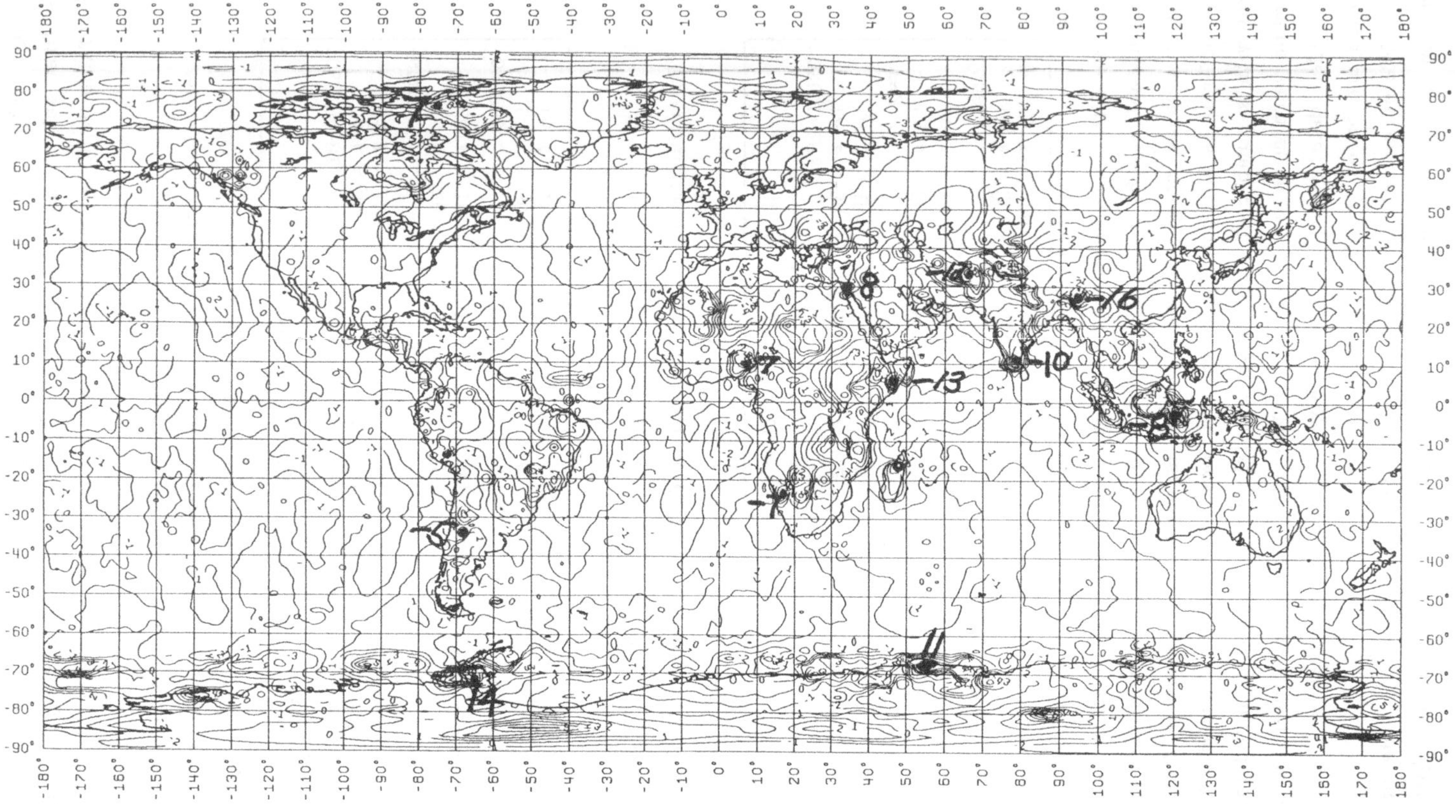

Figure 6
Geoid Undulation Difference Map - OSU86D Minus GPM2
(Contour Interval is 1 m based on a 2°x2° grid)

In 1976 the estimate of the 1°x1° anomaly at ϕ = 5°, λ = 45° was 98 mgal while in the current OSU solution it is -28 mgal. Other such examples exist.

Another reason for some of the large discrepancies is related to how the solutions respond to data gaps if they exist in a solution. Two specific 1°x1° blocks in which gaps may have existed when the GPM2 solutions were carried out are shown in Table 10. Shown in this table are the mean anomalies implied by several potential coefficients sets as well as estimates from other data sources.

Table 10

Comparison of Anomalies for Two 1°x1° Blocks

(units are mgals)

	Block (Northwest Corner)	
Source	ϕ = 30°, λ = 34°	ϕ = 0°, λ = 320°
Terrestrial (1986)	13	0
Altimeter	—	-23
OSU86C*	21	-24
OSU86D*	20	-24
GPM2*	-81	54

* to ℓ = 180

We see from Table 10 the large discrepancies between the OSU86 and GPM2 solutions for two 1°x1° blocks. I believe this is caused by gaps in the data used in the GPM2 solution and the resultant high frequency changes in the anomaly and undulations that take place in this gap region.

Figure 8 shows the geoid undulation differences OSU86C minus OSU86D. The discrepancies, which reach 9 m, occur in the areas where the geophysical anomalies were used in the 86D solution. The small root mean square difference of ±73 cm occurs because the differences are essentially restricted to only the land areas in which the geophysical anomalies have been incorporated in the solution.

A figure similar to Figure 8 showing anomaly differences would show a maximum difference of 116 mgals with a number of differences greater than 90 mgals.

6.2 ANOMALY DEGREE VARIANCE COMPARISONS

Another way to consider high degree solutions is through degree variances, usually anomaly degree variances or unitless potential degree variances. The degree variances provide information on the power contained in the coefficients and is information for which estimates may be made from other sources. Table 11 provides selected anomaly degree variances from several potential coefficient solutions.

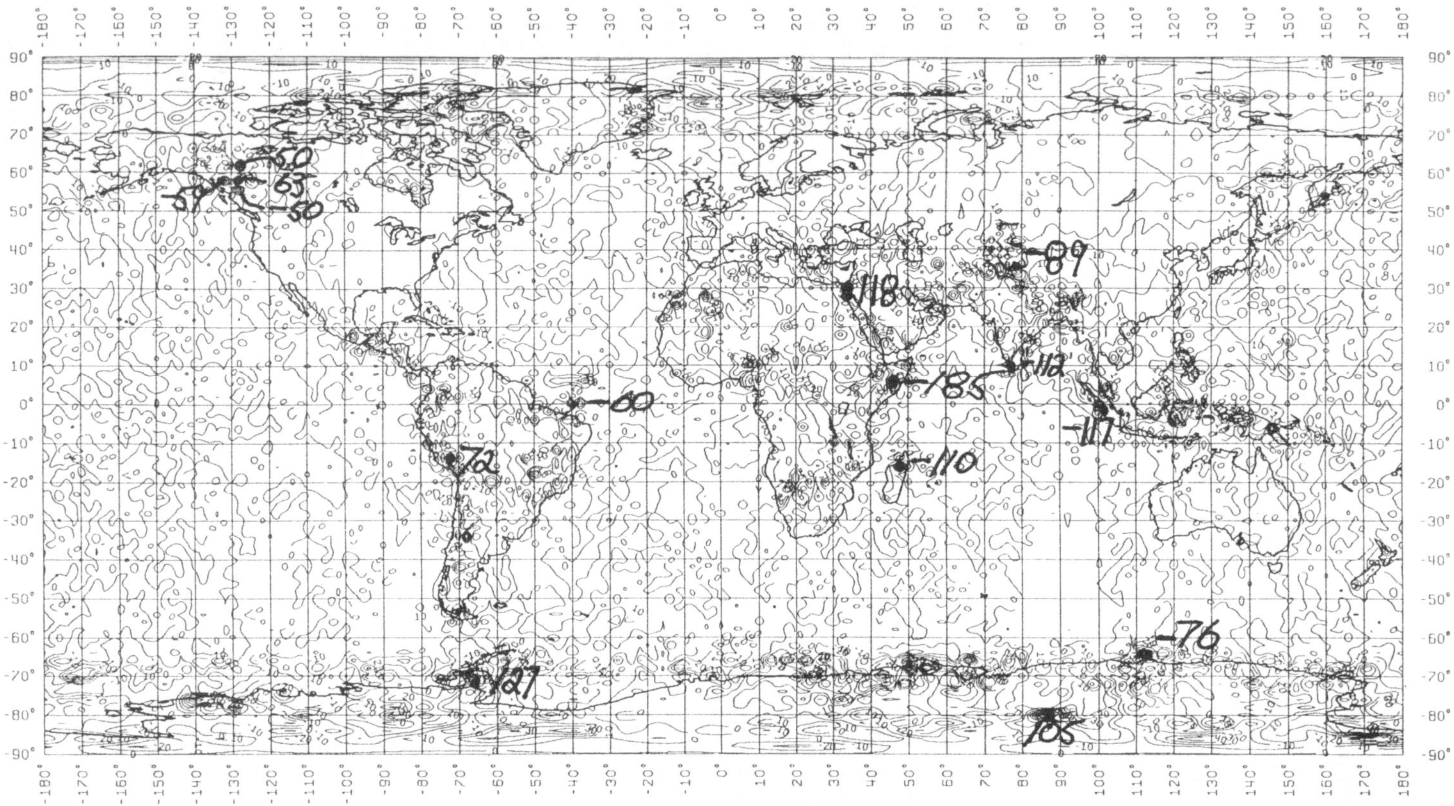

Figure 7
Gravity Anomaly Difference Map - OSU86D Minus GPM2
(Contour Interval is 10 mgal based on a 2°x2° grid)

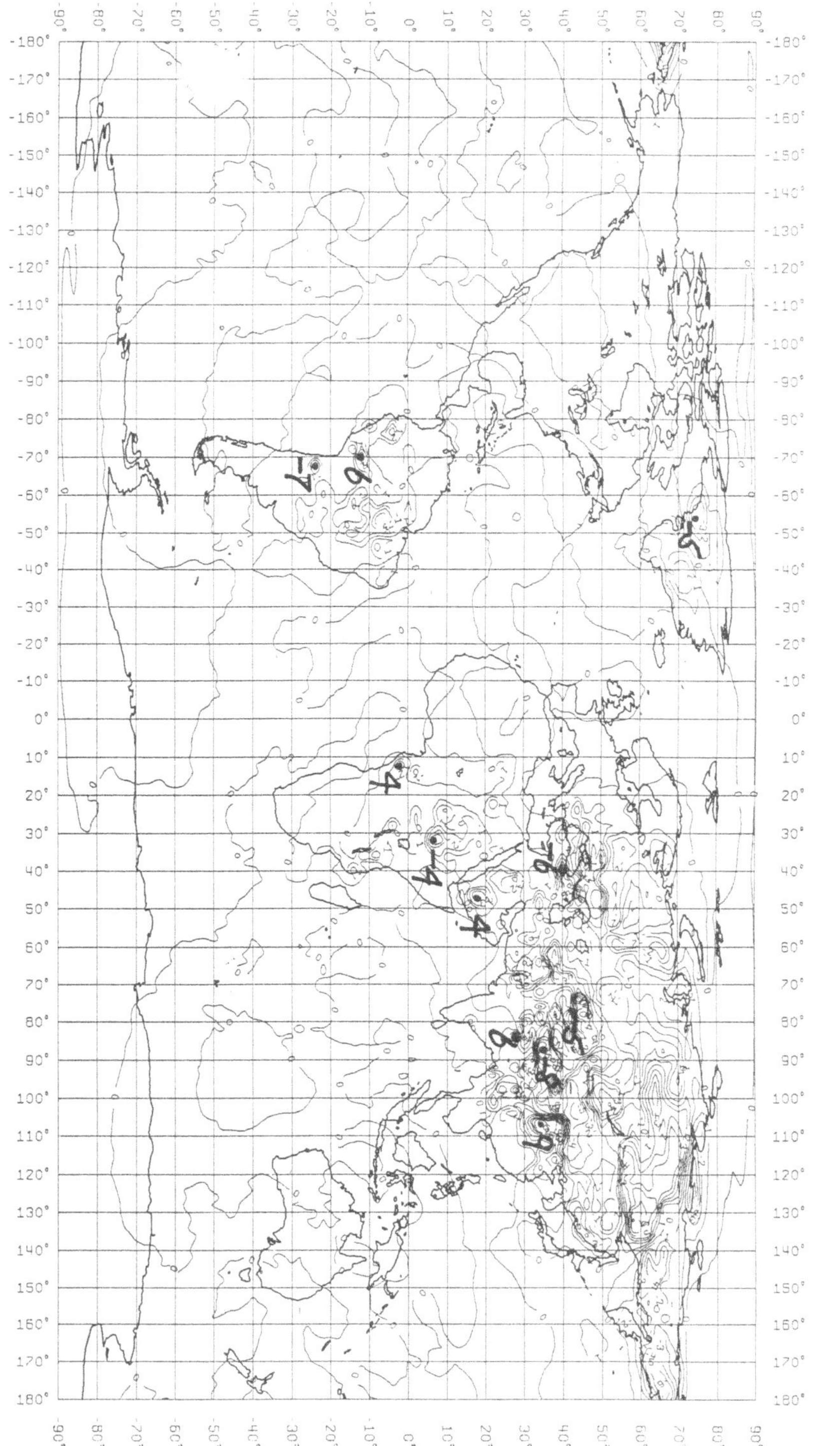

Figure 8
Geoid Undulation Difference - OSU86C Minus OSU86D
(Contour Interval is 1 m based on a 2°x2° grid)

Table 11
Anomaly Degree Variances
(Units are $mgal^2$)

ℓ	OSU81	OSU86C	OSU86D	GPM2	GEM10C
2	7.58	7.59	7.59	7.59	7.56
10	9.77	9.73	9.61	9.49	9.61
20	1.91	2.12	1.94	2.45	3.06
30	2.45	2.58	3.02	2.73	3.88
50	3.62	2.66	2.94	3.29	2.89
75	2.80	2.41	2.50	2.66	1.74
100	2.61	2.23	2.35	2.60	1.40
120	2.60	2.34	2.30	2.66	1.33
150	1.80	1.36	1.48	2.34	1.19
175	1.77	1.05	1.08	2.43	1.11
180	1.65	0.96	1.02	2.62	1.12
200	—	0.57	0.56	2.32	—
230	—	0.31	0.30	—	—
250	—	0.14	0.16	—	—

At degree 180 we see the greatest percent differences between the solutions. In going from OSU81 to OSU86D we see a 38% decrease in the power. The power in GPM2 at degree 175 is 125% greater than that in the OSU86D solution. The additional power in GPM2 is shown at the higher degrees in Figure 9 where GPM2 and OSU86D anomaly degree variances and their accuracy are plotted. Although the differences between the 86D and GPM2 solution at, say degree 175, is 1.35 $mgal^2$, the accuracy of this difference (based on the estimated accuracy of the solution) is on the order of ± 1.5 $mgal^2$, so the difference is not statistically significant.

6.3 DOPPLER UNDULATION COMPARISONS

We next turn to a comparison of the geoid undulations derived from the various geopotential models with undulations derived from Doppler derived positions. If h is the ellipsoidal height of a station derived through Doppler satellite positioning techniques, after coversion to a geocentric system properly scaled (Rapp, 1983), the "Doppler undulation" is:

$$N_0 = h_0 - H \qquad (127)$$

where H is the orthometric height of the point. We compute N from Brun's equation where the disturbing potential is given by equation (6) with r the geocentric distance to the point in question projected to the ellipsoid. In all our comparisons we used the parameters to convert the Doppler system to a geocentric system derived by Boucher and Altamimi (1985). The translation and scale parameters to go from the Doppler system to a geocentric system were used as follows:

$\Delta x = -10.6$ cm

$\Delta y = 69.7$ cm

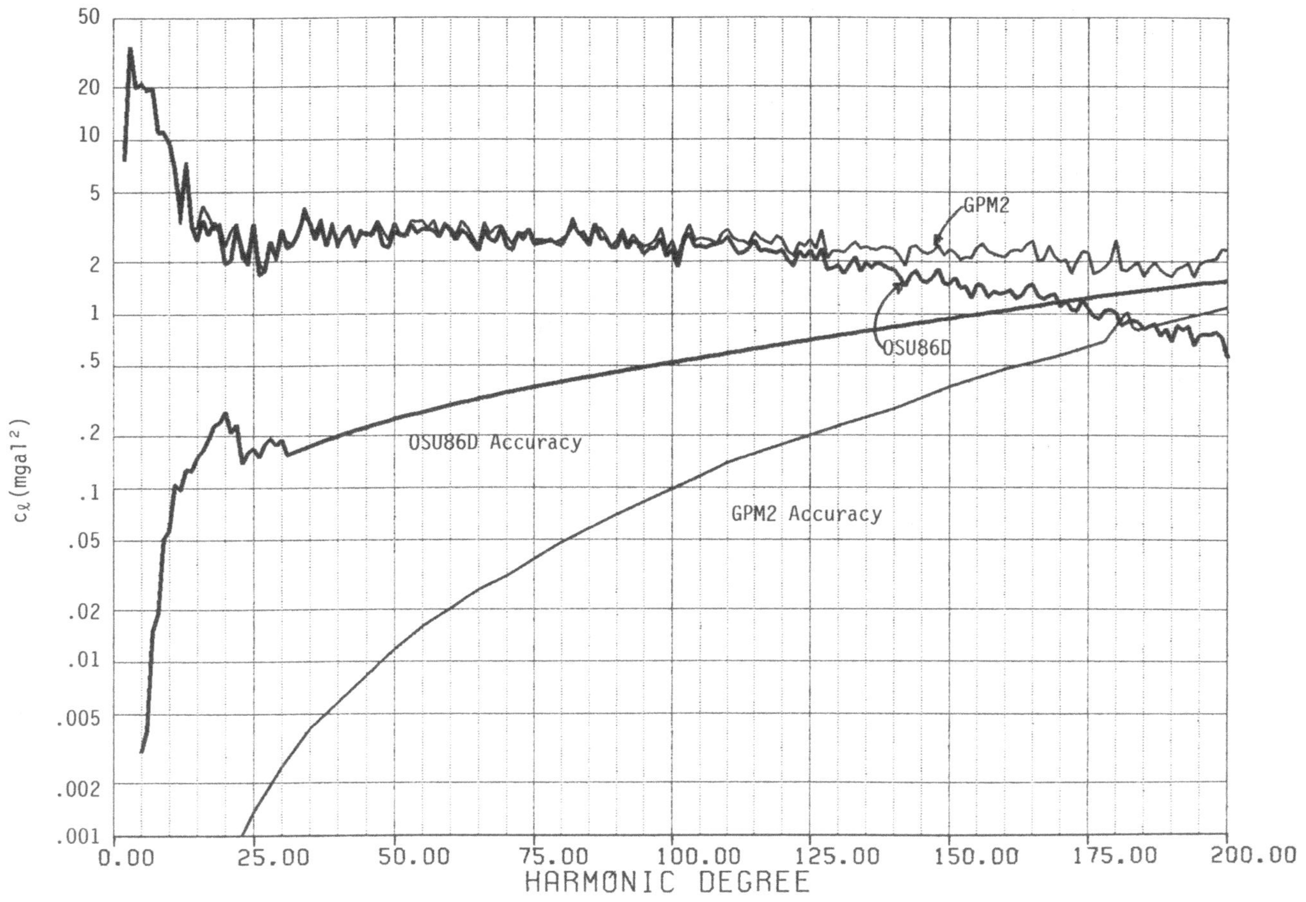

Figure 9
Anomaly Degree Variances and Their Accuracies for the OSU86D and the GPM2 Solution

Δz = 490.1 cm

Δs = − 0.604 ppm

The comparisons made were with two primary sets of Doppler stations. One contained approximately 800 stations and was received from Tscherning. This set was originally obtained from the National Geodetic Survey and contains stations only in North America. The second set contained approximately 2000 stations distributed globally. The coordinates were determined in the years from 1971 to 1985. The number of passes could range from a minimum of 25 to a maximum of 592. In our analysis we made no correction for sun spot effects as suggested by Tscherning and Goad (1985). The overall effect of neglecting this for our large data sets is expected to be only a few cm. In making our comparison we assumed the geoid undulations referred to an equatorial radius of 6378136 m. Comparisons for the Tscherning data set are given in Table 12.

Table 12

Comparison of Doppler Derived Undulations in North America with Corresponding Values Derived from Geopotential Models Complete to Degree 180

Model	Mean Difference	Std Dev Difference	Number of Stations†
OSU78	0.08 m	± 1.62 m	682
OSU81	0.22	1.72	683
GEM10C	-0.08	1.84	652
GPM2	0.18	1.58	695
OSU86C	0.27	1.54	687
OSU86D	0.27	1.55	687

† Stations with residuals greater than 4 meters in absolute value were rejected.

The best solution considering the standard deviation of the difference is the 86C solution. However it is only marginally better than GPM2 results and GPM2 was used at 8 more stations that the 86C solution.

The next set of comparisons are given in Table 13 using the global station set.

Table 13

Comparison of Doppler Derived Undulations on a Global Basis With Corresponding Values Derived from Geopotential Models Complete to Degree 180

Model	Mean Difference	Std Dev Difference	Number of Stations†
OSU78	0.04 m	± 1.76 m	1694
OSU81	0.18	1.76	1721
GEM10C	0.17	1.96	1607
GPM2	0.13	1.64	1735
OSU86C	0.17	1.66	1741
OSU86D	0.21	1.66	1743

† Stations with residuals greater than 4 meters in absolute value were rejected.

In this comparison GPM2 is negligibly better than the OSU86C/D solution. However the OSU86C/D solutions were used at 6 and 8 stations than GPM2.

A subset of the global station set was created for Europe and Australia. The comparisons for the stations are given in Tables 14 and 15.

Table 14

Comparison of Doppler Derived Undulations in Europe With Corresponding Values Derived from Geopotential Models Complete to Degree 180

Model	Mean Difference	Std Dev Difference	Number of Stations†
OSU78	-0.56 m	± 1.76 m	172
OSU81	-0.04	1.51	172
GEM10C	0.14	2.28	172
GPM2	-0.28	1.47	172
OSU86C	-0.47	1.38	172
OSU86D	-0.39	1.42	172

† All stations in the area were used.

Table 15

Comparison of Doppler Derived Undulations in Australia with Corresponding Values Derived from Geopotential Models Complete to Degree 180

Model	Mean Difference	Std Dev Difference	Number of Stations†
OSU78	-1.08 m	± 1.40	114
OSU81	-1.18	1.57	114
GEM10C	-0.96	2.06	114
GPM2	-0.96	1.51	114
OSU86C	-1.13	1.60	114
OSU86D	-1.12	1.60	114

† All stations in the area were used.

From Table 14 and 15 we see comparable accuracy results for GPM2 and the OSU86 solutions. In Europe the 86C solution performs best while in Australia the GPM2 performs best.

These Doppler undulation comparisons are helpful in revealing the relative strengths between solutions. One problem however is that the accuracy of the ellipsoidal height determination is on the order of 80 cm while the accuracy of the gravimetric undulation is on the order of ±1 m. In addition we have other errors related to coordinate system definition (including vertical datum definition and consistency), orthometric height accuracy, and sun spot correlations. With the improvement in the accuracy of our global models it will become more difficult to assess their accuracy with the Doppler comparison.

6.3.1 UNDULATION RESIDUAL CORRELATION WITH ELEVATION

Tscherning (1985, private communication) pointed out that the differences in the Doppler and gravity model undulations were correlated with elevation for the OSU81 solution but not for the GPM2 field. In our research we have computed this correlation for several models for various Doppler data sets by assuming there was a linear correlation with elevation. Results for the global data set and the stations in Europe are shown in Table 16 and 17.

Table 16
Undulation Residual Correlation Using the Global Station Set

Solution	Slope (m/km)	Std Dev. (m/km)	No. of Stations
OSU81	0.73	± 0.16	1721
GPM2	0.00	0.10	1735
OSU86C	0.41	0.12	1741
OSU86D	0.43	0.12	1743

Table 17
Undulation Residual Correlation Using European Station Set

Solution	Slope (m/km)	Std Dev. (m/km)	No. of Stations
OSU81	1.53	± 0.81	172
GPM2	1.87	0.63	172
OSU86C	0.92	0.65	172
OSU86D	1.14	0.69	172

From these results we see that the GPM2 model yields residuals with no elevation correlation for the global data set but significant correlation for the European stations. Tests to be described in a new paper by Rapp and Cruz will indicate the correlation in the non-GPM2 solution arises from the gravity field information in degrees 7 thru 14. Further tests are needed to better understand the apparent correlation and its significance.

6.4 CORRELATION WITH TOPOGRAPHY

Under certain assumptions the topography and its isostatic compensation can be considered to generate a portion of the earth's gravity field. It is appropriate to consider the correlation of the spherical harmonic representation of the topographic/isostatic potential and of the gravitational potential represented by various models. Such correlation, by degree, can be computed from equation (55).

The average correlation for several different models is shown in Table 18.

The topographic/isostatic expansion was basically made using the procedures described in Rapp (1981) where HARMIN was used for carrying out the spherical harmonic developments related to the topography.

Table 18

Average Correlation Coefficients Between Topography and Spherical Harmonic Models of the Earth's Gravity Field for Specified Degree Ranges

	Degree Range	
Model	30 to 180	50 to 180
OSU81	.55	.54
GEM10C	.29	.27
GPM2	.55	.54
OSU86C	.53	.52
OSU86D	.56	.55

Considering Table 18 we note 1) the GEM10C shows poor correlation with topography; 2) the model (86C) that does not use geophysical anomalies has slightly less correlation than the other models that use the geophysical anomalies and 3) the OSU81, OSU86C/D, and GPM2 all show about the same average correlation.

The poor correlation of GEM10C with the topography has been previously pointed out by Tscherning (1984). At the highest degrees (say 176-180) the correlation of GEM10C with topography is 0.1 or less. With the other solution this correlation is on the order of 0.5.

6.5 SATELLITE ORBIT FITS

The accuracy of the calculation of a satellite orbit depends on how well the gravity field of the earth is modeled. One way to test spherical harmonic models is to use them in calculating orbits and examining the observation residuals specifically the root mean square (RMS) residual. In principle the smaller the RMS residual the better the model. This principle is not exact because the satellite reacts, in some cases, to lumped (or resonant) harmonic coefficients.

The tests described in this section were carried out at the Goddard Space Flight Center of NASA. Various gravity models were used in orbit calculations for data in areas whose length ranged from 6 days to 30 days (for Lageos). The RMS residual for several potential coefficient models is given in Table 19.

Table 19

Root Mean Square Residual From Satellite Orbit Fits Using Selected Geopotential Models

	Potential Model			
Satellite	Standard*	86C	86D	GPM2
Starlette	± 47 cm	143	179	737 cm
Seasat	± 0.67 cm/s	1.39	1.46	1.76 cm/s
Oscar	± 1.52 cm/s	1.42	1.42	1.62 cm/s
BEC	± 75 cm	69	175	133 cm
Lageos	± 8 cm	8	8	16 cm
Lageos	± 7 cm	7	7	17 cm
Geos-2	± 134 cm	85	97	347 cm

* see text

The standard field used in Table 19 is a specific field that in some cases is a tailored geopotential model. Specifically the standard models are PGS1331 for Starlette; and GEM10B for Oscar, BEC, and Geos-2. The 86C/D and GPM2 models were used in the fits complete to degree and order 36.

7.0 CONCLUSIONS

This paper has been a review of the various techniques that may be used to combine satellite and terrestrial data to obtain high degree spherical harmonic expansions of the earth's gravitational field. It is clear that a number of different techniques exist each with its own advantages, disadvantages, and approximations. This paper has also included a discussion and comparison of various gravity models including two new OSU solutions. Although the new OSU models are complete to degree 250 the accuracy of the determination suggests that its use be restricted to degree 180 although the higher degree terms seem to be well behaved but probably too smooth. Information does exist in degrees 181 to 250 which should be carefully evaluated before neglecting.

Several problems are clear from the previous discussion that need additional study. For example:

1. How can the error correlation of surface data such as geoid undulations and gravity anomalies be taken into account?

2. What is the best way to combine data of different types (e.g. sea surface heights and terrestrial gravity anomalies) in a high degree solution?

3. What is the smoothing effect on the potential spectrum from the use of the optimal estimation (or least squares collocation procedure)?

4. What is the best way to carry out a high degree expansion using 30′x30′ mean values?

5. Should Doppler derived geoid undulations be incorporated into the geopotential solution and if so, what parameters should be modeled?

6. What is the best way to balance the needs for a gravity model near the earth's surface and in space?

7. Should geophysical anomalies be used at all in the solutions or should their use be restricted to locations where some independent verification is possible?

The answers to these and other similar questions await additional study.

ACKNOWLEDGEMENT

Portions of the research related to the development of the new OSU high degree expansions were supported by the Air Force Geophysics Laboratory under contract F19618-86-K-0016.

REFERENCES

Boucher, C. and Z. Altamimi, Towards an Improved Realization of the BIH Terrestrial Frame, in Proc. of the Int. Conf. on Earth Rotation and Terrestrial Reference Frame, Dept. of Geodetic Science and Surveying, The Ohio State University, Columbus, Vol. 2, 1985.

Colombo, O., Numerical Methods for Harmonic Analysis on the Sphere, Report No. 310, Dept. of Geodetic Science and Surveying, The Ohio State University, 1981.

Cruz, J., Disturbance Vector in Space From Surface Gravity Anomalies Using Complementary Models, Report No. 366, Dept. of Geodetic Science and Surveying, The Ohio State University, Columbus, August, 1985.

Desrochers, G., A Study of the Aliasing Effect on Gravitational Potential Coefficients as Determined from Gravity Data, Report No. 160, Dept. of Geodetic Science, The Ohio State University, Columbus, 1971.

Gleason, D., Partial Sums of Legendre Series via Clenshaw Summation, manuscripta geodaetica, Vol. 10, 115-130, 1985a.

Gleason, D., Using the Fast Fourier Transform and FORTRAN 77 in Gravimetric Spherical Harmonic Expansions, Term Paper, GS 871, Dept. of Geodetic Science and Surveying, The Ohio State University, Columbus, 1985b.

Hajela, D.P. Optimal Estimation of High Degree Gravity Field from a Global Set of 1°x1° Anomalies to Degree and Order 250, Report No. 358, Dept. of Geodetic Science and Surveying, The Ohio State University, 1984.

Jekeli, C., The Downward Continuation to the Earth's Surface of Truncated Spherical and Ellipsoidal Harmonic Series of the Gravity and Height Anomalies, Dept. of Geodetic Science and Surveying, The Ohio State University, Columbus, 1981.

Katsambalos, K., The Effect of the Smoothing Operator on Potential Coefficient Determinations, Report No. 287, Dept. of Geodetic Science, The Ohio State University, Columbus, 1979.

Lerch, F.J., S. Klosko, R. Laubscher, and C. Wagner, Gravity Model Improvement using GEOS-3 (GEM9 and 10), J. Geophys. Res. 84, No. B8, 3897-3916, 1979.

Lerch, F.J., F. Putney, S. Klosko and C. Wagner, Earth Models for Oceanographic Applications (GEM10B and 10C), Marine Geodesy, Vol. 5, 145-187, 1981.

Lerch, F.J., S. Klosko, and G. Patel, A Refined Gravity Model from Lageos (GEM-L2), Geophys. Res. Lett. 9, No. 11, 1263-1266, Nov 1982.

Lerch, F.J., Error Spectrum of Goddard Satellite Models for the Gravity Field, in Geodynamics Branch, Annual Report 1984, NASA Technical Memorandum 86223, 2-2 to 2-9, August 1985.

Meissl, P., A Study of Covariance Functions Related to the Earth's Disturbing Potential, Report 151, Dept. of Geodetic Science, The Ohio State University, Columbus, 1971.

Moritz, H., Advanced Pysical Geodesy, Abacus Press, Kent, U.K., 1980.

Paul, M.K., Recurrence Relations for Integrals of the Associated Legendre Functions, Bulletin Geodesique, Vol. 52, 177-190, 1978.

Pellinen, L.P., Effects of the Earth Ellipticity on Solving Geodetic Boundary Value Problems, in Proc. of Eighth Symposium on Mathematical Geodesy, held in Como, Italy, 1981; Florence, Italy, 1983.

Petrovskaya, M.S., and N.K. Lobkova, On Refinement of Relations Between Stokes' Constants and Surface Gravitational Data, manuscripta geodaetica, Vol. 7, 21-36, 1982.

Priovolos, G., Implementation of Programs Used by Hajela for the Optimal Estimation of High Degree Spherical Harmonic Expansions, unpublished report, Dept. of Geodetic Science and Surveying, The Ohio State University, Columbus, December 1985.

Rapp, R.H., Analytical and Numerical Differences Between Two Methods for the Combination of Gravimetric and Satellite Data, Bollettino di Geofisica Teorica ed Applicata, Vo. XI, No. 41-42, pp. 108-118, 1969.

Rapp, R.H., A Global 1°x1° Anomaly Field Combining Satellite, GEOS-3 Altimeter, and Terrestrial Data, Dept. of Geodetic Science, Report No. 278, The Ohio State University, Columbus, 1978.

Rapp, R.H., Potential Coefficient Determinations from Anomalies Given on a Bounding Sphere as Derived by Least Squares Collocation, Bollettino di Geodesia e Scienze Affini, No. 2-3, 1978.

Rapp, R.H., The Earth's Gravity Field to Degree and Order 180 Using SEASAT Altimeter Data, Terrestrial Gravity Data, and Other Data, Report No. 322, Dept. of Geodetic Science and Surveying, The Ohio State University, Columbus, Dec 1981.

Rapp, R.H., Degree Variances of the Earth's Potential, Topography and its Isostatic Compensation, Bulletin Geodesique, 56, 2, 84-94, 1982.

Rapp, R.H., Geoid Undulation Computations for Doppler Positioning Requirements, in Proc. of the 43rd Annual Meeting, American Congress on Surveying and Mapping, Washington DC, 1983.

Rapp, R.H., The Determination of High Degree Potential Coefficient Expansions from the Combination of Satellite and Terrestrial Gravity Information, Report No. 361, Dept. of Geodetic Science and Surveying, The Ohio State University, Columbus, 1984.

Rapp, R.H., Gravity Anomalies and Sea Surface Heights Derived from a Combined Geos3/Seasat Altimeter Data Set, J. Geophys. Res., 91, B5, 4867-4876, 1986.

Rapp, R.H. and J. Cruz, Development and Comparison of High Degree Spherical Harmonic Models of the Earth's Gravity Field, paper presented at the Symposium, "Definition of the Geoid", Florence, Italy, May 1986.

Reigber, C., G. Balmino, H. Müller, W. Bosch, and B. Moynat, GRIM Gravity Model Improvement Using LAGEOS (GRIM3-L1), J. Geophys. Res., Vol. 90, B11, Sept. 1985.

Rizos, C., An Efficient Computer Technique for the Evaluation of Geopotential from Spherical Harmonic Models, Avst. J. Geodesy, Photogrammetry, and Surveying, No. 31, 161-169, December, 1979.

Rummel, R. and R.H. Rapp, The Influence of the Atmosphere on Geoid and Potential Coefficient Determinations From Gravity Data,, J. Geophys. Res., 81, 5639-5642, 1976.

Singh, A., On Numerical Evaluation of Normalized Associated Legendre Functions, unpublished report, Dept. of Geodetic Science and Surveying, The Ohio State University, Columbus, 1982.

Singh, A., Some Problems of Numerical Evaluation of Normalized Associated Legendre Functions Through a Digital Computer, M.S. thesis, Dept. of Mathematics, The Ohio State University, Columbus, 1984.

Sjöberg, L., A Recurrance Relation for the β_n Function, Bulletin Geodeseque, Vol. 54, No. 1, 69 1980.

Sjöberg, L., Least Squares Combination of Satellite and Terrestrial Data in Physical Geodesy, Annales de Geophysique, Vol. 37, 25-30, 1981.

Sjöberg, L., On the Convergence Problem for the Spherical Harmonic Expansion of the Geopotential at the Surface of the Earth, Bollettino di Geodesia e Scienze Affini, No. 3, 1980.

Tscherning, C.C., The Role of High-Degree Spherical Harmonic Expansions in Solving Geodetic Problems, in Proc. of the IAG Symposia (1983, IUGG), Dept. of Geodetic Science and Surveying, The Ohio State University, 1983.

Tscherning, C.C., R.H. Rapp, and C. Goad, A Comparison of Methods for Computing Gravimetric Quantities from High Degree Spherical Harmonic Expansion, manuscripta geodaetica, Vol. 8, 249-27i2, 1983.

Tscherning, C.C., On the Long-Wavelength Correlation Between Gravity and Topography, in Proc. of 5th Int. Symp, "Geodesy and Physics of the Earth", Vol. 2, 134-143, Zentralinstitut Für Physik Der Erde, Potsdam, 1984.

Tscherning, C.C. and C. Goad, Correlation Between Time Dependent Variations of Doppler - Determined Height and Sunspot Numbers, J. Geophys. Res. 90, B6, 4589-4596, 1985.

Weber, G. and H.-G. Wenzel, Estimation of Error Properties, in Validation of SEASAT-1 Altimetry Using Ground Truth in the North Sea Region, Deutsche Geodatische Kommission, Reihe B: No. 263, Frankfurt An Main, 1982.

Wenzel, H-G., Hochaufloesende Kugelkenktionsmodelle fuer das Gravitationspotential der Erde. Wiss. Arb. Fachrichtung Vermessungswesen der Universitaet Hannover 1985.

GLOBAL TOPOGRAPHIC-ISOSTATIC MODELS

by

H. Sünkel

Institute of Theoretical Geodesy
Division of Mathematical Geodesy
Technical University Graz
Rechbauerstraße 12
A-8010 Graz, Austria

Lecture Notes in Earth Sciences, Vol. 7
Mathematical and Numerical Techniques in Physical Geodesy
Edited by H. Sünkel

INTRODUCTION

According to Newton's law of gravitation, the source of the earth's gravitational field is the mass distribution within the earth's surface (disregarding the attraction due to extra-terrestrial bodies). Consequently, would we know the mass density distribution of the entire earth's body, the gravity field were known both inside and outside the surface of the earth. Since we do not have direct access to that kind of complete source information, we are forced to measure its effect in terms of gravity field quantities. Vice versa, if the surface of the earth would be completely covered by gravity field data, we could, according to potential theory, determine the external gravity field of the earth without any knowledge about its underlying density distribution.

Density data is source information, gravity field data is effect information. These two data types don't compete with each other, they do rather supplement each other. In geodesy, density data is used to smooth the gravity field with the goal to obtain a reduced gravity field with little power and which is statistically homogeneous and isotropic as much as possible.

It seems to be a generally valid law of nature that near events affect us more than remote events. This is true for human life, for politics, natural and man-made disasters to mention a few, and also in the density distribution - to - gravity field relation: the geometrical and physical structure of the earth's crust controls the local irregularities of the gravity field which are hard to catch without sufficiently dense local effect information which, in turn, is expensive to obtain and often entirely unaccessible. Therefore, the best possible use of available source information in terms of digital terrain / digital density models offers itself.

In the sequel the focus is put on exactly the earth's crust, the determination of its most likely shape and its use for the purpose of gravity field smoothing on a local to regional scale.

1. AN ISOSTATIC EARTH MODEL

The theory of isostasy has a well-documented history of more than a hundred years. Its fundamentals are even considerably older and can be traced back to Archimedes. From geodetic evidences some kind of isostatic equilibrium had to be postulated. Although in some limited areas the Pratt/Hayford system seemed to prevail, the Airy/Heiskanen system is now generally believed to model the complex reality better.

The geodetic interest in isostasy was considerably fueled both by the access to high speed computers and the release of a worldwide digital terrain model. Harmonic coefficients of the topographic-isostatic potential have been calculated by several colleagues up to degree and order 36, based on $5^{\circ} \times 5^{\circ}$ mean topographic information, both in linear approximation (Khan, 1973) and in the non-linear mode (Lachapelle, 1975). At that time much higher resolutions could not be obtained because of missing data, but in particular because of excessive computer time requirements. With the design of the fast Fourier algorithm on the sphere by Colombo (1981) and the public release of his programs, high resolution models became suddenly feasible and have been computed complete up to degree 180 by Rapp (1982a) based on the worldwide digital topographic model of 64800 $1^{\circ} \times 1^{\circ}$ mean elevations provided by DMAAC in 1979 and using the Airy/Heiskanen model in linear approximation. Rapp claimed that correlation studies between his geopotential solution of 1981 and the topographic-isostatic potential suggest a considerably deeper level of compensation of about 50 km rather than the usually used value of 30 km. Rapp's linear approximation has been extended recently, taking into account also second and third order terms, by Rummel (private communication).

Tscherning (1985) compared existing solutions, argued against Rapp's 1982 solution because of the too large and therefore, hard to sell depth of compensation, and suggested the use of much smaller compensation depths in the range between 15 and 35 km which he found "optimal" in terms of maximally reducing the power of the observed geopotential.

On the occasion of a research leave at the Department of Geodetic Science the author had the opportunity to investigate that problem from scratch. The goal was to estimate the parameter(s) of the most likely compensation model which is in best possible agreement with the observed geopotential and also geophysically acceptable, and the calculation of the harmonic coefficients of the topographic-isostatic potential complete up to degree and order 180, based on that model (Sünkel, 1985).

Recently a new global $1^{o} \times 1^{o}$ digital terrain model has been established at our Institute at the Technical University Graz, based on the above referenced DMAAC DTM for land areas and the SYNBAPS 5'x5' DTM, released by the National Geophysical Data Center of NOAA, Boulder, Colorado. This model, referred to as TUG 86 - DTM has been used recently for a recalculation of the harmonic coefficients of the topographic-isostatic potential complete up to degree and order 180, employing a least-squares collocation estimate of the two parameters of the Vening Meinesz compensation model (compensation depth, smoothing parameter); it is referred to as TIC 86. This new model matches the OSU 81 gravity field much better than the TIC 85 model, indicated in the frequency domain by the smoothing coefficients per degree, but also by comparing it to observed mean gravity anomalies.

2. HARMONIC ANALYSIS OF THE TOPOGRAPHIC-ISOSTATIC POTENTIAL

The topographic-isostatic potential of Airy/Heiskanen type T_{TI} is defined as the potential of all mass disturbances relative to an ideal crustal layer of uniform density ρ_0 and thickness D, superimposed upon underlying material of equally uniform density, ρ_1. Denoting the mass disturbance by $\delta\rho$, the potential T_{TI} is well-known to be given by

$$T_{TI}(P) = G \iiint_V \ell^{-1}(P,Q)\, \delta\rho\,(Q)\, dv(Q) \tag{2-1}$$

with G ... Newton's gravitational constant,
$\ell(P,Q)$ ... space distance between P and Q,
dv ... volume element.

T_{TI} is certainly harmonic outside the earth's surface and T_{TI}'s spherical harmonic series is certainly convergent outside a sphere completely enclosing the earth. Outside that sphere we may use the convergent series representation of ℓ^{-1},

$$\ell^{-1}(P,Q) = \sum_{n=o}^{\infty} \frac{r_Q^n}{r_P^{n+1}} P_n(\cos\psi_{PQ}) \qquad (2\text{-}2)$$

with r ... modulus of the radius vector,
ψ_{PQ} ... spherical distance between P and Q,
P_n ... Legendre polynomial of degree n

(Heiskanen & Moritz, 1967 (HM), p.33). Decomposing P_n in terms of

$$P_n(\cos\psi_{PQ}) = \frac{1}{2n+1} \sum_{m=-n}^{n} \bar{R}_{nm}(P)\,\bar{R}_{nm}(Q) \qquad (2\text{-}3)$$

with the fully normalized spherical harmonics $\bar{R}_{nm}$,

$$\bar{R}_{nm}(P) = \sqrt{2^{1-\delta_{mo}}(2n+1)\frac{(n-m)!}{(n+m)!}}\, P_{nm}(\cos\theta_P) \begin{cases} \cos m\lambda_P & \text{for } m \leq 0 \\ \sin m\lambda_P & \text{for } m > 0 \end{cases} \qquad (2\text{-}4)$$

(δ_{ij} ... Kronecker symbol, P_{nm} ... Legendre function, θ , λ ... polar distance, longitude), the topographic-isostatic potential is represented by the harmonic series

$$T_{TI}(P) = G\sum_{n=o}^{\infty} r_P^{-(n+1)} \frac{1}{2n+1} \sum_{m=-n}^{n} \bar{R}_{nm}(P) \iiint_V \delta\rho(Q) r^n(Q) \bar{R}_{nm}(Q)\, dv(Q) \qquad (2\text{-}5)$$

with $dv(Q) = r^2(Q)\, dr(Q)\, d\sigma(Q)$
and the spherical surface element $d\sigma$.

Let us now investigate the volume integral, confining ourselves to the usual spherical approximation. Then its contribution due to the topographical masses outside the geoid and ocean water inside the geoid is

$$\iint_\sigma \int_{r=R}^{R+H} \delta\rho(Q) r^{n+2}(Q)\, dr(Q)\, \bar{R}_{nm}(Q)\, d\sigma(Q), \qquad (2\text{-}6a)$$

and its contribution due to the isostatically compensating masses is

$$\iint_\sigma \int_{r=R-D+kH}^{R-D} \delta\rho(Q) r^{n+2}(Q)\, dr(Q)\, \bar{R}_{nm}(Q)\, d\sigma(Q), \qquad (2\text{-}6b)$$

with R ... mean earth radius,
D ... depth of compensation level,
H ... topographical height (+),
ocean bottom height (-),
k ... compensation "factor".

(The compensation "factor" k, which is in the non-planar case actually a function dependent on H and D , will be discussed later.) In (2-6a) the function

$\delta\rho = \rho_o =$ constant for topographic masses,
$\delta\rho = \rho_w - \rho_o =$ constant for ocean masses,

in (2-6b)

$\delta\rho = \rho_o - \rho_1 =$ constant for compensating masses

is used (ρ_w = density of ocean water). The integration of (2-6a,b) with respect to r is straightforward and yields

$$\frac{R^{n+3}}{n+3} \iint_\sigma \delta\rho\,(Q) \left[\left(1 + \frac{H(Q)}{R}\right)^{n+3} - 1\right] \bar{R}_{nm}(Q)d\sigma(Q) \qquad (2\text{-}6a)'$$

and

$$\frac{R^{n+3}}{n+3} \left(1 - \frac{D}{R}\right)^{n+3} \iint_\sigma \delta\rho(Q)\left[\left(1 + \frac{kH(Q)}{R-D}\right)^{n+3} - 1\right] \bar{R}_{nm}(Q)d\sigma(Q), \qquad (2\text{-}6b)'$$

respectively.

The compensation "factor" k

This concept of isostasy is based on the principle of mass balance: surplus + deficit = 0. In the simple Airy/Heiskanen isostatic model this enables us to derive the geometry of the compensation surface (roots/antiroots) such that an exact 1:1 relation is postulated between a topographic height H and the corresponding compensation height kH. This mass balance principle, applied to rock topography and its compensation, yields

$$\rho_o\, R^3 \left[\left(1 + \frac{H}{R}\right)^3 - 1\right] + (\rho_o - \rho_1)(R-D)^3 \left[1 - \left(1 + \frac{kH}{R-D}\right)^3\right] = 0, \qquad (2\text{-}7)$$

and after elementary manipulations

$$\frac{kH}{R-D} = \left\{1 - \left(1 - \frac{D}{R}\right)^{-3} k_o \left[\left(1 + \frac{H}{R}\right)^3 - 1\right]\right\}^{\frac{1}{3}} - 1 \quad \ldots \text{ rock;} \tag{2-7a}$$

for the ocean and its compensation we obtain a similar expression,

$$\frac{kH}{R-D} = \left\{1 - \left(1 - \frac{D}{R}\right)^{-3} k_o \left(1 - \frac{\rho_w}{\rho_o}\right) \left[\left(1 + \frac{H}{R}\right)^3 - 1\right]\right\}^{\frac{1}{3}} - 1 \quad \ldots \text{ ocean} \tag{2-7b}$$

with $k_o := \dfrac{\rho_o}{\rho_1 - \rho_o}$.

It is instructive to investigate the planar case. For this purpose we evaluate kH in a power series with respect to H,

$$kH = -\left(1 - \frac{D}{R}\right)^{-2} k_o H + 0\left(\frac{H^2}{R}\right),$$

and if we let the radius R become infinite, we obtain

$$k = -k_o c_H \tag{2-8}$$

with $c_H = 1$ for $H \geq 0$ and $c_H = 1 - \frac{\rho_w}{\rho_o}$ for $H < 0$.

It is obvious that the compensation factor k is constant in the planar model and independent of both H and D. For standard densities $\rho_o = 2.67$, $\rho_1 = 3.27$, $\rho_w = 1.03$ gcm^{-3} , we obtain the familiar values of -4.45 and -2.73 for rock and ocean compensation, respectively (HM p.136). In the spherical model the compensation "factor" k is no longer a constant; it is a function which depends to a small extent on both the height H and the depth of the compensation level D according to (2-7a,b) with (2-8) as the leading term. In order to simplify, we introduce "normalized" heights $\bar{H}$ and $\bar{H}'$ according to

$$\bar{H} := \frac{H}{R} \tag{2-9}$$

$$\bar{H}' := \frac{kH}{R-D}$$

and the earth's mass M by

$$M \doteq \frac{4\pi R^3}{3} \rho_M \tag{2-10}$$

with the mean earth density $\rho_M = 5.517\ \mathrm{gcm}^{-3}$. With this notation, the topographic-isostatic potential in its harmonic series representation is given by

$$\boxed{T_{TI}(P) = \frac{GM}{r_P} \sum_{n=1}^{\infty} \left(\frac{R}{r_P}\right)^n \sum_{m=-n}^{n} \bar{T}_{nm}\, \bar{R}_{nm}(P)} \tag{2-11}$$

with the harmonic (Fourier) coefficients of the topographic-isostatic potential

$$\bar{T}_{nm} := \bar{T}_{nm}^{(t)} + \bar{T}_{nm}^{(c)} , \tag{2-12a}$$

$$\bar{T}_{nm}^{(t)} := \frac{3\rho_o}{\rho_M} \frac{1}{(2n+1)(n+3)} \frac{1}{4\pi} \iint_\sigma c_H \left[\left(1+\bar{H}(Q)\right)^{n+3} - 1\right] \bar{R}_{nm}(Q)\, d\sigma(Q), \tag{2-12b}$$

$$\bar{T}_{nm}^{(c)} := \frac{3\rho_o}{\rho_M} \frac{1}{(2n+1)(n+3)} \left(1-\frac{D}{R}\right)^{n+3} k_o^{-1} \frac{1}{4\pi} \iint_\sigma \left[\left(1+\bar{H}'(Q)\right)^{n+3} - 1\right] \bar{R}_{nm}(Q)\, d\sigma(Q) \tag{2-12c}$$

(Here "t" stands for topography and "c" for its compensation.) Note that the summation in (2-11) starts with n = 1 due to the mass equality condition; the reader is invited to check that $T_{oo}^{(t)} + T_{oo}^{(c)}$ is indeed equal to zero (hint: use (2-7a,b)).

3. APPROXIMATIONS FOR THE HARMONIC COEFFICIENTS OF THE TOPOGRAPHIC-ISOSTATIC POTENTIAL

3.1 Linear Approximation

Given a worldwide digital terrain model (DTM) representing the geometry of the solid earth surface, H(Q), standard numerical integration techniques have been used in the past (Khan , 1973 ; Lachapelle, 1975). Due to the excessive computation requirements, only low degree coefficients (up to degree and order 36) using a 5° x 5° DTM, could be determined. A couple of years ago, Colombo developed a fast Fourier algorithm for the sphere which can be efficiently applied if the integrand is independent of any degree n. Unfortunately, this is not the case with

(2-12b,c). However, due to the smallness of $\bar{H}$ and $\bar{H}'$ compared to 1, a linear approximation seemed to be justified, yielding integrands

$$\frac{1}{n+3}\left[(1+\bar{H})^{n+3}-1\right] \doteq \bar{H},$$
$$\frac{1}{n+3}\left[(1+\bar{H}')^{n+3}-1\right] \doteq \bar{H}', \tag{3-1}$$

which are obviously independent of the degree n. Therefore, in the linear approximation, the FFT algorithm applied to $\bar{H}$ and $\bar{H}'$, respectively, can be and has been used (Rapp, 1982), providing the Fourier coefficients

$$\frac{1}{4\pi}\iint_\sigma c_H\bar{H}(Q)\bar{R}_{nm}(Q)d\sigma(Q), \tag{3-2}$$
$$\frac{1}{4\pi}\iint_\sigma \bar{H}'(Q)\bar{R}_{nm}(Q)d\sigma(Q),$$

in a very efficient way.

In this linear approximation case, the product $c_H H$ is usually called "equivalent rock topography" which, loosely speaking, trades in rock for water. The products $\rho_o c_H H$ and $\rho_o k_o^{-1} H'$ can consequently be interpreted as surface layer densities at zero level and at the level of compensation. Therefore, in the linear approximation, the topographic-isostatic potential is represented in terms of a double layer potential.

The linear approximation of (2-12b) yields

$$\bar{T}^{(t)}_{nm} \doteq \frac{3\rho_o}{R\rho_M}\,\frac{1}{(2n+1)}\,\frac{1}{4\pi}\iint_\sigma c_H H(Q)\bar{R}_{nm}(Q)d\sigma(Q); \tag{3-3a}$$

the linear term of (2-7b) is

$$\bar{H}' \doteq -\left(1-\frac{D}{R}\right)^{-3} k_o c_H \bar{H}$$

yielding

$$\bar{T}^{(c)}_{nm} \doteq -\frac{3\rho_o}{R\rho_M}\,\frac{1}{(2n+1)}\left(1-\frac{D}{R}\right)^n \frac{1}{4\pi}\iint_\sigma c_H H(Q)\bar{R}_{nm}(Q)d\sigma(Q). \tag{3-3b}$$

Adding $\bar{T}_{nm}^{(t)}$ and $\bar{T}_{nm}^{(c)}$ according to (2-12a) the harmonic series of the topographic-isostatic potential's linear approximation is given by

$$T_{TI}(P) \doteq \frac{4\pi G R \rho_0}{r_P} \sum_{n=1}^{\infty} \left(\frac{R}{r_P}\right)^n \frac{1}{2n+1} \left[1 - \left(1 - \frac{D}{R}\right)^n\right]$$

$$\cdot \sum_{m=-n}^{n} \bar{R}_{nm}(P) \frac{1}{4\pi} \iint_\sigma c_H H(Q) \bar{R}_{nm}(Q) d\sigma(Q) \qquad (3\text{-}4)$$

Exactly the same expression can be obtained if rock/ocean and compensation masses are considered as single layers at zero level and at the level of compensation, respectively.

Equation (3-4) is well suited to discuss the two extreme cases of $D = 0$ and $D = R$. It is obvious that for $D = 0$ the topographic-isostatic potential in its linear approximation vanishes identically. For the second case, $D = R$, the topographic-isostatic potential reduces to the topographic potential with vanishing zero degree harmonic (due to the mass balance). Neither of these two cases is realistic, as we know from geodetic and geophysical evidence. If we consider "reasonable" values of $0 < D \ll R$, we may (at least for low degrees) approximate

$$\left(1 - \frac{D}{R}\right)^n \doteq 1 - n\frac{D}{R}; \qquad (3\text{-}5)$$

we observe that the main terms in (3-4) cancel each other and obtain the approximation

$$T_{TI}(P) \doteq \frac{4\pi G D R \rho_0}{r_P} \sum_{n=1}^{\infty} \left(\frac{R}{r_P}\right)^n \frac{n}{2n+1} \sum_{m=-n}^{n} \bar{R}_{nm}(P) \frac{1}{4\pi} \iint_\sigma c_H H(Q) \bar{R}_{nm}(Q) d\sigma(Q) \qquad (3\text{-}6)$$

which can be further simplified with

$$\frac{n}{2n+1} \doteq \frac{1}{2} \qquad (3\text{-}7)$$

and the harmonic representation of the equivalent rock topography,

$$c_H H(P) = \sum_n \sum_m \bar{R}_{nm}(P) \frac{1}{4\pi} \iint_\sigma c_H H(Q) \bar{R}_{nm}(Q) d\sigma(Q), \qquad (3\text{-}8)$$

we find at zero level

$$T_{TI}(P) \doteq 2\pi \, GD\rho_o c_H H(P) \tag{3-9}$$

(cf. also HM, p. 149). This approximative formula is a very useful rule of thumb which provides a fairly good estimate for a standard Airy/Heiskanen isostatic model. From this simple expression we conclude that the topographic-isostatic potential is approximately linearly dependent on the depth of compensation and also linearly on the height of the equivalent rock topography.

3.2 Higher Order Approximations

Rapp's (1982a) argument that the efficient FFT algorithm is only applicable in the linear approximation, which we have shown is the double layer formulation, is no longer valid if we use the binomial expansion of the left-hand side of (3-1),

$$(1 + \bar{H})^{n+3} - 1 = \sum_{j=1}^{n+3} \binom{n+3}{j} \bar{H}^j \tag{3-10}$$

(analogous for $\bar{H}'$), yielding exact expressions (within spherical approximation) of the harmonic coefficients of the rock/ocean topography and its isostatic compensation:

$$\bar{T}^{(t)}_{nm} = \frac{3\rho_c}{\rho_M} \frac{1}{(2n+1)(n+3)} \sum_{j=1}^{n+3} \binom{n+3}{j} \frac{1}{4\pi} \iint_\sigma c_H \bar{H}^j(Q) \bar{R}_{nm}(Q) d\sigma(Q) \tag{3-11a}$$

$$\bar{T}^{(c)}_{nm} = \frac{3\rho_o}{\rho_M} \frac{1}{(2n+1)(n+3)} \left(1 - \frac{D}{R}\right)^{n+3} k_o^{-1} \sum_{j=1}^{n+3} \binom{n+3}{j} \frac{1}{4\pi} \iint_\sigma c_H \bar{H}'^j(Q) \bar{R}_{nm}(Q) d\sigma(Q) \tag{3-11b}$$

There is no reason why FFT could not be applied to any power of $\bar{H}$ and $\bar{H}'$, respectively. (Note that (3-11a,b) are exact.) But do we gain anything by employing FFT, considering the fact that 2(n+3) FFT's have to be performed (one FFT for each power of $\bar{H}$ and $\bar{H}'$), compared to twice a standard numerical integration? In the case of our earth we know that $|\bar{H}|, |\bar{H}'| \ll 1$; therefore, (3-10) converges very quickly and we can safely terminate the binomial expansion at a very low degree without committing a significant error (an upper limit of power $j = 4$ is perfectly sufficient). Consequently, only a small number of FFT's must be calculated; we typically save a tremendous amount of computer time without sacrificing accuracy, and can arbitrarily improve the result by adding another power of $\bar{H}$ and $\bar{H}'$ if we so desire.

3.3 The Second Order Effect

By the second order effect we understand the contribution of the second power of $\bar{H}$ and $\bar{H}'$ to the topographic-isostatic potential. Because of the rapid convergence of (3-10), this second order effect represents with very good approximation the difference between the actual topographic-isostatic potential and its double layer approximation. Is this difference sufficiently small to be safely neglected?

Actually, there are two kinds of second order effects: one is due to the second power of $\bar{H}$ and $\bar{H}'$ in the binomial expansion (3-10), the other is due to the second power of $\bar{H}$ in the compensation function $k = k(H)$ of (2-7a,b).

The second order effect due to rock/ocean topography is straightforward: using the binomial expansion (3-10) in (2-12b) and observing the binomial coefficient

$$\frac{1}{n+3}\binom{n+3}{2} = \frac{n+2}{2},$$

we obtain

$$\delta\bar{T}_{nm}^{(t)} = \frac{3\rho_o}{2\rho_M}\,\frac{n+2}{2n+1}\,\frac{1}{4\pi}\iint_\sigma c_H \bar{H}^2(Q)\bar{R}_{nm}(Q)d\sigma(Q). \tag{3-12a}$$

For the compensation part we develop (2-7a,b) in a series up to the second power of H,

$$\left(1-\frac{D}{R}\right)^3\bar{H}' = -k_o c_H\bar{H}(1+\bar{H}) - k_o^2 c_H^2\bar{H}^2\left(1-\frac{D}{R}\right)^{-3} + O(\bar{H}^3),$$

and using again the binomial expansion (3-10), we get

$$\frac{n+2}{2}\left(1-\frac{D}{R}\right)^3\bar{H}'^2 = \left(\frac{n}{2}+1\right)k_o^2 c_H^2\bar{H}^2\left(1-\frac{D}{R}\right)^{-3} + O(\bar{H}^3)$$

and obtain as second order contribution of the isostatic compensation

$$\delta\bar{T}_{nm}^{(c)} = \frac{3\rho_o}{\rho_M}\,\frac{1}{2n+1}\left(1-\frac{D}{R}\right)^n\frac{1}{4\pi}\iint_\sigma c_H\bar{H}^2\left[\frac{n}{2}\,k_o c_H\left(1-\frac{D}{R}\right)^{-3} - 1\right]\bar{R}_{nm}(Q)d\sigma(Q). \tag{3-12b}$$

Synthesizing $T_{nm}^{(t)}$ and $\bar{T}_{nm}^{(c)}$ according to (2-11), we obtain the second order effect of the topographic-isostatic potential,

$$\delta T_{TI}(P) = \frac{GM}{r_P} \sum_{n=1}^{\infty} \left(\frac{R}{r_P}\right)^n \sum_{m=-n}^{n} \delta \bar{T}_{nm} \bar{R}_{nm}(P) \tag{3-13}$$

with the harmonic coefficients

$$\delta \bar{T}_{nm} := \delta T_{nm}^{(t)} + \delta T_{nm}^{(c)}$$

$$= \frac{3\rho_0}{\rho_M} \frac{1}{2n+1} \frac{1}{4\pi} \iint_\sigma c_H \bar{H}^2 \left\{ \left(\frac{n}{2}+1\right) + \left(1 - \frac{D}{R}\right)^n \left[\frac{n}{2} k_0 c_H \left(1 - \frac{D}{\bar{R}}\right)^{-3} - 1\right] \right\} \bar{R}_{nm} \, d\sigma . \tag{3-14}$$

Using the approximations (3-5) and (3-7), this odd-looking expression can be considerably simplified if we admit an error on the order of 20%, which can be accepted since the second order effect is small compared to the first order effect, yielding

$$\delta \bar{T}_{nm} \doteq \frac{3\rho_0}{4\rho_M} k_0 \frac{1}{4\pi} \iint_\sigma \left[c_H \bar{H}(Q)\right]^2 \bar{R}_{nm}(Q) d\sigma(Q) . \tag{3-14'}$$

With (3-13) we then obtain at zero level

$$\delta T_{TI}(P) \doteq \frac{GM}{4R^3} \frac{3\rho_0}{\rho_M} k_0 \sum_{n=1}^{\infty} \sum_{m=-n}^{n} \bar{R}_{nm}(P) \frac{1}{4\pi} \iint_\sigma \left[c_H H(Q)\right]^2 \bar{R}_{nm}(Q) d\sigma(Q)$$

and replacing M by (2-10), the approximated second order effect on the topographic-isostatic potential is simply

$$\boxed{\delta T_{TI}(P) \doteq \pi G \rho_0 k_0 \left[c_H H(P)\right]^2} \tag{3-15}$$

We observe that under these approximations the second order effect is independent of the depth of compensation, is proportional to the square of the equivalent rock topography's height, and is therefore always positive.

If we use the standard values for ρ_0, ρ_1, and ρ_w, we obtain a rule of thumb formula for the second order effect on the topographic-isostatic geoidal height, using the Bruns formula,

$$\delta N_{TI\,[m]} \doteq 0.2 \, H^{*2}_{[km]} \tag{3-16}$$

with $H^* := c_H H$, the equivalent rock topography. Note the formal similarity to the Molodensky effect (HM, p.328)!

Combining first and second order effects we obtain

$$T_{TI}(P) \doteq 2\pi G D \rho_o H^* \left(1 + k_o \frac{H^*}{2D} \right) . \tag{3-17}$$

from (3-17) we conclude that the second order effect, relative to the total potential, increases with decreasing depth of compensation and vice versa. For areas with extreme topography (both positive and negative), the second order effect may amount to about 50% of the first order effect and can certainly not be neglected.

4. POWER SPECTRUM CONSIDERATIONS

The power spectrum (degree variances) of the topographic-isostatic potential in linear approximation is implied by the harmonic coefficients $\bar{T}^{(t)}_{nm}$ and $\bar{T}^{(c)}_{nm}$ of (3-3a,b),

$$\bar{T}^2_n := \sum_{m=-n}^{n} (\bar{T}^{(t)}_{nm} + \bar{T}^{(c)}_{nm})^2 . \tag{4-1}$$

Denoting the harmonic coefficients of the normalized equivalent rock topography by $\bar{H}^*_{nm}$,

$$\bar{H}^*_{nm} := \frac{1}{4\pi} \iint_\sigma c_H \frac{H(Q)}{R} \bar{R}_{nm}(Q) d\sigma(Q), \tag{4-2}$$

the degree variances of the topographic-isostatic potential $\bar{T}^2_n$ are related to the degree variances of the equivalent rock topography $\bar{H}^{*2}_n$,

$$\bar{H}^{*2}_n := \sum_{m=-n}^{n} \bar{H}^{*2}_{nm} \tag{4-3}$$

by

$$\bar{T}^2_n = \left\{ \frac{3\rho_o}{\rho_M} \frac{1}{2n+1} \left[1 - \left(1 - \frac{D}{R} \right)^n \right] \right\}^2 \bar{H}^{*2}_n \tag{4-4}$$

(cf. Lambeck, 1979, p.592).

Considering $\bar{T}^2_n$ as a function of the compensation depth D, it is obvious that the power spectrum of the topographic-isostatic potential is gaining power with increasing compensation depth and attains a maximum for D = R such that in this extreme case

$\bar{T}_n^2$ reduces to the degree variance of the potential of the topography for $n > 0$ ($\bar{T}_n^2 = 0$ in any case). For moderately large depths and low degrees n we may use the approximation (3-5) and obtain as ratio between topographic-isostatic degree variances, corresponding to the compensation depths D_1 and D_2,

$$\frac{\bar{T}_n^2 (D_1)}{\bar{T}_n^2 (D_2)} \doteq \frac{D_1^2}{D_2^2} \quad . \tag{4-5}$$

From geodetic satellite and surface data we know that the actual anomalous gravitational potential of the earth has much less power than $\bar{T}_n^2$ $(D = R)$, implying that the rock/ocean topography is isostatically compensated to a large extent with a relatively small D. A global standard value of 30 km is generally accepted and frequently used for the purpose of topographic-isostatic reduction of surface data.

Assuming that the earth's anomalous gravitational potential is to a large extent due to the topographic masses and its isostatic compensation at a compensation depth D, we should expect a good agreement between the power spectrum of the earth's anomalous gravitational potential $\bar{V}_n^2$ and the power spectrum $\bar{T}_n^2$ (D), provided that the standard Airy/Heiskanen model describes reality sufficiently well.

This idea was pursued by Rapp (1982a), using the simple double layer model (linear approximation) and the DTM data set consisting of 64800 1° x 1° mean elevation and ocean depth data, provided by DMAAC. His results demonstrate that the standard Airy/Heiskanen compensation model with the standard compensation depth of D = 30 km produces much too little power over the entire frequency range from n = 2 to 180 and is inadequate to explain the observed power spectrum of the anomalous gravitational potential. Later investigations by Rapp and Rummel (private communication), using in addition the terms of second and third order in $\bar{H}$, confirmed the earlier studies.

Guided by the relation (4-4) or (4-5), considerably larger values for D were suggested by Rapp (1982a). Obviously the best agreement between the two power spectra was achieved for a compensation depth of D = 50 km yielding a good match between degree 50 and 150; poor results have been obtained for the range between 20 and 50, but what is particularly astonishing, also in the high frequency range between degree 150 and 180. Moreover, a compensation depth of 50 km is generally considered to be too large by a factor of almost 2 and is geophysically hard to justify.

The South Pacific area has not only a strong appeal to financially sound vacationers but also to hard working geodesists (don't be misled by this coincidence!) because of extreme gravity field structures. Forsberg (1984) studied the topographic-isostatic geoid in the Tonga Trench and Tahiti area using the recently released 5' x 5' "SYNBAPS" ocean depth data set. (I strongly recommend to study Forsberg's report - it is a beautiful document!) His results, which are also based on the standard Airy/Heiskanen compensation model, also suggest much deeper compensation levels than 30 km. Forsberg argues that at smaller depths of 30 km or so conventional isostasy is led ad absurdum by deep ocean trenches which would imply antiroots ending considerably above the actual ocean bottom(!) - an argument in favor of larger compensation depths. On the other hand, the observed average thickness of the crust below the ocean bottom is only about 6-8 km - a fact which would favor a compensation depth smaller than 30 km.

Isn't there a simple way out of that isostatic dilemma?

4.1 Compensation Depth versus Smoothing

The recent studies of Rapp (1982a) and Forsberg (1984), but also an earlier work by Moritz (1968) and Schwarz (1976) have triggered my interest in this problem; but in particular it was the idea of Vening Meinesz (1939) that the actual compensation takes place regionally rather than strictly locally, which made me to numerically investigate the problem of isostasy. But before we shall hit the target in Chapter 5, let's make a small sidestep to collocation.

During the collocation "high noon" period in the mid-seventies, one of the main issues was how to deal with block mean gravity anomalies within a homogeneous and isotropic statistical model environment. Numerical integration of the covariance function was prohibitive; therefore, an approximation had to be employed; through formally replacing the block averaging operator by a homogeneous and isotropic moving average operator of constant weight, the comfort of homogeneity and isotropy can be achieved. According to the convolution theorem, the moving average convolution process in the space domain corresponds to a simple product between the eigenvalues of the moving average operator and the Fourier coefficients of the function to be averaged in the frequency domain. As far as collocation is concerned, this particular function is the kernel function (in least-squares collocation the covariance function) which, smoothed and unsmoothed, has to be represented by a finite and as simple as possible expression in order to render possible a fast calculation of linear functionals. Unfortunately, the eigenvalues of the moving average operator prevent such a closed expression and therefore, they have to be approximated by another spectrum which behaves properly. Schwarz (1976, p. 37 ff.) suggested to replace the moving average eigenvalues with the eigenvalues of an upward continuation operator. Applying this trick means that the statistical properties of mean values at zero level are replaced by the statistical properties of point values at a certain altitude which depends on the block size of the moving average operator. Numerical tests confirmed that this approximation was admissible.

Have we lost our track? No, we have not! The same idea can be carried over to get, at least partially, out of the isostatic dilemma: assuming that the Vening Meinesz concept of regional isostatic compensation is correct, we my approximate the regional isostatic concept of Vening Meinesz at compensation depth D_1 by the strictly local isostatic concept of Airy/Heiskanen at a larger depth D_2, where $(D_2 - D_1)$ depends on the characteristics of the Vening Meinesz smoothing.

After that verbal avalanche, it is time to be a little more specific: confining ourselves to the linear approximation, the Airy/Heiskanen power spectrum is given by (4-4) which we repeat here

$$\bar{T}_n^2 = \left\{\frac{3\rho_0}{\rho_M} \frac{1}{2n+1} \left[1 - \left(1 - \frac{D}{R}\right)^n\right]\right\}^2 \bar{H}_n^{*2} .$$

The regional Vening Meinesz concept is based on a smoothing of the compensation (= root/antiroot) surface. Since we don't know any better, we propose democratic smoothing represented by a homogeneous and isotropic operator B with eigenvalues β_n,

$$B(P,Q) = \frac{1}{4\pi} \sum_{n=o}^{\infty} (2n+1)\,\beta_n P_n(\cos\psi_{PQ}), \tag{4-6a}$$

$$\beta_n = 2\pi \int_{-1}^{1} B(t) P_n(t)\,dt \qquad (t := \cos\psi), \tag{4-6b}$$

with $\beta_o = 1$ to make the integral of the operator B over the unit sphere equal to 1 which is required for mass balance.

The operator B will now be applied to the equivalent rock topography represented by its harmonic coefficients $\bar{H}^*_{nm}$, at the compensation level only. Note: the rock/ocean topography remains unchanged, only the compensating masses and therefore, the compensation root/antiroot surface is smoothed. This smoothing process (which is a simple convolution in the space domain) is represented by a multiplication of $\bar{H}^*_{nm}$ by the eigenvalue β_n,

$$B * \bar{H}^* \longleftrightarrow \beta_n \cdot \bar{H}^*_{nm} \tag{4-7}$$

Therefore, the Vening Meinesz power spectrum in its linear approximation is given by

$$\tilde{\bar{T}}^2_n\,(D_1) = \left\{ \frac{3\rho_o}{\rho_M}\ \frac{1}{2n+1} \left[1 - \left(1 - \frac{D_1}{R}\right)^n \beta_n\right]\right\}^2 \bar{H}^{*2}_n \tag{4-8}$$

At this point we should briefly discuss two extreme cases of smoothing:

a) $\beta_n = 1 \;\forall\; n = 0, 1, \ldots$

In this case the smoothing operator B of equation (4-6a) degenerates into the Dirac distribution and will reproduce the input; consequently there is no smoothing involved. Therefore, this operator represents the standard Airy/Heiskanen model.

b) $\beta_o = 1,\ \beta_n = 0 \;\forall\; n = 1, 2, \ldots$

In this case the smoothing operator B degenerates into the global average operator with equal weight; the compensation masses form a homogeneous layer which is equivalent to a point mass at the origin. Therefore, this operator represents the

extreme case of the Airy/Heiskanen model with $D_2 = R$.
It is evident that the two power spectra $\bar{T}_n^2$ and $\bar{\bar{T}}_n^2$ coincide if the condition

$$\left(1 - \frac{D_1}{R}\right)^n \beta_n = \left(1 - \frac{D_2}{R}\right)^n \tag{4-9}$$

is fulfilled for all n. This implies an Airy/Heiskanen compensation depth D_2,

$$D_2 = R - (R - D_1)\, \beta_n^{\frac{1}{n}} , \tag{4-10}$$

or, in terms of radius of compensation, $R_1 = R - D_1$,

$$R_2 = R_1\, \beta_n^{\frac{1}{n}} . \tag{4-10)'$$

It is obvious that R_2 is constant (independent of the degree n) if and only if

$$\beta_n^{\frac{1}{n}} = b = \text{const.},$$

which implies

$$\beta_n = b^n \tag{4-11}$$

with $b \leq 1$ because of the compressing properties of a smoothing operator. The smoothing operator B corresponding to (4-11) is obtained by an inverse Fourier transform according to (4-6a)

$$B(P,Q) = \frac{1}{4\pi} \sum_{n=o}^{\infty} (2n+1)\, b^n P_n(\cos\psi_{PQ}) \tag{4-12}$$

which is evidently the interior Poisson operator (HM, p.35). With $b = R_2/R_1$ (equation (4-10)' and (4-11)) we finally obtain

$$B(P,Q) = \frac{R_1(R_1^2 - R_2^2)}{4\pi \ell^3(P,Q)} \tag{4-13}$$

with the spatial distance defined by

$$\ell(P,Q) = (R_1^2 + R_2^2 - 2R_1R_2\cos\psi_{PQ})^{1/2}.$$

We conclude: if the Vening Meinesz regional smoothing of the compensation surface is of Poisson type at level R_1 with parameter R_2, we can exactly replace it by the standard Airy/Heiskanen model with compensation level R_2. Or vice versa, using a standard Airy/Heiskanen model at depth D_2 corresponds exactly to the use of a Vening Meinesz model at depth D_1 with the Poisson smoothing (4-13) and parameter D_2.

Now we are already on much safer ground because we are able to justify the formal use of an unusually large compensation depth. However, the crucial question is still open: does the Poisson smoothing reflect physical reality?

5. AN OPTIMAL VENING MEINESZ ISOSTATIC MODEL

If we are talking about "optimal", we have to choose a norm which defines what "good" is. The choice of the norm is not so difficult if we are primarily interested in the energy budget of the residual gravitational potential. In this case we consider a solution as good if the residual potential after removing the topographic-isostatic effect, has little energy compared with the unreduced observed potential. Therefore, it is quite natural to look for an operator, acting on the compensation part, which fulfills this requirement best.

From our considerations in Chapter 4, we conclude that the minimum number of parameters must be 2 if we are after a compensation smoothing operator: one parameter that controls the level of compensation, the second one controlling the smoothing. (Note that also the Poisson smoothing operator (4-13) has two parameters, R_1 and R_2.)

5.1 Estimation of the Parameters of the Smoothing Operator

In linear approximation, the Vening Meinesz topographic-isostatic power spectrum is given by equation (4-8); the operator's two parameters are D and b, where D stands for depth of compensation and b is a smoothing parameter (note that β_n depends on b). As a matter of fact, the individual harmonic coefficients of the topographic-isostatic potential are also controlled in the same way,

$$\tilde{\bar{T}}_{nm} = \tilde{\bar{T}}_{nm}(D,b). \tag{5-1}$$

Extending a suggestion of Tscherning (1985), we perform a least-squares estimation of the model parameters such that the energy of the residual field is minimized.

$$\| \tilde{\bar{T}} - \bar{V} \|^2 = \min. \tag{5-2}$$

Since the low order harmonics of the anomalous gravitational potential are to a large extent due to density disturbances in the upper mantle, and probably due to even deeper sources, it makes no sense (and would only disturb the isostatic concept) to include the low frequency part in the energy budget.

With a homogeneous and isotropic smoothing operator B having eigenvalues β_n, the harmonic coefficients of the topographic-isostatic potential are given by

$$\tilde{\bar{T}}_{nm}(D,b) = \frac{3\rho_o}{\rho_M} \frac{1}{2n+1} \left[1 - \left(1 - \frac{D}{R}\right)^n \beta_n(b)\right] \bar{H}^*_{nm} . \tag{5-3}$$

The decision of the model spectrum $\{\beta_n(b)\}$ can be based on correlation considerations between $\tilde{\bar{T}}^2_n(D,b)$ and $\bar{V}^2_n$, requiring that

$$\tilde{\bar{T}}^2_n \approx \bar{V}^2_n, \tag{5-4}$$

and yielding empirical estimates

$$\hat{\beta}_n = \left(1 - \frac{D}{R}\right)^{-n} \left[1 - \frac{\rho_M}{3\rho_o} (2n+1) \frac{\bar{V}_n}{\bar{H}^*_n}\right] \tag{5-5}$$

which depend on the choice of D. The figures 6.3 a - d will present graphs of this empirical frequency transfer function $\hat{\beta}_n$ for several compensation depths D. As data the Rapp 81 geopotential model and the worldwide $1^\circ \times 1^\circ$ DTM model, supplied by DMAAC, have been used. All empirical frequency transfer functions $\hat{\beta}_n$, at least for geophysically reasonable compensation depths, strongly suggest a Gaussian model of type

$$\hat{\beta}_n(b) = e^{-b^2 n^2} . \tag{5-6}$$

The solid curves represent the best fit of such a Gaussian

model to the empirical frequency transfer function.

Having chosen a model for $\beta_n(b)$, it is a simple matter of least-squares adjustment with 2 parameters to solve for the "best" D and b. If we use non-equal weights p_n which depend on the degree like

$$p_n = \left(\frac{R}{GM}\right)^{-2} \frac{2n+1}{k_n} \tag{5-7}$$

with the model degree variances k_n, and error variances of the harmonic coefficients of the anomalous potential modelled by

$$\sigma_n^2 = \left(\frac{K}{n-1}\right)^2, \; K \; \ldots \text{ constant,} \tag{5-8}$$

according to Jekeli (1979, pp. 13, 14), we can interpret the least-squares solution as a least-squares collocation solution for the model parameters D and b (cf. Moritz, 1980, pp. 160, 161). (Here the data consists of the vector $l := \{\bar{V}_{nm}\}$.)

Because of (5-7) and (5-8) the signal and error covariance matrices C_s and C_n are diagonal, which implies that the energy

$$\sum_{n=n_o}^{N} \left[\left(\frac{R}{GM}\right)^2 \frac{k_n}{2n+1} + \left(\frac{K}{n-1}\right)^2\right]^{-1} \sum_{m=-n}^{n} \left(\tilde{\bar{T}}_{nm}(D,b) - \bar{V}_{nm}\right)^2 \tag{5-9}$$

must be minimized with respect to D and b. With the Taylor linearization of $\tilde{\bar{T}}_{nm}$

$$\tilde{\bar{T}}_{nm}(D,b) = \tilde{\bar{T}}_{nm}(D^{(o)}, b^{(o)}) + \left.\frac{\partial \tilde{\bar{T}}_{nm}(D,b)}{\partial D}\right|_{(D=D^{(o)},\, b=b^{(o)})} (D - D^{(o)}) + \left.\frac{\partial \tilde{\bar{T}}_{nm}(D,b)}{\partial b}\right|_{(D=D^{(o)},\, b=b^{(o)})} (b - b^{(o)}) \tag{5-10}$$

at an appropriate Taylor point (for example D = 30 km and b according to the solid line in Fig. 6.3b), we obtain the coefficients of the design matrix A. The elements of the two column vectors a_1 and a_2 consist of the partials of (5-10)

$$a_1 = \left\{ \frac{\partial \tilde{\bar{T}}_{nm}(D,b)}{\partial D} \bigg|_{(D=D^{(o)},\ b=b^{(o)})} \right\} ,$$

$$a_2 = \left\{ \frac{\partial \tilde{\bar{T}}_{nm}(D,b)}{\partial b} \bigg|_{(D=D^{(o)},\ b=b^{(o)})} \right\} , \tag{5-11}$$

the residual parameter vector X of

$$X^T := [\delta D, \delta b] . \tag{5-12}$$

The least-squares solution is provided by

$$\hat{X} = (A^T \bar{C}^{-1} A)^{-1} A^T \bar{C}^{-1} (l - l^{(o)}) \tag{5-13}$$

with

$$\bar{C} := C_s + C_n \tag{5-14}$$

and

$$l^{(o)} := \tilde{\bar{T}}_{nm}(D^{(o)}, b^{(o)}) . \tag{5-15}$$

Finally we obtain the wanted parameters D and b of our best-fitting Vening Meinesz isostatic model,

$$\hat{D} = D^{(o)} + \delta\hat{D} ,$$

$$\hat{b} = b^{(o)} + \delta\hat{b} .$$

For error considerations we have

$$E_{xx} = (A^T \bar{C}^{-1} A)^{-1} \tag{5-16}$$

as parameter error covariance matrix (for reference see Moritz (1980, Part B).

5.2 Fine-Tuning of Parameter Estimation

When we derived the least-squares estimation of the isostatic parameters D and b, we made two simplifications: firstly, if our Taylor point is far away from reality, we must iterate the estimation process; secondly, in our discussion of the estimation problem and in all our derivations we have generously and tacitly assumed that $\tilde{\bar{T}}_{nm}$ of equation (5-3) is correct.

The first problem will not be dealt with here because it is simple and standard, provided the Taylor point is not too much off the truth. The second problem is probably slightly more delicate, although a relatively simple solution is possible:

Equation (5-3) presupposes a linear relation between the topographic-isostatic potential and the equivalent rock topography. But according to (2-11) and (2-12a,b,c) we know that the relation is highly non-linear. Therefore, the concept of parameter estimation has to be modified accordingly.

This modification is basically due to second and higher order effects which, we know from our discussion in Chapter 3, are small because the topographic height is so much smaller than the radius of the earth. Therefore, an iteration process offers itself again.

Let B be, as before, the smoothing operator applied to the compensation surface, represented by $\bar{H}'$, yielding the smoothed compensation surface $\tilde{\bar{H}}' = B * \bar{H}'$. Then $\tilde{\bar{H}}'$ should be used for the harmonic coefficients of the compensation $\tilde{\bar{T}}^{(c)}_{nm}$ of equation (3-11b). Note, however, that $\bar{H}$ (and not $\tilde{\bar{H}}$) is used for the harmonic coefficients of the topography $\bar{T}^{(t)}_{nm}$, because smoothing is only applied to the compensating masses. The problem is that also higher order powers of $\tilde{\bar{H}}'$ enter into $\tilde{\bar{T}}^{(c)}_{nm}$,

$$\tilde{\bar{T}}^{(c)}_{nm} = \frac{3\,\rho_o}{\rho_M}\,\frac{1}{(2n+1)(n+3)}\left[1-\frac{D}{R}\right]^{n+3} k_o^{-1}\sum_{j=1}^{n+3}\binom{n+3}{j}\frac{1}{4\pi}\iint_\sigma c_H\,\tilde{\bar{H}}'^{\,j}(Q)\,\bar{R}_{nm}(Q)\,d\sigma(Q) \qquad (5\text{-}17)$$

while $\bar{T}^{(t)}_{nm}$ remains unchanged (equation (3-11a)). Since $\tilde{\bar{H}}'$ is a convolution of B and $\bar{H}'$, higher order powers of $\tilde{\bar{H}}$ correspond to convolutions in the frequency domain. Therefore, the smoothing parameter b enters in a rather complicated manner into the higher order terms of $\tilde{\bar{T}}^{(c)}_{nm}$. But fortunately, these terms are only small correction terms to the leading linear term

(5-3); therefore, if we solve for b, it will be sufficient to calculate the higher order terms based on the latest knowledge of b and to consider them formally as constants in the subsequent least-squares estimation process. It is sufficient to iterate this process a few times because of the extraordinarily rapid convergence. The entire procedure can be summarized as follows:

0) $i = 0$

Harmonic analysis of low order powers of topography according to (3-11a); correlation study (5-5) yields initial estimates $\hat{D}^{(o)}$ and $\hat{b}^{(o)}$;

1) $i := i + 1$
smoothing of compensation surface ($\tilde{\bar{H}}'^{(i)} = B^{(i-1)} * \bar{H}'$) using $\hat{D}^{(i-1)}$ and $\hat{b}^{(i-1)}$;

2) Harmonic analysis of low order powers of smoothed compensation surface $\tilde{\bar{H}}'^{(i)}$ according to (5-17) yields a first order term $\tilde{\bar{T}}_{nm}^{(1)}$ which will be considered as a function of D and b, and higher order (correction) terms $\delta\tilde{\bar{T}}_{nm}$ which will be considered as constants in the subsequent step;

3) Adjustment (collocation) solution for D and b with Taylor point ($D^{(i-1)}$, $b^{(i-1)}$) and linearization restricted to the linear term $\tilde{\bar{T}}_{nm}^{(1)}$ of step (2); Stop if $|\hat{D}^{(i)} - \hat{D}^{(i-1)}| < \varepsilon_D$ and $|\hat{b}^{(i)} - \hat{b}^{(i-1)}| < \varepsilon_b$, else go to (1).

The result of this iteration process is both a best estimate for the depth of compensation and for the parameter of the most likely compensation smoothing, and a set of harmonic coefficients of the topographic-isostatic potential corresponding to that Vening Meinesz regional isostatic compensation model.

6. NUMERICAL STUDIES, RESULTS, CONCLUSIONS

6.1 Solution with the DMAAC 79 - DTM

For the numerical investigations originally the global digital terrain model consisting of 64800 $1^{o} \times 1^{o}$ mean elevations and ocean bottom depths (but no information on ice coverage), supplied by DMAAC in 1979, and, representing the earth's gravity field, the Rapp 1981 solution which is complete to degree and order 180, have been used. Recently that DTM model has been merged with the SYNBAPS dataset, which consists of 5' by 5' mean ocean depths, yielding a new global $1^{o} \times 1^{o}$ DTM, called TUG 86 - DTM, and used for the recalculation of a topographic-isostatic earth model.

The goal of this investigation was both the parameter estimation of the most likely isostatic compensation model and the determination of a set of topographic-isostatic coefficients complete up to degree and order 180, based on that compensation model. This task would be formidable without support of the powerful tool of fast Fourier transform on the sphere. Therefore, the unlimited access to Colombo's (1981) outstanding FFT algorithms "HARMIN" for analysis and "SSYNTH" for synthesis was particularly appreciated. The following is a comprehensive documentation of our numerical studies.

In all our investigations, the following parameters have been used:

density:

$\rho_o = 2.67 \ gcm^{-3}$ ... rock topography
$\rho_w = 1.027 \ gcm^{-3}$ ... ocean water
$\rho_1 = 3.27 \ gcm^{-3}$ ... crust
$\rho_M = 5.517 \ gcm^{-3}$ ... mean earth

earth radius:

$R = 6371$ km

potential coefficient degree variance model:

$$k_n = \left(\frac{A \cdot 10^{-6}}{n^B} \right)^2 , \quad A = 4.1, \ B = 1.8$$

(cf. Rapp, 1979, p. 10)

potential coefficient error model:

$$\sigma_n^2 = \left(\frac{K}{n - 1}\right)^2 , \quad K = 0.0581 \ . \ 10^{-6}$$

(cf. Jekeli, 1979, p. 13 ff.)

harmonic series complete up to degree:

$n = 180$

binomial expansion up to power:

$j = 4$

In order to get an idea about the contribution of first and higher order terms to the harmonic coefficients of the topographic-isostatic potential, we provide in Fig. 6.1 a graph of the degree variances, separate for the linear and second order term as well as the composite degree variances. Note that the linear approximation pretends more power over the entire frequency range than the exact topographic-isostatic model actually has. The contribution of the third order term is already so small that it does not show up in the plotting window. Therefore, we have confined our discussion of higher order terms in Chapter 3 to the second order term only.

The dependence of the power spectrum on the choice of the compensation depth D or, if Poisson compensation smoothing is employed, on the corresponding smoothing parameter (cf. Section 4.1), is illustrated in Fig. 6.2 for depths D = 30, 50, 70, and 100 km. The gain of power with increasing compensation depth (or more compensation smoothing) is obvious.

Frequency transfer functions between the Rapp 81 anomalous gravitational power spectrum and the topographic-isostatic power spectrum in linear approximation are given in Fig. 6.3a-d for compensation depths D = 20, 30, 50, and 70 km. The graphs suggest a Gaussian model of the type (5-6). The solid lines represent that best model fit.

By an inverse Fourier transform of the frequency transfer function β_n we obtain the corresponding space transfer function which is essentially the compensation smoothing operator B,

$$B(P,Q) = \frac{1}{4\pi}\sum_{n=o}^{\infty}(2n+1)\beta_n P_n(P,Q). \qquad (6\text{-}1)$$

Using the best fitting Gaussian frequency transfer model, the smoothing operator has been determined for the most important part of its support and is presented in Fig. 6.4a-d normalized to B(0) = 1. The scale of the abscissa is arc deg. of spherical distance. These operators demonstrate that the smoothing of the compensation surface, as implied by our knowledge of the global gravitational field of the earth, is obviously most likely of Gaussian type rather than of moving average, box-shaped type or of Poisson type. The radius of smoothing (theoretically 180 degrees, practically very much smaller) decreases with increasing compensation depth. This should be expected because an increase of compensation depth accounts already for a certain degree of smoothing, although of Poisson type. For geophysically relevant compensation depths the smoothing radius does practically not exceed 2 degrees.

For the least-squares estimation of the two model parameters, compensation depth D and smoothing parameter b, the initial (Taylor) values

$$D^{(o)} = 30 \text{ km},$$
$$b^{(o)} = 0.0082,$$

suggested by the correlation pattern, have been used. (The smoothing parameter $b^{(o)} = 0.0082$ implies a "correlation length" of the bell-shaped Gaussian frequency transfer function of about (degree) n = 100.) Graphs of both the frequency and the corresponding space domain transfer function(smoothing operator) for these initial values are presented in Figs. 6.3b and 6.4b.

The compensation smoothing $\tilde{\bar{H}}' = B * \bar{H}'$ was naturally performed in the frequency domain. The higher powers of $\tilde{\bar{H}}'$ could have been obtained by spectral convolution in the frequency domain. We found it a lot easier to retransform the spectrum $\tilde{\bar{H}}'_{nm}$ into the space domain using Colombo's Fourier synthesis program SSYNTH and to calculate all powers of $\tilde{\bar{H}}'$ there. The binomial expansion has been terminated at power j = 4. After two iterations of the least-squares parameter estimation process

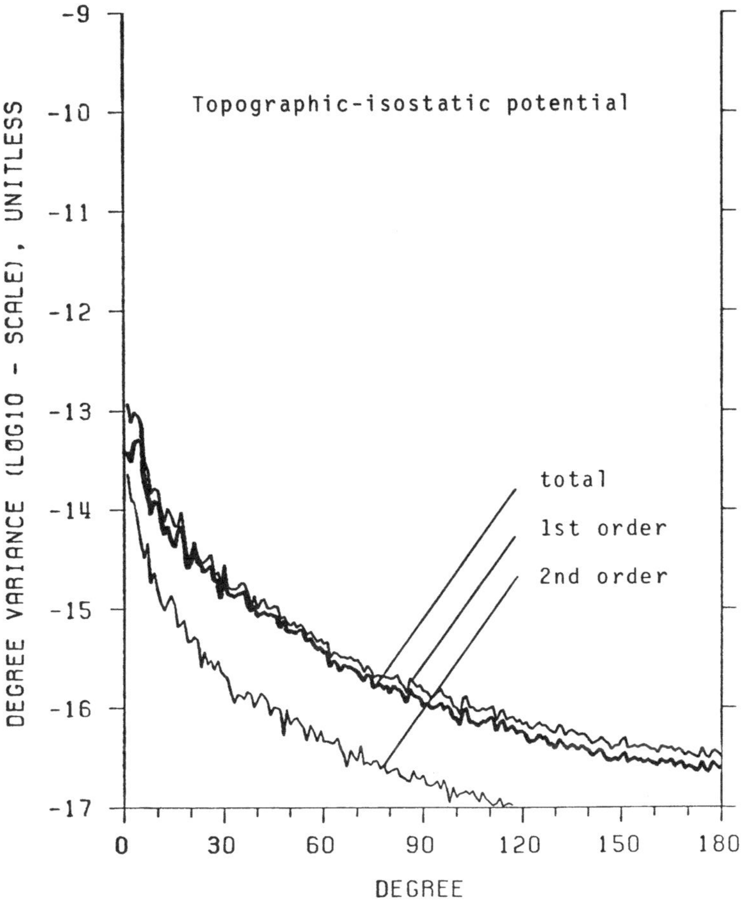

Fig. 6.1 Degree variances, low-order terms
(no compensation smoothing)

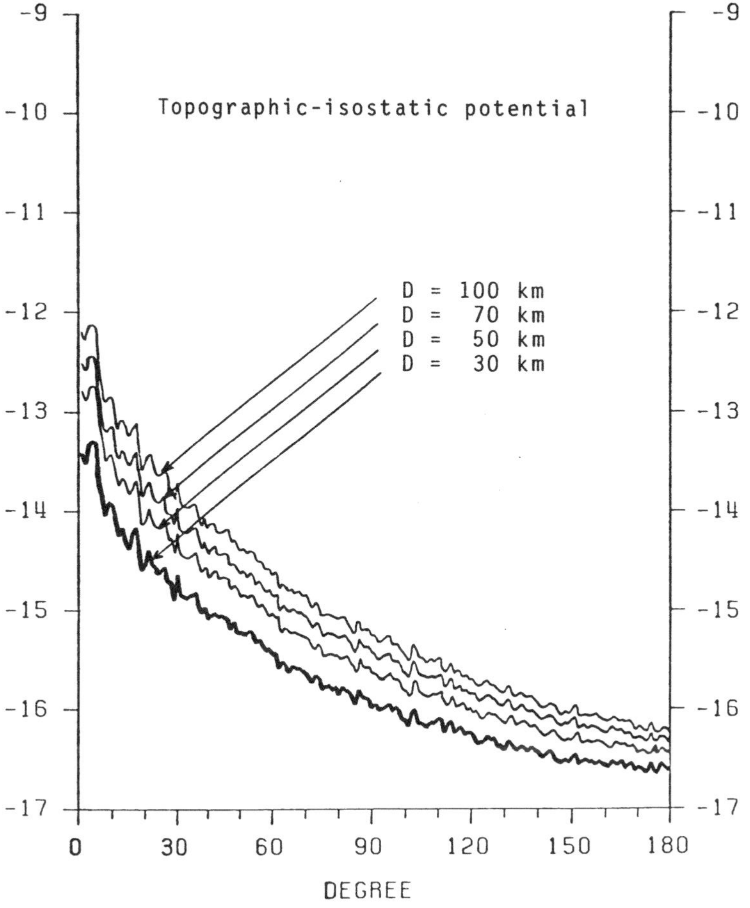

Fig. 6.2 Degree variances, dependent on depth D
(no compensation smoothing)

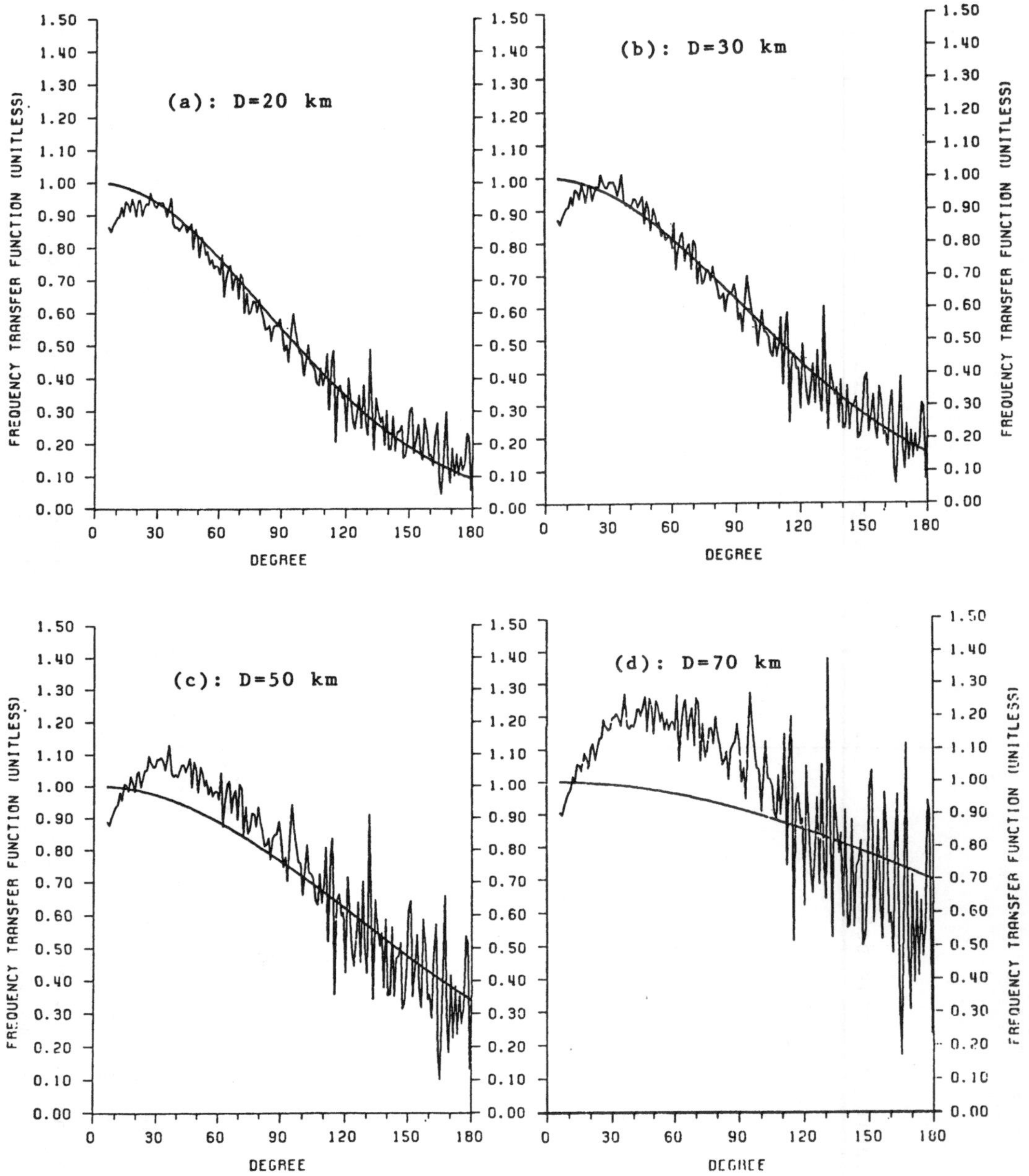

Fig. 6.3a-d Frequency transfer functions: geopotential/topographic-isostatic potential as implied by various compensation depths D .

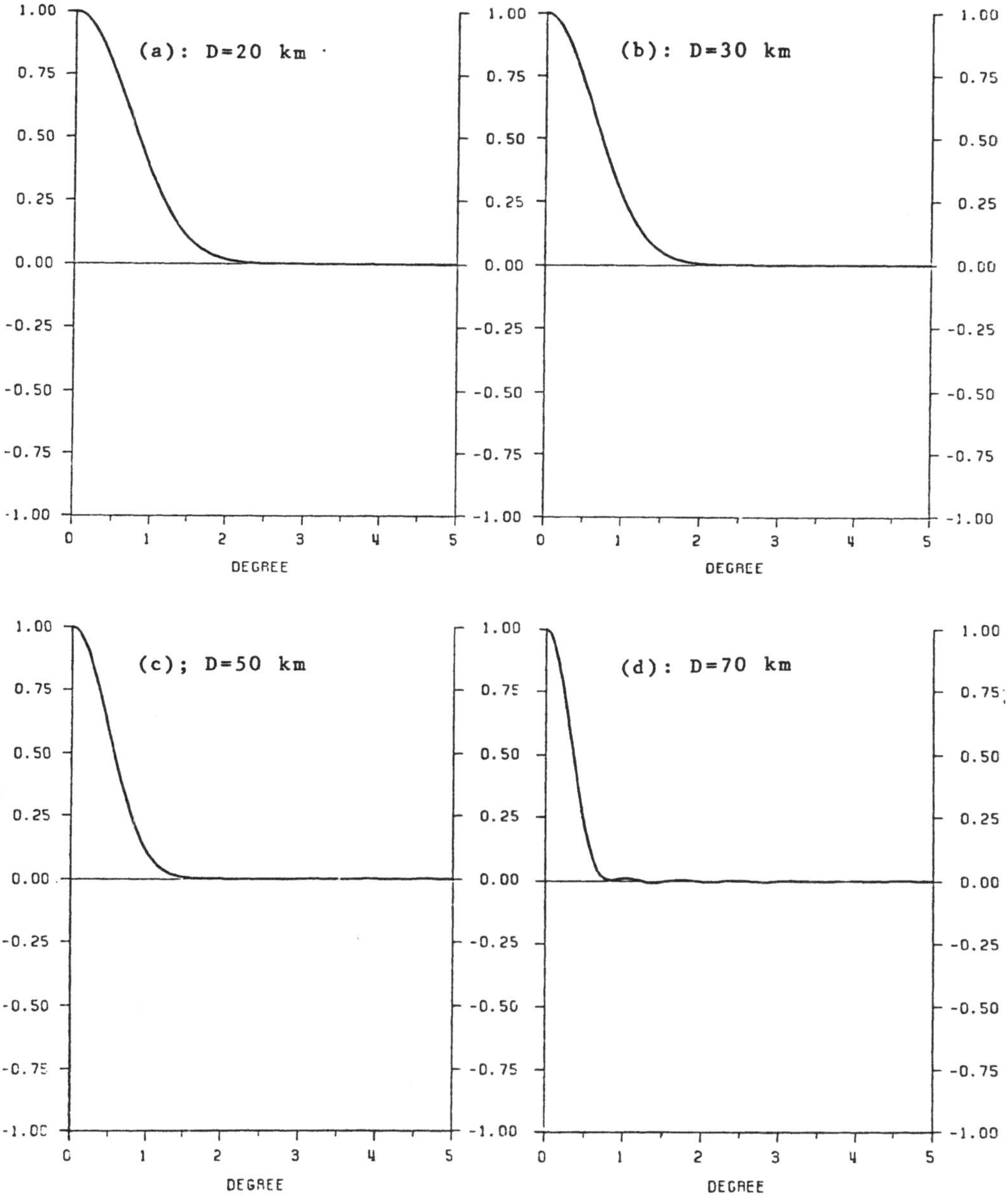

Fig. 6.4a-d Space transfer functions (= smoothing operators), as implied by various compensation depths D , depending on spherical distance (deg.).

no significant change of D and b could be observed. The finally adopted compensation model parameters, which yielded a best possible agreement between gravity field information and a topography-isostasy implied model gravity field on a global scale, were

$$D = 24 \text{ km}$$
$$b = 0.0091 \qquad (6\text{-}2)$$

with a standard deviation of less than 2% (!) each. Note that the very low frequency part of $n < 15$ has been excluded from the parameter estimation process in order to avoid a strong bias of the solution coming from that part of the power spectrum which has the least to do with isostasy. The "best" smoothing operator $B(\psi)$ with $b = 0.0091$, normalized to $B(0) = 1$, is presented in Fig. 6.5 as a function of the spherical distance ψ. Finally we present in Fig. 6.6 both the power spectrum of the Rapp 81 geopotential solution and the power spectrum implied by the "best" Airy/Heiskanen - Vening Meinesz topographic-isostatic model. Because of the above-mentioned exclusion of the very low frequency part a match in that range can never be expected (and if observed, is purely artificial). The very long wavelengths, corresponding to that very low frequency part, are to a great extent due to density disturbances in the earth's mantle and are only superimposed by a relative little topographic-isostatic effect. Therefore, we should primarily focus on the higher and highest frequencies. And indeed, in this range the agreement of the two spectra is remarkable.

6.2 Solutions with the TUG 86 - DTM

The DMAAC 79 - DTM which was used for this study contains several gross errors up to 5000 m (!) which were identified by C.C. Tscherning. Recently we have got access to the 5'x 5' SYNBAPS dataset for ocean areas. This data has been processed into $1^o \times 1^o$ mean ocean depth and used to replace the DMAAC 79 data in ocean areas. This new data set is called TUG 86 - DTM. The histogram in Fig. 6.7 shows the differences between TUG 86 - DTM and the old DMAAC 79 - DTM.

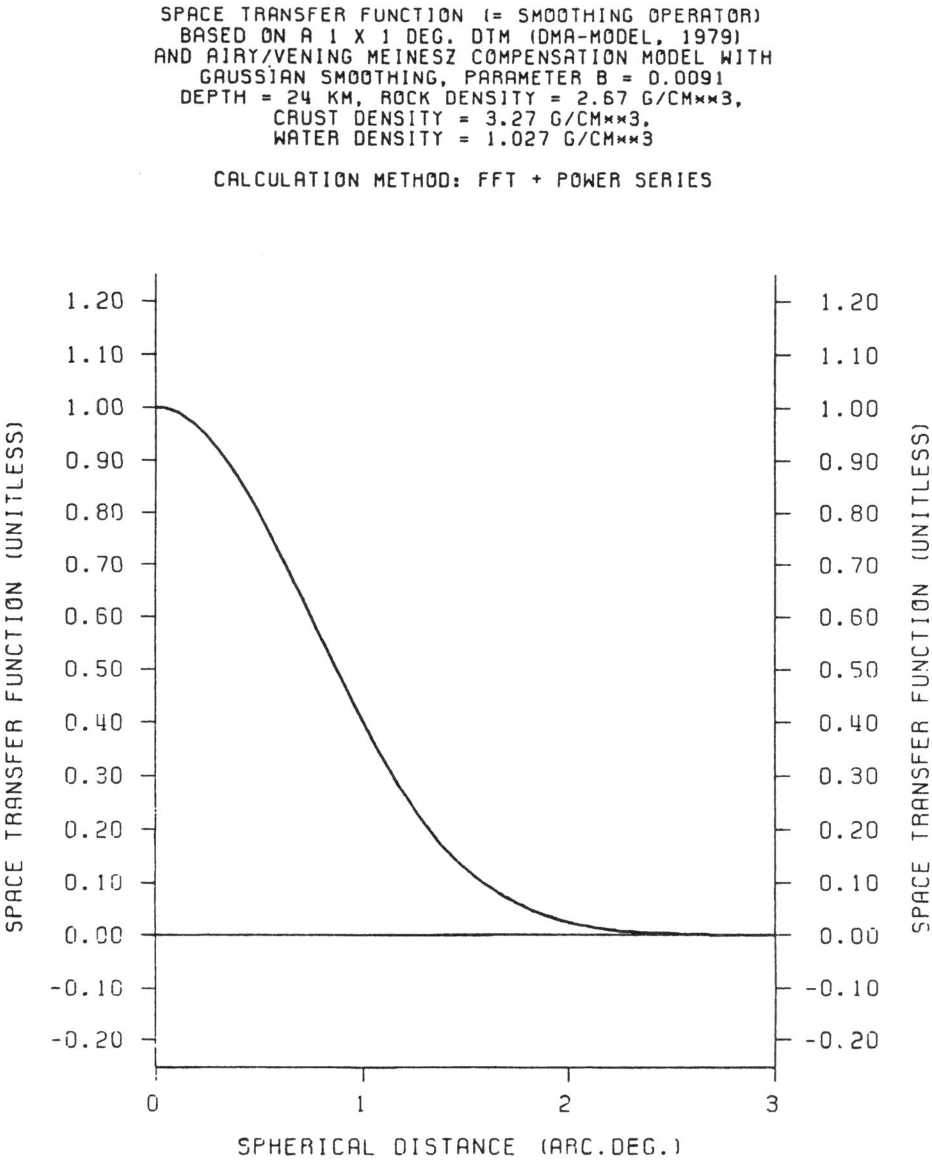
SPACE TRANSFER FUNCTION (= SMOOTHING OPERATOR)
BASED ON A 1 X 1 DEG. DTM (DMA-MODEL, 1979)
AND AIRY/VENING MEINESZ COMPENSATION MODEL WITH
GAUSSIAN SMOOTHING, PARAMETER B = 0.0091
DEPTH = 24 KM, ROCK DENSITY = 2.67 G/CM**3,
CRUST DENSITY = 3.27 G/CM**3,
WATER DENSITY = 1.027 G/CM**3
CALCULATION METHOD: FFT + POWER SERIES
1.20
1.10
1.00
0.90
0.80
0.70
0.60
0.50
0.40
0.30
0.20
0.10
0.00
-0.10
-0.20
SPACE TRANSFER FUNCTION (UNITLESS)
0
1
2
3
SPHERICAL DISTANCE (ARC.DEG.)

Fig. 6.5 Optimal smoothing operator

DEGREE VARIANCES OF THE TOPOGRAPHIC/ISOSTATIC POTENTIAL
BASED ON A 1 X 1 deg. DTM (DMA/MODEL, 1979)
AND AIRY/VENING MEINESZ COMPENSATION MODEL WITH
ROOT/ANTIROOT SMOOTHING OF GAUSSIAN TYPE, A = 0.0091
DEPTH = 24 km, ROCK DENSITY = 2.67 g/cm**3,
CRUST DENSITY = 3.27 g/cm**3,
WATER DENSITY = 1.027 g/cm**3

CALCULATION METHOD: FFT + POWER SERIES
H. SUENKEL, MG/TUG, JUNE 1986

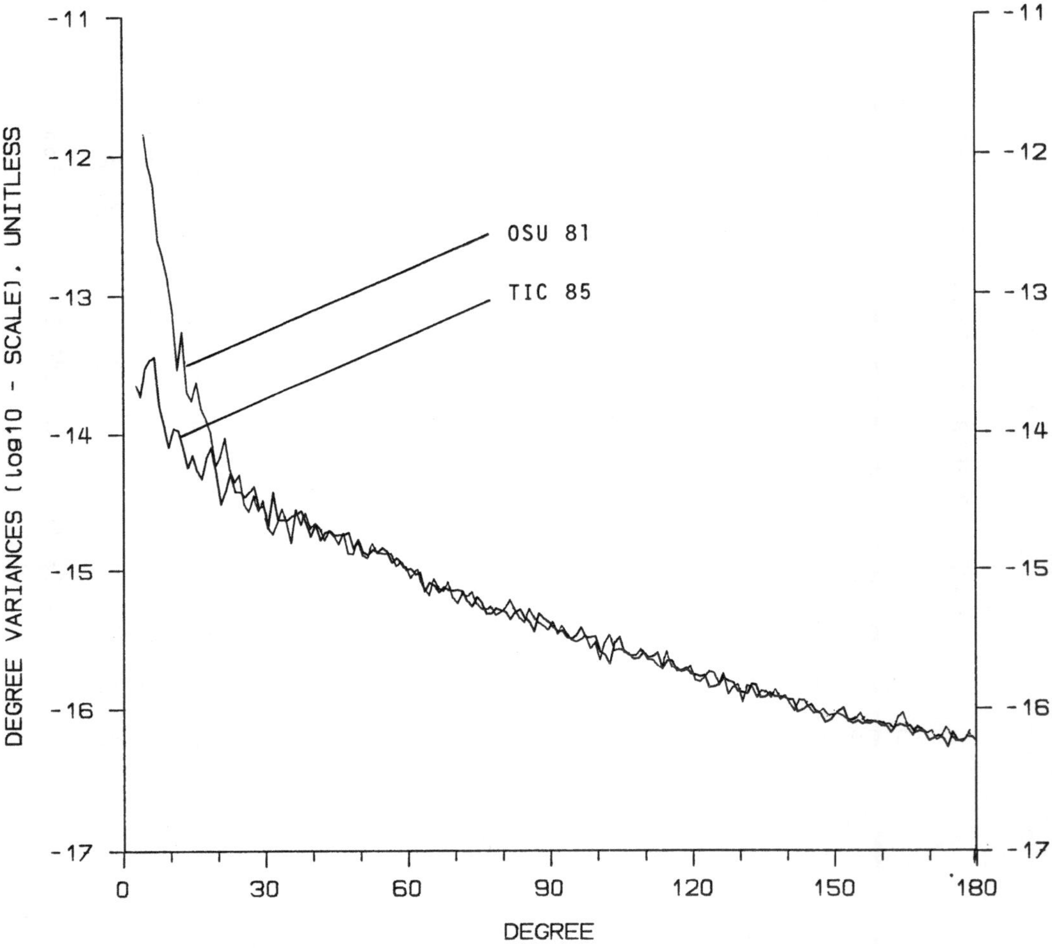

Fig. 6.6 OSU 81 and TIC 85 degree variances

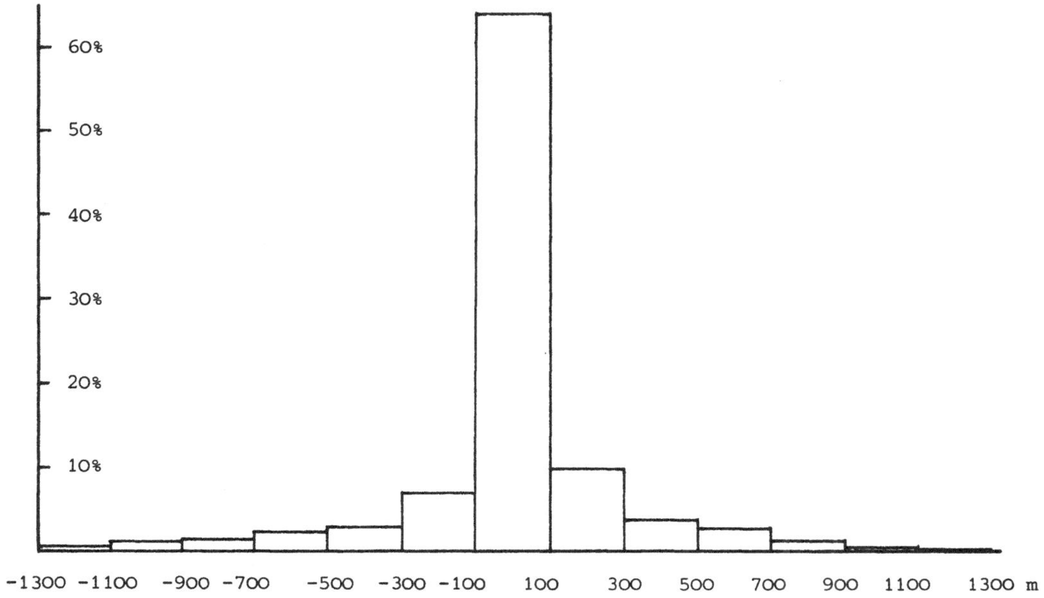

Fig. 6.7 Percentage differences between TUG 86 - DTM and DMAAC 79 - DTM (meters)

Based on TUG 86 - DTM the collocation procedure described above was repeated recently yielding new estimates for the parameters of the isostatic model using both the OSU 81 and Wenzel's GPM2 dataset as gravity field data.

Using the same 2-parameter model with a smoothing operator of Gaussian type and the OSU 81 gravity field model, the iteration process described in section 5.2 yielded the results of Table 6.1 . The same procedure, based on Wenzel's GPM2 earth model yielded the results of Table 6.2 . The difference between these two sets of parameters is insignificant. Compared with the original solution, based on the old DMAAC 79 - DTM, we observe a 7 % deeper compensation level and less smoothing implied by the considerable shorter correlation length of the smoothing operator of about 48 km (Fig. 6.8).

For the sake of interest we also performed a solution without smoothing (1 - parameter solution), yielding the results of Tables 6.3 and 6.4 .

ITR	PARAMETER	STARTING VALUE	CORRECTION	NEW VALUE	S.D.
0	D (M)	24000.0	7138.0	31138.0	424.0
	b	0.009100	-0.004990	0.004110	0.000069
1	D (M)	31138.0	-5583.6	25554.4	319.0
	b	0.004110	0.000243	0.004353	0.000082
2	D (M)	25554.4	9.7	25564.1	307.0
	b	0.004353	-0.000029	0.004324	0.000073
3	D (M)	25564.1	3.9	25568.0	307.0
	b	0.004324	-0.000003	0.004321	0.000073

Table 6.1 Iteration process for the estimation of the topographic-isostatic model parameters D and b using the OSU 81 earth model with n = 15, ..., 180 and the TUG 86 DTM.

ITR	PARAMETER	STARTING VALUE	CORRECTION	NEW VALUE	S.D.
0	D (M)	25570.0	466.8	26036.8	307.1
	b	0.004320	0.000101	0.004421	0.000073
1	D (M)	26036.8	-17.2	26019.6	309.0
	b	0.004421	0.000015	0.004436	0.000073
2	D (M)	26019.6	-4.1	26015.5	310.0
	b	0.004436	0.000002	0.004438	0.000072

Table 6.2 Iteration process for the estimation of the topographic-isostatic model parameters D and b using the GPM2 earth model with n = 15, ..., 180 and the TUG 86 DTM.

ITR	PARAMETER	STARTING VALUE	CORRECTION	NEW VALUE	S.D.
0	D (M)	30000.0	3653.2	33653.2	147.2
1	D (M)	33653.2	128.6	33781.8	153.5
2	D (M)	33781.8	1.6	33783.4	153.7

Table 6.3 Iteration process for the estimation of the topographic-isostatic model parameter D using the OSU 81 earth model with n = 15, ..., 180 and the TUG 86 DTM.

ITR	PARAMETER	STARTING VALUE	CORRECTION	NEW VALUE	S.D.
0	D (M)	33780.0	857.2	34637.2	153.7
1	D (M)	34637.2	12.4	34649.6	155.2
2	D (M)	34649.6	0.1	34649.7	155.2

Table 6.4 Iteration process for the estimation of the topographic-isostatic model parameter D using the GPM2 earth model with n = 15, ..., 180 and the TUG 86 DTM.

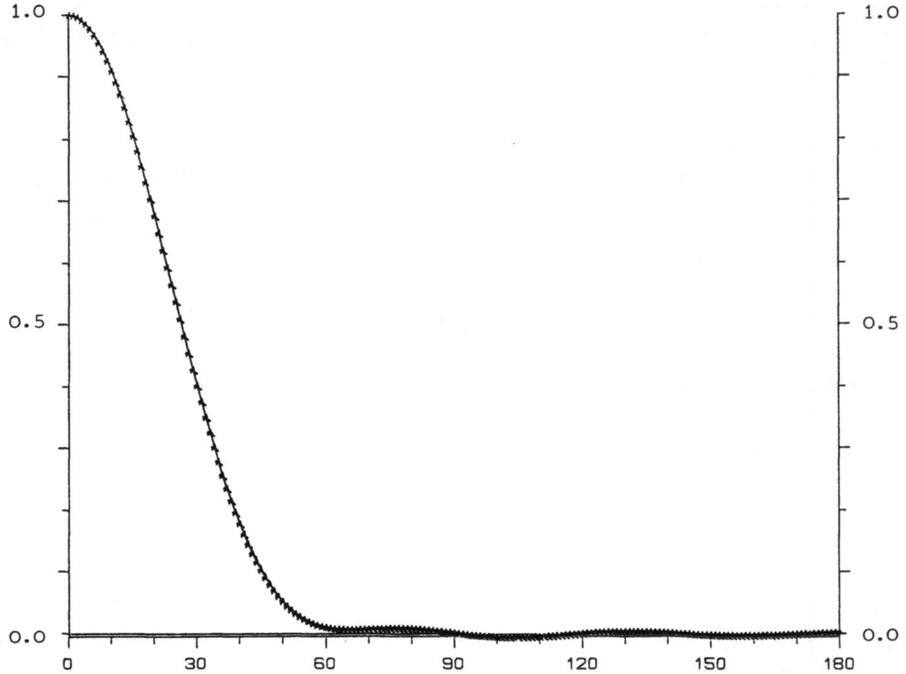

Spherical distance (arcmin.)

Fig. 6.8 Smoothing operator based on the OSU 81 (or the GPM2) earth model and on the TUG 86 DTM.

The considerable deeper compensation level was anticipated because of the explanation given in Section 4.1.

6.3 Correlation and Smoothing

Denoting for the sake of simplicity the fully normalized harmonic coefficients of the topographic - isostatic potential by T_{nm} and the corresponding coefficients of the observed disturbing potential by V_{nm}, the correlation by degree is the inner product of the normalized unit vectors $\{ T_{nm} / \sigma_n^{(T)} \}$ and $\{ V_{nm} / \sigma_n^{(V)} \}$,

$$r_n = \sum_{m=-n}^{n} V_{nm} \; T_{nm} / \sigma_n^{(V)} \; \sigma_n^{(T)} \tag{6-3}$$

with
$$\sigma^{2(V)} := \sum_{m=-n}^{n} V_{nm}^2 ,$$
$$\sigma_n^{2\,(T)} := \sum_{m=-n}^{n} T_{nm}^2 \tag{6-4}$$

As noted by Tscherning (1985), the correlation pattern is not necessarily a good indicator for an agreement or disagreement of two data sets; (note that the correlation coefficient is invariant with respect to a scale factor). For the purpose of geodesy it is of primary importance to obtain smoothing by subtracting one field from the other. In analogy to r_n we can define a smoothing per degree, denoted s_n, as the normalized residual degree variances

$$s_n := \sum_{m=-n}^{n} (V_{nm} - T_{nm})^2 \;/\; \sigma_n^{2\,(V)} \tag{6-5}$$

Ideally the smoothing coefficients should vanish; this would be the case if $T_{nm} = V_{nm}$ for all n, m. Such a perfect smoothing will be never ever achieved because of various reasons: the simple isostatic models are only poor approximations to reality, both the set $\{ V_{nm} \}$ and the used DTM contain noise, thc density assumptions are in some areas not justified (ice coverage, e.g.), etc. All these effects add up to an s_n pattern greater than 0. The other extreme case would be achieved if $T_{nm} = 0$ for all n,m, corresponding to no smoothing at all. (Values of $s_n > 1$ would even indicate a de-smoothing.)

Since the difference between the OSU 81 and GPM2 - derived solutions is so small, we present here only a comparison between TIC 85 (based on OSU 81, DMAAC 79 - DTM) and TIC 86 (based on OSU 81, TUG 86 - DTM) for the 2-parameter model (D, b) both for the correlation coefficients and the smoothing coefficients by degree (Fig. 6.9 - 6.12) .

It is very obvious from these four figures that the TIC 86 solution has a much improved correlation pattern compared with the TIC 85 (and all other existing) solutions, cf. Tscherning, 1985; it does not show any trend to decrease with increasing degree and has an average correlation of about 60%. Particularly striking is the dramatic improvement of the smoothing pattern: the TIC 85 solution (and all other existing solutions, cf. Tscherning, 1985) show a pronounced trend to increase with degree, in some solutions even considerably above the critical threshold 1. TIC 86 makes former solutions look pale: the majority of the smoothness coefficients are located in a very narrow band between 0.5 and 0.7 and do not show any significant trend to increase with degree.

Considering the mean square variation of geoid heights and gravity anomalies, we observe no smoothing in the geoidal heights on a global scale because most of the power of the geoid is coming from the low frequency part of the gravity field which has hardly any relation to isostatic phenomena. The gravity anomalies respond stronger to more local mass distributions, therefore, we should expect some smoothing there; and indeed, subtracting TIC 86 from OSU 81 makes the variance decrease from 586 $mgal^2$ to 421 $mgal^2$ which is a significant improvement over existing solutions and comes very close to the estimate of 420 $mgal^2$ obtained by Tscherning (1985) for compensation depths, optimized per degree.

6.4 Further Improvements

The TIC 86 model has been derived from the TUG 86 - DTM. The quality of this DTM over continental areas is identical to that of the old DMAAC 79 - DTM and therefore rather questionable. A new global DTM which will be derived from the recently released global 5' x 5' ETOPO 5 dataset (NOAA, National Geophysical Data Center, Boulder, Co), called TUG 87 - DTM, can be expected to further improve our results in terms of

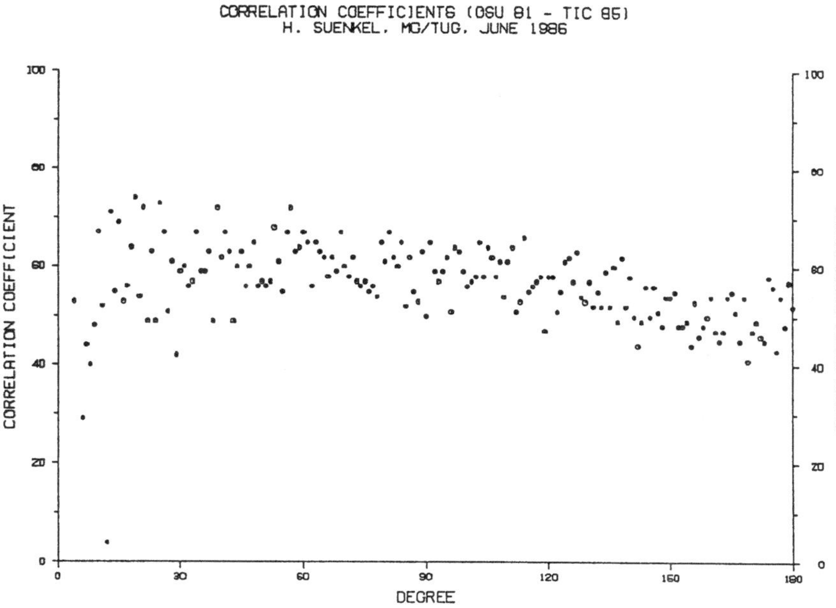

Fig. 6.9 Correlation coefficients OSU 81 versus TIC 85

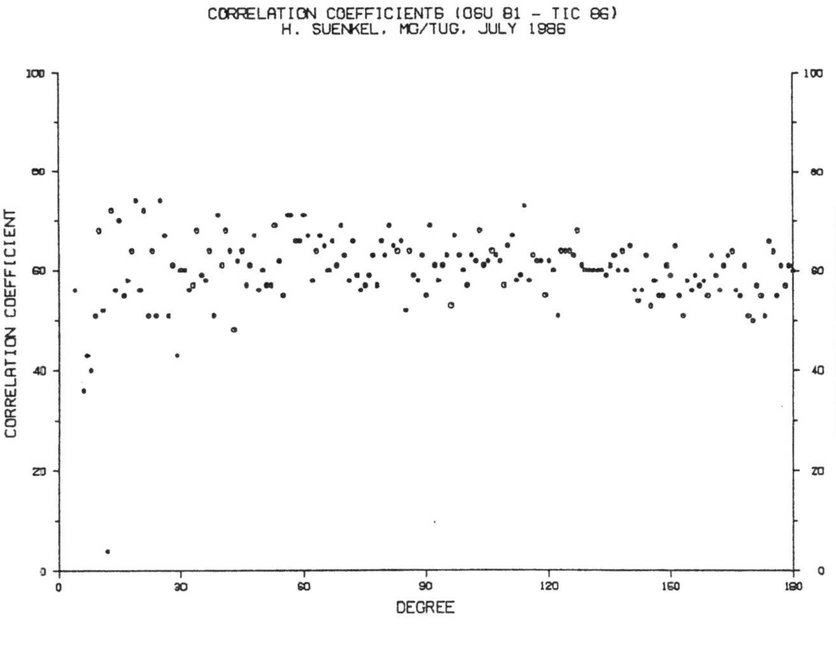

Fig. 6.10 Correlation coefficients OSU 81 versus TIC 86

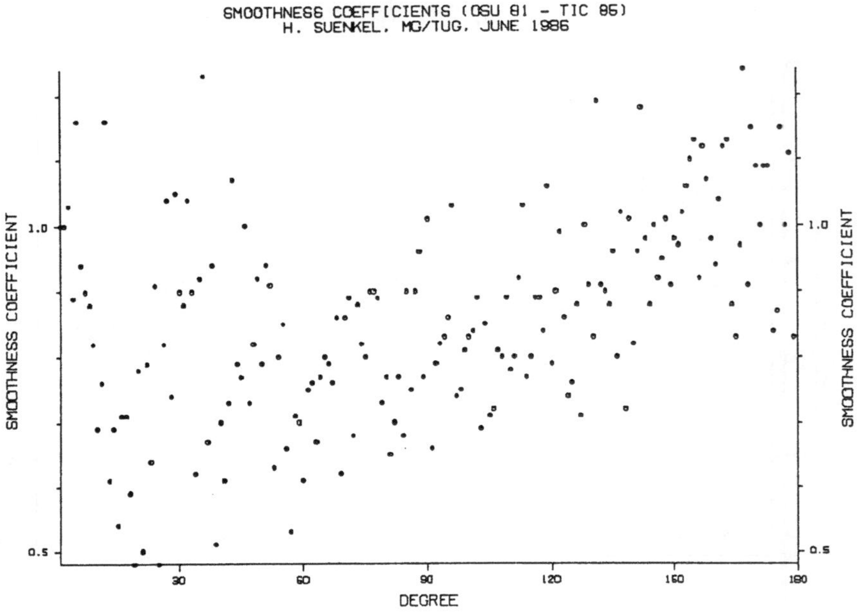

Fig. 6.11 Smoothing coefficients OSU 81 versus TIC 85

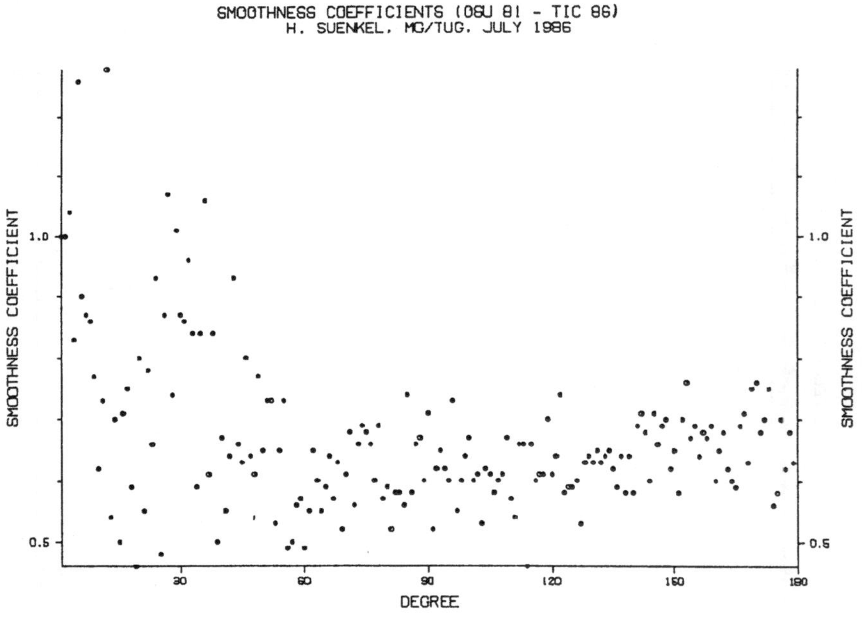

Fig. 6.12 Smoothing coefficients OSU 81 versus TIC 86

yielding higher correlations and lower smoothing coefficients. In addition we intend to use worldwide information about lake depths and ice coverage, the latter one in particular over the Antarctica and over Greenland. These two areas (and also the still strongly isostatically uplifting Scandinavia) are the only major identified areas where a pronounced de-smoothing is observed after the TIC 86 topographic-isostatic reduction of the OSU 81 gravity field. The consideration of ice masses may therefore very well improve our results considerably. A further step is possibly the processing of known lateral density changes within the crust and probably even the use of a smoothing operator which is no longer homogeneous and isotropic.

In any case, the forthcoming TIC 87 model will definitely be a solution complete up to and including degree and order 360 and will be made available upon request by the beginning of 1987.

6.5 Conclusions

Geodesists have been using and are still using the concept of isostasy for filtering purposes only, without bothering very much about its geophysical significance. They are happy with any isostatic model as long as the power of the observed gravity field is sufficiently smooth after a topographic-isostatic reduction. The Vening Meinesz model presented here is probably not capable of smoothing the gravity field dramatically better than vonventional models which have been successfully used in the past, although numerical studies in geodetically pathological areas such as mountainous areas or along ocean trenches and continental margins suggest more optimism. The model presented here should be understood as a bridge between geodesy and geophysics which is more stable and sound than the ones that have been used by geodesists for decades. It is a model which makes us cheat less and therefore, enables us to live scientifically more comfortably.

The compensation smoothing due to Vening Meinesz yields not only a better agreement with observed reality, it also makes Forsberg's argument of antiroots with tops partly above the ocean bottom, a phenomenon observed with the conventional Airy/Heiskanen model, obsolete because the compensation smoothing largely eliminates those singularities. Another argument in favor of a smaller compensation depth is the observed thickness of the crust below the oceans which is on the order of 6 - 8 km. Our model is in good agreement also with this figure. One more argument which speaks in favor of a Vening Meinesz model and against the simple Airy/Heiskanen model is the fact that the strength of the earth's crust is able to support a certain amount of topographic load without local yielding, while in the 1:1 model of Airy/Heiskanen free mobility between vertical mass columns is presupposed - a highly unlikely case. It remains an interesting exercise of applied elasticity theory to figure out whether the kind of compensation smoothing presented here and estimated from global geodetic evidence alone, is compatible with the observed elastic parameters of the crust. Geophysicists are also invited to comment on our geodetic estimates of the depth of compensation.

In any case, we should carefully keep in mind that our model is a global one with only two parameters - it can hardly be simpler. Therefore, it must be considered as a model describing the average behavior of the crust on a global scale. Local deviations from this global model, even large ones, are possible and are well-known from geophysical/geological evidence to exist. Consequently, this model can never be adequate to describe or even explain density patterns which are due to plate tectonics, like slabs sliding down into the mantle at plate margins or the like.

Our topographic-isostatic models received input from a worldwide DTM with a maximum resolution of about 200 km wavelength and partly poor performance in some very large areas. Therefore, we strongly suggest to improve the global DTM both with respect to accuracy and resolution in order to make a computation of an even better model up to degree and order 360 possible.

Topographic-isostatic models of that high resolution are not only important for geophysical research, they are also good and particularly useful for the topographic-isostatic reduction of geodetic gravity field data: using a high resolution model like TIC 86 or better as a global reference, the entire topographic-isostatic reduction problem reduces practically to the processing of small residual topographical masses in a local area which can be done in planar approximation. And for this purpose we have the very powerful FFT algorithm at our disposal which has been so successfully applied by Sideris (1984) for the evaluation of DTM - related integrals, or alternatively, the TC - programs written by Forsberg (1984).

We consider this contribution a small step towards a better understanding of our earth's shell, which A. Wegener once compared with a defendant who declines to answer. The earth scientist, confronted with that defendant, is the judge who has to find the truth from the circumstantial evidences.

ACKNOWLEDGEMENTS

Part of the research related to the development of a global topographic-isostatic earth model was supported by the Air Force Geophysics Laboratory under contract F-19628-82-K-0017. The SYNBAPS 5' x 5' dataset has been made available through the Deutsche Forschungsgemeinschaft under a research cooperation with the FAF University Munich. The new TUG 86 digital terrain model was developed by K. Stubenvoll and M. Wieser and supported by the Austrian Fonds zur Förderung der Wissenschaftlichen Forschung under Project No. P 5481. Part of the computer runs for the development of the new topographic-isostatic earth model TIC 86 have been carried out by M. Hanafy.

REFERENCES

Colombo, O.L.: Numerical Methods for Harmonic Analysis on the Sphere. OSU Report No. 310, 1981.

Forsberg, R.: A Study of Terrain Reductions, Density Anomalies and Geophysical Inversion Methods in Gravity Field Modelling. OSU Report No. 355, 1984.

Heiskanen, W.A. and H. Moritz: Physical Geodesy. Freeman, San Francisco, 1967.

Jekeli, Ch.: Global Accuracy Estimates of Point and Mean Undulation Differences Obtained from Gravity Disturbances, Gravity Anomalies and Potential Coefficients. OSU Report No. 288, 1979.

Khan, M.A.: Earth's Isostatic Gravity Anomaly Field. GSFC, Document No. X-592-73-199, Greenbelt, Md., 1973.

Lachapelle, G.: Determination of the Geoid Using Heterogeneous Data. Mitteilungen der Geodätischen Institute der Technischen Universität Graz, Folge 19, 1975.

Lambeck, K.: Methods and Geophysical Applications of Satellite Geodesy. Rep. Prog. Phys. 42, pp. 547-628, 1979.

Moritz, H.: On the Use of the Terrain Correction in Solving Molodensky's Problem. OSU Report 108, 1968.

Moritz, H.: Advanced Physical Geodesy. Wichmann-Verlag, Karlsruhe, 1980.

Rapp, R.H.: Potential Coefficient and Anomaly Degree Variance Modelling Revisited. OSU Report No. 293, 1979.

Rapp, R.H.: The Earth's Gravity Field to Degree and Order 180 Using SEASAT Altimeter Data, Terrestrial Gravity Data, and Other Data. OSU Report No. 322, 1981.

Rapp, R.H.: Degree Variances of the Earth's Potential, Topography and its Isostatic Compensation. Bull. Géod., No. 56, pp. 84-94, 1982a.

Rapp, R.H.: A Global Atlas of Sea Surface Heights Based on the Adjusted SEASAT Altimeter Data. OSU Report No. 333, 1982b.

Schwarz, K.-P.: Geodetic Accuracies Obtainable from Measurements of First and Second Order Gravitational Gradients. OSU Report No. 242, 1976.

Sideris, M.: Computation of Gravimetric Terrain Corrections Using Fast Fourier Transform Techniques. Publ. No. 20007, Division of Surveying Engineering, The University of Calgary, Alberta, 1984.

Sünkel, H.: An Isostatic Earth Model. OSU Report No. 367, 1985.

Tscherning, C.C.: On the Long-Wavelength Correlation Between Gravity and Topography. Proceedings of the 5th Int. Symposium "Geodesy and Physics of The Earth", Part II, pp. 134-142, Magdeburg, GDR, Veröffentlichungen des Zentralinstituts für Physik der Erde, Nr. 81, Potsdam, 1985.

Vening Meinesz, F.A.: Tables Fundamentales Pour la Réduction Isostatique Régionale. Bull. Géod., No. 63, pp. 771-776, 1939.

THE ERROR MODEL OF INERTIAL GEODESY
A STUDY IN DYNAMIC SYSTEM ANALYSIS

by

K.-P. Schwarz

Division of Surveying Engineering
The University of Calgary
2500 University Drive N.W.
Calgary, Alberta T2N 1N4, Canada

Lecture Notes in Earth Sciences, Vol. 7
Mathematical and Numerical Techniques in Physical Geodesy
Edited by H. Sünkel

1. SCOPE OF THE LECTURE

The lecture uses methods of dynamic system analysis to study the error characteristics of inertial survey systems and their interaction with the anomalous gravity field. Although system analysis is the unifying viewpoint, no attempt has been made to present its theory in a systematic manner. Instead, methods which the author found useful in his own work, will be presented and diversity of techniques will be stressed rather than uniformity of the underlying mathematics. This approach seemed justified in view of the many excellent textbooks on system analysis which have recently been published. They will be referenced for all important details and will thus fill in where the following presentation remains spotty. It is hoped, however, that even this introductory treatment of dynamic system analysis will stimulate interest in a field which, in the author's opinion, is important for future geodetic work.

2. THE MATHEMATICAL FRAMEWORK OF THE PROBLEM

In the following, the formulation of the problems of inertial positioning and inertial gravimetry will be given in a Newtonian framework, i.e. relativistic and quantum mechanical aspects will be ignored. Considering the accuracy of current inertial sensors, such an approach seems to be an appropriate application of Occam's razor.

The representation space for unconstrained Newtonian systems is the Kronecker product

$$E_{31}(\underline{r},t) = E_3(\underline{r}) \times E_1(t) \qquad (1)$$

where

$E_3(\underline{r})$ ... is the three-dimensional homogeneous and isotropic space of coordinates $\underline{r}$, and

$E_1(t)$ ... is the one-dimensional homogeneous space representing time t.

Here as in the sequel, underlined lower case letters denote vectors while underlined upper case letters denote matrices. The definition of $E_3(\underline{r})$ guarantees the existence of inertial frames. The independence of $E_3(\underline{r})$ and $E_1(t)$ in conjunction with the homogeneous character of the latter, ensures the simultaneity of the same event in two inertial frames, i.e. the absoluteness of time. The Galilean equivalence principle requires the invariance of Newton's laws in all inertial frames. Thus, only certain types of coordinate transformations between inertial frames are admissible. This set of linear transformations in $E_{31}(\underline{r},t)$ is called Galilean transformations and is given by the equations

$$\underline{r}' = \underline{R}\,\underline{r} + \underline{v}\,t + \underline{r}_o \qquad t' = t + t_o \tag{2}$$

where

$\underline{R}$ ... is a three-dimensional orthogonal transformation matrix,

$\underline{v}$ ... is a three-dimensional velocity vector,

$(\underline{r}',t')$ ... denotes the coordinate of the new inertial frame with origin $(\underline{r}_o,t_o)$.

The Galilean transformations form a 10-parameter group G_{31} consisting of three rotations, three velocities, and four translations.

Conservative systems are form-invariant under Galilean transformations and can be represented by a system of differential equations of the form

$$m_k\,\ddot{\underline{r}}_k - \underline{f}_k(t,\underline{r},\dot{\underline{r}}) = \underline{0}\;, \tag{3a}$$

where

m_k ... is the mass of the k-th particle,

$\underline{r}_k$ ... is the position vector of the k-th particle in an inertial frame,

$\dot{\underline{r}}_k,\ddot{\underline{r}}_k$ ... are the first and second time derivatives of $\underline{r}_k$, and

$\underline{f}_k$ ... are all forces derivable from a potential W and thus satisfying

$$\underline{f}_k(t,\underline{r},\dot{\underline{r}}) = -\frac{\partial W}{\partial \underline{r}_k} + \frac{d}{dt}\frac{\partial W}{\partial \dot{\underline{r}}_k}\;. \tag{3b}$$

In the following, the inertial measuring unit will be considered as an idealized point mass, so that the subscript k can be omitted. We then have

$$m\,\ddot{\underline{r}} - \underline{f}(t,\underline{r},\dot{\underline{r}}) = \underline{0} \tag{3c}$$

as the fundamental equation and similarly for equation (3b)

$$\underline{f}(t,\underline{r},\dot{\underline{r}}) = -\frac{\partial W}{\partial \underline{r}} + \frac{d}{dt}\frac{\partial W}{\partial \dot{\underline{r}}}\;. \tag{3c}$$

This simplification will not affect subsequent conclusions.

Equation (3c) can be used as a model for inertial navigation as long as all forces acting on the system are of the form (3d). This is certainly the case for gravitational forces, which can be derived from a potential, but not necessarily for all forces acting on accelerometers and gyroscopes which, as e.g. drag forces, may be dissipative in nature. This suggests that equation (3) can be used as a model for error-free navigation, i.e. navigation with perfect sensors in the gravitational field of the Earth. The resulting model equations are usually called mechanization equations. As soon as instrumental errors are taken into account, equation (3c) is not sufficient anymore. It has to be replaced by

$$m\,\ddot{\underline{r}} - \underline{f}(t,\underline{r},\dot{\underline{r}}) = \underline{F}(t,\underline{r},\dot{\underline{r}}) \tag{4}$$

where

$\underline{F}(t,\underline{r},\dot{\underline{r}})$ ... contains all forces which do not satisfy equation (3d).

Equation (4) describes a nonconservative system and is the appropriate model for inertial navigation including system errors. From a theoretical point of view it is, however, a much more difficult model to handle because it is not necessarily form-invariant under Galilean transformations which means that the usual Lagrangian and Hamiltonian approaches are not immediately applicable. The question therefore arises whether the conventional analytic representation can be used for systems of the form (4) which have a specific structure. The answer is in the affirmative for regular holonomic systems of the general form

$$A_{ij}(t,\underline{q},\dot{\underline{q}})\,\ddot{\underline{q}}^{j} + B_{i}(t,\underline{q},\dot{\underline{q}}) = 0 \;, \tag{5}$$

where $\underline{q}$ are generalized coordinates. It should be noted that equation (5) is linear in the accelerations but allows for acceleration coupling through the j-index in the first term. For a more detailed discussion, see Santilli (1978). It appears that equation (5) covers all types of nonconservative forces that may occur in inertial navigation systems. Thus, the conventional Lagrangian or Hamiltonian formulations are applicable for systems of this type.

Equation (5) comprises time-dependent nonlinear couplings in the coordinates and velocities. If they can be neglected and only the linear couplings have to be considered, equation (5) simplifies to the linear system with constant coefficients

$$\underline{A}\,\ddot{\underline{q}} + \underline{B}\,\dot{\underline{q}} + \underline{C}\,\underline{q} = \underline{0} \tag{6}$$

There is some observational evidence (Vassiliou 1984), that the step from equation (5) to equation (6) is an oversimplification in case of inertial positioning.

A more appropriate linearized model for the inertial positioning problem is therefore

$$A\, \underline{\ddot{q}} + \underline{B}\, \underline{\dot{q}} + \underline{C}\, \underline{q} = \underline{d}(t) \tag{7}$$

where $\underline{d}(t)$ is an arbitrary time function. This equation comprises the above as a special case and permits a simple treatment of harmonic excitations. An eigenvector approach to equation (7) will therefore be discussed.

From a theoretical point of view, the approximation of equation (5) by equation (7) is not without difficulty. Classical perturbation theory applied to equation (5) will, in general, lead to non-convergent series because of small denominators. The physics behind the perturbations makes it rather difficult to isolate those regions of phase space where convergence may occur.

However, at the present stage, the use of equation (7) as a model for inertial positioning seems quite appropriate for the following reason. In the survey use of inertial systems frequent zero velocity updates are made. They are used in real time to control the resonances of the system. Thus, in general, perturbations will be kept small and linearization around a reference trajectory is justified.

3. A FIRST LOOK AT THE INERTIAL ERROR MODEL

Inertial survey systems consist in principle of three orthogonally mounted accelerometers whose orientation in space is known at any instant in time. If the basic measuring frame is inertial, the measurements are of the simple form

$$\underline{f}(t) = \underline{a}(t) - \underline{G}(\underline{r},t) \tag{8}$$

where

$\underline{f}(t)$ ... is a 3D specific force vector
$\underline{a}(t)$ ... is a 3D vector of vehicle accelerations
$\underline{G}(\underline{r},t)$ ... is a 3D vector of gravitational accelerations.

Equation (8) is the fundamental equation of inertial geodesy. It implies that the two basic problems of geodesy, positioning and gravity field determination, cannot be separated. To get vehicle accelerations $\underline{a}$, which twice integrated give position differences, $\underline{G}$ must be known; to obtain gravitational acceleration $\underline{G}$, $\underline{a}$ must be given. Data from inertial survey systems are therefore a prime example for the necessity of an integrated approach to geodesy, where both problems are solved simultaneously, see e.g. Hein (1984). If the two problems are treated separately, they will be called inertial positioning and inertial gravimetry, respectively. The latter term then refers to the determination of the gravity disturbance vector. Usually, $\underline{a}$ and $\underline{G}$ are linearized about known values and are iterated if the linear approximation is not sufficient. This is not too difficult if the data can be considered as static in first approximation as is the case with conventional geodetic data. In that case, $\underline{r}$ and $\underline{G}(\underline{r})$ are determined independent of time after appropriate corrections to the actual measurements have been made.

This is not possible with the data coming from inertial survey systems. In that case, the system of differential equations (7) of which equation (8) is a special case, has to be solved for $\underline{r}(t)$ and $\underline{G}(\underline{r},t)$. Errors in the reference trajectory, caused by gravity disturbances and sensor errors, will determine the function $\underline{d}(t)$ on the right-hand side and thus the complexity of the solution. In case of free inertial navigation, an initial value problem has to be solved; in case of aided inertial navigation, it is a multipoint boundary value problem.

Before discussing the error model of a three channel inertial survey system, the simple example of a single axis inertial navigator will be used to explain the basic steps in the derivation of equation (7). The solution of the resulting differential equation for different forms of d(t) will show some important differences between the treatment of dynamic and static data. The discussion of the single axis system can be found in Faurre (1971) and Schwarz (1983a), among others. The system consists of one accelerometer on a stable platform constrained to move in the direction of its sensitive axis. This implies that the accelerometer is nominally orthogonal to the gravity vector and that platform errors can be represented by the error model of a single degree of freedom gyro.

The first step in the derivation of equation (7) is the modeling of the sensor errors. Assuming the usual models as e.g. given in Britting (1971) but excluding the stochastic part, the accelerometer measurement is of the form

$$f_a = (1 + s_a)\, f + b \tag{9a}$$

where

f_a ... is the specific force measured by the accelerometer
s_a ... is the accelerometer scale factor
b ... is the accelerometer bias.

Similarly, the platform rotation can be expressed as

$$\omega_p = (1 + s_g)\omega + d_g \tag{9b}$$

where

ω_p ... is the sensed platform rotation with respect to inertial space
s_g ... is the scale factor of the gyroscope
d_g ... is the gyro drift.

In addition to sensor errors initial errors in

position $\rightarrow \delta r_o$
velocity $\rightarrow \delta v_o$
attitude $\rightarrow \delta \varepsilon_o$

have to be considered. They represent the difference between the given and the true value.

Finally, the difference between the actual gravity vector and the modeled gravity vector has to be taken into account. If a normal gravity field is used, we have

$$\delta g = g_a - \gamma_a = \xi g \tag{9c}$$

where

g_a ... gravity vector component along accelerometer axis
γ_a ... normal gravity vector component along accelerometer axis
ξ ... deflection of the vertical along accelerometer axis
g ... magnitude of gravity vector.

The second step in the derivation of equation (7) is the definition of a set of variables which characterize the physical process under consideration. This is usually done by expressing the process in state variable form, i.e. by using the standard form of a system of first-order differential equations as model. This step will be discussed in some detail in the next chapter. Here, the more direct way will be taken which consists of combining equations (9a) to (9c) to obtain a second-order differential equation for the positioning error δr of the reference trajectory.

The third step, closely linked with the second, is the linearization of the process about the reference trajectory. This is done by setting

$$g = \gamma$$

$$f_a = \ddot{r}_{Ref.}$$

which results in

$$\delta\ddot{r} = f_a - \ddot{r} + \delta g \quad . \tag{10a}$$

Using equation (9a), we obtain

$$\delta\ddot{r} = (1 + s_a)f + b - \ddot{r} + \delta g \tag{10b}$$

which considers the accelerometer errors but not platform rotation and attitude errors. An attitude error $\delta\varepsilon$ leads to a specific force measurement

$$f = \ddot{r}\cos\delta\varepsilon - g\sin\delta\varepsilon$$

which for small $\delta\varepsilon$ can be approximated by

$$f \approx \ddot{r} - g\,\delta\varepsilon \quad .$$

Introducing this into equation (10b) results in

$$\delta\ddot{r} = (1 + s_a)(\ddot{r} - g\delta\varepsilon) + b - \ddot{r} + \delta g \quad . \tag{11a}$$

Neglecting second-order terms in the errors leads to

$$\delta\ddot{r} = b + s_a\ddot{r} - \gamma(\delta\varepsilon - \xi) \tag{11b}$$

The attitude error $\delta\varepsilon$ is a function of the initial attitude and position error, of platform scale factor and drift. It can be written as

$$\delta\varepsilon = \delta\varepsilon_o - \frac{\delta r_o - \delta r}{R} + \frac{1}{R}\int_{\Delta t} s_g\, v\, dt + \int_{\Delta t} d_g\, dt \tag{11c}$$

Introducing this into equation (11b) and neglecting again second-order terms results in

$$\delta\ddot{r} + \omega_s^2\,\delta r = b - \gamma(\delta\varepsilon_o - \xi) + \omega_s^2\,\delta r_o + s_a\ddot{r}$$
$$- \omega_s^2 \int_{\Delta t} s_g\, v\, dt - g \int_{\Delta t} d_g\, dt\,, \qquad (12a)$$

where

$$\omega_s^2 = \gamma/R \qquad (12b)$$

and R is the mean radius of the Earth. This equation has the standard form of a forced vibration

$$\delta\ddot{r} + \omega_s^2\,\delta r = d(t). \qquad (13)$$

The solution of equation (12) can therefore be discussed in the general context of the theory of vibrations. The simplest case

$$d(t) = 0 \qquad (14a)$$

is that of a free, undamped vibration. The solution can in this case be found by solving the eigenvalue problem

$$(\lambda^2 + \omega_s^2)\underline{u} = 0$$

which results in a pair of imaginary eigenvalues

$$\lambda_{1,2} = \pm\sqrt{-1}\,\omega_s\,.$$

The solution is thus of the form

$$\delta r(t) = A\cos\omega_s t + B\sin\omega_s t \qquad (14b)$$

where

$$A = \delta r_o \qquad B = \delta v_o/\omega_s\,. \qquad (14c)$$

This means that the system is stable and oscillates at its natural frequency ω_s.

In our case, ω_s is described by equation (12b) and is almost constant over the Earth's surface. Using the approximation

$$\gamma \approx kM/R$$

we have

$$\omega_s = kM/R^3 = \text{constant}$$

which depends on

k ... universal gravitational constant

M ... mass of the Earth

R ... mean radius of the Earth.

The term ω_s is called the Schuler frequency and its oscillation period is about 84.3 minutes when the above approximations are used. The error behaviour of an inertial system without sensor errors is thus dominated by the Schuler period and has amplitudes which are dependent on the initial errors. This is an important result because it shows that, at least in this simple case, the position and velocity errors are rather systematic in nature and thus highly predictable. The question therefore arises whether or not this is also true for the more realistic case $d(t) \neq 0$.

Before turning to this question, a brief remark on Schuler's principal will be made. In a seminal paper in 1923, Schuler (1923) pointed out that if a mechanical device with a natural oscillation period of 84.3 minutes could be built, such a device would be insensitive to horizontal accelerations near the surface of the Earth. This principle has found its electronic implementation, known as Schuler tuning, in inertial platform systems. It can be visualized in a thought experiment by interpreting equation (13) together with equations (12b) and (14a) as the equation of a simple pendulum. The pivot of the pendulum is on the surface of a spherical earth, while its mass, due to the pendulum length R, rests in the centre of mass of the earth. Movement of the pivot on the surface will not excite oscillations of the hypothetical pendulum making it thus insensitive to horizontal accelerations.

The case of forced oscillations, i.e. the case

$$d(t) \neq 0 \tag{15a}$$

is more difficult to treat. The specific form of d(t) determines whether or not the system remains stable and has Schuler-type oscillations. A discussion of some typical functions d(t) are given in textbooks like Magnus (1976) or Fischer and Stephan (1981), and can be applied to the different terms on the right-hand side of equation (12). To keep the example simple, it will be assumed that b, s_a, s_g, v, d_g, and ξ are constant and that the initial acceleration a_i is constant and lasts only for a period Δt very much shorter than 84 minutes. In that case, equation (12) can be solved. The result is

$$\delta r(t) = \delta r_o + (\delta v_o + s_a\, a_i\, \Delta t)\, \frac{\sin \omega_s t}{\omega_s}$$
$$+ (\frac{b}{\gamma} - \delta\varepsilon_o + \xi_o)\, R(1 - \cos \omega_s t) \tag{15b}$$
$$- (s_g v + R\, d_g)(t - \frac{\sin \omega_s t}{\omega_s}) \, .$$

The solution shows four interesting characteristics:

a. The position error is bounded for δr_o, δv_o, $\delta\varepsilon_o$, b, ξ_o, and s_a; it shows, however, unlimited error growth for s_g and d_g. This means that only aided inertial navigation will be stable. This is even more apparent from a similar analysis of the height channel. In its simplest form it leads to

$$\delta h(t) = e^{\sqrt{2}\, \omega_s t}\, \delta h_o \, . \tag{16}$$

Thus, an initial error in height δh_o will grow very quickly out of bounds. Update measurements are therefore a stringent requirement for inertial geodesy.

b. The position error is still dominated by Schuler-like oscillations, at least for surveys of normal duration. This means that a large part of this error is predictable and can thus be eliminated in real time if update measurement of sufficient accuracy are available. The use of real-time estimates in the system equations corresponds to a damping of the system. The error analysis should therefore consider viscous or structural damping terms in the differential equations depending on whether velocity or position updates have been made.

c. Different types of system errors produce the same characteristic in $\delta r(t)$. Thus, the effects of δv_o and s_a, b and $\delta\varepsilon_o$, s_g and d_g are not separable on the basis of δr-measurements. Similarly, the initial deflection of the vertical ξ_o cannot be separated from an initial attutude error or an accelerometer bias. This indicates that most of the sensor errors are not directly observable. They are lumped together as coefficients of a relatively small number of base functions. By using these characteristic base functions, a simple error model can be built.

d. Zero velocity update measurements will not be affected by scale factor errors s_a and s_g because $v = 0$. This means that the effect of scale factor errors will only show up in coordinate updates.

Since, in normal survey operations, coordinate updates are rare while zero velocity updates are frequent, a two step procedure seems to be most appropriate. In the first step, the combined effect of δr_o, δv_o, $\delta\varepsilon_o$, b, ξ_o, and δ_g is eliminated by the use of zero velocity measurements. In the second step, s_a and s_g are determined from the coordinate updates.

Table 1 gives the magnitude of the position errors caused by the different error sources in equation (15b).

Input error		Position error after		Error growth
Type	Size	1 hour	2 hours	
Initial position error δr_o	3 m	3 m	3 m	zero
Initial velocity error δv_o	3 mm s^{-1}	-2 m	1 m	bounded
Initial attitude error $\delta\varepsilon_o$	5"	192 m	291 m	bounded
Accelerometer bias b	10 mgal	81 m	122 m	bounded
Gyro scale factor s_g	10^{-4}	6 m	10 m	unbounded
Gyro drift d_g	15" h^{-1}	564 m	878 m	unbounded

Table 1: Position error due to sensor and initial errors

They show a somewhat different pattern from similar values for inertial navigation. This is due to the fact that in comparison the accuracy of the initial values in inertial geodesy is extremely high, while the duration of the survey is quite short. With current system accuracies, the largest position errors are caused by gyro drift and by the combined effect of initial attitude error and accelerometer bias.

The foregoing analysis has shown that in case of a single axis navigator the error behaviour is dominated by Schuler-type oscillations and is therefore systematic in nature. It is thus possible to eliminate the total error by regular update measurements. Individual error sources are, however, difficult to separate because different error sources show the same signature in the system output. In addition, not every system output is sensitive to all error sources. Thus, a combination of update measurements is necessary to eliminate the total error. The instability of the basic differential equation does not cause a practical problem as long as a proper mix of regular update measurements is applied.

An extension of these results to the three-dimensional case will be attempted in the next chapters. The basic technique will be the same although more emphasis will be put on the derivation of the standard state space model. In addition, stochastic error terms d(t) and a more realistic gravity model δg will be considered.

4. DERIVATION OF THE 3D DYNAMIC ERROR MODEL

The derivation of the error model for the three-dimensional case will follow the pattern outlined in the last chapter:

- modeling of sensor errors
- process representation by state variables
- linearization about reference trajectory.

The presentation will be rather concise and reference is made to Britting (1971) and Wong (1982) for all details. To simplify the discussion, the derivation will only be given for the local-level mechanization. Other mechanizations and their relation to the one discussed here can again be found in Britting (1971).

The error model for the accelerometer triad of local-level mechanization is

$$\underline{f}_a = \underline{b} + \underline{B}\,\underline{f} + \underline{e}_f \tag{17a}$$

where

$\underline{f}_a$ ... is the measured specific force vector
$\underline{b}$ ... is a vector of accelerometer biases
$\underline{e}_f$ ... is the accelerometer random uncertainty,

and $\underline{B}$ is of the form

$$\underline{B} = \underline{I} + \underline{S} + \underline{E}_\varepsilon + \underline{E}_\alpha \tag{17b}$$

with the scale factor matrix $(\underline{I} + \underline{S})$

$$\underline{I} + \underline{S} = \begin{bmatrix} 1+s_E & 0 & 0 \\ 0 & 1+s_N & 0 \\ 0 & 0 & 1+s_U \end{bmatrix} , \qquad (17c)$$

the orthogonal platform attitude error matrix $\underline{E}_\varepsilon$

$$\underline{E}_\varepsilon = \begin{bmatrix} 0 & \varepsilon_U & -\varepsilon_N \\ -\varepsilon_U & 0 & \varepsilon_E \\ \varepsilon_N & -\varepsilon_E & 0 \end{bmatrix} , \qquad (17d)$$

and the non-orthogonal misalignment matrix of the accelerometer triad $\underline{E}_\alpha$

$$\underline{E}_\alpha = \begin{bmatrix} 0 & -\alpha_{EU} & \alpha_{EN} \\ \alpha_{NU} & 0 & -\alpha_{NE} \\ -\alpha_{UN} & \alpha_{UE} & 0 \end{bmatrix} . \qquad (17e)$$

The subscripts {E,N,U} stand for east, north, and up. The vector of accelerometer errors

$$\underline{\delta f}_a = \underline{f}_a - \underline{f} \qquad (18a)$$

is thus of the form

$$\underline{\delta f}_a = \underline{b} + (\underline{S} + \underline{E}_\varepsilon + \underline{E}_\alpha)\underline{f} + \underline{e}_f \; . \qquad (18b)$$

The error model for the platform rotation of a local-level IMU is given by

$$\underline{\omega}_p = \underline{E}\;\underline{\omega} + \underline{e}_\omega \qquad (19a)$$

where

$\underline{\omega}_p$ ... is the actual angular velocity of the platform with respect to the inertial reference frame,

$\underline{\omega}$ ... is the true angular velocity of the platform with respect to the inertial reference frame,

$\underline{e}_\omega$ ... is the uncertainty in the angular velocity of the platform, mainly generated by uncompensated gyro errors,

and $\underline{E}$ is of the form

$$\underline{E} = (\underline{I} + \underline{T} + \underline{E}_\beta) \qquad (19b)$$

with the gyro torque scale factor matrix

$$\underline{I} + \underline{T} = \begin{bmatrix} 1+t_x & 0 & 0 \\ 0 & 1+t_y & 0 \\ 0 & 0 & 1+t_z \end{bmatrix} , \qquad (19c)$$

and the misalignment matrix $\underline{E}_\beta$ describing the gyro nonorthogonalities

$$\underline{E}_\beta = \begin{bmatrix} 0 & -\beta_{xz} & \beta_{xy} \\ \beta_{yz} & 0 & -\beta_{yx} \\ -\beta_{zy} & \beta_{zx} & 0 \end{bmatrix} . \qquad (19d)$$

The vector of platform rotation errors

$$\underline{\delta f}_\omega = \underline{\omega}_p - \underline{\omega} \qquad (20a)$$

is thus

$$\underline{\delta f}_\omega = (\underline{T} + \underline{E}_\beta)\, \underline{\omega} + \underline{e}_\omega \quad . \qquad (20b)$$

In contrast to the model used previously, no drift terms appear explicitely in equations (19). Their effect is contained in the term $\underline{e}_\omega$, i.e. they are considered as stochastic variables. This is certainly the more correct approach considering that platform drifts are lumped parameters depending on a variety of error sources and that drift compensation is a major task in factory calibration. There are, however, good practical reasons to split the total $\underline{e}_\omega$-term into two parts. One which accounts for linear drift and another which accounts for nonlinear drift and white noise. In general, both parts will be correlated stochastic processes.

Initial errors in position, velocity, and attitude are defined as before and are denoted by $\underline{\delta r}_o$, $\underline{\delta v}_o$, $\underline{\delta \varepsilon}_o$, respectively. The gravity disturbance vector is

$$\underline{\delta g} = \underline{g} - \underline{\gamma}_h = \begin{bmatrix} g\,\eta \\ g\,\xi \\ \delta_g \end{bmatrix} \qquad (21)$$

where

$\underline{g}$ ... is the gravity vector

$\underline{\gamma}_h$ ... is the normal gravity vector at h

ξ, η ... are the deflections of the vertical in N-S and E-W direction

δ_g ... is the gravity disturbance.

The second step, the construction of a state variable representation, defines those variables which are essential in characterizing the error process. The underlying idea is that the state variable model provides an explanation of the observed data. It explains the internal workings of the black box called system errors. In our case, the system is known for errorless data. It is given by the so-called mechanization equations. The obvious approach is therefore to represent the system errors as a perturbation of the nominal trajectory given by these equations.

The starting equation is the specific force equation for a local-level system which is of the form

$$\underline{f} = \underline{\omega} \times (\underline{\omega} \times \underline{r}) + \dot{\underline{\omega}} \times \underline{r} + 2\underline{\omega} \times \dot{\underline{r}} + \ddot{\underline{r}} - \underline{G} \tag{22}$$

where

$$\underline{\omega} = \begin{bmatrix} -\dot{\phi} \\ \cos\phi\ \dot{\mu} \\ \sin\phi\ \dot{\mu} \end{bmatrix}$$

$$\dot{\phi} = \frac{v_N}{(M + h)} \qquad \dot{\mu} = \omega_E + \dot{\lambda}$$

$$\dot{\lambda} = \frac{v_E}{(N + h)\cos\phi}$$

$$M = \frac{a(1 - e^2)}{(1 - e^2 \sin^2\phi)^{3/2}} \qquad N = \frac{a}{(1 - e^2 \sin^2\phi)^{1/2}}$$

and

ϕ, λ, h ... are the ellispoidal latitude, longitude, and height

v_N, v_E ... are velocities in north and east direction

ω_E ... is the rotation rate of the earth

M, N ... are the meridian and prime vertical radius of curvature, respectively

a, e ... are the semimajor axis and the first eccentricity of the ellipsoid

$\times$ ... denotes the cross product.

For a concise derivation of the formula and a discussion of the coordinate systems involved, see Schwarz (1983b). The actual mechanization equations are a transformation of equation (22). They are derived in detail in Farrell (1976). They replace the three second-order differential equations in $\underline{r}$ by three first-order equations in $\underline{v}$ and are of the form

$$\underline{f} = \dot{\underline{v}} + \underline{\rho} \times \underline{v} - \underline{g} \tag{23}$$

where

$$\underline{v} = \begin{bmatrix} v_E \\ v_N \\ v_U \end{bmatrix} \qquad \underline{\rho} = \begin{bmatrix} -\dot{\phi} \\ \ell \cos\phi \\ \ell \sin\phi \end{bmatrix} \qquad \underline{g} = \begin{bmatrix} \eta g \\ \xi g \\ g \end{bmatrix}$$

$$\ell = 2\omega_E + \dot{\lambda}$$

and $\underline{g}$ is again the gravity vector. This equation will be used to derive the perturbation equations.

The choice of state variables for a specific process is not unique. One would, however, expect that two different sets of state variables will lead to the same results, i.e. biases between representations should not occur. System theory has put this principle into mathematical form by defining a canonical realization of a process. It has the property that the model depends only on the data and is not dependent on any external bias. This property can be tested for any state variable model by investigating reachability and observability. A system which is both reachable and observable is called canonical. The concepts of reachability and observability are rather involved for the general nonlinear case. They take, however, a rather simple form in the linear case and a few more details will be given later. One important result is that for linear systems canonical is equivalent to minimal, i.e. for a given set of data the canonical representation is the one with the smallest possible number of states. Combined with the isomorphism between different canonical realizations of the same data set, this leads to a simple transformation between canonical state vectors. It appears that a number of questions in the area of state space modeling of inertial data could be brought closer to an answer by applying some of the recent advances in partial realization theory to these problems. An excellent review of recent developments can be found in Casti (1985).

In the following, the canonical state vector discussed in Britting (1971) will be taken as the basis for the subsequent development. It consists of three states each, for attitude errors $\underline{\delta\varepsilon}$, position errors

$\underline{\delta r}$, and velocity errors $\underline{\delta v}$. We thus have

$$\underline{x} = \{\underline{\delta\varepsilon},\ \underline{\delta r},\ \underline{\delta v}\}^T . \tag{24}$$

The third step, the linearization of equation (23) with respect to $\underline{x}$ is straightforward in principle although somewhat involved in detail. The reference trajectory is obtained by setting

$$\underline{g} = \underline{\gamma} \tag{25a}$$

$$\underline{\dot{v}}_R = \underline{f}_a - \underline{\rho}_R \times \underline{v}_R + \underline{\gamma} \tag{25b}$$

where the subscript R indicates reference values. Using vector differentiation, we obtain

$$\underline{\delta\dot{v}} = \underline{\delta f}_a + \underline{v}_R \times \underline{\delta\rho} - \underline{\rho}_R \times \underline{\delta v} + \underline{\delta g} \tag{26a}$$

where $\underline{\delta f}_a$ is given by equation (18b) and $\underline{\delta\rho}$ is of the form

$$\underline{\delta\rho} = \underline{R}\ \underline{dr}_1 \tag{27}$$

with

$$\underline{R} = \begin{bmatrix} -\dot{\mu}\sin\phi & \cos\phi\,\frac{d}{dt} \\ -\frac{d}{dt} & 0 \\ -\dot{\mu}\cos\phi & -\sin\phi\,\frac{d}{dt} \end{bmatrix} \qquad \underline{dr}_1 = \begin{bmatrix} \delta\phi \\ \delta\lambda \end{bmatrix}$$

Combining equations (18b), (26a), and (27) leads to

$$\underline{\delta\dot{v}} = -\underline{\rho} \times \underline{\delta v} + \underline{w}\{\underline{b},\ \underline{s},\ \underline{\delta\varepsilon},\ \underline{\alpha},\ \underline{dr},\ \underline{dg},\ \underline{e}_f\} \tag{26b}$$

which gives one set of equations in the standard form, i.e. as a system of first-order differential equations in the error terms. Note that the attitude errors $\underline{\varepsilon}$ in equation (18b) are the same as the $\underline{\delta\varepsilon}$ in equation (24). The computation of the terms $\underline{w}\{\cdot\}$ is quite involved and reference is again made to Britting (1971) for all details. Similar equations can be derived for $\underline{\delta\dot{\varepsilon}}$ and $\underline{\delta\dot{r}}$ which after reordering lead to

$$\begin{bmatrix} \underline{\delta\dot{\varepsilon}} \\ \underline{\delta\dot{r}} \\ \underline{\delta\dot{v}} \end{bmatrix} = \underline{F} \begin{bmatrix} \underline{\delta\varepsilon} \\ \underline{\delta r} \\ \underline{\delta v} \end{bmatrix} + \underline{w} \left\{ \begin{array}{c} (\underline{t},\ \underline{\beta},\ \underline{e}_\omega) \\ (\underline{s},\ \underline{\alpha},\ \underline{e}_f,\ \underline{\delta}_g) \end{array} \right\} \tag{26c}$$

This is the simplest form of the state space model describing the error behaviour of a three-dimensional inertial survey system. The dynamics matrix for this specific formulation is given in Table 2. For a derivation of the matrix and a more detailed discussion of individual terms, see Wong (1982).

$$\underline{F} = \begin{bmatrix} 0 & \ell \sin\phi & -\ell\cos\phi & 0 & 0 & 0 & -1 & 0 & 0 \\ -\ell\sin\phi & 0 & -\dot{\phi} & -\ell\sin\phi & 0 & 0 & 0 & \cos\phi & 0 \\ \ell\cos\phi & \dot{\phi} & 0 & \ell\cos\phi & 0 & 0 & 0 & \sin\phi & 0 \\ 0 & 0 & 0 & 0 & 0 & 0 & 1 & 0 & 0 \\ 0 & 0 & 0 & 0 & 0 & 0 & 0 & 1 & 0 \\ 0 & 0 & 0 & 0 & 0 & 0 & 0 & 0 & 1 \\ f_U/r & 0 & -f_E/r & 0 & 0 & 0 & 0 & -\ell\sin 2\phi & -2\dot{\phi}/r \\ 0 & -f_U\sec\phi/r & f_N\sec\phi/r & 0 & 0 & 0 & (\dot{\lambda}+2\omega_e)\tan\phi & 0 & -(\dot{\lambda}+2\omega_e)/r \\ -f_N & f_E & 0 & 0 & 0 & 2\gamma/r-k2 & 2r\dot{\phi} & 2r\ell\cos^2\phi & 0 \end{bmatrix}$$

where $k2 = 2\alpha\gamma/r$ is a damping constant and the state vector x is defined by

$$\underline{x} = (\varepsilon_E, \varepsilon_N, \varepsilon_{UP}, d\phi, d\lambda, dh, d\dot{\phi}, d\dot{\lambda}, d\dot{h})^T$$

Table 2: Dynamics matrix and state vector of equation (24)

Equation (26c) has the general form

$$\dot{\underline{x}} = \underline{F}\,\underline{x} + \underline{G}\,\underline{u} \tag{27a}$$

which is the standard linear state model of system theory. In engineering applications the individual terms are often called

$\underline{x}$ ... state vector
$\underline{F}$ ... dynamics matrix
$\underline{u}$ ... input vector (white noise)
$\underline{G}\,\underline{u}$ ... forcing function (random).

This system has the output

$$\underline{y} = \underline{H}\,\underline{x} + \underline{e} \tag{27b}$$

where

$\underline{y}$ ... system output (observable)
$\underline{H}$ ... design matrix
$\underline{e}$ ... measurement noise.

The matrices $\{\underline{F}, \underline{G}, \underline{H}\}$ describe the internal model of the linear dynamical system. System theory provides the mathematical tools to determine $\{\underline{F}, \underline{G}, \underline{H}\}$ or, in a more general setting, functions $\{f, g, h\}$ which satisfy certain mathematical properties. The model which has been most extensively applied and for which suitable algorithms are available is the one given by equations (27) where $\underline{F}$, $\underline{G}$, and $\underline{H}$ are constant coefficient matrices. It will be exclusively used in the following. An in-depth discussion of more general linear cases is given in Casti (1977).

The question naturally arises whether in our case equation (27) is a realistic model. Table 2 shows that $\underline{F}$ is not a constant coefficient matrix but is dependent on accelerations. This difficulty can be overcome by computing $\underline{F}(\Delta t)$ for time intervals Δt of constant acceleration and by summing the sequence of $\underline{F}(\Delta t)$-matrices. Nonlinearities in the model, as e.g. the dependance of $\underline{F}$ on $\underline{x}$, can be kept at an acceptable level by real-time updating of $\underline{F}$. The dynamics in standard survey applications seems to be benign enough to justify such a procedure, even on theoretical grounds; for some numerical results, see Vassiliou and Schwarz (1985). The most stringent modeling requirement is in our case the white noise property of $\underline{u}$ in equation (27a). As an inspection of the last term in equation (26c) shows, it is not very likely that the stochastic processes in $\underline{w}$ can all be generated by sending white noise through an appropriately chosen matrix $\underline{G}$. Therefore an extension of the state vector (24) has frequently been chosen as an alternative. The additional states are described by a spectral density function and are, in the majority of cases, low-order Gauss-Markov processes. They are chosen because they can be easily incorporated in the state variable model.

A typical case is the inclusion of three platform drifts $\underline{\delta\omega}$ in the state vector (24) which then takes the form

$$\underline{x}_{12} = \{\underline{\delta\varepsilon}, \underline{\delta r}, \underline{\delta v}, \underline{\delta\omega}\}^T \tag{28a}$$

The reasoning in this case is that in surveying applications the platform commands of a local-level mechanization are dominated by the earth rotation and can be considered constant in first approximation. Multiplied by the constant elements of $\underline{T}$ and $\underline{E}_\beta$, they will produce constant terms in the forcing functions of the attitude equations. Constant terms in these equations correspond to a linear drift in the attitude angles while the small variations of these terms can be interpreted as nonlinear drifts. It is therefore an accepted practice to lump the effects of $\underline{t}$, $\underline{\beta}$, and $\underline{e}_\omega$ into three linear drift rates $\underline{\delta\omega}$. They will absorb the linear part of the platform rotation errors. The remaining part $\underline{\delta e}_\omega$ can usually be described in good approximation by the stochastic properties of $\underline{G}\ \underline{u}$.

The dynamics matrix belonging to the state vector $\underline{x}_{12}$ can be written as

$$\underline{F}_{12} = \begin{bmatrix} \underset{(9,9)}{\underline{F}} & \underset{(3,3)}{I} \\ \underset{(3,9)}{\underline{0}} & \underset{(9,3)}{\underline{0}} \end{bmatrix} \tag{28b}$$

with $\underline{F}$ defined as in Table 2. The corresponding state variable model has the form

$$\begin{bmatrix} \underline{\delta\dot\varepsilon} \\ \underline{\delta\dot r} \\ \underline{\delta\dot v} \\ \underline{\delta\dot\omega} \end{bmatrix} = \underline{F}_{12} \begin{bmatrix} \underline{\delta\varepsilon} \\ \underline{\delta r} \\ \underline{\delta v} \\ \underline{\delta\omega} \end{bmatrix} + G \begin{bmatrix} \underline{0} \\ \underline{0} \\ \underline{e}_f \\ \underline{\delta e}_\omega \end{bmatrix} + \underline{\delta g} + C \begin{bmatrix} s \\ \alpha \end{bmatrix} . \tag{28c}$$

Since the last term on the right-hand side does not affect $\underline{\delta\dot v}$ at zero velocity, it can be excluded when zero velocity measurements are considered and a good reference trajectory is available.

The gravity disturbance vector has been written separately because it can be treated in different ways. In areas where a good local gravity field approximation is available, it can be used as a control function with the remaining uncertainty to be modeled in $\underline{G}\ \underline{u}$. In areas where no gravity information beside the normal model is available, it can be included in the state vector, and thus the estimation, with an appropriate spectral density function. In cases where it is considered as a nuisance parameter, i.e. a disturbance in the inertial positioning problem, it can be modeled into $\underline{G}\ \underline{u}$ and its effect can be thus eliminated.

In the following, the state variable model (27) with state vector (24) and (28a) will be discussed from two different points of view. In the next chapter, an essentially deterministic approach will be taken via an eigenvalue-eigenvector analysis. In chapter 6, the standard solution of equation (27) using Kalman filtering will be discussed which provides an optimal estimate of $\underline{x}$ in case of correct stochastic assumptions.

5. EIGENSYSTEM ANALYSIS - THE DETERMINISTIC POINT OF VIEW

This chapter extends the discussion of chapter 3 to the state variable models developed in the last chapter. The task is thus the solution of the eigenvector equation

$$(\underline{F} - \lambda_i \underline{I})\underline{v}_i = \underline{0} \qquad (29)$$

for systems (26c) and (28b). Here λ_i are the eigenvalues and $\underline{v}_i$ the corresponding eigenvectors. There should be no difficulty to keep the eigenvector $\underline{v}_i$ and the velocity vector $\underline{v}$ apart by looking at the context. Equation (29) solves only the homogeneous part of model (27), i.e. it gives only information on the free vibrations of the system. Forcing terms can be introduced as before and will give a more complete picture of the system response. The forcing terms $\underline{G}\ \underline{u}$ have not to be stochastic which is a major difference to the filtering approach in the next chapter.

The approach outlined here is related to modal analysis which is often applied to the determination of vibrations in mechanical systems, specifically large engineering structures. However, two important differences have to be kept in mind. In modal analysis the system vibrations are derived from measurements, i.e. we have a problem of parametric model identification. Identification is a secondary point in our analysis where the eigenvalues and eigenvectors are determined analytically from the given matrix $\underline{F}$. This makes for a somewhat simpler problem. On the other hand, modal analysis usually makes symmetry assumptions on the matrices in the basic differential equations. They are not applicable in our case because an active system, not a passive one, has to be treated. A good introduction into the standard cases encountered in modal analysis is given in Natke (1983).

The eigenvalue analysis of models (26c) and (28b) is given in detail in Vassiliou and Schwarz (1985) for the zero-acceleration case and will only be summarized here. Since difficulties can be expected for the height channel, the solution is first given for the seven state variable case excluding δh and $\delta\dot{h}$. The characteristic determinant is in this case

$$\det(\lambda\underline{I}-\underline{F}_7) = \lambda(\lambda^2+\omega_e^2)(\lambda^4+2\lambda^2(2\omega_e^2\sin^2\phi+\omega_s^2)+\omega_s^4) \tag{30a}$$

which leads to the eigenvalues

$$\begin{aligned}
\lambda_1 &= j\ \omega_e\\
\lambda_2 &= -j\ \omega_e\\
\lambda_3 &= j\ \theta_1\\
\lambda_4 &= -j\ \theta_1\\
\lambda_5 &= j\ \theta_2\\
\lambda_6 &= -j\ \theta_2\\
\lambda_7 &= 0
\end{aligned} \tag{30b}$$

where $j = \sqrt{-1}$

$$\begin{aligned}
\theta_1 &= \{\omega_s^2 + 2\omega_e\sin\phi(\omega_e\sin\phi - [\omega_e^2\sin^2\phi + \omega_s^2]^{1/2}\}^{1/2}\\
\theta_2 &= \{\omega_s^2 + 2\omega_e\sin\phi(\omega_e\sin\phi + [\omega_e^2\sin^2\phi + \omega_s^2]^{1/2}\}^{1/2} \quad .
\end{aligned} \tag{30c}$$

The zero eigenvalue is due to the indeterminacy of the longitude origin and can, without loss of generality, be disregarded in the following discussion. The three pairs of imaginary eigenvalues indicate that the solution is stable if the longitude origin is fixed. It is governed by three dominant frequencies which are the

- Schuler rate ω_s with a period of 1.4 h
- Earth rate ω_E with a period of 24 h
- Foucault rate ω_F with a period of 24 h/sinϕ.

The frequencies ω_s and ω_F are obtained from equation (30c). θ_i (i=1,2) is rewritten as

$$\theta_i = \omega_s\{1 + 2\alpha^2\sin^2\phi \pm 2\alpha\sin\phi(1 + \alpha^2\sin^2\phi)^{1/2}\}^{1/2}$$

where

$$\alpha = \frac{\omega_e}{\omega_s} \cong 0.058$$

and where ($\pm$) indicates the two solutions. Considering the size of α, the square root in the bracket can be approximated by

$$(1 + \alpha^2\sin^2\phi)^{1/2} \cong 1 + \frac{1}{2}\alpha^2\sin^2\phi$$

which results in

$$\theta_i = \omega_s\{1 \pm 2\alpha\sin\phi + 2\alpha^2\sin^2\phi \pm \alpha^3\sin^3\phi\}^{1/2} \quad .$$

Using the same approximation for the whole bracket gives

$$\theta_i = \omega_s\{1 \pm \alpha\sin\phi + \alpha^2\sin^2\phi - \frac{1}{2}\alpha^3\sin^3\phi\} \quad .$$

Neglecting terms with α^n ($n \geq 2$), we get the usual approximation, see e.g. Britting (1971, p. 128)

$$\theta_i = \omega_s(1 \pm \alpha\sin\phi) \quad . \tag{30d}$$

The dominant term in this equation is the Schuler frequency ω_s. The additional term gives rise to a beat frequency ω_F with a period of $(2\pi/\omega_e \sin\phi)$. Figure 1 shows the three dominant oscillations and the total effect which characterizes the error behaviour of an inertial system without height channel. It is clear that for survey missions of one or two hours duration, the Schuler-type oscillations will again be the dominant effect while the other two oscillations will basically add a constant and a linear term.

Using the solution (30) as a basis, model (24) can now be solved from the characteristic determinant

$$\det(\lambda\underline{I} - \underline{F}) = \lambda\{\lambda^2-(2\omega_s^2 - k_2)\}[\lambda^4+2\lambda^2(2\omega_e^2\sin^2\phi+\omega_s^2)+\omega_s^4]. \tag{31a}$$

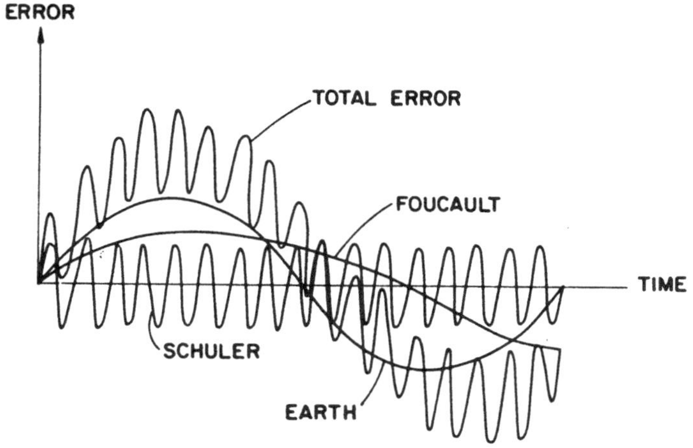

Figure 1: Dominant error frequencies of two-dimensional inertial navigator

The eigenvalues λ_1 to λ_7 are the same as before, thus the same error behaviour can be expected for ϕ, λ and their corresponding velocities. The two additional eigenvalues are

$$\lambda_{8,9} = \pm \left\{ \begin{array}{ll} (2\omega_s^2 - k_2)^{1/2} & \text{for } 2\omega_s^2 \geq k_2 \\ j(2\omega_s^2 - k_2)^{1/2} & \text{for } k_2 > 2\omega_s^2 \end{array} \right\} \qquad (31b)$$

Two cases have to be distinguished. If $2\omega_s^2 \geq k_2$, one eigenvalue with a positive real part is obtained and the system becomes unstable. If $2\omega_s^2 < k_2$, another pair of imaginary roots is obtained and the system will remain stable. The factor k_2 regulates the amount of damping in the vertical accelerometer. This shows that an undamped inertial survey system is unstable due to the height channel, a result already obtained in the single axis case. Equation (31b) shows also how this problem can be overcome without sacrificing measuring sensitivity. By introducing outside information on heights and weighting it properly, a stable solution can be achieved. In inertial surveying, outside height information is provided by zero velocity updates during vehicle stops. This corresponds to viscous damping of the error system. In aircraft applications the inertial output is often stabilized by precise barometric measurements. In future, GPS range or range rate measurements will be used for this purpose. This corresponds to structural damping of the error system. No corresponding study on the change in eigenvalues due to the damping of the horizontal channels is known to the author.

The analysis of system (28b) leads to the characteristic determinant

$$\det(\lambda \underline{I} - \underline{F}_{12}) = \lambda^4 \{\lambda^2 - (2\omega_s^2 - k_2)\} [\lambda^4 + 2\lambda^2 (2\omega_e^2 \sin^2\phi + \omega_s^2) + \omega_s^4] \tag{32}$$

which, in addition to the eigenvalues obtained previously, has three zero eigenvalues. Thus, the system is unstable.

The physical explanation is simple in this case. A platform drift, however small, will generate unbounded error growth for $t \to \infty$.

Mathematically, this instability can be eliminated by replacing matrix (28b) by

$$\underline{F}_{12} = \left[\begin{array}{c|c} \underline{F}(9,9) & \underline{I}\ (3,3) \\ & \underline{0}\ (6,6) \\ \hline \underline{0}\ (3,3) & -\beta_i \underline{I}\ (3,3) \end{array}\right] \tag{33a}$$

where $i = 1,2,3$ and the β_i are small positive constants. The characteristic determinant is

$$\begin{aligned} \det(\lambda \underline{I} - \underline{F}_{12}) &= \lambda(\lambda + \beta_1)(\lambda + \beta_2)(\lambda + \beta_3) \{\lambda^2 - (2\omega_s^2 - k_2)\} \\ &\cdot [\lambda_4 + 2\lambda^2 (2\omega_e^2 \sin^2\phi + \omega_s^2] \end{aligned} \tag{33b}$$

with eigenvalues

$$\begin{aligned} \lambda_{10} &= -\beta_1 \\ \lambda_{11} &= -\beta_2 \\ \lambda_{12} &= -\beta_3 \,. \end{aligned} \tag{34}$$

Physically, this means that the platform drifts are not constant but change in a stochastic manner governed by a first-order Gauss-Markov process. The constants β_i determine the correlation length of the corresponding correlation function. Although the total drift model has certainly components of this type, linear drifts cannot be excluded and, thus, the above instabilities remain. In practice, the problem is again solved by updating. The constant drift is determined by position updates taken at appropriate time intervals, and the correlated drift component is modelled by equation (33).

The above eigenvalue analysis provides considerable insight into the expected error behaviour of an inertial system. Although the solution will somewhat change when appropriate forcing functions are introduced, it can be expected that the dominant oscillations will remain the same, as long as the duration of the survey does not exceed a few hours and the sensor errors can be considered as constant for this time period. Thus, a simple system of base functions can be used to model velocity and position errors. It consists of trigonometric functions in $\omega_S t$ and $\omega_E t$, a constant and a linear term. Such a system has been proposed in Schwarz (1983c) and has been successfully implemented by Vassiliou (1984). The coefficients of these base functions are determined from an adjustment of the output data (27b) which may be velocity updates or coordinate updates; for a more detailed discussion of the method, see Schwarz (1985). The procedure gives excellent results when applied to each channel individually. A combined solution for all three channels gives poorer results. This may be due to a constraint exercised by the neglected damping terms in the horizontal channels. This should be further investigated using long term measurements. Another interesting phenomenon is the presence of small oscillations in the residuals after the adjustment. They have frequencies ($k\omega_S$) where k is an integer, see Vassiliou (1984), and may be due to neglected nonlinearities in the dynamic sensor models, see Möhlenbrinck (1986), or to a sampling problem. Some well designed experiments with high data rates are needed in this case to decide on the source of these oscillations.

Although the insight gained by an eigenvalue analysis leads immediately to useful applications, a theoretically more satisfying approach is a full eigenvalue-eigenvector analysis. It provides the solution to the homogeneous part of equation (27a) in the form

$$\underline{x}(t) = \underline{c}^T e^{\underline{\Lambda} t} \underline{V} \qquad (35)$$

where

$\underline{\Lambda}$ ... is the eigenvalue matrix

$\underline{V}$ ... is the eigenvector matrix

$\underline{c}$ ... is a vector of normalizing constants.

For a discussion of this formula, see Braun (1983). The determination of $\underline{V}$ has been performed in Schwarz and Vassiliou (1986) for the model (26c). The derivation is rather involved and the resulting analytical expressions for the eigenvectors are somewhat lengthy and will not be repeated here. An immediate application of formula (35) seems difficult, however, because of the complexity of the analytical manipulations necessary to form the right-hand side. Due to the non-symmetry of $\underline{F}$, the computation of eigenrows is required for the normalizing factors $\underline{c}$, see Fawzy and Bishop (1976) for a concise discussion. The multiplication of the three factors on the right-hand side of formula (35) will thus be extremely cumbersome. Considering that this is only the first, though very important step towards a general solution, this problem may have to await the availability of more powerful analytical formula manipulation programs before the next step can be taken.

6. KALMAN FILTERING - THE STOCHASTIC VIEWPOINT

This chapter reviews first some system theoretical results on realization which give the general background for the Kalman filtering problem. Since the presentation will again be very brief, reference is made to Casti (1977) for all details. The filtering problem will then be formulated and an application oriented derivation of the main equations be given. For alternative derivations and a full discussion of filtering aspects, textbooks like Bryson and Ho (1969) or Jazwinski (1970) should be consulted. Among the many excellent publications of Kalman on the subject, his seminal paper in 1960 and his review in 1978 are of special interest. The rest of the chapter will be devoted to implementation aspects providing a link to the labs connected to this lecture. Among the application oriented textbooks Biermann (1977), Gelb (1978), and Brown (1983) are specifically mentioned.

Realization theory treats the question of deriving an internal model $\Sigma = (\underline{F}, \underline{G}, \underline{H})$ from an external input-output map. As has been mentioned in chapter 4, the concepts of reachability and observability are essential in defining a canonical representation. Their application to linear constant systems will now be discussed. Let Ω be the set of admissable inputs, Γ the set of outputs, and $X = R^n$ the space of states. The reachability problem can then be formulated: Given a fixed initial state $\underline{x}_o$, which states may be reached in finite future time by applying input sequences from Ω? If the entire space X is reachable, Σ is said to be completely reachable. Clearly reachability is a concept of system control and, for Σ given in internal form, is only determined by $\underline{F}$, $\underline{G}$, and the set $\underline{\Omega}$. For linear constant systems $\Sigma = (\underline{F}, \underline{G}, -)$ the following theorem exists: Σ is completely reachable if and only if the matrix

$$C = [G/FG/F^2G/\dots F^{n-1}G] \qquad (36)$$

has rank n.

The problem of observability is dual to the reachability problem and can be formulated in the following way: Given the observation $\underline{H}\ \underline{x}$ and knowledge of all future inputs $\underline{u}$, are all future states $\underline{x}$ uniquely determined? If the answer is in the affirmative, we say the state $\underline{x}$ is observable. If the entire space X is observable, then Σ is said to be completely observable. The concept of observability involves only the matrices $\underline{F}$ and $\underline{H}$. For linear constant systems $\Sigma = (\underline{F},-,\underline{H})$ the following theorem exists: Σ is completely observable if and only if the matrix

$$0 = [H^T/F^TH^T/\ldots/(F^T)^{n-1}H^T] \tag{37}$$

has rank n.

Equations (36) and (37) provide simple test criteria for reachability and observability. They can thus be used to find the canonical model from an arbitrary state variable representation ($\underline{F}$, $\underline{G}$, $\underline{H}$) by asserting complete reachability and observability. Since, in case of a linear system, the canonical model is also minimal, this has not only the advantage of excluding external biases but also gives a state vector of minimal length. Finally, a theorem on the algebraic equivalence of different realizations will be quoted because it gives the set of admissable state vector transformations. The constant linear system $\Sigma = (\underline{F}, \underline{G}, \underline{H})$ is algebraically equivalent to the system $\bar{\Sigma} = (\bar{F}, \bar{G}, \bar{H})$ if and only if there exists a nonsingular matrix T such that

$$\underline{\bar{F}} = \underline{T}\ \underline{F}\ \underline{T}^{-1} \qquad \underline{\bar{G}} = \underline{T}\ \underline{G} \qquad \underline{\bar{H}} = \underline{H}\ \underline{T}^{-1}\ . \tag{38}$$

This means that Σ and $\bar{\Sigma}$ differ only by a base change in X.

The above properties are of a fundamental nature and ascertain the consistency of the internal model Σ with the external input-output map. They relate therefore directly to the discussion on the choice of a state variable model in chapter 4. These properties have to be satisfied before any additional constraint can be applied to the system. Such constraints are e.g. given by optimality criteria. One of them, namely the minimization of the mean square deviation between true state and estimated state is behind the Kalman filter concept. The basic problem is as follows. Consider $\underline{x}(t)$ as a random process generated by the model (27a) and $\underline{y}(t)$ as an observed signal of this process. The functions $\underline{u}(t)$ and $\underline{e}(t)$ are independent, white, Gaussian noise processes with zero means and covariance matrices

$$E\{\underline{u}(t),\ \underline{u}(\tau)\} = \underline{C}^{u}(t)\delta(t-\tau)$$

$$E\{\underline{e}(t),\ \underline{e}(\tau)\} = \underline{C}^{e}(t)\delta(t-\tau)$$

$$E\{\underline{u}(t),\ \underline{e}(\tau)\} = 0 \quad \text{for all } t \text{ and } \tau$$

where

$\delta(\cdot)$ ... is the Dirac delta function
$E\{\cdot\}$ ... is the mathematical expectation
$\underline{C}^{u}$ ... is symmetric and positive semidefinite
$\underline{C}^{e}$ ... is symmetric and positive definite .

In addition,

$$E\{\underline{x}(t_o),\ \underline{x}(t_o)\} = C_o^{x} .$$

The filtering problem can then be stated:

Given known values of $\underline{y}(s)$ in the interval $t_o \leq s \leq t$, find an estimate $\hat{\underline{x}}(t_1)$ of $\underline{x}(t_1)$ of the form

$$\hat{\underline{x}}(t_1) = \int_{t_o}^{t_1} \underline{A}(t_1,s)\ \underline{y}(s)\ ds \tag{39a}$$

where $\underline{A}$ is an $(n \times p)$ matrix and $\hat{\underline{x}}(t_1)$ has the property

$$E\{\underline{x}(t_1) - \hat{\underline{x}}(t_1)\ ,\ \underline{x}(t_1) - \hat{\underline{x}}(t_1)\}^2 \rightarrow \min. \tag{39b}$$

An interpretation of this problem, often used in engineering applications, runs as follows: Consider $\dot{\underline{x}} = \underline{F}\ \underline{x}$ as the mathematical idealization of the original system and $\underline{G}\ \underline{u}$ as a stochastic input accounting for the uncertainties in the model. The filtering problem is then, to get a best estimate of the state vector at time t_1 from noisy measurements $\underline{y}$ in the time interval (t_o,t_1), using the statistics of the noise processes $\underline{u}$ and $\underline{e}$. If best is defined by equation (39b), this leads directly to the Kalman filter.

A number of different approaches to the solution of this problem have been published. The one discussed in the following emphasizes aspects of the filter which are important in applications. It thus is closely linked to the rest of the chapter. A full discussion of this approach is e.g. given in Gelb (1978) or Brown (1983).

The solution of equation (27a) for constant coefficient matrices $\underline{F}$ and $\underline{G}$ is obtained by standard results from differential equation as

$$\underline{x}(t) = \underline{\Phi}(t,t_o)\underline{x}_o + \int_{t_o}^{t} \underline{\Phi}(t,\tau)\ \underline{G}\ \underline{u}(\tau)\ d\tau \tag{40}$$

where

$\underline{\Phi}$ ... is the transition matrix

$\underline{x}_o$ ... is the vector of initial values

and the whole equation is often called the matrix superposition integral. $\underline{\Phi}$ is given by the matrix exponential

$$\underline{\Phi}(t,t_o) = e^{\underline{F}(t-t_o)} . \tag{41}$$

Equation (40) shows that for a system without input the shape of the error curves is completely determined by $\underline{\Phi}$ and the amplitudes by $\underline{x}_o$. This corresponds to the case of free vibrations in the previous chapter. The explicit relationship is given by

$$\underline{\Phi}(t,t_o) = \underline{V}\ e^{\underline{\Lambda}(t-t_o)}\ \underline{V}^{-1} \tag{42}$$

where

$\underline{V}$ ... is the eigenvector matrix

$\underline{\Lambda}$ ... is the diagonal eigenvalue matrix.

The Kalman filtering equations will now be derived by combining the discrete form of equation (40)

$$\underline{x}_k = \underline{\Phi}_{k,k-1}\ \underline{x}_{k-1} + \underline{w}_{k,k-1} \tag{43}$$

where

$$\underline{x}_k = \underline{x}(t_k)$$

$$\underline{\Phi}_{k,k-1} = \underline{\Phi}(t_k,t_{k-1})$$

$$\underline{w}_{k,k-1} = \int_{t_{k-1}}^{t_k} \underline{\Phi}(t_k,\tau)\ \underline{G}_k\ \underline{u}(\tau)\ d\tau$$

with the discrete output equation

$$\underline{y}_k = \underline{H}_k\ \underline{x}_k + \underline{e}_k . \tag{44}$$

The combination will be done in such a way that the resulting estimate is a linear combination of equations (43) and (44), that it is unbiased, and optimal in the sense defined above.

The linearity requirement stipulates that the estimate is of the basic form

$$\hat{\underline{x}}_k(+) = \underline{K}_k' \, \hat{\underline{x}}_k(-) + \underline{K}_k \, \underline{y}_k \tag{45}$$

where

$\hat{\underline{x}}(-)$ ... is the estimate of $\underline{x}_k$ before update (predicted)

$\hat{\underline{x}}(+)$ ... is the estimate of $\underline{x}_k$ after update (filtered)

$\underline{K}_k'$, $\underline{K}_k$ ... are arbitrary weighting matrices.

Thus, the required estimate is obtained by weighting the predicted estimate and the update measurements in a suitable fashion. Equation (45) describes a rather broad class of linear estimates. The required estimate is obtained by making specific choices for $\underline{K}_k'$ and $\underline{K}_k$. This is done in two steps. First, a relation betweek $\underline{K}_k'$ and $\underline{K}_k$ is established by requesting unbiasedness of the estimate. Second, a specific matrix $\underline{K}_k$ is chosen by requesting optimality of the estimate.

Unbiasedness of the estimate is expressed by

$$E\{\hat{\underline{x}}_k(+)\} = \underline{x}_k \ , \tag{46}$$

i.e. by stipulating that the expectation value of the estimates $\hat{\underline{x}}(+)$ is equal to $\underline{x}$ itself. Inserting equation (44) into (45) results in

$$\hat{\underline{x}}(+) = \{\underline{K}_k' + \underline{K}_k \, \underline{H}_k\} \, \underline{x}_k + \underline{K}_k' \, \underline{e}_k(-) + \underline{K}_k \, \underline{e}_k \tag{47}$$

where

$$\underline{e}_k(-) = \underline{x}_k - \hat{\underline{x}}_k(-) \quad \text{with } E\{\underline{e}_k(-)\} = \underline{0}$$

is the estimation error in the predicted state vector $\hat{\underline{x}}_k(-)$. Denoting the estimation error in $\hat{\underline{x}}_k(+)$ in a similar fashion by

$$\underline{e}_k(+) = \underline{x}_k - \hat{\underline{x}}_k(+) \quad ,$$

equation (48) can be rewritten as

$$\underline{e}_k(+) = \{\underline{K}_k' + \underline{K}_k \, \underline{H}_k - \underline{I}\} \, \underline{x}_k + \underline{K}_k' \, \underline{e}_k(-) + \underline{K}_k \, \underline{e}_k \ .$$

Applying now the condition (46) in the form

$$E\{\underline{e}_k(+)\} = \underline{0} \ ,$$

results in

$$\underline{K}_k' + \underline{K}_k \, \underline{H}_k - \underline{I} = \underline{0}$$

and thus in

$$\underline{K}_k' = \underline{I} - \underline{K}_k \underline{H}_k \ . \tag{48}$$

Backsubstituting equation (48) into (45) gives the linear unbiased estimate

$$\hat{\underline{x}}_k(+) = \hat{\underline{x}}_k(-) + \underline{K}_k\{\underline{y}_k - \underline{H}_k \hat{\underline{x}}_k(-)\} \ . \tag{49}$$

Equation (48) gives the required relation between $\underline{K}_k$ and $\underline{K}_k'$. It determines the structure of (49) which now depends only on one weighting matrix $\underline{K}_k$. Fixing the structure of $\underline{K}_k$ by the optimality criterion will give the final estimate.

Optimality is defined by

$$\text{trace } \{C_k^x(+)\} = \text{minimum} \tag{50}$$

where

$$C_k^x(+) = E\{\underline{e}_k^T(+) \ e_k(+)\} \tag{51}$$

is the error covariance matrix of $\hat{\underline{x}}_k(+)$. Performing the minimization (50) leads to

$$\underline{K}_k = \underline{C}_k^x(-) \ \underline{H}_k^T\{\underline{H}_k \ C_k^x(-) \ \underline{H}_k^T + C_k^e\}^{-1} \tag{52}$$

where

$$\underline{C}_k^x(-) = E\{\underline{e}_k^T(-) \ \underline{e}_k(-)\}$$

and

$$\underline{C}_k^e = E\{e_k^T \ e_k\}.$$

$\underline{K}_k$ is called the Kalman gain matrix. Its use transforms the linear unbiased estimate (49) into a best linear unbiased estimate.

The Kalman filtering algorithm is obtained by combining formulas (43), (44), (49), and (52). It consists of two distinct sets of formulas, the prediction equations and the update equations. The first set is used when no measurements of type (44) are available. The prediction equations are of the form

$$\hat{\underline{x}}_k(-) = \underline{\Phi}_{k,k-1} \ \hat{\underline{x}}_{k-1} \tag{53a}$$

$$\underline{C}_k^x(-) = \underline{\Phi}_{k,k-1} \ \underline{C}_{k-1}^x \ \underline{\Phi}_{k,k-1}^T + C_{k,k-1}^w \ . \tag{53b}$$

The derivation of these equations is straightforward. Equation (53a) comes directly from equation (43) and equation (53b) is obtained by standard covariance propagation. It should be noted that on the right-hand side of these equations the symbols (+) and (-) have been omitted because the prediction interval does seldom agree with the update interval. To preserve numerical accuracy, $\underline{C}_k^x(-)$ must be computed every few seconds, in some cases even milliseconds. The time interval between update measurements is often, however, of the order of a few hundred seconds. Thus $\hat{\underline{x}}_{k-1}$ and $\underline{C}_{k-1}^x$ present estimates based on the last update but propagated in time. If (k-1) is an update they will have the symbol (+), otherwise (-).

The update equations are of the form

$$\hat{\underline{x}}(+) = \hat{\underline{x}}(-) + \underline{K}\{\underline{y} - \underline{H}\ \hat{\underline{x}}(-)\} \tag{54a}$$

$$\underline{C}^x(+) = \{\underline{I} - \underline{K}\ \underline{H}\}\ \underline{C}^x(-) \tag{54b}$$

$$K = \underline{C}^x(-)\ \underline{H}^T\{\underline{H}\ \underline{C}^x(-)\ \underline{H}^T + \underline{C}^e\}^{-1} \tag{54c}$$

They express the new estimate of the state vector $\hat{\underline{x}}(+)$ in terms of the old estimate $\hat{\underline{x}}(-)$ and the measurement $\underline{y}$. The new estimate, the measurement, and the old estimate refer to the same instant t_j. The subscript has therefore been omitted. The second term on the right of equation (54a) shows how the measurement $\underline{y}$ is used for the update. It is compared to its predicted value $\underline{H}\ \hat{\underline{x}}(-)$ and the difference between the two is multiplied by the Kalman gain $\underline{K}$. Thus, the effect of the measurement on the estimate $\hat{\underline{x}}(+)$ is determined by $\underline{K}$. It uses the covariance matrices $\underline{C}^x$, $\underline{C}^e$, and $\underline{C}^w$ to determine the weight of the measurement $\underline{y}$. If these matrices are not judiciously chosen $\underline{K}$ may become so small that $\underline{y}$ will have no effect on $\hat{\underline{x}}(+)$. This so-called divergence of the Kalman filter is one of the problems to be considered in filter implementation. Equation (54a) also shows that an update will not change the state vector in case $\{\underline{y} - A\ \hat{\underline{x}}(-)\} = 0$. In this case, the prediction agrees with the measurement. Usually, this means that the prediction is working well and there is no reason for change. The derivation of formulas (54a) and (54c) was outlined previously. Equation (54b) is obtained by covariance propagation using formulas (51), (48) and (52).

A major advantage of Kalman's formulation of the dynamic estimation problem is the fact that equations (54) are recursive. They give the optimal estimate of the state vector based on all measurements up to the time t_j without the necessity to store these data.

The information on previous measurements is contained in the state vector and its covariance matrix. Thus, besides the update measurements and its covariance matrix no other information is needed.

Kalman filtering as a typical real-time procedure gives optimal estimates based on all measurements available at time t_j. If a new measurement becomes available at time t_{j+1}, all previous estimates can be improved by the new information. The process of improving previous estimates by a new measurement is called smoothing. If the optimality criterion (39b) is employed again, we speak of optimal smoothing. Since smoothing goes backward in time it is a typical post-mission process. It should be noted that optimal smoothing differs from Kalman filtering only in terms of data completeness. It does not use a different methodology.

There are different forms of the optimal smoothing algorithm. The most intuitive approach considers optimal smoothing as a weighted average of a forward and a backward filter. It was proposed by Fraser and Potter (1969) and is discussed for survey applications in Schwarz (1980). However, this approach is not the most efficient one from a computational point of view. Therefore the algorithm proposed by Rauch et al. (1965) will be given here. It relates the Kalman estimates $\hat{\underline{x}}_j(+)$ directly to the smoothed estimates $\hat{\underline{x}}_j^s$. The formulas are

$$\hat{\underline{x}}_j^s = \hat{\underline{x}}_j(+) + \underline{D}_{j+1,j}\{\hat{\underline{x}}_{j+1}^s - \hat{\underline{x}}_{j+1}(-)\} \qquad (55a)$$

$$\underline{C}_j^s = \underline{C}_j^x(+) + \underline{D}_{j+1,j}\{\underline{C}_{j+1}^s - \underline{C}_{j-1}^x(-)\}\, \underline{D}_{j+1,j}^T \qquad (55b)$$

$$\underline{D}_{j+1,j} = \underline{C}_j^x(+)\, \underline{\Phi}_{j+1,j}(\underline{C}_{j+1}^x(-))^{-1} \qquad (55c)$$

The smoothing algorithm goes backward in time. If the last update measurement has been made at time t_N, then obviously for this point

$$\hat{\underline{x}}_{j+1} = \hat{\underline{x}}_N^s = \hat{\underline{x}}_N(+) \qquad (56a)$$

$$\underline{C}_{j+1}^s = \underline{C}_N^s = \underline{C}_N^x(+) \qquad (56b)$$

because the smoothed estimate is equal to the filtered estimate. The difference between the smoothed state vector and the predicted state vector can thus be formed for t_N. The difference is multiplied by the weight matrix $\underline{D}_{j+1,j}$ and is added to the updated state vector at the next point backwards in time. The weight matrix is again derived using optimality criteria. It differs from the Kalman gain matrix because it does not refer to the same point in time but relates the estimates at time t_{j+1} and t_j. The quantities required for optimal estimation are the state vectors and covariance matrices before and after filtering and the transition matrices between update points. Since optimal smoothing is a post-mission procedure, off-line storage of these quantities is sufficient.

The implementation of the Kalman filter algorithm starts usually with the computation of the transition matrix Φ: Rewriting equation (41) as a series in $\underline{F}$, one obtains

$$\underline{\Phi}(t,t_o) = \sum_{n=0}^{\infty} \frac{1}{n!} \underline{F}^n (t-t_o)^n \tag{57}$$

which can be used to determine $\underline{\Phi}(t,t_o)$ either numerically or analytically. The numerical approach is preferred today because it makes for very flexible programming and does not require major changes when the dimension of $\underline{x}$ increases. It uses a truncated form of the series (57) to determine the transition matrix, see e.g. Gelb (1978). If $(t-t_o)$ is chosen small enough, the simple approximation

$$\underline{\Phi}(t,t_o) = \underline{I} + \underline{F}(t-t_o)$$

will often be sufficient. This is the case in inertial survey applicatins where the internal integration loop is as small as 16 milliseconds.

Although a numerical computation of the transition matrix is usually preferred in production type problems,an analytical approach is often superior in error analysis. In simple cases it can be obtained by an analytical development of the series (57) for a number of terms. Closed expressions can then be deduced by inspection of the first few terms in the series. However, for larger matrices $\underline{F}$ this approach becomes extremely awkward.

It is often replaced by the inverse Laplace transform technique which gives a solution by forming

$$\underline{\Phi}(t,t_o) = L^{-1}\{(s\underline{I} - \underline{F})^{-1}\} \tag{58}$$

for individual elements of the inverse matrix on the right-hand side. Extensive tables of inverse Laplace transforms L^{-1} exist, see e.g. McCullum and Brown (1965), and it is often possible to transform the elements of the inverse matrix in such a way that they correspond to a table entry. However, the analytical inversion of $(s\underline{I} - \underline{F})$ is a problem, especially for larger matrices $\underline{F}$.

A third method to arrive at an analytical expression for Φ is by way of equation (42). Note, that $\underline{F}$ must have distinct eigenvalues in this case. In case of multiple eigenvalues the situation is more complicated and reference is made to textbooks like Hochstadt (1975) or Braun (1983). Although this approach is the most elegant, the practical difficulties of deriving $\underline{V}$ and $\underline{\Lambda}$ in analytical form are again considerable.

A detailed numerical example of Kalman filtering for a simple case is given in Schwarz (1983a). It also contains some discussion if implementation aspects like algorithm stability, causes for Kalman filter divergence, reliability of the computed covariances, and determination of station cross-covariances. Many more details are discussed in the application oriented textbooks mentioned before. Kalman filtering has been implemented in Litton's LASS. Unfortunately, the detailed filter implementation has not been published in the open literature. Applications of Kalman filtering and optimal smoothing to problems of inertial geodesy can be found among others in Schwarz (1980), Wong (1982), Wong and Schwarz (1983), Wei and Brockstein (1983), Bose and Baussus von Luetzow (1986), Wong and McMillan (1986).

7. CONCLUDING REMARKS

The system of differential equations which describes the error model of inertial geodesy, has been discussed from two different points of view. One is deterministic in nature and uses the theory of vibrations as a starting point, while the other is stochastic in principle and results in the well-known Kalman filtering and optimal smoothing equations. Theoretically, the two approaches are closely related because they both solve the same system of differential equations, albeit with different sets of assumptions for the forcing functions. Practically, they are quite different. One provides important insights into the spectral behaviour of the error system while the other gives an optimal space domain solution for a well-defined set of stochastic assumptions. They emphasize therefore different aspects of the problem and provide complementary insights.

The deterministic approach has advantages as an analysis and model identification tool. It shows clearly the dominant oscillations of the system, allows the study of the system for different types of excitations, gives detailed analytical expressions for the transformation of various input signals, and provides a simple approach to study updating procedures and their effect on system stability. This type of analysis is usually restricted to simplified models. It gives, however, insights which can be applied to more complex situations. A case in question is the analysis of gravity disturbances given in Schwarz (1986). It shows that the effect of these disturbances can be almost completely eliminated by suitable update measurements but that it is very difficult to separate them from instrumental errors. This explains why even crude methods of zero velocity updating work well for the positioning problem. It also explains why more sophisticated methods are needed for the gravimetry problem. In terms of model identification, the approach has the advantage that the dominant frequencies of the system are usually known from the eigenvalue analysis and can be represented by a small number of simple base functions. The spectral analysis of a given data set will then show whether other than the expected frequencies are contained in the data. Since covariance assumptions are not critical in this case, and the main periods are known in advance, an optimal experimental design is possible. The treatment of transient signals is problematic in this approach, as is the consideration of cross-coupling effects between channels. The method has its main strength in post-mission analysis, although some of its results are also useful in real-time estimation.

The stochastic approach has advantages in situations where reliable real-time estimation is required without an explicit identification of individual frequencies. The error structure contained in the homogeneous part of the system of differential equations is optimally resolved if the statistical assumptions on $\underline{u}$ are correct. Thus, there is an interplay between the choice of the state vector and the white noise characteristics of $\underline{u}$ which can be used for model identification. This is important in situations where sampling rates are high and statistical tests can be performed with some reliability, as e.g. in INS/GPS integration. The standard form of the state variable model makes numerical model identification simple because augmentation or reduction of the state vector are easy to do. However, outside information on the states to be added, must be available. Furthermore, only the overall performance of the filter can be tested. The significance of individual states is difficult to establish. On the other hand, because of the stochastic a priori information, transient signals are absorbed rather well. The choice of the covariance structure is critical, however, and major approximations are usually made at this point. If the covariances are not correct, results will be distorted and the interpretation of individual state vector components will become ambiguous. In geodesy, where the state vector and its covariance matrix is needed, the standard algorithms of Kalman filtering are an asset because they supply both. The main strengths of the method are a standard form which is well adapted to many practical problems, optimal real-time performance if the model assumptions are satisfied, and a capability for data compaction by the way of its recursive structure.

There are a number of interesting avenues for future research. On the theoretical side, the reachability and observability criteria of realization theory should be applied to the different state variable models that have been proposed for the processing of inertial data. This is a fundamental step which would help to identify states which are weakly observable with the current update data. Similarly, the duality between the stochastic filtering problem and the corresponding deterministic control problem could be used to get further insight into the model. Established results on the modal control problem can be used to show the relationship of the two approaches more clearly. Finally, the effect of updating on the eigenvalues and the stability of the system should be further studied. On the practical side, a thorough comparison of the two approaches is needed in terms of accuracy, reliability, and data sensitivity. Some preliminary results have been published in Schwarz (1983b), Vassiliou (1984), and Forsberg et al (1986) but they consider specific aspects of each approach and are not general enough. Since the INS/GPS integration will give much higher sampling rates than the standard zero velocity update procedure, the comparison should be directed towards the new data type. Finally, a data analysis of the error damping effect due to zero velocity and coordinate updating would complement the theoretical studies in a meaningful way and give much needed parameters for future applications.

Acknowledgements: The paper was written while the author was visiting scientist at the Sonderforschungsbreich 228 "Hochgenaue Navigation" in Stuttgart working on problems of inertial data modeling. His appreciation is expressed to Profs. Linkwitz and Grafarend for making time and facilities of the Sonderforschungsbereich available and to a number of colleagues who by discussion and critique helped to clarify the concepts.

REFERENCES

Bierman, G.F., Factorization methods for discrete sequential estimation. Academic Press, New York, 1977.

Bose, S. and H. Baussus von Luetzow, Improved determinations of deflections of the vertical from astrogeodetic-inertial measurements. Proc. Third Int. Symp. on Inertial Technology for Surveying and Geodesy, UCSE Publ. No 60005, Calgary, 1986.

Braun, M., Differential equations and their aplications. Springer Verlag, New York, 1983.

Britting, K.R., Inertial navigation systems analysis, Interscience, New York, 1971.

Brown, R.G., Introduction to random signal analysis and Kalman filtering. John Wiley and Sons, New York, 1983.

Bryson, A.E. and Y.C. Ho, Applied optimal control. Blaisdell Publishing Co., Waltham, Mass., 1969.

Casti, J.L., Dynamical systems and their applications: Linear theory. Academic Press, New York, 1977.

Casti, J.L., Nonlinear system theory, Academic Press Inc., Orlando, 1985.

Farrell, J.L., Integrated Aircraft Navigation. Academic Press, New York, 1976.

Faurre, P., Navigation inertielle optimale et filtrage statistique. Dunod, Paris, 1971.

Fawzy, I. and R.E.D. Bishop, On the dynamics of linear nonconservative systems. Proc. Royal Society, London, A 352, 25-40, 1976.

Fraser, D.C. and J.E. Potter, The optimum linear smoother as a combination of two optimum linear filters. IEEE Trans. on Autom. Control, AC-14, 4, 387-390, 1969.

Fischer, K. and W. Stephan, Schwingungen. Birkhäuser Verlag, Basel 1981.

Forsberg, R., A. Vassiliou, K.P. Schwarz and R.V.C. Wong, Inertial gravimetry - a comparison of Kalman filtering and post-mission adjustment techniques. Proc. Third Int. Symp. on Inertial Technology for Surveying and Geodesy, UCSE Publ. No 60005, Calgary, 1986.

Gelb, A., Applied optimal estimation. The M.I.T. Press, Cambridge, Mass., 1974. Fourth printing 1978.

Hein, G.W., The local gravity field in the concept of integrated geodesy. Proc. Beijing Int. Summer School on Local Gravity Field Approximation. UCSE Publ. No 60003, Calgary, 1984.

Hochstadt, H., Differential equations. Dover, New York, 1975.

Jazwinski, A.H., Stochastic processes and filtering theory. Academic Press, New York, 1960.

Kalman, R.E., A new approach to linear filtering and prediction problems. Journal of Basic Engineering. Transactions of the ASME, 82 D, 35-45, 1960.

Kalman, R.E., A retrospective after twenty years: From the pure to the applied. Proc. AGU Chapman Conf. on Applications of Kalman Filter to Hydrology, Hydraulics, and Water Resources, Pittsburgh, 1978.

Magnus, K., Schwingungen. Teubner Studienbücher, Stuttgart, 1976.

McCullum, P.A. and B.F. Brown, Laplace Transform Tables and Theorems. Holt, Rinehart and Winston, New York, 1965.

Möhlenbrink, W., Nonlinearities in the dynamic model of inertial sensors. Proc. Third Int. Symp. on Inertial Technology for Surveying and Geodesy, UCSE Publ. No 60005, Calgary, 1986.

Natke, H.G., Einführung in die Theorie und Praxis der Zeitreihen und Modalanalyse. Friedr. Vieweg & Sohn, Braunschweig, 1983.

Rauch, H.E., F. Tung and C.T. Striebel, Maximum likelihood estimates of linear dynamic systems. AIAA J., 3, 8, 1445-1450, 1965.

Santilli, R.M., Foundations of theoretical mechanics I. Springer Verlag, New York, 1978.

Schuler, M., Die Störung von Pendel- und Kreiselapparaten durch die Beschleunigung des Fahrzeugs. Physikalische Zeitschrift, 24, 344-350, 1923.

Schwarz, K.P., Error propagation in inertial positioning. The Canadian Surveyor, 34, 3, 265-276, 1980.

Schwarz, K.P., Kalman filtering and optimal smoothing. In Papers for the CIS Adjustment and Analysis Seminars, Ottawa, 1983a.

Schwarz, K.P., Inertial surveying and geodesy. Rev. Geoph. and Space Phys., 21, 4, 878-890, 1983b.

Schwarz, K.P., Inertial adjustment models - A study of some underlying assumptions. In Geodesy in Transition, Publication 60002, Division of Surveying Engineering, The University of Calgary, 1983c.

Schwarz, K.P., A unified approach to post-mission processing of inertial data. Bulletin Géodésique, 59, 1, 33-54, 1985.

Schwarz, K.P.,Geoid profiles from an integration of GPS satellite and inertial data. Paper presented at the Int. Symp. on the Definition of the Geoid, Florence, May, 26-30, 1986.

Schwarz, K.P. and A.A. Vassiliou, An eigenvector base for the error model in inertial surveying. Proc. Third Int. Symp. on Inertial Technology for Surveying and Geodesy, UCSE Publ. No 60005, Calgary, 1986.

Vassiliou, A.A., Processing of unfiltered data from an inertial survey system of local level type. Manuscripta Geodaetica, 9, 4, 231-306,1984.

Vassiliou, A.A. and K.P. Schwarz, Eigenvalues of the dynamics matrix used in inertial geodesy. Manuscripta Geodaetica, 10, 3, 213-221, 1985.

Wei, S.Y. and A. J. Brockstein, Offline optimal adjustment of inertial system survey data. Proc. Fall Conv. ACSM-ASP, Salt Lake City, Utah, Sept. 1983.

Wong, R.V.C., A Kalman filter-smoother for an inertial survey system of local level type. Publ. 20001, Div. Surv. Eng., The Univ. of Calgary, 1982.

Wong, R.V.C. and S.C. McMillan, Precise aircraft positioning with an integrated inertial navigation system. Proc. Third Int. Symp. on Inertial Technology for Surveying and Geodesy, UCSE Publ. No 60005, Calgary, 1986.

Wong, R.V.C. and K.P. Schwarz, Dynamic positioning with an integrated GPS-INS: Formulae and base line tests. UCSE Publ. No 30003, Calgary, 1983

INTEGRATED GEODESY
STATE-OF-THE-ART 1986 REFERENCE TEXT

by

G.W. Hein

Institute of Astronomical and Physical Geodesy
University FAF Munich
Werner-Heisenberg-Weg 39
D-8014 Neubiberg, F.R. Germany

Lecture Notes in Earth Sciences, Vol. 7
Mathematical and Numerical Techniques in Physical Geodesy
Edited by H. Sünkel

1. INTRODUCTION

It was *Torben Krarup* from the Danish Geodetic Institute in 1973 who had the idea - wisely far-sighted that the following decades present geodesy and surveying a large amount of data from new observation techniques. Thus, in view of electronic computers those data have to be processed using certain model strategies. And they should not be processed like in classical geodesy: a variety of model-based approaches without interaction between and often not taking advantage of other observation types which could also contribute to the solution of the problem. As a consequence, an *integrated data processing* was needed as far as possible using only one *unified* model for a rigorous threedimensional geodesy: *integrated geodesy*, as it was named by T. Krarup, was born. He created the adjustment model able to determine positions and the gravity potential (and its functionals) in *one* step using all available observations. The underlying model developed Krarup years before *(Krarup 1969)*: the *least-squares collocation* (see also *Moritz 1973*).

Thus, integrated geodesy is not a research direction in physical geodesy. It is *the* philosophy to use geodetic (and other) observations for solving our problems nowadays. And it is not a break with regard to the traditional approaches applied in the past. They are still used - at least implicitly to built-up our covariance models - and they are systematically combined and incorporated in the integrated model. But what has changed is the way of extracting information from the observations: we are using the *operational* way. This has led Erik Grafarend to name this approach according to Bridgeman: *operational geodesy*, see also *Grafarend (1978a), Moritz (1978)*.

The readiness to create combined models was, in particular, influenced by the fast development of *high-speed computers* where the solution of large systems of equations was becoming possible. In addition, the availability of 3d-geocentric coordinates from *new satellite techniques*, accelerated the set-up of 3d-networks.

With the advent of the *Global Positioning System (GPS)* it first seemed that this revolutionizing positioning technique made physical geodesy and gravity field considerations superfluous in the working-day of a surveyor. However, just the opposite happened! With this new observation technique a challenge was presented to physical geodesists: to determine *gravity potential differences* (or geoid height differences) with an accuracy which could compete together with 3d-baseline vectors from GPS the centuries-old spirit levelling technique. Thus, again there is a need to consider geometrical and dynamical observations in *one* approach.

These lecture notes summarize the state-of-the-art in the application of the integrated geodesy philosophy in the various branches of geodesy. The different concepts are outlined generally. For the reader who is interested to go into details extensive reference is made to relevant work.

<u>*REFERENCES.*</u> *Some historical notes: It was already Bruns (1878) who proposed in 1878 a rigorous threedimensional geodesy. Marussi (1949) favoured the use of natural (holonomic) coordinates astronomical latitude* Φ *and longitude* Λ *and the gravity potential* W *. Dufour (1962) developed an iterative solution between the two normal equation systems belonging to the horizontal and the vertical coordinates. Wolf (1963a,b) presented a 3d-approach where he introduced five unknowns per station,* (x, y, z, Φ, Λ) *. The inclusion of the observations of spirit levelling by means of orthometric heights was proposed by Heitz (1963) and demonstrated by Preusser (1977).*

2. THE PRINCIPLE OF INTEGRATED GEODESY

We begin with the derivation of the general principle of operational geodesy and follow more or less the notation used in *(Moritz 1980)*. Every geodetic measurement l can be expressed as a nonlinear functional depending on one or several position vectors $\underline{x} = (x,y,z)$ in space and on the gravity field of the earth, symbolically written

$$l = F(\underline{x},W) \tag{2-1}$$

where W is the gravity potential

$$W = V + \omega^2(x^2+y^2)/2 \quad . \tag{2-2}$$

V is the potential of the gravitational force and ω is the angular velocity of the earth's rotation. By (x,y,z) or (x_1,x_2,x_3) we denote the Cartesian coordinates of the geocentric system $E^{\bullet}$ defined as follows: the origin is at the earth's center of mass, the z-axis coincides with the (mean) rotation axis, and the x-axis goes through the (mean) Greenwich meridian. We presume further that the observations are corrected for time-dependent effects, therefore we do not consider geodynamical phenomena in this approach.

The scope of geodesy is now to determine the coordinates of material points on the surface of the earth (and in space) <u>and</u> the gravity potential including its functionals by the relation (2-1). Since our measurements are nonlinear functionals we have to introduce approximate values for the linearization process.

$$\underline{x} = \underline{x}_0 + \delta\underline{x} \tag{2-3}$$

$$W(\underline{x}) = U(\underline{x}) + T(\underline{x}) \tag{2-4}$$

$\underline{x}_0$ is the approximate position vector, $U(\underline{x})$ is some kind of trial value for the actual gravity potential at $\underline{x}$. This could be a so-called normal potential belonging to an arbitrary ellipsoid or any other reference potential, e.g. a low order harmonics expansion derived from satellite geodesy. For the sake of comparison of the results with those of other computations one would use one of the adopted reference systems of the International Association of Geodesy. As done hitherto we will call $T(\underline{x})$ the disturbing potential and we will assume further that both, $T(\underline{x})$ and $\delta\underline{x} = (\delta x, \delta y, \delta z)^T = (\delta x_1, \delta x_2, \delta x_3)^T$ are small quantities, which allow us to work with a linear model.

Thus we get for (2-1)

$$l = F(\underline{x}_0 + \delta\underline{x}, U+T) \tag{2-5}$$

Applying Taylor's theorem at $\underline{x}_0 = [x_1^0, x_2^0, x_3^0]^T$ and restricting ourselves to the first-order terms in the series it follows

$$l = F(\underline{x}_0, U) + \sum_{i=1}^{3} F_{x_i}(\underline{x}_0, U)\delta x_i + L(T) \tag{2-6}$$

where $L(T)$ is a linear operator applied on the function T. By the substitution

$$\delta l = l - F(\underline{x}_0, U) \tag{2-7}$$

$$a_i = \frac{\partial F}{\partial x_i}(\underline{x}_0, U) \tag{2-8}$$

we get the general linear observation equation of the form

$$\delta\underline{l} = \underline{a}^T\delta\underline{x} + L(T) \tag{2-9}$$

where the first expression of the right hand side represents the *metric coordinates* part and the second the (functionals of the) *disturbing potential.*

We give some examples of observation equations of type (2-9) and select for further considerations the main observations of physical geodesy. The ones determining the gravity vector on the earth's surface are astronomical latitude Φ, astronomical longitude Λ and gravity g. In addition, gravity potential differences ΔW (derived from spirit levelling observations dh and gravity via $\Delta W = \int g\, dh$) will be discussed.

We start with the nonlinear relation *(Moritz 1980, p. 233)*

$$\begin{bmatrix} W_x \\ W_y \\ W_z \end{bmatrix} = -g \begin{bmatrix} \cos\Phi\cos\Lambda \\ \cos\Phi\sin\Lambda \\ \sin\Phi \end{bmatrix} \tag{2-10}$$

and linearize at $P_0(\underline{x}_0)$ the left hand side according to the principle of integrated geodesy. Thus, we get

$$\begin{bmatrix} W_x \\ W_y \\ W_z \end{bmatrix} = \begin{bmatrix} U_x \\ U_y \\ U_z \end{bmatrix}^0 + \underline{M}\,\delta\underline{x} + \text{grad } T \tag{2-11}$$

where $\underline{M} = (\partial^2 U/\partial x_i\, \partial x_j)$, $(i = \{1,2,3\},\ j = \{1,2,3\})$ is the famous *Marussi tensor* of second-order derivatives of the approximate (normal) potential U ,

$$\underline{M} = \begin{bmatrix} U_{xx} & U_{xy} & U_{xz} \\ U_{xy} & U_{yy} & U_{yz} \\ U_{xz} & U_{yz} & U_{zz} \end{bmatrix} . \tag{2-12}$$

On the other hand we linearize the right hand side of (2-10) by introducing $(j := \gamma)$

$$\Phi = \phi^0 + \delta\Phi \tag{2-13}$$

$$\Lambda = \lambda^0 + \delta\Lambda \tag{2-14}$$

$$g = j^0 + \delta g \tag{2-15}$$

resulting in

$$\begin{bmatrix} W_x \\ W_y \\ W_z \end{bmatrix} = \begin{bmatrix} U_x \\ U_y \\ U_z \end{bmatrix}^0 + \underline{Q} \begin{bmatrix} j\,\delta\Phi \\ j \cos\phi\, \delta\Lambda \\ \delta g \end{bmatrix} \tag{2-16}$$

where $\underline{Q}$ is defined by

$$\underline{Q} = \begin{bmatrix} \sin\phi\cos\lambda & \sin\lambda & -\cos\phi\cos\lambda \\ \sin\phi\sin\lambda & -\cos\lambda & -\cos\phi\sin\lambda \\ -\cos\phi & 0 & -\sin\phi \end{bmatrix}^0 . \tag{2-17}$$

Comparing now (2-16) with (2-11) yields $(\underline{Q}^T = \underline{Q}^{-1})$

$$\begin{bmatrix} j\,\delta\Phi \\ j\cos\phi\,\delta\Lambda \\ \delta g \end{bmatrix} = \underline{Q}^T\underline{M}\,\delta\underline{x} + \underline{Q}^T \text{grad } T . \tag{2-18}$$

This is the fundamental relation to derive the observation equations for astronomical latitude, astronomical longitude and gravity in the integrated model. Recall that (2-18) is still expressed in the global geocentric reference frame.

The derivation of the observation equation for potential differences $\Delta W_{ij} = W_j - W_i$ is simple. Considering the approximate value

$$\Delta W^0_{ij} = U(\underline{x}_{0,j}) - U(\underline{x}_{0,i}) = \Delta W_{ij} - \delta W_{ij} \qquad (2\text{-}19)$$

we get

$$\delta W_{ij} = [U_x\ U_y\ U_z]^0_j\ \delta\underline{x}_j - [U_x\ U_y\ U_z]^0_i\ \delta\underline{x}_i + T_j - T_i \qquad (2\text{-}20)$$

or

$$\delta W_{ij} = \underline{j}^T_j\ \delta\underline{x}_j - \underline{j}^T_i\ \delta\underline{x}_i + T_j - T_i \qquad (\underline{j} := \underline{\gamma})\ . \qquad (2\text{-}21)$$

REFERENCES. *For the detailed structure of the observation equations of integrated or operational geodesy see Eeg and Krarup (1973), Dermanis (1985), Grafarend (1975, 1978a,b, 1981), Hein (1982a,b, 1985a), Krarup (1978), Moritz (1978, 1980), Zaiser (1984). Reilly (1980, 1984) used a somewhat different approach to combine gravity data with other (so-called) geometrical observations.*

3. THE CLASSICAL APPROACH TO PHYSICAL GEODESY

3.1 Approximations

In order to work in a spherical system with coordinates spherical latitude b, longitude l and r we apply the corresponding transformation from the geocentric to the spherical system with respect to the potential part of (2-18)

$$\text{grad}_x\ T = \begin{bmatrix} T_x \\ T_y \\ T_z \end{bmatrix} = \begin{bmatrix} -\sin b \cos l & -\sin l & \cos b \cos l \\ -\sin b \sin l & \cos l & \cos b \sin l \\ \cos b & 0 & \sin b \end{bmatrix} \begin{bmatrix} T_b/r \\ T_l/(r \cos b) \\ T_r \end{bmatrix} \qquad (3\text{-}1)$$

or

$$\text{grad}_x\ T = \underline{R}(b,l) \begin{bmatrix} T_b/r \\ T_l/(r \cos b) \\ T_r \end{bmatrix}\ . \qquad (3\text{-}2)$$

Inserting (3-2) into (2-18) results in

$$\begin{bmatrix} \delta\Phi \\ \delta\Lambda \\ \delta g \end{bmatrix} = \begin{bmatrix} j & 0 & 0 \\ 0 & j \cos\Phi & 0 \\ 0 & 0 & 1 \end{bmatrix}^{-1} \underline{Q}^T\underline{M}\ \delta\underline{x} + \underline{Q}^T\underline{R} \begin{bmatrix} T_b/(r\cdot j) \\ T_l/(r\cdot j \cos b) \\ T_r \end{bmatrix} \qquad (3\text{-}3)$$

where

$$\underline{Q}^T\underline{R} = \begin{bmatrix} -\cos(\phi-b) & 0 & \sin(\phi-b) \\ 0 & -\dfrac{1}{\cos\phi} & 0 \\ \sin(\phi-b) & 0 & -\cos(\phi-b) \end{bmatrix} . \tag{3-4}$$

Remember that ϕ is the latitude of the introduced *geodetic* coordinate system consistent with the reference system for the model or normal potential. It is therefore often called "normal latitude". The difference $(\phi-b)$ will be certainly small, say, less than 10" in general, so that the following approximation is justified.

$$\underline{Q}^T\underline{R} \doteq \begin{bmatrix} -1 & 0 & 0 \\ 0 & -\dfrac{1}{\cos\phi} & 0 \\ 0 & 0 & -1 \end{bmatrix} \tag{3-5}$$

Considering (3-5) in (3-3) yields $(\phi \doteq b)$

$$\begin{bmatrix} \delta\Phi \\ \delta\Lambda \\ \delta g \end{bmatrix} = \begin{bmatrix} j & 0 & 0 \\ 0 & j\cos\phi & 0 \\ 0 & 0 & 1 \end{bmatrix}^{-1} \underline{Q}^T\underline{M}\,\delta\underline{x} + \begin{bmatrix} -T_b/(r\cdot j) \\ -T_l/(r\cdot j\cos^2 b) \\ -T_r \end{bmatrix} \tag{3-6}$$

or, expressing the most right part in the form of so-called *disturbances*

$$\begin{bmatrix} \delta\Phi \\ \delta\Lambda \\ \delta g \end{bmatrix} = \begin{bmatrix} j & 0 & 0 \\ 0 & j\cos\phi & 0 \\ 0 & 0 & 1 \end{bmatrix}^{-1} \underline{Q}^T\underline{M}\,\delta\underline{x} + \begin{bmatrix} \delta\phi \\ \delta\lambda \\ \bar{\delta g} \end{bmatrix} \tag{3-7}$$

where

$$\delta\phi = -\frac{1}{rj}\frac{\partial T}{\partial b} = \xi \tag{3-8}$$

$$\delta\lambda = -\frac{1}{rj\cos^2 b}\frac{\partial T}{\partial l} = \frac{\eta}{\cos b} \tag{3-9}$$

$$\bar{\delta g} = -\frac{\partial T}{\partial r} \tag{3-10}$$

$\xi = \delta\phi$ and $\eta = \delta\lambda\cos b$ are the components of the deflections of the vertical in north-south and east-west direction, respectively.

We are now transforming the first term on the right hand side of (3-7) into the local astronomical system with coordinates $\underline{x}^* = (x^*\ y^*\ z^*)^T$ using the transformations

$$\delta\underline{x} = \underline{R}\,\delta\underline{x}^* \tag{3-11}$$

and

$$M = \underline{Q}\,\underline{M^*}\,\underline{Q}^T \tag{3-12}$$

where $\underline{M^*}$ is the matrix of local second-order derivations of the normal gravity potential referring to the local astronomical triad (x^*,y^*,z^*) ,

$$\underline{M^*} = \begin{bmatrix} U_{x^*x^*} & U_{x^*y^*} & U_{x^*z^*} \\ U_{y^*x^*} & U_{y^*y^*} & U_{y^*z^*} \\ U_{z^*x^*} & U_{z^*y^*} & U_{z^*z^*} \end{bmatrix} . \tag{3-13}$$

Thus we get after considering that

$$\begin{aligned} \underline{R} &\doteq -\underline{Q} \\ \underline{Q}^T\underline{Q} &= \underline{I} \end{aligned} \tag{3-14}$$

the final expression (equivalent to eq. (2-9) $\delta\underline{l} = \underline{a}^T\,\delta\underline{x} + L(T)$)

$$\begin{bmatrix} \delta\Phi \\ \delta\Lambda \\ \delta g \end{bmatrix} = \begin{bmatrix} -j^{-1}U_{x^*x^*} & -j^{-1}U_{x^*}U_{y^*} & -j^{-1}U_{x^*z^*} \\ -(j\cos\phi)^{-1}U_{x^*y^*} & -(j\cos\phi)^{-1}U_{y^*y^*} & -(j\cos\phi)^{-1}U_{y^*z^*} \\ -U_{x^*z^*} & -U_{y^*z^*} & -U_{z^*z^*} \end{bmatrix} \begin{bmatrix} \delta x^* \\ \delta y^* \\ \delta z^* \end{bmatrix} + \begin{bmatrix} \delta\phi \\ \delta\lambda \\ \bar{\delta g} \end{bmatrix} \tag{3-15}$$

or, in short

$$\delta\underline{g} = \underline{\tilde{M}^*}\,\delta\underline{x^*} + \text{grad } T . \tag{3-16}$$

Note, that in classical physical geodesy δz^* is considered as the so-called *height anomaly*.

Looking now more in detail to the design matrix $\underline{a}^T = \underline{\tilde{M}^*}$ of (3-15) we can write

$$\begin{bmatrix} \delta\Phi \\ \delta\Lambda \\ \delta g \end{bmatrix} = \begin{bmatrix} -k_1 & -t_1 & \kappa_1 \\ -t_1\sec\phi & -k_2\sec\phi & \kappa_2\sec\phi \\ j\,\kappa_1 & j\,\kappa_2 & \partial j/\partial z^* \end{bmatrix} \begin{bmatrix} \delta x^* \\ \delta y^* \\ \delta z^* \end{bmatrix} + \begin{bmatrix} \delta\phi \\ \delta\lambda \\ \bar{\delta g} \end{bmatrix} \tag{3-17}$$

where

$k_1 = j^{-1}U_{x^*x^*}$ is the *curvature* of the *normal* equipotential surface in *north-south* direction,

$k_1 = j^{-1}U_{y^*y^*}$ is the *curvature* of the *normal* equipotential surface in *east-west* direction,

$t_1 = j^{-1}U_{x^*y^*}$ is the *torsion* of the *normal* equipotential surface,

$\kappa_1 = -j^{-1}U_{x^*z^*}$ is the *north-south* component of the *curvature* of the plumb-line in the *normal* gravity field, and

$\kappa_2 = -j^{-1}U_{y^*z^*}$ is the *east-west* component of the *curvature* of the plumb-line in the *normal* gravity field.

Using an ellipsoidal or spherical normal gravity field we approximate the quantities above by

parameter	*normal gravity field*	
	ellipsoidal	*spherical*
k_1	$1/\rho_1$	$1/r$
k_2	$1/\rho_2$	$1/r$
t_1	0	0
κ_1	$\frac{\partial \ln j}{\rho_1 \, \partial\phi}$	0
κ_2	$\frac{\partial \ln j}{\rho_2 \, \partial\phi}$	0
$\partial j/\partial z^*$	$\partial j/\partial z^*$	$-2j/r$ (where $U = GM/r$)

.

We can rewrite (3-17) in *ellipsoidal* approximation

$$\begin{bmatrix} \delta\Phi \\ \delta\Lambda \\ \delta g \end{bmatrix} = \begin{bmatrix} -1/\rho_1 & 0 & \frac{\partial \ln j}{\rho_1 \, \partial\phi} \\ 0 & -1/(\rho_2 \cos\phi) & 0 \\ \frac{j \, \partial \ln j}{\rho_1 \, \partial\phi} & 0 & \partial j/\partial z^* \end{bmatrix} \begin{bmatrix} \delta x^* \\ \delta y^* \\ \delta z^* \end{bmatrix} + \begin{bmatrix} \delta\phi \\ \delta\lambda \\ \bar{\delta} g \end{bmatrix} \tag{3-18}$$

where ρ_1, ρ_2 are the two axes of the ellipsoid. In *spherical* approximation (3-15) reads

$$\begin{bmatrix} \delta\Phi \\ \delta\Lambda \\ \delta g \end{bmatrix} = \begin{bmatrix} -1/r & 0 & 0 \\ 0 & -1/(r\cos\phi) & 0 \\ 0 & 0 & -2j/r \end{bmatrix} \begin{bmatrix} \delta x^* \\ \delta y^* \\ \delta z^* \end{bmatrix} + \begin{bmatrix} -T_b/(r \cdot j) \\ -T_l/(r \cdot j \cdot \cos^2 b) \\ -T_r \end{bmatrix} \tag{3-19}$$

The same transformations and approximations we have applied to the gravity vector $\underline{g}$ with observations Φ, Λ, g , we use now in modifying the potential difference (2-20) considering (3-11). Thus, we get

$$\delta W_{0j} = -\Delta W_0 + [U_{x^*} \; U_{y^*} \; U_{z^*}] \; \delta \underline{x}_j^* + T(\underline{x}_j) \tag{3-20}$$

where $P_i := P_0$ and $\Delta W_0 = [U_x \; U_y \; U_z] \; \delta \underline{x}_i^0 + T^0(\underline{x}_i)$. In spherical approximation we have $U_{x^*} = U_{y^*} = 0$, $U_{z^*} = -j$, so that (3-20) changes into

$$\delta W_0 = -\Delta W_0 + [0 \;\; 0 \;\; -j] \; \delta \underline{x}^* + T \tag{3-21}$$

or, in detail,

$$\delta W_0 = -\Delta W_0 - j \; \delta z^* + T \tag{3-22}$$

3.2 Approximate coordinates

For the linearization applied in the last chapter to the gravity vector and the potential we need the corresponding *approximate* values. Those quantities can be derived from the measurements themselves unless it is possible to get horizontal and vertical coordinates from a map or another source. Thus, three conditions using three of the four observations (3-3), (3-22) can be used to derive the approximate coordinates. The two famous choices are the so-called

Marussi mapping, leading to the *Marussi telluroid*, using the natural coordinates potential, latitude and longitude,

$$\begin{matrix} \delta W_0 = U(\underline{x}^0) - U_0(\underline{x}_0^0) & & \delta W_0 = 0 & & \\ \Phi(\underline{x}) = \phi(\underline{x}^0) & \text{equi-valent to} & \delta\Phi = 0 & \text{in} & \begin{pmatrix}3\text{-}22\\3\text{-}7\end{pmatrix} \\ \Lambda(\underline{x}) = \lambda(\underline{x}^0) & & \delta\Lambda = 0 & & \end{matrix} \tag{3-23}$$

and the *gravimetric* mapping leading to the *gravimetric telluroid* using

$$\begin{matrix} \Phi(\underline{x}) = \phi(\underline{x}^0) & & \delta\Phi = 0 & & \\ \Lambda(\underline{x}) = \lambda(\underline{x}^0) & \text{equi-valent to} & \delta\Lambda = 0 & \text{in (3-7)} & \\ g(\underline{x}) = j(\underline{x}^0) & & \delta g = 0 & & . \end{matrix} \tag{3-24}$$

This means, that the observations belonging to points $P(\underline{x})$ of the physical earth's surface are set equal to the corresponding quantities derived from a chosen model or normal gravity field U at $P(\underline{x}^0)$ of the approximate figure of the earth, the telluroid.

3.3 Solution

The linearized observation equations for the potential, astronomical latitude and longitude, and gravity contain two types of unknowns, the coordinates and the functionals $L(T)$ of the gravity disturbing potential T. Let us solve now (3-16) for the geometric unknowns $\delta \underline{x}^*$, assuming that $\underline{\tilde{M}}^*$ is invertible,

$$\delta \underline{x}^* = \underline{\tilde{M}}^{*-1}(\delta \underline{g} - \text{grad } T) \tag{3-25}$$

and insert it into (3-21)

$$\delta W_0 = -\Delta W_0 + \underline{j}^T \underline{\tilde{M}}^{*-1}(\delta \underline{g} - \text{grad } T) + T \tag{3-26}$$

or, rearranged,

$$T - \underline{j}^T \underline{\tilde{M}}^{*-1} \text{grad } T = \delta W_0 + \Delta W_0 - \underline{j}^T \underline{\tilde{M}}^{*-1} \delta \underline{g} \quad . \tag{3-27}$$

This equation is the *fundamental boundary condition* of the *linearized Molodensky problem* assuming that the surface (telluroid) is known.

Let us have a more detailed look on the corresponding formulas in spherical approximation. From (3-22) we get the well-known *Bruns'* equation

$$\delta z^* = \frac{-\delta W_0 - \Delta W_0 + T}{j} \tag{3-28}$$

and from (3-19) using (3-28) we derive the *fundamental equation of physical geodesy* in *spherical* approximation

$$\delta g - \frac{2}{r}\,\delta W_0 = \frac{2}{r}\,\Delta W - \left(\frac{2T}{r} - \frac{2}{r}\frac{\partial T}{\partial r}\right) \quad . \tag{3-29}$$

Depending on the type of mapping - see (3-23) and (3-24) - either δg or δW_0 is becoming zero in the equations above. In practical work we use $\delta W_0 = 0$ for the determination of height anomalies $\zeta := \Delta z^*$ (3-28),

$$\zeta = \frac{T}{j} - \frac{\Delta W_0}{j} \tag{3-30}$$

The right hand side of (3-27) or the left hand side of (3-29) can be now considered as new (derived) observable in order to solve for T and its linear functionals using a *Molodensky* or *Stokes* type of solution, respectively. Afterwards one can go back and insert $L(T)$ in (2-9), or in detail, into (3-19), (3-20) to solve for the geometrical part in the observation equations, the coordinates. In case that the approximate coordinates used in the mapping were not sufficiently accurate to deal with the boundary value problem linearly, the whole approach has to be iterated using improved (approximate) coordinates as starting values.

<u>REFERENCES.</u> *The solution of the classical Molodensky boundary value problem is nicely described by Moritz (1980). The Bruns' transformation in 3d-coordinates and the different mappings can be found in (Grafarend 1980). The way from the observational model to gravity parameter estimation is discussed by Rummel (1985). Numerical aspects of the different mappings are outlined by Bakker (1983).*

4. THE INTEGRATED APPROACH

4.1 The general solution

We recall that the linearization of all types of geodetic observations results in the general system of linear equations (2-9)

$$\delta\underline{l} = \underline{a}^T\delta\underline{x} + L(T) \quad . \tag{4-1}$$

Whereas in classical physical geodesy some of the observations were used to eliminate first the coordinate unknowns $\delta\underline{x}$ to get a system containing only functionals $L(T)$ of the gravity disturbing potential T on the right hand side of (4-1), the integrated geodesy approach solves for both simultaneously. Looking onto (4-1) as a *mixed* model where the part $\underline{a}^T\delta\underline{x}$ is considered to be *deterministic* ("fixed"), and the gravity disturbing field to be *stochastic*, we can identify (4-1) as a *general collocation model,*

$$\delta\underline{l} = \underline{A}\delta\underline{x} + \underline{R}\,\underline{t} + \underline{n} \tag{4-2}$$

where $\underline{A} := \underline{a}^T$ and $\underline{R}$ are the design matrices of the coordinate vector $\delta\underline{x}$ and the vector $\underline{t}$ of linear functionals of the disturbing potential T, respectively,

$$\underline{t} = \left[T\,,\ -\frac{\partial T}{\partial r}\,,\ -\frac{\partial T}{\partial b\ j\ r}\,,\ -\frac{\partial T}{\partial l\ r\cos b}\,,\ \frac{\partial^2 T}{\partial x_i \partial x_j}\right]^T \quad . \tag{4-3}$$

As one can see from the observation equations, T is only present in (4-3) if potential differences (see 2-20) are observed. Second-order derivatives of T appear when dealing with gradiometer or torsion-balance data. $\underline{n}$ is the vector of observational noise. Eq. (4-2) can be solved using the hybrid minimum condition

$$\underline{n}^T\underline{C}_{nn}^{-1}\,\underline{n} + \underline{t}^T\underline{K}_{tt}^{-1}\,\underline{t} = \min \tag{4-4}$$

where $\underline{C}_{nn}$ and $\underline{K}_{tt}$ are the covariance matrices of $\underline{n}$ and $\underline{t}$. In case of non-correlated observations, $\underline{C}_{nn}$ is a diagonal matrix containing the error variances of the observations. The covariance matrix $\underline{K}_{tt}$ can be derived from a (global) covariance model, see e.g., *Tscherning (1976).*

The solution of the system (4-2) by using the minimum norm (4-4) was first proposed by *Eeg and Krarup (1973).* Its justification by looking onto the problem as the regularization of an improperly posed problem was discussed by *Moritz (1980, p.238 f.).*

Although in my opinion still some theoretical work has to be done with respect to the choice-of-norm problem in integrated geodesy, I try to give some heuristic explanations for the application of (4-4).

Depending on the contents of the (pseudo-)stochastic vector $\underline{t}$ (4-3) we minimize the disturbing potential and first- and second-order derivatives of the disturbing potential T. In contrast to the exact mapping (3-23) or (3-24) applied in classical physical geodesy we use more than just three conditions for the solution taking into consideration more information from *all available* observations. Thus, a minimum principle has to take place since only three conditions have to be chosen. Consequently, we get some kind of mixture between Marussi and gravimetric type of mapping for the (local) telluroid. In case, that no potential differences and gradiometric data are considered in the solution, e.g., only astronomic and/or gravity observations are carried out, the minimization of $\underline{t}^T \underline{K}_{tt}\, \underline{t}$ corresponds to some extent to a gravimetric mapping (2-18), since only first-order derivatives appear in $\underline{t}$.

The solution of a general collocation model of type (4-2) can be found, e.g., in *(Moritz 1980, p. 111 f.)*. The unknown *coordinates* are given by

$$\delta\hat{\underline{x}} = (\underline{A}^T \underline{D}^{-1} \underline{A})^{-1} \underline{A}^T \underline{D}^{-1} \delta\underline{l} \tag{4-5}$$

and the *functionals of the disturbing potential* at the observation points are

$$\hat{\underline{t}} = \underline{K}_{tt}\, \underline{R}^T \underline{D}^{-1} (\delta\underline{l} - \underline{A}\delta\hat{\underline{x}}) \quad . \tag{4-6}$$

Since in the collocation model an interpolation of the stochastic part is implicitly built in, (4-6) can be also used for the determination of *any other functional* $\underline{s}$ of the gravity disturbing potential at *any other station* when knowing the corresponding crosscovariances $\underline{K}_{st}$

$$\hat{\underline{s}} = \underline{K}_{st}\, \underline{R}^T \underline{D}^{-1} (\delta\underline{l} - \underline{A}\delta\hat{\underline{x}}) \; . \tag{4-7}$$

The error statistics are given by

$$\underline{E}_{xx} = (\underline{A}^T \underline{D}^{-1} \underline{A})^{-1} \tag{4-8}$$

and

$$\underline{E}_{ss} = \underline{K}_{ss} - \underline{K}_{st}\, \underline{D}^{-1} [\underline{I} - \underline{A}(\underline{A}^T \underline{D}^{-1} \underline{A})^{-1} \underline{A}^T \underline{D}^{-1}]\, \underline{K}_{ts} \tag{4-9}$$

where $\underline{I}$ is the identity matrix and

$$\underline{D} = \underline{C}_{nn} + \underline{R}\, \underline{K}_{tt}\, \underline{R}^T \quad . \tag{4-10}$$

For the estimation process above there are certain properties necessary, as

$$\bar{E}\{\underline{n}\} = 0 \tag{4-11}$$

$$\bar{E}\{\underline{t}\} = 0 \tag{4-12}$$

$$\bar{E}\{\underline{s}\} = 0 \tag{4-13}$$

where $\bar{E} = EM$ is the *total average*, and E is the expectation operator and M describes a homogeneous and isotropic average over the sphere.

4.2 Implications and numerical considerations

Let us figure out first the advantages of the integrated determination of threedimensional coordinates and the functionals of the gravity disturbing potential. Geometrical and physical geodesy are no longer separated. Thus, the historically grown separate set-up of horizontal and vertical networks is no longer justified (in particular also due to modern 3d-satellite techniques). The threedimensional approach was for a long time rejected since the high-precision results of spirit levelling could not be used in combination with the other so-called geometrical measurements. The integrated geodesy adjustment presented here allows this combination leading as a consequence to precise heights. Moreover, gravity observation have also an influence on the coordinates, and, in particular, stabilize the vertical component in case of weak or even no other vertical information. Astronomical observations which had to be carried out at the network vertices for the use in plumbline corrections can be replaced by low-cost surface gravity measurements which - and this holds also for all other types of data - can be situated anywhere due to the internal interpolation properties of collocation. Refraction and other parameters can be built-in into the deterministic coordinate unknown vector $\delta\underline{x}$.

In the derivation of the general principle of integrated geodesy we were dealing with a global geocentric reference frame. The question now is: Can we apply this approach also locally? The answer is definitely yes! Transformations can be easily built-in to work with local - even with planar - covariance models and coordinates, see e.g., *Barzaghi et al (1984), Hein and Landau (1983)*. As a consequence, the functionals of the disturbing potential in $\underline{t}$ and the derived telluroid are *relative*.

Numerical difficulties may arise when dealing with large data sets. Two matrix inversions have to be performed whereby the matrix $\underline{D}$ (4-10) is of the order of the number of observations and $(\underline{A}^T\underline{D}^{-1}\underline{A})$ is as large as the number of coordinate unknowns is. Since we do not want to adjust the positions the gravity observations were taken (their coordinates can be of low precision), we fix the corresponding coordinates in the solution. Although this keeps the coordinate vector $\delta\underline{x}$ small, we still have the problem of increasing the matrix $\underline{D}$. To speed-up the determination of covariances for the definition of $\underline{K}_{tt}$ the corresponding values can be computed and stored in a preprocessing step on a certain grid which can serve as basis for subsequent final determination of covariances by simple interpolation methods *(Sünkel 1979)*. If one arranges the gravity data on a equally spaced grid assuming equal weights for all observations and using an isotropic and stationary covariance function the resulting covariance matrix has the form

$$\underline{K}_{tt} = \begin{bmatrix} K_0 & K_{-1} & K_{-2} & \cdots & K_{-N} \\ K_1 & K_0 & K_{-1} & \cdots & K_{-(N-1)} \\ K_2 & K_1 & K_0 & \cdots & K_{-(N-2)} \\ \vdots & \vdots & \vdots & & \vdots \\ K_N & K_{N-1} & K_{N-2} & \cdots & K_0 \end{bmatrix} . \qquad (4\text{-}14)$$

Thus, $\underline{K}_{tt}$ has a so-called *block-Toeplitz* structure where the (i,j)th submatrices themselves are *simple Toeplitz* matrices fulfilling the condition $K_{ij} = K_{i-j}$ which means that they are a function of $(i-j)$. The inversion of such Toeplitz matrices can be done very efficiently using computations proportional to N^2 instead of N^3 when applying conventional inversions. In case of block-Toeplitz matrices the computational time is of the order of n^2m^3 instead of n^3m^3 where n denotes the number of rows and m the number of columns of the grid. In addition, the saving in memory is considerable. Let us give an example. For 10 000 gravity observations on a 100×100 grid

$$(n\,m + 1) * \frac{n\,m}{2} = (10\,000 + 1) * 5000 = 50\,005\,000$$

values have to be computed and stored for the covariance matrix in the conventional approach instead of

$$n \cdot [(m-1)\frac{m}{2} + 1] = 495\,100$$

values in the Toeplitz approach.

If data are regularly distributed on a sphere the block-Toeplitz matrix above becomes a so-called *block circulant matrix* consisting of several circulant matrices of type

$$\underline{K}_C = \begin{bmatrix} K_0 & K_1 & K_2 & \cdots & K_{N-1} \\ K_{N-1} & K_0 & K_1 & \cdots & K_{N-2} \\ K_{N-2} & K_{N-1} & K_0 & \cdots & K_{N-3} \\ \vdots & \vdots & \vdots & & \vdots \\ K_1 & K_2 & K_3 & & K_0 \end{bmatrix} \qquad (4\text{-}15)$$

having the property that each row i is equal the row (i-1) rotated by one element in the column. Circulant matrices become diagonal matrices under discrete Fourier transform. Therefore they can be easily inverted in frequency domain. Using Fast Fourier Transform (FFT) techniques the solution requires only N log N elementary operations compared to N^3 with conventional inversion methods. A special theorem exists *(Gray 1971)* which proves that each infinite Toeplitz matrix is asymptotically equivalent to a circulant matrix. Thus, for large gravity data sets it is allowed and possible to use FFT-techniques for the inversion of the covariance matrix.

Although abovementioned methods speed-up the numerical computations considerably, one can again think - as in classical geodesy done, see chapter 3.3 - about using a stepwise procedure to solve the integrated model (2-9). The motivation to split-up the solution comes due to the fact, that on the one hand some users are only interested in coordinates, on the other hand the gravity-related observations need to be considered but having only limited influence, say in the centimeter to decimeter range, on coordinates. Thus, the idea is to split-up the observations into a vector $\underline{l}_1$ containing only the more or less geometrical measurements like e.g., distances s, angles Ω, directions R, zenith distances Z, azimuths A and astronomical latitude Φ and longitude Λ,

$$\underline{l}_1 = (\Phi, \Lambda, A, Z, R, \Omega, s) \tag{4-16}$$

and to perform with it a common least-squares adjustment with respect to 3d-coordinates. Afterwards those coordinates are corrected iteratively for the influence of the gravity field by the vector $\underline{l}_2$ of gravity-related observations, like e.g., gravity g, gravity differences Δg, potential differences ΔW, gradiometer observations $W_{x_i^* x_j^*}$,

$$\underline{l}_2 = (g, \Delta g, \Delta W, W_{x_i^* x_j^*}) \quad . \tag{4-17}$$

If finally desired, also functionals $\underline{t}$ of the disturbing potential can be determined. This procedure was proposed by *Hein (1982b)* and realized by *Landau et al (1985)*. The main elements in the derivation are:

- the stepwise collocation approach, see e.g. *(Moritz 1980, p. 144 f.)*
- the split-up of the matrix $\underline{D}^{-1}$ into

$$\underline{D}^{-1} = \underline{C}_{nn}^{-1} - \underline{C}_{nn}^{-1} \underline{R}(\underline{I} + \underline{K}_{tt} \underline{R}^T \underline{C}_{nn}^{-1} \underline{R})^{-1} \underline{K}_{tt} \underline{R} \underline{C}_{nn}^{-1} \quad , \tag{4-18}$$

- the iterative computation of $\underline{n}$, $\underline{t}$, $\underline{P}^{-1} = (\underline{A}^T \underline{D}^{-1} \underline{A})^{-1}$

REFERENCES. *An operational program for integrated geodesy, called OPERA was developed by Hein and Landau (1983). Its extension with respect to the stepwise approach can be found in (Landau et al, 1985). Numerical computations are carried out by Bäumker (1984), Bäumker et al (1983), Engler et al (1982), Grist (1984), Hein et al (1984), Torge and Wenzel (1978), Zaiser (1984, 1986). Grafarend et al (1985) included a priori information in the operational adjustment. As a starting point for covariance models see Tscherning and Rapp (1974), Tscherning (1976), Sünkel (1979). The inversion of covariance matrices by Fast Fourier Methods is discussed by Eren (1980), Gray (1971), Rino (1970). Barzaghi et al (1985) transformed the whole framework of integrated geodesy into a local reference coordinate system.*

5. THE ROLE OF THE GRAVITY FIELD IN VIEW OF THE PRESENCE OF NEW SATELLITE TECHNIQUES: NEED FOR INTEGRATED APPROACHES ?

5.1 General considerations

With the revolutionizing surveying technology provided by the NAVSTAR satellites of the *Global Positioning System* (GPS) one might pose the question whether integrated geodesy approaches are still needed in the future. The new observation type is of *purely geometrical* character (neglecting the fact that the satellite orbit is a function of the gravity field): *threedimensional Cartesian baseline components* over 50 - 100 km (or more) are derived with an accuracy in the 1 ppm (part per million: 10^{-6}) level, in the near future perhaps good to 0.1 ppm when precise satellite orbits better than 2 m in accuracy are available. Thus, a threedimensional geometrical control can be built-up solely by that type of measurements.

Remember, however, that vertical coordinates should in most cases - at least implicitly - answer the question: Where does water flow? This question cannot be answered using a geometrically defined vertical coordinate. On the contrary, dynamical quantities are needed, namely *gravity potential differences*. As a consequence, the fundamental relation

$$H = h - N \tag{5-1}$$

where H is an orthometric height, h is an ellipsoid height and N is the corresponding geoid undulation, plays a fundamental role in *vertical positioning* as well as in *gravity field (or geoid) determination* in the near future.

Rewriting (5-1) in difference-form considering the fact that the observations to the Global Positioning System yield only precise *relative* coordinate information, results in

$$H_j - H_i = (h_j - h_i) - (N_j - N_i) \tag{5-2a}$$

or

$$N_j - N_i = (h_j - h_i) - (H_j - H_i) \tag{5-2b}$$

where the indices i and j stand for the stations P_i, P_j between the relative ellipsoidal height difference $(H_j - H_i)$ was computed from GPS baseline components. Assuming that the orthometric height difference $(H_j - H_i)$ is measured or derived from a vertical network, then the geoidal height difference can be determined by (5-2b).

The accuracy of the geoid height differences according to (5-2b) relies in practice highly on the quality of the used orthometric heights. Remember that in most countries either no orthometric corrections were applied (depending on the order of the net) or, if, normal gravity was often used instead of observed real gravity in those formulas (→ normal orthometric corrections). In addition, the application of non-rigorous or even non-adjustment strategies at all, might have led to considerable network deformations. Both types of error influence ΔN_{ij} in a nearly one-to-one correspondence.

Assuming that geoid heights (or better geoid height differences) can be computed with high accuracy, GPS-derived ellipsoidal heights can be used to determine orthometric heights. For the determination of geoid heights the following approaches are possible:

(i) Integration over (mean) gravity anomalies using Stokes' formula

(ii) Collocation using heterogeneous data (deflections of the vertical, gravity anomalies, etc.)

(iii) Mixed approaches of (i) and (ii)

Prerequisite for all kind of computations is a dense coverage with gravity in the area of consideration.

Whether orthometric height or geoid height difference determination by (5-2) is the primary goal, in any case *geometrical* and *physical* aspects are closely connected. Using separate strategies for the computation of the one or the other and/or assuming them as known implies that their full error behaviour enters in the target function of (5-2a) or (5-2b).

Here the integrated geodesy approach can provide a useful strategy to treat this problem in one model providing both quantities (5-2a,b). In addition, geoid height and/or orthometric height differences can be used in the solution with known variance-covariance structure.

5.2 Example 1

Let us assume that certain preprocessing of GPS observations provides Cartesian coordinate differences, $\Delta \underline{x}_{ij}^{obs} = (\underline{x}_j - \underline{x}_i)^{obs}$ as pseudo-observations associated with a covariance matrix, $\underline{C}_{nn} = \underline{C}_{\Delta x \Delta x}$ in a *nearly conventional terrestrial* system, e.g.

WGS 72. The basic relationship to a national reference frame is then given by

$$\Delta \underline{x}_{ij}^{obs} = (1 + k) \, \underline{R}(\underline{\varepsilon}) \, \Delta \underline{x}_{ij} \tag{5-3}$$

where k is a small scale change and $\underline{R}(\underline{\varepsilon})$ is a rotation matrix for small Eulerian rotations $\varepsilon_1, \varepsilon_2, \varepsilon_3$ given in second approximation by

$$\underline{R}(\underline{\varepsilon}) = \begin{bmatrix} 1 - \frac{\varepsilon_2^2}{2} - \frac{\varepsilon_3^2}{2} & \varepsilon_3 + \varepsilon_1\varepsilon_2 & \varepsilon_1\varepsilon_3 - \varepsilon_2 \\ \varepsilon_3 & 1 - \frac{\varepsilon_1^2}{2} - \frac{\varepsilon_3^2}{2} & \varepsilon_2\varepsilon_3 + \varepsilon_1 \\ \varepsilon_2 & -\varepsilon_1 & 1 - \frac{\varepsilon_1^2}{2} - \frac{\varepsilon_2^2}{2} \end{bmatrix} \tag{5-4}$$

Since the users are mainly interested in ellipsoidal latitude ϕ, and ellipsoidal longitude λ, and orthometric height H we insert the following two relations in (5-3).

(I)

$$\underline{x} = \underline{x}(\phi,\lambda,h) = \begin{bmatrix} (N_\phi + h) & \cos\phi \cos\lambda \\ (N_\phi + h) & \cos\phi \sin\lambda \\ (\frac{b^2}{a^2} N_\phi + h) & \sin\phi \end{bmatrix} \tag{5-5}$$

with the east-west radius of curvature of the ellipsoid with axes a, b,

$$N_\phi = \left[\frac{a^2}{a^2\cos^2\phi + b^2\sin^2\phi}\right]^{0.5} \tag{5-6}$$

(II)

$$h = H + N \doteq H + T/j \quad . \tag{5-7}$$

The geoid height N is in (5-7) replaced by the (simple) Bruns' formula (3-30).

In order to get the linear observation equation of (5-3) considering (5-4) to (5-6) we linearize (5-3) using a Taylor expansion.

$$\begin{aligned} \delta\underline{b}_{ij} = \Delta\underline{x}_{ij}^{obs} - \Delta\underline{x}_j^0 = {} & \frac{\partial \underline{b}_{ij}}{\partial k} k + \frac{\partial \underline{b}_{ij}}{\partial \underline{\varepsilon}} \delta\underline{\varepsilon} + \frac{\partial \underline{b}_{ij}}{\partial \phi_j} \delta\phi_j - \frac{\partial \underline{b}_{ij}}{\partial \phi_i} \delta\phi_i \\ & + \frac{\partial \underline{b}_{ij}}{\partial \lambda_j} \delta\lambda_j - \frac{\partial \underline{b}_{ij}}{\partial \lambda_i} \delta\lambda_i + \frac{\partial \underline{b}_{ij}}{\partial H_j} \delta H_j - \frac{\partial \underline{b}_{ij}}{\partial H_i} \delta H_i \\ & \frac{\partial \underline{b}_{ij}}{\partial T_j} \delta T_j - \frac{\partial \underline{b}_{ij}}{\partial T_i} \delta T_i \end{aligned} \tag{5-8}$$

where

$$\frac{\partial \underline{b}_{ij}}{\partial k} = \underline{R}(\underline{\varepsilon})\Delta\underline{x}^0_{ij} \doteq \Delta\underline{x}^{obs}_{ij} \tag{5-9}$$

$$\frac{\partial \underline{b}_{ij}}{\partial \underline{\varepsilon}} = (1+k)\begin{bmatrix} \varepsilon_2\ \Delta x + \varepsilon_3\ \Delta z & \varepsilon_1\ \Delta y - \varepsilon_2\ \Delta x - \Delta z & \Delta y - \varepsilon_1\ \Delta z \\ \Delta z - \varepsilon_1\ \Delta y & \varepsilon_3\ \Delta z & \varepsilon_2\ \Delta z - \varepsilon_3\ \Delta y - \Delta x \\ -\Delta y - \varepsilon_1\ \Delta z & \Delta x - \varepsilon_2\ \Delta z & 0 \end{bmatrix} \tag{5-10}$$

with

$$\begin{aligned} \Delta x &= x_j - x_i \\ \Delta y &= y_j - y_i \\ \Delta z &= z_j - z_i \\ \underline{x} &= (x,y,z)^T \quad . \end{aligned} \tag{5-11}$$

Neglecting small terms (the products with small rotational angles ε) in the matrix on the right hand of (5-10) we can approximate (5-10) by

$$\frac{\partial \underline{b}_{ij}}{\partial \underline{\varepsilon}} \doteq \begin{bmatrix} 0 & -\Delta z & \Delta y \\ \Delta z & 0 & -\Delta x \\ -\Delta y & \Delta x & 0 \end{bmatrix} \doteq (\delta \underline{R}_x)_{ij} \quad . \tag{5-12}$$

We have further in (5-8) the expressions *(Eissfeller et al 1985, Eissfeller 1986)*

$$\frac{\partial \underline{b}_{ij}}{\partial \phi} = \begin{bmatrix} -(M + H + \frac{T}{j})\ \sin\phi\ \cos\lambda \\ -(M + H + \frac{T}{j})\ \sin\phi\ \sin\lambda \\ (M + H + \frac{T}{j})\ \cos\phi \end{bmatrix}, \quad \phi = \{\phi_1,\phi_2\} \tag{5-13}$$

$$\frac{\partial \underline{b}_{ij}}{\partial \lambda} = \begin{bmatrix} -(N - H + \frac{T}{j})\ \cos\phi\ \sin\lambda \\ (N + H + \frac{T}{j}\ \ \cos\phi\ \cos\lambda \\ 0 \end{bmatrix}, \quad \lambda = \{\lambda_1,\lambda_2\} \tag{5-14}$$

$$\frac{\partial \underline{b}_{ij}}{\partial H} = \begin{bmatrix} \cos\phi\ \cos\lambda \\ \cos\phi\ \sin\lambda \\ \sin\phi \end{bmatrix}, \qquad H = \{H_1,H_2\} \tag{5-15}$$

$$\frac{\partial \underline{b}_{ij}}{\partial T} = \frac{\partial \underline{b}_{ij}}{\partial H}, \qquad \begin{aligned} T &= \{T_1,T_2\} \\ H &= \{H_1,H_2\} \end{aligned} \tag{5-16}$$

where M is the north-south radius of curvature of the ellipsoid. By (5-7) we have now an observation of type

$$l = f(k, \underline{\varepsilon}, \phi, \lambda, H, T) \tag{5-17}$$

or in matrix form the expression (4-2)

$$\underline{l} = \underline{A}\,\underline{x} + \underline{R}\,\underline{t} + \underline{n} \tag{5-18}$$

where $\underline{x} = (k, \underline{\varepsilon}, \phi, \lambda, H)$, $\underline{R} = \underline{I}$, $\underline{t} = (T)$. Note, that the terms with respect to H and T are linear dependent. However, when applying the hybrid minimum norm (4-4) the defect is removed. For the solution see paragraph 4.1.

In the approach above we are now able to combine GPS-derived coordinate information in the form of (5-8) with all other types of gravity field functionals, e.g., gravity data and deflections of the vertical, as well as *known* horizontal coordinates, orthometric heights and geoid heights. For the observation equations of the first see *Hein (1982a)*. Since GPS precise coordinates are relative, it is of great advantage to introduce known coordinates and geoid heights when densifying networks.

In (5-8) approximate values for T_i have to be introduced in order to deal with δT_i . This can be done using different strategies. One possibility is to determine it by using $T^0 = j(h - H)$ and known orthometric heights H and GPS ellipsoidal heights. Another way is splitting up T^0 in three different parts

$$T^0 = T_{180} + T_{terrain} + T_{trend}$$

where T_{180} is the part coming from an earth model (e.g. GEM 10 C, $n = m = 180$), $T_{terrain}$ is the part computed by a digital terrain model and (known or assumed) density, and T_{trend} is a local trend function.

5.3 Example 2

The second method starts from the philosophy that a vertical network has to be densified and local deformations as well as available accuracy estimates of the network quantities should be taken into account.

For that reason we deal with the observation equation for GPS baseline components leaving it in the form of $\underline{l} = \underline{A}\,\underline{x} + \underline{n}$ with covariance matrix $\underline{C}_{nn}$, in detail

$$\Delta\underline{x}_{ij}^{obs} - \Delta\underline{x}_{ij}^{0} = \delta\underline{x}_j - \delta\underline{x}_i + (\delta\underline{R}_x)_{ij}\,\delta\underline{\varepsilon} + \Delta\underline{x}_{ij}^{obs}\,k \tag{5-19}$$

eventually considering the transformation $\underline{x} = (x,y,z)^T \rightarrow (\phi,\lambda,h)^T$ in the corresponding quantities of (5-18), but dealing *only* with ellipsoidal *geometrical* heights h .

In order to take advantage of available orthometric heights H with certain accuracy estimates ($\rightarrow \underline{C}_{\Delta H \Delta H}$) - looking consequently onto the determination of new orthometric heights as some kind of interpolation approach - we are developing an observation equation for H which can be considered as "directly measured quantity" or pseudo-observation. This approach was proposed by *Hein (1985)*.

Starting from Helmert's definition for orthometric heights *(Heiskanen and Moritz 1967, p. 167)*

$$H(P) = \frac{C(P)}{g(P) + \alpha H(P)} \tag{5-20}$$

where

C is the geopotential difference, $C(P) = W(Q) - W(P)$, where $W(Q)$ belongs to point Q on the geoid,

g is the actual gravity at surface point P , and

α is a coefficient derived from the normal gravity field,

$$\alpha = -\left(\frac{1}{2}\frac{\partial j}{\partial h} + 2\pi G\rho\right) \qquad \textit{(see Heiskanen and Moritz 1967, p. 167)} \tag{5-21}$$

where

$\frac{\partial j}{\partial h}$ is the normal vertical gravity gradient,

G is the gravitational constant, and

ρ is the normal density, 2.67 g/cm^3 .

Inserting numerical values in eq. (5-21), α is determined to be

$$\alpha = 0.0424 \text{ gal km}^{-1} . \tag{5-22}$$

In the linearization process the coordinates $\underline{x}_Q$ of the foot point of H on the geoid (Q) come into the picture. Since the corresponding surface coordinates $\underline{x}_P$ and $\underline{x}_Q$ are linear dependent, $\underline{x}_Q$ on the geoid is approximated by

$$\underline{x}_Q \doteq \underline{x}_P + H\, \underline{n}_P \tag{5-23}$$

where $\underline{n}_P$ is the unit vector normal to the geopotential surface at P ,

$$\underline{n}_P = \begin{bmatrix} \cos\Phi_P & \cos\Lambda_P \\ \cos\Phi_P & \sin\Lambda_P \\ \multicolumn{2}{c}{\sin\Phi_P} \end{bmatrix} \tag{5-24}$$

Φ, Λ are the astronomical latitude and longitude of the unit gravity vector at P. Relation (5-23) holds for the case that the curvature and the torsion of the plumb-line are set to zero. Finally, a linear observation equation of type (4-2) or (5-18) is derived.

For the detailed form and the corresponding coefficients see *Hein (1985b)*.

Again, with those observation equations all kind of other data, especially gravity field data, can be combined in the general *integrated geodesy adjustment* model.

REFERENCES. For the details of the approaches presented above see Eissfeller et al (1985), Eissfeller (1986), Hein (1985b). The role of the gravity field and the geoid within the integrated approach was discussed by Hein (1982c, 1983, 1985a, 1986b). Seven theses to set-up a modern network were presented for discussion by Hein (1986b).

6. THE OBSERVATION EQUATIONS OF SATELLITE GEODESY IN THE INTEGRATED MODEL

For the derivation of the observation equations of satellite geodesy it is necessary to get the geocentric position and velocity vector of the considered satellite. Those vectors are a solution of a system of differential equation of second order. Both, position and velocity vector can be expressed as a nonlinear functional of the gravity potential.

6.1 Equations of motion and orbit integration

Let $\underline{x}(t)$ be a vector in an inertial reference system describing the orbit of a satellite. Neglecting relativistic effects the motion of a satellite is given in such a system by Newton's second law,

$$\underline{\ddot{x}}(t) = \nabla_x W(\underline{x}(t),t) + \underline{f}(\underline{x}(t),\underline{\dot{x}}(t),t) \tag{6-1}$$

where

t	is the time parameter,
$\underline{x}(t)$	is the position vector of the satellite,
$\underline{x}(t) = d\underline{x}/dt$	is the velocity vector of the satellite,
$\underline{\ddot{x}}(t) = d^2\underline{x}/dt^2$	is the acceleration vector of the satellite,
$\nabla_x W(\underline{x}(t),t)$	is the gradient of the earth's gravity potential referring to the inertial coordinate system, and
$\underline{f}(\underline{x}(t),\underline{\dot{x}}(t),t)$	is the vector of all other resultant accelerations acting on the satellite.

The explicit functional relationship of the gradient $\nabla_x W(\underline{x}(t),t)$ with time is due to the fact, that the gravity potential W refers to an earth-fixed coordinate system moving relatively to the inertial system in time.

Eq. (6-1) is a nonlinear vector differential equation of second order. It corresponds to a nonlinear system of three scalar differential equation of second order.

For the solution of the problem the following decomposition is used:

$$W(\underline{x},t) = W_0(\underline{x}) + W_1(\underline{x},t) \tag{6-2}$$

Corresponding to (6-2) we get for the gradient of the gravity potential

$$\nabla_x W(\underline{x},t) = \nabla_x W_0(\underline{x}) + \nabla_x W_1(\underline{x},t) \tag{6-3}$$

where

$$W_0(\underline{x}) = GM/|\underline{x}| \tag{6-4a}$$

and

$$\nabla_x W_0(\underline{x}) = -GM\ \underline{x}/|\underline{x}|^3 \ . \tag{6-4b}$$

GM is the product of the gravitational constant G and the mass M of the earth.

Thus, the decomposition (6-2) results in a radial symmetrical part (6-4a) of the gravity potential and a small disturbance W_1 . Furthermore, we note that the corresponding resultant acceleration term $\nabla_x W_1(\underline{x},t)$ is small in comparison to (6-4b).

Inserting (6-3) in (6-1) we get

$$\ddot{\underline{x}}(t) = \nabla_x W_0(\underline{x}) + \nabla_x W_1(\underline{x},t) + \underline{f}(\underline{x},\dot{\underline{x}},t) \ . \tag{6-5}$$

Adding the two most right terms in (6-5), and expressing the corresponding acceleration vector by $\underline{a}$,

$$\underline{a}(\underline{x},\dot{\underline{x}},t) = \nabla_x W_1(\underline{x},t) + \underline{f}(\underline{x},\dot{\underline{x}},t) \tag{6-6}$$

we have

$$\ddot{\underline{x}} = \nabla_x W_0(\underline{x}) + \underline{a}(\underline{x},\dot{\underline{x}},t) \ . \tag{6-7}$$

Thus, (6-7) can be considered as a *perturbed homogeneous* problem, where the perturbing part $\underline{a}(\underline{x},\dot{\underline{x}},t)$ is small in comparison to $\nabla_x W_0(\underline{x})$.

The classical solution starts with the solution of the homogeneous problem, which possesses six integration constants. Afterwards the special solution of the perturbed problem can be derived by the method of variation of constants.

According to (6-7) the homogeneous problem is given by

$$\ddot{\underline{x}} = -GM\ \underline{x} / |\underline{x}|^3 \tag{6-8}$$

This problem is known as the so-called *Kepler* problem.

The position vector $\underline{x}$ of the so-called *Kepler ellipse* is of the following form,

$$\underline{x} = \underline{x}(\underline{u},t) \tag{6-9}$$

where $\underline{u}$ is the vector of six integration constants (orbital elements)

$$\underline{u} = [u_1, \ldots, u_6]^T \ . \tag{6-10}$$

In order to get a special solution of the perturbed problem (6-7), the constants u_i in (6-10) are considered as functions of time, e.g. $u_i = u_i(t)$, $i = \{1,\ldots,6\}$.

Following *Bucerius (1966, p. 171 ff.)* a differential equation system of first order can be found for the determination of the unknown parameters $u_i(t)$, $i = \{1,\ldots,6\}$.

Replacing the integration constants in (6-9) by the functions $u_i(t)$ we get

$$\underline{x} = \underline{x}(\underline{u}(t),t) \ . \tag{6-11}$$

In order to get the above-mentioned differential equation system for $\underline{u}(t)$, the position vector $\underline{x}$ has to be differentiated twice with respect to time and to be inserted in (6-7).

For the differentiation the following abbreviations are used:

$$\underline{X} := \frac{\partial \underline{x}}{\partial \underline{u}} \tag{6-12a}$$

$$\dot{\underline{X}} := \frac{\partial \dot{\underline{x}}}{\partial \underline{u}} \tag{6-12b}$$

Thus, the first and second order derivatives of (6-11) are given by

$$\dot{\underline{x}} = \underline{X}\ \dot{\underline{u}} + \partial \underline{x}/\partial t \tag{6-13a}$$

$$\ddot{\underline{x}} = \dot{\underline{X}}\ \dot{\underline{u}} + \partial^2 \underline{x}/\partial t^2 \tag{6-13b}$$

where

$$\underline{X}\ \dot{\underline{u}} = 0 \ . \tag{6-13c}$$

The condition (6-13c) may be chosen due to the fact, that the special solution of (6-7) is determined uniquely by three scalar functions (three components of the position vector $\underline{x}$). Thus, only three independent functions $u_i(t)$ can be determined.

Inserting (6-13b) in (6-7), results in

$$\underline{\dot{X}}\ \underline{\dot{u}} + (\partial^2\underline{x}/\partial t^2) = \nabla_X W_0(\underline{X}) + \underline{a}(\underline{x},\underline{\dot{x}},t) \qquad (6\text{-}14a)$$

where

$$(\partial^2\underline{x}/\partial t^2) - \nabla_X W_0(\underline{x}) = 0 \ . \qquad (6\text{-}14b)$$

Eq. (6-14b) expresses the fact that $\partial^2\underline{x}/\partial t^2$ just solves the homogeneous problem (6-8).

Using (6-13c) and (6-14a) we get a nonlinear differential equation system of first order for the determination of the vector $\underline{u}(t)$.

$$\underline{\dot{u}} = \begin{bmatrix} \underline{X} \\ \underline{\dot{X}} \end{bmatrix}^{-1} \begin{bmatrix} \underline{0} \\ \underline{a} \end{bmatrix} \qquad (6\text{-}15)$$

Since the first three components of the vector on the right hand side are zero-elements, the following simplification might be reasonable.

$$\underline{Y} = \begin{bmatrix} \underline{X} \\ \underline{\dot{X}} \end{bmatrix}^{-1} [0,0,0,1,1,1]^T \qquad (6\text{-}16)$$

Thus, the following differential equation system may be obtained

$$\underline{\dot{u}}(t) = \underline{Y}(\underline{u}(t),t)\ \underline{a}(\underline{u}(t),t) \ . \qquad (6\text{-}17)$$

The solution is given by

$$\underline{u}(t) = \underline{u}_0 + \int_{t_0}^{t} \underline{Y}(\underline{u}(t),t)\ \underline{a}(\underline{u}(t),t)\ dt \qquad (6\text{-}18)$$

where

$$\underline{u}_0 = \underline{u}(t_0) \ .$$

Since $\underline{a}$ is considerable small in comparison to $\nabla_X W_0(\underline{x})$ (6-7) - for low satellites we have approximately $|\underline{a}|/\nabla_X W_0 \doteq 0.002$ - *Picard's iteration method by successive approximations* is suited for the solution of (6-18).

6.2 General form of observation equations

Every satellite observation l_i can be considered as nonlinear functional depending on the threedimensional position vector $\underline{x}$ and on the velocity vector $\dot{\underline{x}}$ of the ground station G and of the satellite position Q , respectively.

$$l_i = l_i(\underline{x}_Q, \dot{\underline{x}}_Q, \underline{x}_G, \dot{\underline{x}}_G) \tag{6-19}$$

The position vector $\underline{x}_G$ of the ground station can also be expressed in the earth-fixed coordinate system $\underline{y}_G$.

Introducing approximate coordinate and velocity vectors $\underline{x}_Q^0, \dot{\underline{x}}_Q^0, \underline{x}_G^0, \dot{\underline{x}}_G^0$ by the decompositions,

$$\begin{aligned}
\underline{x}_Q &= \underline{x}_Q^0 + \delta\underline{x}_Q \\
\dot{\underline{x}}_Q &= \dot{\underline{x}}_Q^0 + \delta\dot{\underline{x}}_Q \\
\underline{x}_G &= \underline{x}_G^0 + \delta\underline{x}_G \\
\dot{\underline{x}}_G &= \dot{\underline{x}}_G^0 + \delta\dot{\underline{x}}_G
\end{aligned} \tag{6-20}$$

considering the corresponding approximate observations l_i^0 ,

$$l_i = l_i^0 + \delta l_i \tag{6-21}$$

and expanding δl_i in a Taylor series neglecting higher-order terms, leads to

$$\delta l_i = \left.\frac{\partial l_i}{\partial \underline{x}_Q}\right|_{\underline{x}_Q^0} \delta\underline{x}_Q + \left.\frac{\partial l_i}{\partial \dot{\underline{x}}_Q}\right|_{\dot{\underline{x}}_Q^0} \delta\dot{\underline{x}}_Q + \left.\frac{\partial l_i}{\partial \underline{x}_G}\right|_{\underline{x}_G^0} \delta\underline{x}_G + \left.\frac{\partial l_i}{\partial \dot{\underline{x}}_G}\right|_{\dot{\underline{x}}_G^0} \delta\dot{\underline{x}}_G \; . \tag{6-22}$$

Since the position vector $\underline{x}$ and the velocity vector $\dot{\underline{x}}$ of the satellite is a function of dynamical parameters, e.g., parameters of the acceleration model and of the gravity potential, the linearization process should result in a linear functional of the form (see also *Moritz 1980, p. 226 ff.*)

$$\delta l_i = \underline{b}_i^T \, \delta\underline{y}_G + \underline{a}_i^T \, \delta\underline{p} + L(T) \tag{6-23}$$

where

- $\delta\underline{y}_G$ — is the vector of residual coordinate unknowns of the ground station in the earth-fixed reference frame,
- $\delta\underline{p}$ — is a vector of residual dynamical parameters, e.g., parameters of the acceleration model,

$L(t)$ is a linear operator of the residual (or disturbing) potential, and

$\underline{a}_i$, $\underline{b}_i$ are coefficient vectors of the unknowns $\delta\underline{y}_G$, $\delta\underline{p}$.

The linear relation between $\delta\underline{x}_Q$, $\delta\underline{\dot{x}}_Q$ in (6-22) and $\delta\underline{p}$, $L(t)$ in (6-23) can only be derived by considering the coordinate vector $\underline{u}$ resulting from the successive approximations (6-18).

Decomposing $\underline{u}(t)$ into an approximate vector $\underline{u}^0(t)$ and a residual (linear) vector $\delta u(t)$,

$$\underline{u}(t) = \underline{u}^0(t) + \delta\underline{u}(t) \tag{6-24}$$

leads to

$$\underline{x}_Q = \underline{x}_Q(\underline{u}^0(t),t) + \frac{\partial\underline{x}_Q}{\partial\underline{u}}(\underline{u}^0(t),t)\;\delta\underline{u}(t) \tag{6-25}$$

$$\underline{\dot{x}}_Q = \underline{\dot{x}}_Q(\underline{u}^0(t),t) + \frac{\partial\underline{\dot{x}}_Q}{\partial\underline{u}}(\underline{u}^0(t),t)\;\delta\underline{u}(t) \quad . \tag{6-26}$$

Using the abbreviations $\underline{X}$, $\underline{\dot{X}}$ (6-12) we get

$$\underline{x}_Q(t) = \underline{x}_Q(\underline{u}^0(t),t) + \underline{X}(\underline{u}^0(t),t)\;\delta\underline{u}(t) \tag{6-27a}$$

$$\underline{\dot{x}}_Q(t) = \underline{\dot{x}}_Q(\underline{u}^0(t),t) + \underline{\dot{X}}(\underline{u}^0(t),t)\;\delta\underline{u}(t) \quad . \tag{6-27b}$$

Thus, we have for $\delta\underline{x}_Q$, $\delta\underline{\dot{x}}_Q$ in (6-20) the relation

$$\delta\underline{x}_Q = \underline{X}\;\delta\underline{u} \tag{6-28a}$$

$$\delta\underline{\dot{x}}_Q = \underline{\dot{X}}\;\delta\underline{u} \quad . \tag{6-28b}$$

The vector $\underline{p}$ (or its residual vector $\delta\underline{p}$) forms together with the ground station coordinates the *deterministic unknown* part of model (6-23). It is defined by the introduced acceleration model.

According to (6-6) the resulting acceleration vector $\underline{a}$ is given by

$$\underline{a} = \nabla_X W_1 + \underline{f} \quad .$$

The linearizing principle of integrated geodesy consists now of two steps:

$$\text{(i)} \quad W_1 = U_1 + T \tag{6-29}$$

$$\text{(ii)} \quad \underline{p} = \underline{p}^0 + \delta\underline{p} \tag{6-30}$$

U_1 in (6-29) is the so-called *normal (or model) potential* U reduced for the radial-symmetrical part $U_0 = GM//\underline{x}/$, consequently $U_1 = U - U_0$, and T is the gravity *disturbing potential* of the earth. $\underline{p}^0$ is the vector of approximate parameters of $\underline{p}$.

Thus, the gradient of the gravity potential W_1 (6-29) is

$$\nabla_x W_1 = \nabla_x U_1 + \nabla_x T \tag{6-31}$$

or using the formerly introduced notations of *Hein (1982a, p. 43)*,

$$\underline{g}_1 = \underline{j}_1 + \delta\underline{g} \tag{6-32}$$

where

$$\underline{g}_1 = \nabla_x W_1 \tag{6-33a}$$

$$\underline{j}_1 = \nabla_x U_1 \tag{6-33b}$$

$$\delta\underline{g} = \nabla_x T \quad . \tag{6-33c}$$

Introducing (6-32) in (6-6) yields

$$\underline{a} = \underline{g}_1 + \underline{f} = \underline{j}_1 + \underline{f} + \delta\underline{g} \quad . \tag{6-34}$$

Note, that the vector $\delta\underline{g}$ in (6-34) is already of first order.

Next, we insert (6-34) into the successive approximation of (6-18), and linearize it. Thus, we get

$$\underline{u}_{k+1}(t,\underline{p}) = \underline{u}_{k+1}(t,\underline{p}^0) + \delta\underline{u}_{k+1}(t,\underline{p}^0) \tag{6-35}$$

where

$$\underline{u}_{k+1}(t,\underline{p}^0) =$$

$$\underline{u}_0(t,\underline{p}^0) + \int_{t_0}^{t} \underline{Y}(\underline{u}_k(t,\underline{p}^0),t)\ [\underline{j}_1(\underline{u}_k(t,\underline{p}^0),p^0,t) + \underline{f}(\underline{u}_k(t,\underline{p}^0),\underline{p}^0,t)]\ dt \tag{6-35a}$$

$$\delta\underline{u}_{k+1}(t,\underline{p}^0) =$$

$$\frac{\partial}{\partial\underline{p}}\left[\underline{u}_0(t,\underline{p}) + \int_{t_0}^{t} \underline{Y}(\underline{u}_k(t,\underline{p}),t)\ [\underline{j}_1(\underline{u}_k(t,\underline{p}),p,t) + \underline{f}(\underline{u}_k(t,\underline{p}),p,t)]\ dt\right]_{(\underline{p}^0)} \delta\underline{p}$$
$$+ \int_{t_0}^{t} \underline{Y}(\underline{u}_k(t,p^0),t)\ \delta\underline{g}(\underline{u}_k(t,\underline{p}^0),\underline{p}^0,t)\ dt \tag{6-35b}$$

By $\delta\underline{u}(t)$ we are able to determine the variations $\delta\underline{x}_Q$, $\delta\dot{\underline{x}}_Q$ (6-28a,b) of the position and velocity vector of the satellite Q . These variations are needed to derive the linear observation equations. In addition we need the variations $\delta \underline{x}_G$, $\delta \underline{\dot{x}}_G$ of the position and velocity vector of the ground station G in the inertial reference frame.

$$\delta \underline{x}_G = \frac{\partial \underline{x}_G(t)}{\partial \underline{p}} \delta \underline{p} + \underline{R}(t)\, \delta \underline{y}_G \tag{6-36a}$$

$$\delta \underline{\dot{x}}_G = \frac{\partial \underline{\dot{x}}_G(t)}{\partial \underline{p}} \delta \underline{p} + \underline{\dot{R}}(t)\, \delta \underline{y}_G \tag{6-36b}$$

where

$$\underline{R}(t) = \underline{N}(t_0)\, \underline{P}(t_0)\, \underline{P}^T(t)\, \underline{N}^T(t)\, \underline{R}_3(-\Theta(t))\, \underline{S}(t) \quad . \tag{6-36c}$$

The matrix $R(t)$ is due to precision (P), nutation (N), earth rotation (R_3) and polar motion (S).

6.3 Examples for satellite observation equations

We will show two examples for observation equations. For all others the reader is referred to *Eissfeller and Hein (1986).*

Distance observations

The distance d between a ground point G and a satellite Q is given by

$$d(t) = |\underline{x}_Q(t) - \underline{x}_G(t)| \quad . \tag{6-37}$$

Decomposition of $d(t)$ using

$$d(t) = d^0(t) + \delta d(t) \tag{6-38}$$

results in

$$\delta d(t) = \frac{\partial d(t)}{\partial \underline{x}_Q(t)} \left\{ \left[\frac{\partial \underline{x}_Q(t)}{\partial \underline{u}(t)} \frac{\partial \underline{u}(t)}{\partial \underline{p}} - \frac{\partial \underline{x}_G(t)}{\partial \underline{p}} \right] \delta \underline{p} - \underline{R}(t)\, \delta \underline{y}_G + \frac{\partial \underline{x}_Q(t)}{\partial \underline{u}(t)} \int_{t_0}^{t} \underline{Y}(t)\, \delta \underline{g}(t)\, dt \right\} \tag{6-39}$$

Doppler observations (doppler shift)

The doppler shift $\Delta f(t)$ depends on the velocity $\dot{S}(t)$ of a satellite Q relative to a ground station G in the direction of $\underline{u}(t)$. f is the frequency.

$$\Delta f(t) = -\frac{f}{c}\dot{S}(t) \tag{6-40}$$

where

$$\dot{S}(t) = \frac{[\underline{x}_Q(t) - \underline{x}_G(t)]^T[\dot{\underline{x}}_Q(t) - \dot{\underline{x}}_G(t)]}{S(t)} \quad . \tag{6-41}$$

Decomposition of $\Delta f(t)$ using

$$\Delta f(t) = \Delta f^0(t) + \delta\Delta f(t) \tag{6-42}$$

results in

$$\begin{aligned}
\delta\Delta f(t) = \\
& - \frac{f}{c}\left\{\frac{\partial\dot{S}(t)}{\partial\underline{x}_Q(t)}\left[\frac{\partial\underline{x}_Q(t)}{\partial\underline{u}(t)}\frac{\partial\underline{u}(t)}{\partial\underline{p}} - \frac{\partial\underline{x}_G(t)}{\partial\underline{p}}\right]\right. \\
& \qquad \left. + \frac{\partial\dot{S}(t)}{\partial\dot{\underline{x}}_Q(t)}\left[\frac{\partial\underline{x}_Q(t)}{\partial\underline{u}(t)}\frac{\partial\underline{u}(t)}{\partial\underline{p}} - \frac{\partial\dot{\underline{x}}_G(t)}{\partial\underline{p}}\right]\right\}\delta\underline{p} \\
& + \frac{f}{c}\left[\frac{\partial\dot{S}(t)}{\partial\underline{x}_Q(t)}\underline{R}(t) + \frac{\partial\dot{S}(t)}{\partial\dot{\underline{x}}_Q(t)}\dot{\underline{R}}(t)\right]\delta\underline{y}_G \\
& - \frac{f}{c}\left[\frac{\partial\dot{S}(t)}{\partial\underline{x}_Q(t)}\frac{\partial\underline{x}_Q(t)}{\partial\underline{u}(t)} + \frac{\partial\dot{S}(t)}{\partial\dot{\underline{x}}_Q(t)}\frac{\partial\dot{\underline{x}}_Q(t)}{\partial\underline{u}(t)}\right]\int_{t_0}^{t}\underline{Y}(t)\,\delta\underline{g}(t)\,dt
\end{aligned} \tag{6-43}$$

It is also possible to find a similar equation for observed doppler counts.

REFERENCES. *Basic reference text for satellite geodesy is Kaula (1966). Satellite geodesy in the integrated model is presented by Eissfeller and Hein (1986); short summaries of that can be found in Eissfeller and Hein (1984), Hein (1985c). Orbit improvement in this context was discussed by Eissfeller (1985). In particular, GPS observations were treated in the integrated model by Eissfeller et al (1985), Hein and Eissfeller (1985a, 1986), Landau (1986). The vertical datum problem to be solved by integrated adjustment was proposed by Hein and Eissfeller (1985b).*

7. GEOPHYSICAL DATA IN THE INTEGRATED MODEL

Although we consider in the collocation model the gravity field of the earth as the realization of a *stochastic* process we may not forget that *deterministic* phenomena are responsible for building it up. The *Newton integral* represents the physical relationship between the gravity potential W and those quantities characterizing the material of the earth's body, namely *density* contrasts,

$$W(P) = G \int_{\sigma_E} \frac{\rho(Q)}{|P - Q|} d\sigma(Q) + \Phi(P) \tag{7-1}$$

where Φ is here the centrifugal potential, by ρ the density is denoted, and G is the gravitational constant. The integration has to be performed over the whole volume σ_E of the earth E.

The reason for not having used the deterministic relation (7-1) throughout whole physical geodesy to construct the gravity potential lies in the fact that in most regions the available density information is insufficient to achieve the necessary accuracy. However, geophysical surveys are increasing, and on the other hand, why not using the density information - although insufficient - to *smooth* the gravity field when dealing with it numerically. Similar goals we have in mind when taking out of the free-air gravity anomalies the height-dependent term in case of interpolation, for example. And let us go one step further: Geophysics is using (besides other techniques) *seismics* to get density contrasts within the earth. Why not also using *seismic velocities* and *seismological displacement* vectors in the integrated approach? Again, our purpose is *not* to solve geophysical problems, our intention is only to recover the gravity field more and more deterministically, leaving the stochastic part as small as possible.

A simple consideration of density is possible, when starting from the *generalized Poisson equation*

$$\nabla^2 W(\underline{x}) = -4\pi G \rho(\underline{x}) + 2\omega^2 \tag{7-2}$$

where ∇^2 is the Laplace operator, ω is the angular velocity of the earth's rotation. We linearize by decomposing ρ into an approximate value ρ_0 and a density disturbance or anomaly $\delta\rho$,

$$\rho = \rho_0 + \delta\rho \tag{7-3}$$

and use (2-4) in connection with (2-2) which leads to the linear observation equation

$$\delta\rho(\underline{x}) = -(4\pi G)^{-1} \nabla^2 T = -(4\pi G)^{-1} \sum_{i=1}^{3} \partial^2 T / \partial x_i^2 \tag{7-4}$$

where the *normal* or *model density* ρ_0 is given by

$$\rho_0(\underline{x}) = -(4\pi G)^{-1} \nabla^2 U = -(4\pi G)^{-1} \sum_{i=1}^{3} \partial^2 U / \partial x_i^2 \quad . \tag{7-5}$$

In the same way we may use *seismic velocities* v. Assuming homogeneous material along the travel path of the seismic waves characterized by (constant) *Lamé* parameters λ, μ and density ρ we have the relation, here for example using the *longitudinal* (compression, P-) wave velocity v_P ,

$$v_P^2 = (\lambda + 2\mu)/\rho \quad . \tag{7-6}$$

Thus, the linearized observation equation is of the form

$$v_P = v_{P_0} + \frac{\partial v_P}{\partial \lambda}\,\delta\lambda + \frac{\partial v_P}{\partial \mu}\,\delta\mu + \frac{\partial v_P}{\partial \rho}\,\delta\rho \quad . \tag{7-7}$$

If λ , μ is (empirically) known in the area under investigation, and only ρ is considered a variable, we get using (7-4) for (7-7) the expression

$$\delta v_P = (8\pi G)^{-1}\rho_0^{-1.5}(\lambda + 2\mu)^{0.5}\;\nabla^2 T \tag{7-8}$$

where

$$\delta v_P = v_P - v_{P_0} \tag{7-9}$$

and

$$v_{P_0} = [(\lambda + 2\mu)/\rho_0]^{0.5} \quad . \tag{7-10}$$

Similarly, we can use the shear (S-) wave velocities. Geophysicists might argue that the relations above are too simplified for a real world. However, we always should have in mind that we do not want to solve geophysical problems but like to use any information which can contribute to a smoothing of the gravity field.

The crucial point in using the observation equations above is the definition of a *model density* ρ_0 consistent with the introduced normal potential U ,

$$U(P) = G\int_{\sigma_E} \frac{\rho_0(Q)}{|P - Q|}\,d\sigma(Q) \quad . \tag{7-11}$$

Whereas U can be uniquely defined from the right hand side of (7-10), is the inverse relation $\rho_0 = f(U)$ ambigious (*inverse problem* of geophysics). Additional constraints have to be used to get a unique and physically reasonable solution.

A somewhat more sophisticated approach to consider density in a geodetic model starts from the Newton integral (7-1) and uses the decomposition (7-3) leading to

$$W(P) = G\int_{\sigma_E} \frac{\rho_0(Q)}{|P - Q|}\,d\sigma(Q) + G\int_{\sigma_E} \frac{\delta\rho(Q)}{|P - Q|}\,d\sigma(Q) \tag{7-12}$$

where the second integral on the right hand side of (7-12) may be considered as consistent with the disturbing potential T . When, in addition, we separate the local region of the earth with volume σ from the rest $\sigma_E - \sigma$ of the earth, we get

$$T(P) = G \int_{\sigma_E - \sigma} \frac{\delta\rho(Q)}{|P - Q|} d\sigma(Q) + G \int_{\sigma} \frac{\delta\rho(Q)}{|P - Q|} d\sigma(Q) \quad . \tag{7-13}$$

The first integral representing the contribution of the outer zone can be approximated using a spherical harmonic expansion, for example, where the separation between local and global is found by spectral analysis of the observations in the frequency domain. Thus, the effect of known density anomalies in the *local* volume σ on the disturbing potential is

$$\delta T_\sigma(P) = G \sum_i \frac{\delta\rho(Q_i)}{|P - Q_i|} \Delta\sigma_i \quad , \tag{7-14}$$

or, in matrix form ($\underline{B}$ is then an integral operator)

$$\underline{t} = \underline{B}\, \delta\underline{\rho} \quad . \tag{7-15}$$

The critical point in this approach remains the same as before: the construction of a model or normal density $\rho_0 = f(U)$. In addition, the separation of the right hand side integrals of (7-13) is problematic.

With respect to seismological *displacement vectors* $\underline{u}$ we start from the partial differential equation describing the *seismic wave motion* in an *inhomogeneous* medium (Eq. (7-6) is a homogeneous solution) with time t

$$\begin{aligned} \rho \frac{\partial^2 u}{\partial t^2} &= \nabla[(\lambda + 2\mu)\nabla\cdot\underline{u}] - \nabla\times\mu\, \nabla\times\underline{u} \\ &+ 2(\nabla\mu\cdot\nabla)\underline{u} - 2(\nabla\mu)\nabla\cdot\underline{u} \\ &+ 2(\nabla\mu)\times\nabla\times\underline{u} \quad . \end{aligned} \tag{7-16}$$

ρ, λ, μ are irregular functions of position and, therefore, may be treated as *randomly distributed functions* in space. Thus, this gives us the key for the application of the concept of stochastic processes and corresponding covariance functions. The decompositions

$$\underline{u} = \underline{u}_0 + \delta\underline{u} \tag{7-17}$$

and (7-3) leads after taylorization to the vector displacement variation

$$\delta\underline{u} = \nabla\delta\Phi + \nabla\times\delta W \tag{7-18}$$

where

$\delta\Phi = \nabla\cdot\delta\underline{u}$ are the compression (P-) waves, (7-19)

$\delta W = \nabla\times\delta\underline{u}$ are the shear (S-) waves, (7-20)

in detail,

$$\delta\Phi(P) = -\frac{1}{4\pi\rho_0\omega^2}\int_\sigma \frac{(\nabla\times\underline{f})\, e^{ik\alpha|P - Q|}}{|P - Q|}\, d\sigma(Q) \tag{7-21}$$

$$\delta\underline{w}(P) = \frac{1}{4\pi\rho_0\omega^2}\int_\sigma \frac{(\nabla\times\underline{f})\, e^{ik\beta|P - Q|}}{|P - Q|}\, d\sigma(Q) \tag{7-22}$$

with

$$\alpha^2 = (\lambda_0 + 2\mu_0)/\rho_0 \tag{7-23}$$

$$\beta^2 = \mu_0/\rho_0 \tag{7-24}$$

$$\underline{f} = -\frac{\partial^2\underline{u}_0}{\partial t^2}\,\delta\rho \quad . \tag{7-25}$$

u_0 is harmonic and time-dependent through $\exp(-i\omega t)$

$$\underline{u}_0(P,t) = \exp(-i\omega t)u_0(P) \tag{7-26}$$

$$\rightarrow \quad \underline{f}\,(P,t) = \omega^2\exp(-i\omega t)\ \delta\rho(P)\ \underline{u}_0(P) \quad . \tag{7-27}$$

Inserting (7-19) and (7-20) into (7-16) gives the final observation equation for *seismological displacement variations* $\delta\underline{l} = \delta\underline{u}$, yielding a relation of type

$$\delta\underline{l} = f(\delta\underline{\rho}\ ,\ \nabla\underline{\rho}) \tag{7-28}$$

which simply expresses the reasonable physical fact, that the seismic wave motion is not only a function of density but also of its gradient. As a consequence we have to supplement the (pseudo-)stochastic vector $\underline{t}$ (4-3) in the integrated model (4-2) by $(\delta\underline{\rho}\ ,\ \nabla\underline{\rho})$, minimizing the last besides $\underline{t}$ by applying the hybrid minimum condition (4-4).

REFERENCES. For further details on the approach above see Eissfeller et al (1986), Hein (1984). A good reference for density models is (Bullen, 1975). Corresponding covariance functions are presented by Tscherning (1977) based on the ideas of Krarup (1970). A new approach, called "mixed collocation" was proposed by Sanso and Tscherning (1982).

8. SOME REMARKS ON A FOURDIMENSIONAL INTEGRATED GEODESY

In the following we like to outline a general model for a *4d-integrated geodesy* where we consider besides a time-dependent position vector a time-dependent gravity potential.

As a basic starting point the geodetic observation may be expressed as follows:

$$l(t) = F\left(\underline{x}(\underline{a},t)\ ,\ W(\underline{x}(\underline{a},t)\ ,\underline{b},t)\right) \tag{8-1}$$

where

$l(t)$ is the time dependent observation,

$x(\underline{a},t)$ is the time dependent position vector in the standard geocentric reference system. The parameters $\underline{a}$ model the position change with time.

$W(\underline{x}(\underline{a},t),\ \underline{b},t)$ is the time and position dependent gravity potential. W is a function of position and independently a function of time as well. The time variation of the potential is modelled by parameters $\underline{b}$.

F denotes a non-linear functional which, in simple terms, operates upon $\underline{x}$ and W to produce the real number l .

For the purposes of linearizing equation (8-1) we shall now introduce the following decompositions:

$$\underline{x}(\underline{a},t) = \underline{x}_0 + \Delta\underline{x}_0 + \Delta\underline{x}(\underline{a},t) \tag{8-2}$$

where

$\underline{x}_0$ is the approximate position vector at time t (the reference epoch),

$\Delta\underline{x}_0$ is the time independent increment to the approximate position vector at t_0 , *to give the true position vector at the reference epoch.*

$\Delta\underline{x}(\underline{a},t)$ is the time dependent increment to the reference epoch position vector to give the position at some time t .

In addition, we decompose the gravity potential as follows:

$$W(\underline{x}(\underline{a},t),\ \underline{b},t)) = U(\underline{x}(\underline{a},t)) + T(\underline{x}(\underline{a},t)) + \tau(\underline{x}(\underline{a},t),\ \underline{b},t) \tag{8-3}$$

where

$U(\underline{x}(\underline{a},t))$	is some approximate reference (normal) potential which is a function only of position (although position may vary with time).
$T(\underline{x}(\underline{a},t))$	is the usual disturbing potential which again is position dependent.
$\tau(\underline{x}(\underline{a},t), \underline{b},t)$	Here we explicity introduce the time dependence of the gravity potential by this additional disturbing component which is function of both position and time.

The Linearization Process

Firstly look at equation (8-3) expressing the decomposition of the gravity potential. Into this expression we may substitute equation (8-2) for the position vector so that we obtain:

$$W(\underline{x}(\underline{a},t), \underline{b},t) = U(\underline{x}_0 + \Delta\underline{x}_0 + \Delta\underline{x}(\underline{a},t)) + T(\underline{x}_0 + \Delta\underline{x}_0 + \Delta\underline{x}(\underline{a},t)) + \tau((\underline{x}_0 + \Delta\underline{x}_0 + \Delta\underline{x}(\underline{a},t)), \underline{b},t) \tag{8-4}$$

Applying a first order Taylor expansion to the terms on the right hand side of this expression, then

$$\begin{aligned} W(\underline{x}(\underline{a},t), \underline{b},t) = \; & U(\underline{x}_0) + \frac{\partial U}{\partial \underline{x}} \Delta\underline{x}_0 + \frac{\partial U}{\partial \underline{x}} \Delta\underline{x}(\underline{a},t) + \\ & T(\underline{x}_0) + \frac{\partial T}{\partial \underline{x}} \Delta\underline{x}_0 + \frac{\partial T}{\partial \underline{x}} \Delta\underline{x}(\underline{a},t) + \\ & \tau(\underline{x}_0,\underline{b},t) + \frac{\partial \tau}{\partial \underline{x}} \Delta\underline{x}_0 + \frac{\partial \tau}{\partial \underline{x}} \Delta\underline{x}(\underline{a},t) \; . \end{aligned} \tag{8-5}$$

Deleting the second order-terms from this expression, and rearranging the order of the remaining terms, gives

$$W(\underline{x}(\underline{a},t), \underline{b},t) = U(\underline{x}_0) + \frac{\partial U}{\partial \underline{x}} \Delta\underline{x}_0 + T(\underline{x}_0) + \frac{\partial U}{\partial \underline{x}} \Delta\underline{x}(\underline{a},t) + \tau(\underline{x}_0,\underline{b},t) \; . \tag{8-6}$$

Note in equation (8-6) the first three terms correspond to the linearization of the usual time invariant gravity potential and the last two allow for the time dependency.

As a further step in the linearization process we differentiate the equations of position (8-3) and potential (8-6) with respect to the parameters $\underline{a}$ and $\underline{b}$. This allows for the fact that the function modelling the change in position and potential with time may not be linear with respect to their parameters $\underline{a}$ and $\underline{b}$. We introduce the following decomposition of these parameter vectors.

$$\underline{a} = \underline{a}_0 + \Delta\underline{a} \qquad \text{and} \qquad \underline{b} = \underline{b}_0 + \Delta\underline{b} \tag{8-7}$$

Substitution into equation (8-2) and linearization by the Taylor expansion yields

$$\underline{x}(\underline{a},t) = \underline{x}_0 + \Delta\underline{x}_0 + \Delta\underline{x}(\underline{a}_0 + \Delta\underline{a},t)$$

i.e.
$$\underline{x}(\underline{a},t) = \underline{x}_0 + \Delta\underline{x}_0 + \Delta\underline{x}(\underline{a}_0,t) + \frac{\partial\Delta\underline{x}}{\partial\underline{a}}\Delta\underline{a} \tag{8-8}$$

This equation (8-8) is the *linearized position vector.*

Performing similar substitutions and expansion upon equation (8-6) for the gravity potential, we obtain

$$W(\underline{x}(\underline{a},t),\ \underline{b},t) = U(\underline{x}_0) + \frac{\partial U}{\partial \underline{x}}\Delta\underline{x}_0 + T(\underline{x}_0) + \frac{\partial U}{\partial \underline{x}}\Delta\underline{x}(\underline{a}_0 + \Delta\underline{a},t) + \tau(\underline{x}_0,\ \underline{b}_0 + \Delta\underline{b},t) \tag{8-9}$$

$$W(\underline{x}(\underline{a},t),\ \underline{b},t) = U(\underline{x}_0) + \frac{\partial U}{\partial \underline{x}}\Delta\underline{x}_0 + T(\underline{x}_0) + \frac{\partial U}{\partial \underline{x}}\Delta\underline{x}(\underline{a}_0,t) + \frac{\partial U}{\partial \underline{x}}\frac{\partial\Delta\underline{x}}{\partial\underline{a}}\Delta\underline{a} + \tau(\underline{x}_0,\underline{b}_0,t) + \frac{\partial\tau}{\partial\underline{b}}\Delta\underline{b} \tag{8-10}$$

This equation (8-10) is the *linearized gravity potential.*

A substitution of the linear equations (8-8) and (8-10) into the general model of equation (8-1) can now take place and the final step of the linearization with respect to the functional F performed.

Substituting (8-8) and (8-10) into equation (8-1) gives

$$l(t) = F[(\underline{x}_0 + \Delta\underline{x}_0 + \Delta\underline{x}(\underline{a}_0,t) + \frac{\partial\Delta\underline{x}}{\partial\underline{a}}\Delta\underline{a})\ ,\ (U(\underline{x}_0) + \frac{\partial U}{\partial \underline{x}}\Delta\underline{x}_0 + T(\underline{x}_0) + \frac{\partial U}{\partial \underline{x}}\Delta\underline{x}(\underline{a}_0,t) + \frac{\partial U}{\partial \underline{x}}\frac{\partial\Delta\underline{x}}{\partial\underline{a}}\Delta\underline{a} + \tau(\underline{x}_0,\underline{b}_0,t) + \frac{\partial\tau}{\partial\underline{b}}\Delta\underline{b})]\ . \tag{8-11}$$

And now linearizing via the Taylor expansion gives

$$
\begin{aligned}
l(t) = F(\underline{x}_0, U(\underline{x}_0)) &+ \frac{\partial F}{\partial \underline{x}} \Delta\underline{x}_0 + \frac{\partial F}{\partial \underline{x}} \Delta\underline{x}(\underline{a}_0, t) + \frac{\partial F}{\partial \underline{x}} \frac{\partial \Delta\underline{x}}{\partial \underline{a}} \Delta\underline{a} + \\
&\frac{\partial F}{\partial W} \frac{\partial U}{\partial \underline{x}} \Delta\underline{x}_0 + \frac{\partial F}{\partial W} T(\underline{x}_0) + \frac{\partial F}{\partial W} \frac{\partial U}{\partial \underline{x}} \Delta\underline{x}(\underline{a}_0, t) + \\
&\frac{\partial F}{\partial W} \frac{\partial U}{\partial \underline{x}} \frac{\partial \Delta\underline{x}}{\partial \underline{a}} \Delta\underline{a} + \frac{\partial F}{\partial W} \tau(\underline{x}_0, \underline{b}_0, t) + \frac{\partial F}{\partial W} \frac{\partial \tau}{\partial \underline{b}} \Delta\underline{b} \qquad (8\text{-}12)
\end{aligned}
$$

Now collecting coefficients of common unknowns we get:

$$
\begin{aligned}
l(t) = F(\underline{x}_0, U(\underline{x}_0)) &+ \left(\frac{\partial F}{\partial \underline{x}} + \frac{\partial F}{\partial W} \frac{\partial U}{\partial \underline{x}}\right) \Delta\underline{x}_0 + \left(\frac{\partial F}{\partial \underline{x}} + \frac{\partial F}{\partial W} \frac{\partial U}{\partial \underline{x}}\right) \Delta\underline{x}(\underline{a}_0, t) \\
&+ \frac{\partial F}{\partial W} T(\underline{x}_0) + \frac{\partial F}{\partial W} (\underline{x}_0, \underline{b}_0, t) \\
&+ \left(\frac{\partial F}{\partial \underline{x}} \frac{\partial \underline{x}}{\partial \underline{a}} + \frac{\partial F}{\partial W} \frac{\partial U}{\partial \underline{x}} \frac{\partial \Delta\underline{x}}{\partial \underline{a}}\right) \Delta\underline{a} + \frac{\partial F}{\partial W} \frac{\partial \tau}{\partial \underline{b}} \Delta\underline{b} \qquad (8\text{-}13)
\end{aligned}
$$

This is the proposed *linear model for 4d-integrated geodesy*. The unknowns in the model may be classified as follows.

(i)	*Positional*	$\Delta\underline{x}_0$	is the time independent increment to the assumed position at time t_0 .
		$\Delta\underline{x}(\underline{a}_0,t)$	is the time dependent increment to the actual position vector at t_0 to give position at t .
(ii)	*Potential*	$T(\underline{x}_0)$	is the time invariant disturbing potential.
		$\tau(\underline{x}_0,\underline{b}_0,t)$	is the time dependent disturbing potential.
(iii)	*Model parameters*	$\Delta\underline{a}$	are the increments to the parameters modelling the time variation of the position vector
		$\Delta\underline{b}$	are the increments to the parameters modelling the time variation of the gravity potential.

REFERENCES. Collier et al (1986), Hein (1984). Space time geodesy is also extensively researched by the Stuttgart school, see e.g., Grafarend (1982), Grafarend and Sanso (1982).

Acknowledgement. The development of integrated geodesy with its state-of-the-art as it was tried to outline here would not be becoming possible without the tremendous efforts and ideas of my coworkers Bernd Eissfeller and Herbert Landau. Philipp Collier from the University of Melbourne, Australia, assisted in working out the 4d-integrated model during his stay at my institute this year spring. Many individuals from the Special Study Group 1.73 of the International Association of Geodesy contributed with their work and hints. The German Research Foundation (Deutsche Forschungsgemeinschaft) supported the developments financially with several research grants in the last years. The careful typing was carried out by Mrs. Grandl.

REFERENCES

BAKKER, G. (1983), *Some Computational Aspects of the Mapping Problem in Threedimensional Geodesy.* Quaterniones Geodaesiae 4, 35-57

BÄUMKER, M. (1984), *Zur dreidimensionalen Ausgleichung von terrestrischen und Satellitenbeobachtungen.* In: Wiss. Arbeiten der Fachrichtung Vermessungswesen der Universität Hannover, Nr. 130

BÄUMKER, M., D. Egge, H.W. Schenke, H.W. Wenzel (1983), *Common Adjustment of Doppler and Terrestrial Networks.* Geodezja 79, 17-33, Cracōw

BARZAGHI, R., B. Betti, F. Sanso (1985), *Integrated Geodesy: A Purely Local Approach.* Manuscript available from the authors

BRUNS, H. (1878), *Die Figur der Erde.* Publ. Königl. Preuss. Geod. Inst., P. Stankiewicz Buchdruckerei, Berlin

BUCERIUS, H. (1966), *Vorlesungen über Himmelsmechanik, Band 1.* Bibliographisches Institut Mannheim

BULLEN, K.E. (1975), *The Earth's Density.* Chapman and Hall, London

COLLIER, P., B. Eissfeller, G.W. Hein and H. Landau (1986), *On a Fourdimensional Integrated Geodesy.* Manuscript, in print

DERMANIS, A. (1985), *Optimization Problems in Geodetic Networks with Signals.* In: Grafarend, E.W. and F. Sansò (eds.), Optimization and Design of Geodetic Networks, Springer-Verlag, 221-256

DUFOUR, H. (1962), *Eléments fondamentaux de la Géodésie tridimensionelle.* Paris

EEG, J., T. Krarup (1973), *Integrated Geodesy.* The Danish Geodetic Institute, Internal Report No. 7, Kopenhagen

EISSFELLER, B., G.W. Hein (1984), *The Observation Equations of Satellite Techniques in the Model of Integrated Geodesy.* In: Proc. of the International Symposium on Space Techniques for Geodynamics, Sopron, Hungary, July 9-13, 1984, (2) 119-129

EISSFELLER, B. (1985), *Orbit Improvement Using Local Gravity Field Information and Least-Squares Prediction.* Manuscripta Geodaetica 10, 91-101

EISSFELLER, B., H. Landau, G.W. Hein, A.D. Bottse (1985), *The Processing of GPS Baseline Vectors in Conventional Geodetic Networks Using Gravity Field Information and Least-Squares Collocation.* Presented at the 7th International Symposium on Geodetic Computations, Cracow, Poland, June 18-21, 1985

EISSFELLER, B. (1986), *The Estimation of Orthometric Heights from GPS Baseline Vectors Using Gravity Field Information and Least-Squares Collocation.* In: Landau, H., B. Eissfeller and G.W. Hein, GPS Research 1985 at the Institute of Astronomical and Physical Geodesy, Heft 19 der Schriftenreihe des Universitären Studiengangs Vermessungswesen der Universität der Bundeswehr München, 107-126, University FAF Munich

EISSFELLER, B., G.W. Hein (1986), *A Contribution to 3d-Operational Geodesy. Part 4: The Observation Equations of Satellite Geodesy in the Model of Integrated Geodesy.* Heft 17 der Schriftenreihe des Universitären Studiengangs Vermessungswesen der Universität der Bundeswehr München, University FAF Munich

EISSFELLER, B., K. Hehl, G.W. Hein and H. Landau (1986), *On the Incorporation of Geophysical Data in an Integrated Model.* Manuscript, in print

ENGLER, K., E. Grafarend, P. Teunissen, J. Zaiser (1982), *Test Computations of Threedimensional Geodetic Networks with Observables in Geometry and Space.* In: Proc. of the International Symposium on Geodetic Networks and Computations of the IAG, Munich, Aug. 31 to Sept. 5, 1981, Deutsche Geodätische Kommission, Reihe B, Nr. 258/VII, 119-141

EREN, K. (1980), *Spectral Analysis of GEOS-3 Altimeter Data and Frequency Domain Collocation.* The Ohio State University, Department of Geodetic Science, Report 297, Columbus

GRAFAREND, E.W. (1975), *Threedimensional Geodesy I. The Holonomity Problem.* Zeitschrift für Vermessungswesen 100, 269-281

GRAFAREND, E.W. (1978a), *Operational Geodesy.* In: Approximation Methods in Geodesy, Eds. Moritz and Sünkel, Herbert Wichmann Verlag, Karlsruhe 1978, 235-284

GRAFAREND, E.W. (1978b), *Dreidimensionale geodätische Abbildungsgleichungen und die Näherungsfigur der Erde.* Zeitschrift für Vermessungswesen 103, 132-140

GRAFAREND, E.W. (1980), *The Bruns Transformation and a Dual Setup of Geodetic Observational Equations.* NOAA Technical Report NOS 85 NGS 16, Rockville, Md.

GRAFAREND, E.W. (1981), *Die Beobachtungsgleichungen der dreidimensionalen Geodäsie im Geometrie- und Schwereraum - ein Beitrag zur operationellen Geodäsie.* Zeitschrift für Vermessungswesen 106, 411-429

GRAFAREND, E.W. (1982), *Six Lectures on Geodesy and Global Geodynamics.* In: Moritz, H. and H. Sünkel (Eds.), Geodesy and Global Geodynamics, Mitteilungen der geodätischen Institute der Technischen Universität Graz, Folge 41, 531-672

GRAFAREND, E.W. (1983), *Spacetime Telluroid Versus Spacetime Geoid or the Bruns-Love Dialogue.* In: Commemorative volume on the occasion of Erik Tengström's 70th birthday. Report No. 19 of the Department of Geodesy of the University of Uppsala, 105-126

GRAFAREND, E.W., F. Sanso (1984), *The Multibody Space-Time Geodetic Boundary Value Problem and the Honkasalo Term.* Geophys. J. R. ostr. Soc. 78, 255-275

GRAFAREND, E.W., H. Kremers, W. Lindlohr (1985), *Threedimensional Operational Adjustment of Geodetic Observations of Terrestrial Type Including Prior Information of the Unknowns.* Presented at the 7th International Symposium on Geodetic Computations, Cracow, June 18-21, 1985

GRAY, R.M. (1971), *Toeplitz and Circulant Matrices: A Review.* Technical Report No. 6502-1, Stanford University, Stanford

GRIST, S.N. (1984), *Computing Models in Spatial Geodesy.* Ph.D. Thesis, University of Nottingham, Department of Civil Engineering

HEIN, G.W. (1982a), *A Contribution to 3D-Operational Geodesy. Part 1: Principle and Observational Equations of Terrestrial Type.* In: Proc. of the International Symposium on Geodetic Networks and Computations of the International Association of Geodesy, Munich, Aug. 31 to Sept. 5, 1981, Deutsche Geodätische Kommission, Reihe B, Nr. 258/VII, 31-64, München 1982

HEIN, G.W. (1982b), *A Contribution to 3D-Operational Geodesy. Part 2: Concepts of Solution.* In: Proc. of the International Symposium on Geodetic Networks and Computations of the International Association of Geodesy, Munich, Aug. 31 to Sept. 5, 1981, Deutsche Geodätische Kommission, Reihe B, Nr. 258/VII, 65-85, München 1982

HEIN, G.W. (1982c), *The Geoid as Part of Integrated Geodesy.* Presented at Symposium No. 4b of the General Meeting of the International Association of Geodesy, Tokyo, Japan, May 7-20, 1982

HEIN, G.W. (1983), *Erdmessung als Teil einer integrierten Geodäsie - Begründung, Stand und Entwicklungstendenzen.* Zeitschrift für Vermessungswesen 108, 93-104

HEIN, G.W. (1984), *Earth Deformation Analysis in the Context of Integrated Geodesy.* Presented at the International Symposium on Recent Crustal Movements of the Pacific Region, Febr. 9-14, 1984, Wellington, New Zealand

HEIN, G.W. (1985a), *The Local Gravity Field in the Concept of Integrated Geodesy.* In: Schwarz, K.P. (ed.), Local Gravity Field Approximation, Beijing International Summer School, Aug. 21 - Sept. 4, 1984, 107-216, The University of Calgary

HEIN, G.W. (1985b), *Orthometric Height Determination Using GPS Observations and the Integrated Geodesy Adjustment Model.* NOAA Technical Report NOS 110 NGS 32, U.S. Department of Commerce

HEIN, G.W. (1985c), *Satellitenmessungen und Integrierte Geodäsie.* Heft 15 der Schriftenreihe des Wiss. Studiengangs Vermessungswesen der Universität der Bundeswehr München, 193-197, University FAF Munich

HEIN, G.W. (1986a), *How Does an Integrated Geodetic Network Look Like? - Seven Theses on the Set-up of a Modern Geodetic Network.* Presented at the 3rd Symposium on Geodesy in Africa on the subject "Toward an Integrated Geodetic Network for Africa", Yamoussoukrou, Côte d'Ivoire, April 10-17, 1986

HEIN, G.W. (1986b), *The Role of GPS Data in Gravity Field Approximation or The Role of the Gravity Field in GPS Surveys.* Presented at the International Symposium on the Definition of the Geoid, May 26-30, 1986, Florence, Italy

HEIN, G., B. Eissfeller (1985a), *Integrated Modelling of GPS-Orbit and Multi-Baseline Components.* In: Proc. First International Symposium on Precise Positioning with the Global Positioning System, Rockville, Md., April 15-19, 1985, 263-272

HEIN, G.W., B. Eissfeller (1985b), *Vertical Datum Definition by Integrated Geodesy Adjustment.* In: Proc. Third International Symposium on the North American Vertical Datum, Rockville, Md., April 21-26, 1985, 121-135

HEIN, G.W. and B. Eissfeller (1986), *The Basic Observation Equations of Carrier Phase Measurements to the Global Positioning System Including General Orbit Modelling.* In: Landau, H., B. Eissfeller and G.W. Hein, GPS Research 1985 at the Institute of Astronomical and Physical Geodesy. Heft 19 der Schriftenreihe des Universitären Studiengangs Vermessungswesen der Universität der Bundeswehr München, 7-45, University FAF Munich

HEIN, G.W. and H. Landau (1983), *A Contribution to 3D-Operational Geodesy. Part 3: OPERA - A Multi-Purpose Program for Operational Adjustment of Geodetic Observations of Terrestrial Type.* Deutsche Geodätische Kommission, Reihe B, Nr. 264, München

HEIN, G.W., H. Landau, K. Egreder (1984), *Erste Erfahrungen zur integrierten geodätischen Netzausgleichung.* Zeitschrift für Vermessungswesen 109, 75-86

HEISKANEN, W.A., H. Moritz (1967), *Physical Geodesy.* W.H. Freeman and Co., San Francisco and London

HEITZ, S. (1973), *Ein dreidimensionales Berechnungsmodell für Punktbestimmungen mit Berücksichtigung orthometrischer Höhen.* Zeitschrift für Vermessungswesen 98, 479-485

KAULA, W.M. (1966), *Theory of Satellite Geodesy.* Blaisdell Publ. Comp., Waltham, Mass.

KRARUP, R. (1969), *A Contribution to the Mathematical Foundation of Physical Geodesy.* The Danish Geodetic Institute, Meddelelse No. 44, Kopenhagen

KRARUP, T. (1970), *On the Determination of a Reasonable Distribution of Masses Corresponding to a Potential Harmonic Outside a Closed Surface.* Letter to the members of IAG Special Study Group 4.31

KRARUP, R. (1978), *Integrated Geodesy.* Lecture Notes, Summer School Erice 1978

LANDAU, H. (1986), *GPS Baseline Vectors in a Threedimensional Integrated Adjustment Approach.* In: Landau, H., B. Eissfeller and G.W. Hein, GPS Research 1985 at the Institute of Astronomical and Physical Geodesy, Heft 19 der Schriftenreihe des Universitären Studiengangs Vermessungswesen der Universität der Bundeswehr München, 7-45, University FAF Munich

LANDAU, H., G.W. Hein, B. Eissfeller (1985), *A Stepwise Approach for the Integrated Geodesy Adjustment Model.* Presented at the 7th International Symposium on Geodetic Computations, Cracôw, June 18-21, 1985

MARUSSI, A. (1949), *Fondaments de géometrie différentielle absolue du champ potentiel terrestre.* Bulletin Géodesique 14, 411-439

MORITZ, H. (1973), *Least-Squares Collocation.* Deutsche Geodätische Kommission, Reihe A, Nr. 75, München

MORITZ, H. (1978), *The Operational Approach to Physical Geodesy.* The Ohio State University, Department of Geodetic Science, Report 277, Columbus

MORITZ, H. (1980), *Advanced Physical Geodesy.* Herbert Wichmann Verlag, Karlsruhe

PREUSSER, A. (1977), *Ein dreidimensionales Berechnungsmodell für geodätische Punkt- und Geoidbestimmungen.* Deutsche Geodätische Kommission, Reihe C, Nr. 238, München

REILLY, W.I. (1980), *Three-dimensional Adjustment of Geodetic Networks with Incorporation of Gravity Field Data.* Dept. of Scientific and Industrial Research, Geophysics Division, Report No. 160, Wellington

REILLY, W.I. (1984), *Three-dimensional Adjustment of Geodetic Networks: Examples from Southern New Zealand and West Germany.* Zeitschrift für Vermessungswesen 109, 279-292

RINO, C.L. (1970), *The Inversion of Covariance Matrices by Finite Fourier Transforms.* IEEE Transactions on Information Theory, 230-232

RUMMEL, R. (1985), *From the Observational Model to Gravity Parameter Estimation.* In: Schwarz, K.P. (ed.), Local Gravity Field Approximation, Beijing International Summer School, Aug. 21 - Sept. 4, 1984, 67-106, The University of Calgary

SANSÒ, F. and C.C. Tscherning (1982), *Mixed Collocation: a Proposal.* Quaterniones Geodaesiae 3, 1-15, Thessaloniki

STUMPFF, K. (1974), *Himmelsmechanik III.* VEB Deutscher Verlag der Wissenschaften, Berlin

SÜNKEL, H. (1979), *A Covariance Approximation Procedure.* The Ohio State University, Department of Geodetic Science, Report No. 286, Columbus

TORGE, W., H.G. Wenzel (1978), *Dreidimensionale Ausgleichung des Testnetzes Westharz.* Deutsche Geodätische Kommission, Reihe B, Nr. 234, München

TSCHERNING, C.C. (1976), *Covariance Expressions for Second and Lower Order Derivatives of the Anomalous Potential.* The Ohio State University, Department of Geodetic Science, Report No. 225, Columbus

TSCHERNING, C.C. (1977), *Models for the Auto- and Crosscovariances between Mass Density Anomalies and First and Second Order Derivatives of the Anomalous Potential of the Earth.* Proc. Third International Symposium Geodesy and Physics of the Earth, 261-268, Weimar 1976, Potsdam

TSCHERNING, C.C., R. Rapp (1974), *Closed Covariance Expressions for Gravity Anomalies, Geoid Undulations, and Deflections of the Vertical Implied by Anomaly Degree Variance Models.* The Ohio State University, Department of Geodetic Science, Report No. 208, Columbus

WOLF, H. (1963a), *Dreidimensionale Geodäsie. Herkunft, Methodik und Zielsetzung.* Zeitschrift für Vermessungswesen 88, 109-116

WOLF, H. (1963b), *Die Grundgleichungen der Dreidimensionalen Geodäsie in elementarer Darstellung.* Zeitschrift für Vermessungswesen 88, 225-233

ZAISER, J. (1984), *Ein dreidimensionales geometrisch-physikalisches Modell für konventionelle geodätische Beobachtungen - Beobachtungsfunktionale, Parameterschätzung und Deformationsanalyse.* Deutsche Geodätische Kommission, Reihe C, Nr. 298, München

ZAISER, J. (1986), *Begründung, Beobachtungsgleichungen und Ergebnisse für ein dreidimensionales geometrisch-physikalisches Modell der Geodäsie.* Zeitschrift für Vermessungswesen 111, 190-197